HISTORY, PHILOSOPHY AND SOCIOLOGY OF SCIENCE

Classics, Staples and Precursors

HISTORY, PHILOSOPHY AND SOCIOLOGY OF SCIENCE

Classics, Staples and Precursors

Selected By

**YEHUDA ELKANA
ROBERT K. MERTON
ARNOLD THACKRAY
HARRIET ZUCKERMAN**

THE
NATURAL SCIENCES

An Introduction to the
Scientific Philosophy
of to-day

BY

BERNHARD BAVINK

ARNO PRESS
A New York Times Company
New York — 1975

Reprint Edition 1975 by Arno Press Inc.

Reprinted from a copy in
  The Columbia University Library

HISTORY, PHILOSOPHY AND SOCIOLOGY OF SCIENCE:
Classics, Staples and Precursors
ISBN for complete set: 0-405-06575-2
See last pages of this volume for titles.

Manufactured in the United States of America

————◆————

Library of Congress Cataloging in Publication Data

Bavink, Bernhard, 1879-1947.
  The natural sciences.

  (History, philosophy, and sociology of science)
  Reprint of the translation of Ergebnisse und Prob-
leme der Naturwissenschaften, 4th ed., published in
1932 by the Century Co., New York.
  Bibliography: p.
  1. Science--Philosophy. I. Title. II. Series.
Q175.B32 1975   501   74-26248
ISBN 0-405-06578-7

THE
NATURAL SCIENCES

THE
NATURAL SCIENCES

An Introduction to the
Scientific Philosophy
of to-day

BY

BERNHARD BAVINK

Translated from
the 4th German Edition,
with additional notes
and Bibliography
for English readers, by

H. STAFFORD HATFIELD

NEW YORK
THE CENTURY COMPANY
1932

Printed in Great Britain by the Camelot Press Limited
London and Southampton

PREFACE

IN writing a few words of introduction for this English translation of my book, I should like first of all to express my thanks to the publishers for their courage in venturing to bring out in English, in despite of the general world crisis, so extensive a book by a foreign author. In the next place I wish to say that I count it a peculiar honour and pleasure to be able to make with this book some return, as it were, to the world of English speech for the many important works of a similar character which English and American authors have given to the German public. I need only mention Faraday, Maxwell, Lord Kelvin, Tait and Snyder in the past, and, in more recent times, Eddington and Jeans. The task of expressing in simple form to a wide public the great results of scientific research, and of attempting to evaluate from a general and philosophic point of view the significance of these results, and of the great problems still unsolved, is one that has always appealed to the leaders of English and American science as *nobile officium*, a worthy and a noble duty, and foundations such as the famous Gifford lectures have been instituted with no other purpose. In Germany, unfortunately, such a tradition is not as strong as could be wished, although there has been some improvement in this respect in recent times, and when occasionally German men of science, e.g. Haeckel, have embarked upon more comprehensive philosophic discussions the result has not always been particularly fortunate.

A German who wished not only to study science but also to examine the great general and philosophic questions at the back of his special subject was thrown largely upon the reading of such books as I have indicated, and I remember with gratitude the great impression which they made upon me as a young student. Here was something like a true *'universitas literarum'* which was long, and I fear still is, little more than a phrase with us Germans. However, as a result of rapid development and inevitable specialisation, young men in all countries, England and America included, now find themselves so immersed in their own special branches of science that they scarcely have time or energy

v

to devote to the study of more comprehensive problems. Already before the World War Germany was the farthest advanced along the path of specialisation. It is not surprising, perhaps, that in this country the need was, and is, most strongly felt of an escape from this state, which becomes unendurable in the long run – a need of supplementing the specialist's knowledge by a more comprehensive consideration of the problems of science, without which the former is little more than a kind of scientific trade. It is only on some such basis that I can explain the astonishing success which this book of mine has had in Germany, although, or perhaps because, the author is neither engaged in research (not even in his special subject of physics) nor is a professional philosopher. There is, apparently, an outstanding need of a real synthesis on broad philosophic lines of the inexhaustible material of natural science, and I believe and trust that this need is felt in England and America almost as urgently as in Germany.

It was clear to me from the start that this need could be met in a way that would satisfy men of science only if the presentation dealt in the first place with the scientific results and not the philosophic problems. This book therefore deals with *Inductive Philosophy*, so to speak: the philosophic questions grow naturally out of the results and problems of science. I confidently believe that this method is the one that will commend itself to English and American scientists.

Finally I should like to express my best thanks to the translator for the trouble which he has taken with his task, and for submitting the manuscript for my perusal. This has made it possible for me to improve and amplify in one or two places the text of the last German edition. I have completely recast the section dealing with the present state of our knowledge of primitive man. I shall be very grateful for any corrections or suggestions from English or American readers.

THE AUTHOR

TRANSLATOR'S NOTE

THE writing of this Note concludes a most interesting task.

I have done my best to adapt the very full notes and bibliography for the use of English readers, by replacing German references by English where this was possible. Dr. Baviňk's bibliography is given in full, with translations substituted for foreign works where these exist. I have added a supplementary bibliography of some recent English works representing diverse opinions.

Dr. Bavink read the MS., and I am much obliged to him for many suggestions and addenda. I am also greatly indebted to Professor H. Munro Fox and Dr. H. I. Coombs for their valuable assistance in the biological and biochemical sections.

H. STAFFORD HATFIELD

CONTENTS

PART I. FORCE AND MATTER

PART II. COSMOS AND EARTH

PART III. MATTER AND LIFE

PART IV. NATURE AND MAN

ILLUSTRATIONS

xi

PART I

FORCE AND MATTER

WHEN man awakes to reflection concerning himself and the world around him, the first piece of fundamental knowledge which offers itself to him is the difference between living and dead bodies. He draws the boundary between these two classes, wrongly, it is true, to our present ideas, but he always draws it. The distinguishing mark of the non-living is for him the fact that the body in question cannot move 'on its own account.' Such bodies require to be moved by human or animal muscular power; or by other causes acting in nature, which, by analogy, have likewise been called *forces*. Apart from changes of position, non-living substances are able to undergo all sorts of 'changes of state'; they are able to change their form, their temperature, and so on, but they may also be transformed into quite different substances, as when iron rusts, copper forms verdigris, and so on. Changes of this sort are also, by a further extension of the concept of force, referred to the action of certain 'forces of nature,' and we can thus summarise our first and simplest experience of the non-living world as follows:

The bodies of the non-living world consist of various kinds of material, and are kept in a state of continual motion and change by the action of forces.

When the question is considered more deeply, these facts lead us to three problems:

1. What is substance or matter, and how are various kinds of matter really different from one another?

2. What is force, and how do the various kinds of force differ from one another?

3. How does force act upon matter?

These three questions may be regarded as the fundamental problems of all science of non-living nature – or in our present day terms, physics and chemistry. It can in fact be shown that they are not yet finally solved even to-day, although their three-fold character appears to be only provisional (see below).

Bs   1

But while the scientist, at least since the foundation of empirical natural science by Galileo, only attempts to approach these problems of his step by step, the philosopher takes another course. He is interested by the general question which is behind these problems. The general descriptions 'matter,' 'force,' already point for him to a final unity behind them, and assuming this, he arrives at the conviction that in order to imagine anything at all happening in the world, there must be a fundamental substance, to which, with which, or in which something happens. And that, secondly, everything that happens in the world must be regarded as a change in the 'states,' or 'accidents' of this substance or substances; and thirdly, that force must always be regarded as the active cause of such changes of state.

According as the philosopher leans in his theory of knowledge to idealism or realism, these three statements are then regarded either as statements concerning the real world itself, or as necessary forms of our thought concerning the world (the latter being for example Kant's view). And up to the present day, many philosophers are of the opinion that they in fact state, in one way or the other, the last truth concerning nature (so far as we leave life out of consideration for the present). We shall show in this book that this opinion is vulnerable in more than one respect, and that everything rather points to there being here, not three separate problems, but in truth only one; and that our present day science of inorganic nature has already approached very closely the final solution of this single problem. We can only, in following out our programme, touch quite by the way on historical matters ; we shall rather build up our point of view systematically, and for this purpose we shall begin with that science which has to do with the problem of matter : chemistry.

I

THE FUNDAMENTAL FACTS OF CHEMISTRY

Present day chemistry knows of about three quarters of a million single definite substances,[1] such as water, sulphur, common salt, iron, soda, sugar, indigo, vanilla, urea, etc., etc. Every educated person also knows, even if he is otherwise almost totally ignorant of chemistry, this much at least: that all substances can be decomposed by chemistry into certain fundamental substances or 'elements,' and that they can for the most part also be built up again from these elements by combination, naturally with the exception of those, such as iron and sulphur, which are themselves elements.

Most people also know that the great majority of the 92 known elements are metals (e.g., iron, copper, silver, zinc, sodium, potassium, magnesium, aluminium) and that besides these there are 14 non-metals, namely hydrogen, oxygen, nitrogen, carbon, fluorine, chlorine, bromine, iodine, sulphur, selenium, phosphorus, arsenic, boron, silicon (three others, tellurium, antimony and germanium being on the border between metals and non-metals). It will also be known to many that modern chemistry has added 6 very peculiar gases, helium, neon, argon, krypton, xenon and radium-emanation, which are distinguished from all other elements by complete incapacity for chemical combination.

Of these 92 elements, only about two dozen play a practical part,[2] the others are so rare that they are only of interest to science. The whole of living matter consists of substances which mostly contain only the first four elements named, H, O, N, C, while very many contain only H, O, and C. Many organic substances, as for example most proteins, also contain sulphur and phosphorus; and compounds of other elements such as iodine, chlorine, magnesium, iron, etc., also occur here and there in plants and animals. On the other hand, the solid crust of the earth consists for the most part of the silicates of a few metals, namely aluminium, magnesium, calcium, iron, sodium and potassium, together with carbonate of lime (calcium carbonate, $CaCO_3$) as the only other main constituent, so that here also a small number of elements supply almost the whole requirement.

These are the broad lines of chemistry as it has resulted from 250 years of work. In order to understand the following discussion of fundamentals, we must interpolate a glance at history. Our present day concepts, now known to every school-child, of element, compound, and decomposition, go back as far as the great English physicist and chemist Robert Boyle (1667–1691). He was the first really to state that it is needful to investigate the elementary constituents into which the substances found in nature can be decomposed.

On the other hand, the whole of the Middle Ages, and also the beginning of modern times, were ruled by the doctrines of Aristotle, who set out from a totally different position. The Stagirite's profound doctrine is based upon the difference between 'substance' and 'form,' which he obviously arrived at by observing the activities of craftsmen and artists. Both by their formative activity make matter (Latin *materia*, wood for building; Greek *hyle*, wood) into what alone interests us. Without this activity, matter is only unformed and indifferent mass. But since there are undoubtedly very many different kinds of matter, Aristotle further teaches us that these different kinds depend for their existence upon a formative process which has already been exerted upon a general matter, in itself without qualities. In accordance with this doctrine, for a detailed description of which the reader must consult text books of the history of philosophy,[3] it should be possible to change any one kind of matter into any other, if we could but discover what formative principle must act upon the substance in question in order to produce the required effect. This idea lay ready to hand, for the simple reason that it is actually possible, by the use of such forces of nature as heat, to carry out what we call to-day chemical transformations. 'Elements' in the sense of this doctrine are then not fundamental substances which cannot be further decomposed (our meaning of the word), they are the purest representatives of the single formative principles; water is the combination of cold with moistness, air the combination of cold with dryness, etc. (From our point of view the elements of the ancients are thus hybrids, as it were, of state of aggregation and temperature).

The alchemists of the Middle Ages attempted, in accordance with this doctrine, to transform base metals into noble metals. This was obviously possible on the basis of Aristotle's view; and we know to-day that Aristotle actually anticipated our latest knowledge of the constitution of matter, which has only become clear to the science of to-day. In those days, however, such experiments were bound to fail on account of the impossibility of producing the transmutation of the chemical elements by the means of ordinary chemistry.

Boyle has the credit of having perceived this fact, and thus brought into being our present day concept of an element. In order to estimate truly the greatness of this service, we must realise how great a strain Boyle laid upon simple and direct thinking when he stated that, for example, the two substances, sulphur and mercury are present as such in the red mineral cinnabar. Cinnabar 'consists' of them, just as a house consists of bricks, mortar, and beams of wood, or a piece of cloth of threads. This assertion is entirely paradoxical, since nothing can be perceived of the properties of the two elements as soon as they are combined together. But the acceptance of this paradox gradually disentangled the confusion of the numerous single discoveries already made by the alchemists, and of other discoveries which were added to them. It is true that a further hundred years were necessary before Lavoisier finally put the system of the elements upon its true basis, for until then an error, the phlogiston theory of Stahl, had led investigation far astray. Lavoisier was the first to begin to carry out Boyle's programme; in a further 150 years, chemistry worked out the system of the elements practically completely. Of the 92 elements from the lightest to the heaviest (hydrogen to uranium), only two, an alkali metal and a halogen, are missing.

But how do we know this? How can we say in advance what elements must exist, since according to Boyle's proposition they are supposed to be fundamental constituents, which can only be discovered by experience? This brings us to a new set of facts which again put matters in an essentially new light.

The famous Swedish chemist Berzelius, who at the beginning of the 19th century performed a great service by setting up the system of chemical elements, and himself discovered quite a number of new ones, already noticed that certain relations of similarity exist between single groups of elements, and show themselves particularly in the numerical values of the atomic weights – a term which we shall presently explain. But it was only in the year 1869 that the German Lothar Meyer, and the Russian Mendeléeff, succeeded almost simultaneously in extending these relationships, hitherto regarded more as a play with figures than a serious scientific problem, and thus arranging the whole of the elements hitherto known (fully 60 in number) in a system of such relations of similarity. Since this system, the so-called *Periodic System* of the elements, plays a fundamental part in our whole physico-chemical knowledge of to-day, we must describe it a little more closely, assuming for a moment the fundamental concept of atomic weight.

Mendeléeff and Meyer noticed that when the elements are arranged in the order of their atomic weights, beginning with hydrogen, $H = 1$, and ending with uranium, $U = 238$, that elements similar to one another recur at definite intervals. If we leave hydrogen and also helium ($He = 4$) out of account, the first two 'horizontal rows' of the Periodic System, each consisting of eight elements, are as follows :

$$Li = 7 \quad Be = 9 \quad B = 11 \quad C = 12 \quad N = 14 \quad O = 16 \quad F = 19 \quad Ne = 20$$
$$Na = 23 \quad Mg = 24 \quad Al = 27 \quad Si = 28 \quad P = 31 \quad S = 32 \quad Cl = 35 \quad A = 40$$

The correspondence between the elements which are below one another here is almost complete. Upon these two so-called small periods, two larger ones of 18 elements each follow, namely,

K, Ca, | Sc, Ti, V, Cr, Mn, Fe, Co, Ni, Cu, Zn, | Ga, Ge, As, Se, Br, Kr
Rb, Sr, | Y, Zr, Nb, Mo, Ma, Ru, Rh, Pd, Ag, Cd, | In, Sn, Sb, Te, I, X

in which the correspondence is again almost perfect. Furthermore, the first two elements of these series (K, Ca, and Rb, Sr, metals of the alkalis and alkaline earths) correspond to the two first of the small periods, and the last six to the last six of the small periods, with great exactness. Hence the two 18-periods are best regarded as two 8-periods in which the ten elements enclosed by a line are interpolated. The table continues in a similar, though more complicated, way, and the reader may refer to the arrangement given in note 106.

What we have given here is however the present day arrangement of the Periodic System. In those days (1869), and up to a short time ago, most investigators attempted to carry a period of 8 elements right through ; this arrangement is still found to-day in many books on chemistry.   I even retained it in previous editions of this book, since it was then usual.   Furthermore, in Mendeléeff's time the system was far from complete. He was obliged to leave gaps at many points, that is to say, to assume that elements still unknown must exist to fill them, and it is one of the greatest triumphs of the human spirit of investigation, that elements foreseen by him, and characterised in their properties on the basis of his system, were afterwards discovered and found to confirm his predictions in all essential respects.   Above all, this is true of the elements which were later called by their discoverers in accordance with their own nationality, Scandium, Gallium and Germanium.   Later too, the discovery of missing elements also confirmed the predictions made from the periodic system, although complications arose.   The whole column of noble gases came much later.   Furthermore, we need to reverse the order at two points (K, A; Te, I), that is, to change the order given by the atomic weight, in order to bring about correspondence.   How these difficulties, which in earlier times as Ramsay said 'made it a painful pleasure to study the Periodic System' are solved to-day, we shall see later on.   But they could naturally not prevent the system, taken as a whole, from already appearing as a discovery of the first rank.

The conclusion which had necessarily to be drawn was that some inner connection, completely unknown to us, must exist between these 70 or 80 substances, and that therefore they might possibly be regarded as combinations in various patterns of one and the same fundamental substance, and hence that the old philosophic idea of the world-substance might be true.   In actual fact, chemists interested in philosophy drew this conclusion in many cases, but their more empirically minded contemporaries mostly showed little sympathy, since nothing concrete could be found in this general idea.   It needed the newer discoveries, which brought for the first time a true transmutation of elements into the region of practical experience.

We can only deal with these discoveries in detail at a much later point.   It must suffice here to point out that radium, for example, changes in the course of long periods of time, slowly but continuously, through a number of intermediate stages into lead, and that helium is also formed in the course of this process, both being undoubtedly elements having a definite place in the Periodic System, just as radium itself, which finds its correct position as alkaline earth metal in the column with calcium, strontium and barium.   Here therefore we have without any doubt a true transformation of an element, and after the discovery of radio-activity, many such cases became known.   Accordingly, no further doubt remains that our chemical elements, hitherto regarded as unchangeable, are not simply fundamental and invariable substances, but that at least between certain groups of them, and probably between them all, a genetic connection must exist, which, it is true, is not yet entirely cleared up to-day.

When this knowledge spread fairly rapidly at the beginning of the century, numerous voices were heard, which are not entirely silent even to-day, proclaiming with great emphasis the 'complete breakdown of chemical science,' or its 'bankruptcy,' or other pleasant things of the kind; and the public which loves sensation and is flattered when it finds that know-all science can also make bad mistakes, seized greedily on such statements, often indeed coming from authoritative scientific quarters, which ought to have known better.   What had really happened ?   Common salt continues to consist of sodium and chlorine, water of oxygen and hydrogen, and all substances we know of on earth are composed of one or more of the 92 'elements.'   In this respect the whole result of chemical investigation has not changed in the least.

The new fact is, that the result of these investigations shows our previous chemistry to be an intermediate stage to a step further on.   In other words, a new floor has been added to the structure of our knowledge, or its foundation has been carried down a floor further.   If we call the 92 substances in question ' elements of the first order,' and the fundamental substances of which they are again made up 'elements of the second order,' everything is correctly stated, and there can be no question of 'bankruptcy' or even 'revolution' in chemistry.   For no sensible chemist has ever had any doubt, that a smaller number of simpler elements might be found to make up the so-called chemical elements.   But chemistry as an empirical science stood, as was its simple duty so long as no experimental foundation for further hypothesis existed, by its 92 elements, saying that these could not be further decomposed by hitherto available means. It would have been a piece of quite unjustifiable dogmatism for the chemist to have gone further and maintained that the substances could not, under any circumstances, be decomposed, or transformed into one another. Such negative dogmatism would have taken him beyond the region of experience, which could simply bear witness that they had never hitherto been decomposed. It is possible that such negative dogmatism has appeared in scientific works, as for example in the one time well known materialistic work *Force and Matter* by L. Büchner. Such errors are human, and the opposite type, of which we spoke above, has also occasionally occurred in scientific work. But such single voices are not science.

Science had not merely not questioned, but often enough actually considered, before the discovery of radio-activity and since the discovery of the Periodic System, that an inner connection, still unknown to us, might exist between the so-called elements. But there was no reason in this fact for using any other term than the word 'element,' since for the moment, it covered the known facts completely; and although we have progressed a whole stage further, there is nothing that obliges us to abandon this term to-day. It is simply a matter of a word, of the most practical method of description ; and obviously the most practical thing to do is to keep unchanged the term by which chemical elements have hitherto been designated. The fundamental substances of the second order may then be otherwise described, in any way which best suits their nature. We shall see that they have already received a suitable appellation.

The incorrigible scholastic may, however, object that this is sophistry. Chemistry in the past has actually designated the 92 substances as elements, because this word was connected with the concept of a fundamental, untransformable material. Either these 92 substances are such elements or they are not. It was earlier supposed that they were, but now we know that this is not the case. We can only reply that the scholastic is forgetting that science does not deal like philology with the meaning of words, but with things which it designates by suitable words. What the word 'element' once meant for the philosophers of antiquity and of the Middle Ages (the etymology is not clear) is of no moment for the scientist, and has only an historical interest for him as soon as he has made use of this word for a well defined class of things present to his experience.

This is actually the case with the 92 substances in question. Into them all other substances can be transformed in accordance with Boyle's programme. And they show themselves, not by the Periodic System alone, but in numerous other ways, to be the next stage of decomposition lying behind the three-quarters of a million known substances, which in any case would have to have a special name, even if it had only just been discovered. It is entirely a matter by itself that these substances can again probably be decomposed and transformed into one another by using means totally different from existing chemical means; this fact has never been seriously questioned by chemistry. The scholastic thus ascribes to the word 'element' an absolute sense, which it could never have in the chemist's language, since he is only expressing an empirical judgment. By making use of this ascription of meaning quite foreign to chemistry, he is attempting to prove self-contradiction or revolution.

It was unavoidable that we should make clear by means of this example at the very beginning how foolish the great majority of such criticism of science is which finds so much public interest to-day. Any one who supposes that scholastic word-juggling of this description can inflict a wound, shows that he has not the slightest conception of the spirit of science. If a revolution is to take place, it needs to be proved that science has actually been guilty of essential errors. This is possible, and has actually happened often enough, especially in new sciences. In such cases it is right to speak of revolution and the like. In the present case no revolution has taken place; for what the new discovery might have refuted: the assertion, namely, that the 92 substances are absolutely unchangeable elements, has actually never been made by science; while what has been asserted, namely that all other substances may be decomposed into the 92, has not only not been refuted, but on the contrary, has been confirmed beyond all doubt.

False critics have committed still greater follies in respect of the two further concepts of chemistry now to be discussed.

II

MOLECULES AND ATOMS

We must again refrain from going too far into the history of atomic science.[4] But it is generally known that it arose in ancient times, and that the doctrine of its most famous representative in antiquity, Democritus, was regarded as the antithesis of the doctrine of Aristotle.[5] Democritus is regarded as the father of what was later called materialism, and this is true in so far as he appears to have taught (his teaching is only known to us through the writings of his opponents) that the formation of the various bodies and forms in the world is to be explained as arising from the action of blind forces, such as weight, upon the atom. This doctrine is evidently irreconcilable with that of Aristotle, which assumed in its formative principle the 'entelechy,' a force acting constructively according to a plan. But it is doubtful in what sense Democritus himself meant his doctrine to be construed. It is certain that his later disciple, the Roman Lucretius, taught a fairly undisguised form of materialism.[5]

As long as the doctrine of atoms was customarily regarded only from this general philosophical standpoint, which was often enough the case until modern times, and even well into the last century, it was simply impossible to come to any agreement concerning its value or otherwise. But matters are in an entirely different position as soon as such a concept gains a real footing in empirical science, and is thus used to describe matters of experience, or at least expressly introduced as an explanation of such matters. This has been the case with the doctrine of atoms since the beginning of the last century, when the English chemist Dalton and the Italian physicist Avogadro (1808–1811) succeeded in explaining by means of the atomic theory the two most important fundamental laws of chemistry, the law of multiple proportions, and the law of simple gas volumes.[6]

From that time the atomic theory never again disappeared from science, although physics was for very long doubtful about it. But in this science also, it finally gained a footing when in

the years 1856 to 1857 the two German physicists, Krönig and
Clausius, succeeded in referring the fundamental laws of heat to
the hypothesis that heat consists in the movement of molecules
or atoms (see below).   Curiously enough, it was precisely in the
field of physics that a strong reaction against it once more
occurred.   Towards the end of the 19th century, the whole
body of physicists, and also many chemists, were of the opinion
that the atomic theory was certainly only an accessory notion,
which had no real value in itself as knowledge.   It was 'a scaffold'
which can be taken down when the building is finished (Bucherer),
a picture or model (Hertz), which we are able to make of phenomena, since they behave in many respects as if matter consisted
of atoms.   Indeed, there are a few voices of this kind still to be
heard to-day, although in the meantime the true existence of
atoms has been put beyond any doubt, and there can no longer
be any question of a mere 'as if.'   We shall return later to the
deeper epistemological background behind this rejection of
atomic theory.   For the present it is our task to state the present
condition of the atomic theory, in which we shall be obliged to
be somewhat dogmatic, since a full statement of the basis would
take us at many points too far into special science.

The essential point of modern atomic theory may be summarised
by Ostwald's statement that 'matter has a grained structure,'
which means to say that it does not, as would appear to direct
observation, fill the space it takes up fully and uniformly (homo-
geneously), but instead, possesses a structure of dimensions lying
below the limits of even microscopical visibility, this structure
being something like a network of fine meshes, or a piece of sand-
stone, or a cloud of mist particles.   If we imagine a water surface
covered with very many small waves, this also is inhomogeneous,
without therefore being discontinuous, that is to say interrupted.

We must guard against supposing this statement to be simply
equivalent to the other, that matter consists of single particles
separated from one another by empty spaces.   In the same way
a photograph of the heavens, taken out of focus, would show,
not single points of light separated from one another, but a
diffused sheen, which would be more concentrated at certain
points, without its being possible to distinguish sharp boundaries
with ease.   Such might also be the case with matter, although
it is true that the simpler and more obvious picture of separated

particles is usually taken as a basis. But we must at this point be clear about the fact that this special form of the picture may be abandoned without the essential point, the inhomogeneous, grained occupation of space suffering in any way.

In order to find our way about this world of the smallest dimensions, it is general in present day physics to make use of the unit of length 1 micron, μ = 0.001 millimetre, and 1 milli-micron, $\mu\mu$ = 0.001 μ = 0.000001 millimetre; a unit ten times smaller is also frequently used, namely, the so-called Ångstrom unit = 0.1 $\mu\mu$. The limit of visibility for the ordinary microscope lies about at 0.2 μ = 200 $\mu\mu$. By means of the ultramicroscope of Siedentopf and Zsigmondy, we are able to recognise the existence, but not the form or size, of particles of a size about 4 to 5 $\mu\mu$, assuming that they are far enough apart, and are sufficiently different optically from the surrounding medium. Now the 'grained' structure of matter which we are considering, is only a little finer than these extreme obtainable limits of direct visibility. The distance apart of molecules or atoms in ordinary bodies is about 0.1 $\mu\mu$ or 1 ångstrom unit. How this is measured will be seen later.

Since the concept of a molecule has become a little variable in recent times, we shall do best to start from chemistry, the central idea of which is the atom. There are as many different kinds of atoms as there are chemical elements (see however pages 129 ff.), that is to say 92. Every defined substance, such as water, common salt, sugar, or the like, contains the atoms of its elements in a quite definite numerical relationship to one another, which is given by the chemical formula (in the cases named, H_2O, NaCl, $C_{12}H_{22}O_{11}$). The weights of the different kinds of atoms are different from one another. The so-called atomic weights of chemistry are relative numbers, which tell us how many times each kind of atom is heavier than the lightest of all, the hydrogen atom, H.

These 'relative atomic weights' were being found with ever increasing accuracy by chemists, even from the beginning of last century, by means which cannot be given here in detail.[6] Every year, the International Commission on Atomic Weights produces a revised table, which to-day usually undergoes little change. The exactitude of the determination usually extends as far as the fourth figure, but in many particularly important cases the fifth figure has been reached. In this book we shall only reckon with round numbers, unless it is precisely the deviation from these which we are considering. Since the atomic weight of oxygen is not exactly 16, but 15.88, and since most atomic weights are not determined directly by reference to hydrogen, but to oxygen, it is general to-day to follow Ostwald's proposal, and put $O = 16$ instead of $H = 1$. The table printed at the end of the book is based upon this system. H then becomes, instead of 1, 1.008. This figure is of importance in connection with an investigation to be discussed later.

So far, there is no difficulty in carrying out the idea. The matter becomes somewhat more complicated when we proceed to discuss the concept of a molecule. The most obvious notion, which is also that introduced by Dalton with the atomic theory into chemistry, is clearly to regard the complex of atoms given by the chemical formula of a compound, such as H_2O, NaCl, etc. as the smallest portion of the substance in question which can exist by itself, and this was the view held throughout the whole of the 19th century. These complexes of atoms were called molecules, and it was tacitly assumed that these are present as such in the substances in question, and that water therefore, to keep to it as an example, consists of single molecules H_2O, which again consist of the three atoms. We have reason for assuming that this is actually the case with gases and in part also with liquids. In the case of gases there can be no doubt that such molecules or complexes according to the chemical formula exist, and that these molecules fly about freely in space like gnats in a swarm, whereby they continually collide with one another, but between collisions pursue a 'free path' (see page 91). Avogadro drew from the chemical facts in 1811 the following conclusion, which also follows necessarily from the physical facts of heat, that:

All gases contain an equal number of molecules in equal volumes at the same temperature and pressure (Avogadro's rule).

If this whole doctrine is accepted, the facts further compel us to ascribe to many free elementary gases, such as hydrogen, oxygen, chlorine, molecules which are not simply equal to an atom of the elements in question, but contain two or even more atoms of like kind. Gaseous hydrogen consists of molecules H_2, which just as much represent a chemical combination as say the molecule of gaseous hydrochloric acid, HCl. The difference that in one case the two atoms combined into a molecule are of the same kind, and in the other of different kinds, is of no essential importance. This is not true however for all elementary gases. The noble gases (helium, argon etc.) are monatomic, as are also the vapours of metals, such as sodium or mercury. The various facts upon which these results are based cannot be stated here; the reader will find them in any text book of chemistry. If the atomic theory is once accepted, these conclusions follow of necessity.

In the case of solid and liquid bodies, matters are considerably more complicated. Let us first take the opposite extreme of the gas, namely the solid. Here the molecules or atoms are fixed as a whole. When they move, and they do so in all bodies according to the kinetic theory of heat (see below), they can only oscillate to and fro about a middle position. They are retained in this by the forces exerted upon them by neighbouring particles. In the case of liquids, these forces are, on account of the greater distance apart, already so much smaller that they are not sufficient to keep the molecules in one position, but a particle in the interior of a liquid never gets right outside the range of action of these forces.

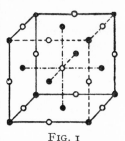

FIG. I

Structure of common salt. • = Na atom, o = Cl atom

For this reason, the particles of a liquid are capable of displacement to any extent relatively to one another, that is, a sensible restoring force (elasticity) does not occur in them. Nevertheless they still possess cohesion, although this is small as compared with solid bodies, and their particles do not separate freely from one another as do those of gases. We are, unfortunately, still without exact information concerning the details of the facts just given in broad outline, since we still know too little with certainty concerning the forces exerted upon one another by the individual parts.

It is therefore rather difficult to make out to what extent the chemical molecules, such as H_2O or the like, have a real existence in solid bodies. In certain definite cases we know with fair certainty that this concept loses its meaning. Common salt NaCl, for example, is crystallised in cubic form when solid. In accordance with the latest results of X-ray investigation, which will be discussed in detail later, it is now settled that the state of the crystal is that of a regular arrangement of its parts, a so-called space lattice, which in the case of common salt has the form shown in figure 1 (the sodium and chlorine atoms are shown by small circles and dots respectively).   We see at once that there is no sense here in speaking of single molecules NaCl, and that this formula merely states that in the whole crystal, which contains many billions of atoms, there is always one chlorine atom present for every sodium atom.   In other crystal lattices, how-ever, the single atomic groupings form relatively independent complexes among themselves, so that one might speak of single molecules in such cases.   But when such a crystal evaporates or dissolves in a liquid, it is usually not single atoms, but again definite complexes that leave the space lattice, and move about freely in the solution or gas space.   These are then again chemical molecules in the earlier sense.

Matters are more complicated in the case of highly complex organic substances such as cellulose or starch.   We shall return to these later in another connection.   The reader should not allow himself to be daunted by these difficulties.   They prove no more than that matters are considerably more complex than was thought 40 to 50 years ago.   This has nothing whatever to do with the existence of atoms.   The sole question here is in what way they may be supposed to build up the various forms of matter, solid, liquid and gas.   The molecule is in many cases an inevitable accessory concept, and in others superfluous, and neither too much or too little value should be attached to it.

The latter fact is seen most clearly by such examples as the molecule of acetic acid $C_2H_4O_2$, or that of grape sugar, $C_6H_{12}O_6$, which as we see in both cases consist of a multiple of the simple formula $C\,H_2O$, but nevertheless are not to be regarded as like this simple molecule.   The latter also exists, it is the substance formaldehyde (formalin), well known as a disinfectant. But acetic acid has molecules twice as large, and grape sugar six times as large, facts which can be conclusively demonstrated from certain physico-chemical facts.   We must therefore beware of concluding that the concept of a molecule has no further meaning to-day.   All that is meant is, that the question as to what higher complex units are to be regarded as forming the fine structure of a liquid or solid substance, needs in each case special investigation, and such investigations are continually being made.

An equally difficult chapter of modern investigation consists of a number of other facts, which were regarded 50 years ago as much simpler than they actually are; the *valency* of the elements. As soon as the formulae of the more important groups of atoms had been obtained fairly completely, a peculiar regularity regarding the number of atoms combining together appeared, which may be shortly stated as follows. If we start from the simplest and lightest element, hydrogen, there are first found a number of elements which combine with hydrogen atom for atom, as we see from the formulae HCl, HBr, HI and HF, and further, a number of metals which replace the hydrogen, atom for atom, in these compounds (the so-called halogen acids), thus for example NaCl, NaBr, NaI, NaF. All these atoms, which thus combine with or replace an atom of hydrogen, are called, as is hydrogen itself, monovalent.

There are also other atoms, such as O, S, Ca, Ba, Cu, Zn, etc., which combine with or replace two atoms of hydrogen or other monovalent element. These are called divalent or bivalent.

Examples of these are H_2O, H_2S, $CaCl_2$, $ZnCl_2$, and further CaO, Cl_2O. In exactly the same way, we explain the concepts of trivalent, tetravalent, etc., elements.

Nitrogen in particular is trivalent, carbon tetravalent, as is seen by the example NH_3, NCl_3, CH_4, CCl_4, $CHCl_3$ (chloroform), CO_2, etc.

An idea of this regularity – it is not a law, since there are too many exceptions to the rule – is best gained by imagining on each atom a certain number of points to which other atoms may be attached, which may be represented by hooks, or more simply by lines, thus:

$$H - \qquad -O- \qquad -N- \quad \text{or } N \equiv$$
$$\text{or } O =$$

The rule then says that the atoms usually combine in such numbers that no free lines (valencies) exist. We thus get for example the following diagrams:

$$H-O-H \qquad \begin{matrix} H & & H \\ \ \diagdown & & \diagup \\ & N & \\ & | & \\ & H & \end{matrix} \qquad \begin{matrix} H \\ | \\ H-C-H \\ | \\ H \end{matrix} \qquad O=C=O$$

Now there is actually more in these diagrams than they were originally intended to express. For if the whole idea rests somehow or other upon a real property of the atoms, these formulae give us at least a suggestion of the structure of the molecule, the linkage of the atoms within it; such formulae are called structural or constitutional formulae. In actual fact, this simple fundamental idea has enabled organic chemists to solve, in the course of about 50 years, the hitherto insoluble riddle of organic substances almost completely. On the basis of structural formulae thus discovered, for example, that of indigo:

Cs

we manufacture to-day all the innumerable products of the organic chemical industry, in particular tar products (colours, medicines, etc.).

Of recent times, so many objections have arisen to this simple original form of the doctrine of valency that it has been necessary to change it essentially in order to do justice to newly discovered facts. In place of valencies, we have to-day the so-called co-ordination numbers of Werner, but the whole matter has thus become so complicated, that we cannot deal with it in a book intended for the lay public. Fortunately, organic chemistry, that is the chemistry of carbon compounds, is very little affected by it. Only because we shall have to refer to the latter later on, have we gone into it at all.

To summarise what we have said so far, we may state the position as follows. While the existence of atoms themselves is proved, the question as to how matter in various states is built up from them, cannot yet in all cases be answered with certainty. But in certain cases this construction is fairly well known, and this is true in particular of the simpler crystal lattices and of the structural formulae of organic chemistry, in which a very considerable portion of the truth must certainly be contained, as otherwise the technical practice based upon them would be quite incomprehensible. It is further also certain, as we shall see more clearly below, that the molecules of all bodies are in continual motion, and that this motion is what our nerves experience as heat. Considerable uncertainty exists even to-day concerning the more detailed nature of the forces exerted by the atoms upon one another, by which the state of aggregation is conditioned. We shall see later what is known to-day concerning this question.

Having thus given a short sketch of the present day atomic theory, we may inquire by what right this, after all very hypothetical idea, actually remains a part of science? Upon what is based the assertion, repeatedly made above, that the real existence of atoms is an undoubted fact? Is not this after all something beyond the capacity of a purely empirical science to prove? The answer to these questions is of the greatest general interest as regards the theory of knowledge. The atomic theory has always been rightly considered the pattern of all physical and scientific hypotheses. What is true for it, will therefore be true in the widest degree, and for most other scientific hypotheses.

The following are the reasons which have forced present day physics and chemistry to recognise the *real existence of atoms and molecules* as an undoubted fact.

1. It is a fact that all matter, when reduced to sufficiently thin films, namely, as soon as these films approach the limit 1 $\mu\mu$, assumes fairly suddenly other properties than those which it had in thicker layers. The thinnest films which have recently been produced by mechanical means were the metal foils of 10 $\mu\mu$ thickness, produced by Müller some years ago.[7] In the case of these foils, which still contained about 30 layers of atoms, the properties were, in all essential respects, the same as those of metal in ordinary thickness. Much thinner layers than these can however be produced by other methods, for example by the dissipation of wires traversed by electric currents (every one knows the very thin layer that is formed in this way inside our electric lamps). Such layers behave, as regards their electric conductivity, as if they were full of holes from a certain limit of size downwards. The same is true of other properties, in the case of soap bubbles, in oil films on water, and other cases of the kind. Very numerous investigations of this kind during the last three decades, have agreed in proving that the sudden change of properties occurs close to the limit known for other reasons as the range of molecular action. Scarcely any other explanation than this can be given, that a layer consisting of only one or two molecules is obviously quite different in quality as regards its whole condition of internal connection, from one containing 20 or 100 or more molecules in thickness.[8]

2. In recent years a new department of physico-chemical investigation has been developed, namely, *colloid chemistry*, which one of its chief investigators, Wo. Ostwald, has not inaptly described in a recent work as 'the world of neglected dimensions.' In order to explain this, we must digress for a little. At one time a strict distinction was made between the solubility or insolubility of a solid or liquid in another liquid. Common salt, for example, is soluble in water, sulphur insoluble, but the latter is soluble in carbon bisulphide, which itself is insoluble in water

(not miscible with water). Now in spite of this, it is obviously possible to take a liquid or solid body which is insoluble in a given liquid, and grind or otherwise divide it into particles so extremely fine, that they remain suspended in the liquid in which they are insoluble for quite a long time before they separate by sinking to the bottom of it or rising to the top of it, according to their density. So close a mixture is called a suspension or emulsion. If a solid body is suspended in liquid in a finely divided form, such as carbon in Indian ink, we speak of a suspension, but if one liquid is suspended in another, such as the drops of fat in milk, we have an emulsion.

These mixtures are distinguished from true solutions (such as those of common salt or sugar) in their external appearance, since they are turbid, and further by the fact that the suspended substance always collects after some time either at the top or at the bottom of the liquid as the case may be, or at least can easily be separated by centrifuging (milk) ; and finally by the fact that the ordinary microscope enables us to detect the single suspended particles. However the old maxim: *Natura non facit saltum* once more holds good. About a hundred years ago mixtures of substances were already known which could not be made to fit into this scheme of solution or suspension, and recently innumerable other such cases have been found. Certain reasons would lead us to assign them to the one class, while others would seem to place them in the other class. These mixtures, first investigated by Graham (1861), were called by him colloids or colloidal solutions; gum (Latin: *colla*) being one of the chief examples.

Very extensive investigation in recent years has proved that these *colloids form a completely continuous transition from suspensions and emulsions to true solutions*, it being simply impossible to define a boundary. Slight changes in the conditions of preparation enable the size of the particles dispersed in the liquid to be changed within wide limits. There are colloids which by their turbid appearance exhibit their closeness to suspensions, but in many such cases a slight change in the conditions of preparation results in the same substance forming a colloidal solution which appears perfectly clear, and even seems to be homogeneous under the ordinary microscope, but allows the suspended particles to be detected in the ultramicroscope, in which the illumination is across the field. In other cases, such colloids capable of being detected by the ultramicroscope are continuously connected with others for which our finest optical instruments fail us, but nevertheless allow of a separation by an ingenious method of filtration invented by Bechold, and called 'Ultra-filtration.'

When this method of separation also fails, the particles having been made still a little smaller by a slight change (still entirely continuously), centrifuging may still allow of separation, but we have then already reached mixtures which have always been considered to be true solutions. For example, Bechold was able to separate dextrine solution to a great extent by ultra-filtration, while Lobry de Bruyn was able to produce differences of concentration in sugar solutions by centrifuging (sugar, being heavier, moved outwards), and lately similar effects have even been produced in salt solutions. Svedberg has done most to demonstrate the completeness of continuity.

But if this is the case, it follows that what applies to colloids, must also apply to true solutions. No one can have any doubt in the case of the former, when a small change of temperature or concentration has made the particles so small that they can no longer be seen in the microscope, that they still exist (ultra-filtration being a proof of this fact, if a proof be necessary) and hence it is impossible to doubt that they still exist when a further small change results in a product which has all the properties of a 'true solution.' But this brings us to molecular magnitudes, for the quantitative examination of all these phenomena shows us, that the transition occurs when the particles reach the degree of smallness (at which they are no longer visible) which had been ascribed to molecules for other reasons altogether. I should state expressly, that reading the original work of Svedberg is much more convincing than a mere summary of this kind can ever be.

We may add that the quantitative explanation of such suspensions and colloids has resulted in finding ways and means of determining the absolute number of molecules or atoms in a given quantity of matter. According to Avogadro's law (see above, page 14), the number of molecules of all gases contained in a certain volume, for example a cubic centimetre, is equal under like conditions. Avogadro himself probably regarded as hardly possible that this number could ever be actually discovered, supposing it simply to be enormously great. The Austrian physicist, Loschmidt, was the first to calculate this number in the year 1865, on the basis of the result of the kinetic theory of heat, which we shall discuss later on. *The Loschmidt number N, is the number of molecules in a cubic centimetre of gas at 0°C, and atmospheric pressure (760 mm. of mercury).* Loschmidt's results were not very exact, but the true value of his number is, according to the latest investigations, 27.2 trillions with an uncertainty of about $\frac{1}{2}\%$. One of the most famous, and at the same time simplest investigations which aimed at the determination of this number, was Perrin's counting of suspensions,[9] which belongs to the above discussion concerning suspension. We shall discuss other methods later.

3. The proofs which we owe to recent investigations of radio-activity and allied phenomena are still more convincing, since they admit of direct observation. We cannot yet deal with them in detail, since an understanding of them requires the development of further concepts, and we must be content with the remark that the investigations in question, although they do not allow single atoms or molecules themselves to be directly observed, allow the effects of them to be seen and even photo-graphed (see page 131), so that the path of a single such particle can be directly followed in the surrounding gas space, and the number of electrically charged particles thrown off by a small quantity of radium in a given time can be directly counted. This method again gives us means of determining the absolute size and number of molecules, and the results agree excellently with the Loschmidt number as given above.

4. Just as in the case of the radio-active processes we can count individual charged atoms passing through space, so also can we investigate particles of other substances suspended in gases and electrically charged. This method again leads to a determination of the atomic constants with the same result as before (Ehrenhaft, Millikan; see page 131).

5. In crystallised substances, the molecules or atoms take up positions which, according to all that we know, are arranged according to certain laws, and form a point lattice in space, as we have already described. Now a regular arrangement of this kind of suitable fineness acts upon light waves, as we have known for a very long time, in such a way as to produce what are called *diffraction* effects. Everyone knows the haloes, which

are visible around the sun or moon when the air is filled with fine ice crystals, and likewise the coloured rings around a bright light when seen through a dusty or moist pane of glass.

These phenomena, which have been known for more than one hundred years in optics, and have been exactly investigated and completely explained by the wave theory, should also occur when light waves pass through a regular point lattice such as a crystal. They however require that the size of the light wave (the wave length) should be approximately comparable with the distance apart of the points of the lattice. Now this is not the case with ordinary light waves, the length of which is about 400 to 800 $\mu\mu$ (see page 116), while the distance between the molecules, as we have seen, is about 0.1 $\mu\mu$. The light waves are therefore much too large. They bear much the same relation to the molecules as do the waves of the sea to a coarse sieve. In such a case, theory tells us that no diffraction phenomena are possible. However, we know of a kind of light which has much shorter waves, namely the X-rays. In their case the wave length is less than 1 $\mu\mu$. V. Laue, and his co-workers Friedrich and Knipping, actually succeeded in getting wonderfully sharp diffraction images from X-rays by passing them through crystals; and thus gave us a new and beautiful confirmation, not only of their lattice structure, but also of the molecular theory itself. For further details see page 167.

In this connection we may mention that the mere existence of crystallography, with its fundamental fact of 32 possible groups of crystal forms, is likewise a good confirmation of the molecular theory, since these 32 groups, as Bravais, Sohncke, Schönflies, and others have shown, may also be deduced theoretically as all the possible forms of a regular point lattice.

6. Besides these direct proofs of the actual existence of molecules or atoms, there are others almost equally convincing, which have recently been added to the already existing indirect confirmation of the theory. Particular attention should be directed to certain recent results of the kinetic theory of heat, in particular the kinetic theory of gases. We have already referred to Loschmidt's first calculation of N. This depended upon formulae deduced by Maxwell, van der Waals, O. E. Meyer, and others for the kinetic theory of gases, in which experimentally measureable quantities, such as the co-efficient of internal friction of a

gas, the rate of diffusion, the conductivity for heat, and other quantities are connected with certain molecular magnitudes, above all with the 'mean free path' of a gas molecule, that is, the average distance travelled in freedom by a gas molecule before it again collides with another gas molecule.

Loschmidt was the first to show how one can find N by a quite simple calculation from a free path thus determined.[10]　But his results for N were entirely different according to the gas chosen, the constants of which were used in the calculation (for example, for hydrogen 11, for cyanogen 89 trillions).　Since this is a crass contradiction of Avogadro's law, which itself is placed beyond a doubt for other reasons, it was necessary at that time to regard the whole calculation with great mistrust.　At the very most, the order of magnitude of N might be regarded as fairly probable. It is also easy to recognise that the calculation, as carried out by Loschmidt according to methods originated by his predecessors, could not possibly lead to satisfactory results.　For the calculations all depended upon simplified assumptions concerning the collisions between molecules, the manner in which a molecule fills the space occupied by it, etc.; and these assumptions were bound to be far from reality in the case of molecules of anything like complicated construction.

These assumptions had to be made for want of anything better, and in order to make the application of mathematics at all possible, but it was not surprising that the results were only approximate.　Now it is an obvious question to ask which gases are most likely to give a correct result.　It is self-evident that these must be the monatomic gases, which were only discovered in more recent times, after Ramsay, Rayleigh and others had introduced the noble gases, helium, argon, neon, etc., to us.　And in the case of these we have a really excellent confirmation of the theory. Not only did it appear that numerous relations required by theory hold strictly for these gases, and particularly for helium and argon, relations that are only approximate in the case of other gases, but Loschmidt's method gave numerical values for N almost identical with those already known; Sirk obtained for argon 27.9, Ghose for helium 28.0 trillions.

7. The whole theory thus finds its chief support, less in any single direct proof, however striking this may be, than in the surprising agreement between all the quantitative results obtained

in the most various ways.   It is obviously quite impossible that all these values should be the same as a result of pure chance.

We give here a short summary,[11] which, however, is by no means complete :

AUTHOR	METHOD	N in trillions (10^{18})
O. E. Meyer, Chapman and others after van der Waals	Kinetic gas theory	(1–106)
Sirk	Kinetic gas theory (argon)	27.9
Ghose	Kinetic gas theory (helium)	28
Perrin	Suspensions Brownian motion	(29–31)
Svedberg	Brownian motion	27.7
Ehrenhaft	fine metallic particles	27.6
Millikan	similar	26.4
Rutherford–Geiger	Radio-activity	27.8
Regener	Radio-activity	27.0
Rutherford–Boltwood	Radio-activity	27.0
Mean of values not in brackets...................		27.4

The best recent determinations depend upon the measurements of the elementary electric charge (see page 131) by Millikan and Ehrenhaft. The most reliable value for this at present is 4.772×10^{-10} electrostatic units.   This gives for the Loschmidt number 27.25 trillions, and for the mass of an atom of hydrogen 1.62 quadrillionths of a gram.   This is the number by which we must multiply the 'atomic weights' to obtain the real weight of a single atom of each element.

Having thus become acquainted with the convincing reasons which lead physics and chemistry of to-day to admit the real existence of atoms and molecules, we will consider the philosophical consequences of the fact.

III

THE MEANING AND VALUE OF
PHYSICAL HYPOTHESES

In spite of all that has been said in the last few paragraphs, many readers may still be left with the uncomfortable feeling that we are after all dealing with things that no one has ever really seen, and that, indeed, the theory in question for ever removes them from the possibility of being seen and touched. For light waves, as already indicated on page 23, are naturally much too coarse for the size of the molecules, and the fact that single molecules cannot make any direct impression on the skin will also be readily believed. Also, what is seen on photographs in experiments by Regener, Rutherford, Wilson and others, referred to above and later to be fully described, is not actually the single atom itself, but rather an effect produced by a single rapidly moving atom: a flash of light, a galvanometer deflection, or a chain of drops of mist. Is it still permissible, in spite of this, to maintain that not only these visible effects, but also their assumed causes, the single atoms and their motion, really exist? Are we not thereby going beyond the limits of what has actually been determined by experiment?

The answer to these questions brings us right into the middle of epistemological considerations. We will first state the question quite precisely as follows: are we to regard molecules and atoms as real things in the same sense, and with the same certainty, as bricks or plant cells? In other words, is the judgment that matter consists of atoms, true in exactly the same sense as the judgment: this house consists of bricks, this leaf consists of cells? Our first direct impression leads us to say no, for we can see the bricks and the cells directly but not the atoms. Is this answer a correct one?

Physics teaches us that the visibility of a body depends upon the way in which it behaves towards the rays of visible light which fall upon it, the wave length of these rays lying between 380 and 760 $\mu\mu$. In surroundings of a like refractive index, a body is invisible, since in this case it has no influence on the light waves. Is it therefore not there at all? No, for it can be made visible in other ways, or its presence can otherwise be proved. But when a body is very small, it no longer influences the light waves in such a way that we are able to perceive its form and size. But we are nevertheless able to cause it to produce an effect upon our eye, namely by using the ultramicroscope already referred to, in which it diffracts the light which strikes it from the side. This produces on the eye the impression of a point of light, which however has neither form nor size. Now no one will doubt that such particles, beautifully visible in colloidal solutions of gold, for example, are present, even although we are not able to recognise their form and size, in spite of the fact that in most cases this is a matter of great importance for us.

Let us now go a step further. By adding suitable reagents to this gold solution we are able to throw down a brown precipitate, which then causes the points of light to disappear, and this brown mud can then be transformed by further manipulation into the well-known yellow and shining metal. How did we know that those points of light were particles of gold? Only from a whole chain of observations, which are so closely connected with innumerable earlier experiences and resulting judgments, that a whole book would be required for the complete analysis of one single case.

The same is true in all other cases. All our judgments concerning so-called facts, even those which appear as simple as the cases of the house and the leaf referred to above, turn out upon closer inspection to be an almost inextricable complex of single observations and chains of judgments which connect these observations with other observations. Not a single so-called fact of experimental science is an exception to this position of affairs. Let anyone, for example, try to bring to mind the innumerable single observations and judgments which lead to the result, that the earth is approximately a sphere with a radius of 6370 kilometres. This is surely a fact, or do the whole of geography and geology rest upon an agreement to act as if the earth were such a sphere?

If we were to take up such an attitude – necessary, it is true, in a certain sense in considering the theory of knowledge – we should be clearly obliged to recognise that there are no such things as facts in the ordinary sense of the word. For then, only the single sense impressions would 'really exist,' and strictly speaking only my own sense impressions and these only at the present moment, for we pass beyond direct experience when we assume that memory is trustworthy, and further, that we recognise the existence of minds other than our own.[12] Hence if empiricism is truly logical, it does away with all possibility of science itself.

But as soon as we decide, in order to explain the existence of science, to admit the necessary extension, it is impossible to avoid seeing that all statements concerning reality made by science are fundamentally similar in nature. They are all complicated structures of data of the senses, definitions of concepts, and logical conclusions, and this is just as true of so-called judgments of fact, as of judgments based for the present on supposition (hypothetical judgments). The difference between the latter and the former does not consist in their being quite different in nature, but simply in the fact that we believe and have the right to believe the one to be as good as certain, whereas in the case of the other we are still uncertain whether we may not be wholly or partly mistaken.[13]

From this point of view, the judgment that matter consists of atoms, is in itself in no way different from the judgment that plants consist of cells. Only the degree of complication of the relations which lead to the first judgment is considerably greater, and this judgment is therefore not so simple to reach as is the other.

Furthermore, a deeper insight shows us that no group of sense impressions has any fundamental priority as compared with any other. It is simply custom which leads us, since man is a seeing animal, to regard 'direct visibility' as the simplest means of distinguishing reality from pure supposition. If we were able to see by means of X-rays instead of by ordinary light with its much longer wave length, we should without any doubt be able to see the fine structure of matter just as directly as we are able to see the cell structure of plants in the microscope. In the latter case also, an apparatus is placed in front of the eye, and we do not see the cells directly in the literal sense. However, no one seriously doubts that the picture seen in the microscope is in general similar to reality, and even geometrically similar, if we leave distortion out of account, which itself can be exactly calculated.

In principle, Rutherford is doing no more than this when he interposes between the flying atom and the eye a somewhat more complicated apparatus, consisting of an ionisation chamber together with a galvanometer; only in this case the way is somewhat longer. *All investigation of nature depends upon an enlargement of our senses of this description.*[14] Only the most naïve and common-sense theory of knowledge could blind us to this state of affairs, and persuade us that it is an entirely different matter when we are able to see directly.

That is real which has a real effect. The effects of single atoms can be proved to exist in numerous different ways. Hence these single atoms also really exist. What they are, is another question, which is naturally not answered by the mere fact of their existence, just as little as an observation in the ultramicroscope, which shows us only single points of light, can tell us without the assistance of other and further facts, what kind of thing causes this diffraction of light. We shall see later that the atoms are certainly something entirely different from what they have been supposed to be by certain crude materialists: 'rigid lumps of reality,' which have existed from the beginning of time and will continue to exist for ever. But whatever we may learn about them subsequently will make no difference to the fact that these small units of matter exist as such, and hence represent a completely definite physical concept, which corresponds to a thing which can be found in reality. *Atoms are just as real things as cannon balls or grains of sand, as waves on water or mountains.* Whether the former or the latter analogy affords a better picture of them and their properties, will appear later on.[15]

If we hold fast to these facts, we are then prepared to turn our attention to the more general epistemological questions which lie behind this special question, and it is now necessary to devote a few words to the history of the theory of physical knowledge during the last half century.

During by far the greater part of last century, the relations between science and philosophy were fairly strained, or even non-existent. The responsibilty for this lay with the unfortunate 'nature philosophy,' of the Schelling-Hegel period, which was arrogant enough to thrust aside laborious empiricism and attempt to distil the essence of all things from pure speculation, according to the scheme thesis-antithesis-synthesis. As an example of the nonsense which resulted we may take Hegel's remarkable definition of heat: 'Heat is the self-restoration of matter in its formlessness, its liquidity the triumph of its abstract homogeneity over specific definiteness, its abstract, purely self-existing continuity as negation of negation, is here set as activity.'

It is easy to understand, as Verweyen quite rightly says, that the scientist turned away from such 'science' with a shudder, and that the result was the production in a majority of scientists of the belief, not yet extinguished, that philosophy is nothing but completely useless chatter, in the best case devoid of content, in the worst simply nonsensical, concerning matters about which it knows nothing, and concerning which nothing whatever can be said in any other way, excepting by the slow advance, step by step, of experimental science.

When the breakdown of Hegel's philosophy occurred about the middle of the century, and the academic philosophers determined upon a 'return to Kant,' and hence to exclusive pre-occupation with epistemology, the breach between philosophy and science, at least in Germany, was only the more complete, for the latter was fully occupied with the ever increasing mass of experimental results, while philosophy for its part, moved at a height of abstraction which had no interest for most scientists. At only one point was there direct contact between philosophy and science, namely in the matter of the biological doctrine of development or descent, which immediately triumphed in science after the appearance of Darwin's book in 1859, and then, as we know, influenced not a few scientists very strongly in the sense of a materialistic philosophy. It is characteristic of the academic philosophy of that time that it could find no other means of opposing this tendency than epistemological considerations,[16] for it had abandoned metaphysics, to which the question really belonged, and hence its effect remained limited to a narrow circle of persons with a philosophical training. All in all, this question merely served at that time (about 1860 to 1890) to enlarge the breach between science and philosophy.

The reaction was first brought about by the fact that within the confines of science itself, originally in physics and chemistry, insistent philosophical questions arose; again initially of an epistemological character, and such that they could not be simply dismissed as superfluous by science. Indeed there was no desire to do so, since – and this is one of the strangest paradoxes in the history of philosophy – it was hoped that in this way philosophy might finally be got rid of altogether. The science of those days was empirical to the last degree. It recognised only experience, and again experience, as the true source of knowledge; every kind of speculative activity appeared fundamentally objectionable, and could at the most be regarded as an aid to reaching further experimental results.

Starting from this fundamental idea, several scientists arrived almost simultaneously in the last decades of the century at the idea of examining physics and chemistry to see whether, after all, these sciences did not still contain speculative 'metaphysical' elements, which would then require to be removed as quickly as possible, for 'metaphysical' was considered to be synonymous with superfluous and even injurious. 'The tendency of this essay is anti-metaphysical' – so begins the preface to a work which became of leading importance in this direction: Ernst Mach's *The Development of Mechanics*. Ostwald states his task in quite similar words at the beginning of his work on *Nature Philosophy*. The fact that he had the courage to make use of this ill-famed term was already remarkable, and did not do his work any good. Fundamentally he was pursuing the same aim as Mach, to whom the book is dedicated, though he afterwards put forward, with his theory of energy, something altogether different from what he intended, namely a system of metaphysics. This school of thought is represented in England by W. J. M. Rankine and Karl Pearson.

A similar antimetaphysical tendency, less explicit than in the case of these two but not less clear and plain, is found in innumerable writings[17] of the leaders of science of those days, and even of to-day. Particularly in introductions to text-books of physics and chemistry, in speeches at the openings of congresses and festivals, intended for a larger public, we find innumerable such sentences, as for example, the following, chosen at random from Kirchoff: 'The task of science is to describe phenomena which take place in nature as completely as possible and as simply as possible.' In all these cases, the antimetaphysical tendency in question is chiefly directed against those elements found everywhere in science and called hypotheses, of which the best example is the atomic hypothesis (others being the wave theory of light, the kinetic theory of heat, etc.). Mach and Ostwald regarded it as an ideal, only however to be reached in the distant future, to construct a physics and chemistry free from hypotheses, and gave expression to this in no uncertain manner many times.   Since the atomic hypothesis was unquestionably the prototype of all physico-chemical hypotheses, it was the chief objective in the fight.

Mach has explained at length in many passages in his works – particularly in his theory of heat[18] – why he hopes that atoms, which are inaccessible to direct experience, will soon disappear from physics, and Ostwald occupied himself and many of his scholars for many years with the plan of proving that chemistry can do without atoms, hoping to explain the law of multiple proportions, which was the cause of their introduction into chemistry by Dalton, in another way.   In these circumstances, the opposition of the great majority of physicists and chemists to atomic conceptions was so strong, that men like Boltzmann, whose life work was concerned with developing the atomic theory and particularly the kinetic theory of heat, were actually ignored, or even despised, by their colleagues.   This want of appreciation even contributed to the unhappy ending of Boltzmann's life.

What, then, did these physicists and chemists consider to be the nature of the hypotheses in question ?   The answer to this was given by Hertz in his *Mechanics*, in a form which has been quoted innumerable times: Hypotheses are pictures or models, which we make for ourselves of certain phenomena or groups of phenomena, in order to gain a better general view of them, and to see them more clearly.   An hypothesis is good when this picture or model does not merely represent correctly the charac-teristics of the particular group of experimental facts, but when it further exhibits characteristics which lead us to new facts; when, in other words, 'the consequences of the pictures further prove to be pictures of the consequences.'   Hence Mach distinguishes a 'thought-economy value' and a 'heuristic value' of hypotheses, but both he and Hertz and all their followers quite clearly reject any intrinsic value in real knowledge of these pictures or models. For example, Nernst says in the introduction to his *Theoretical Chemistry* (second edition, 1898): 'The introduction of hypotheses has become necessary in order to reach a deeper knowledge of natural phenomena, which leads to the discovery of new laws ('heuristic value'); the latter can be tested by experiment, and success proves the utility of the hypothesis, but by no means its correctness. . . . The hypothesis is therefore a very valuable aid to science; it is in no way an end in itself (at least for those who are devoted to exact science), but must needs produce the proof of its right to exist by connecting known experimental facts with one another, or assisting us to arrive at new ones.'

It must be expressly stated that this is not at all the peculiar private opinion of Nernst; but that he is merely stating what the great majority of his colleagues at that time thought, and even for the most part still think to-day. It is true that these sentences, which also appear in the new edition of his work, seem almost comical to-day, since what follows them disproves them point for point. For Nernst, no more than Ostwald, has been able to continue to avoid admitting that the 'grained structure of matter' (as the latter says), or in other words the atomic theory, must be regarded to-day as a fact. Thus even Nernst says in the fourth edition (1913): 'In view of such remarkable confirmation of the view which the kinetic theory gives us of the molecular world, we are obliged to admit that this theory is beginning to lose its hypothetical character.'

Nevertheless, in his time the whole earnest endeavour of that generation of investigators was directed towards eliminating from science hypothetical elements of this kind, for the reason that they had been arrived at purely speculatively and were not founded on direct observation. But an endeavour of this kind means, whether we like the word or not, a philosophical activity, in particular in the field of epistemology; and thus these physicists and chemists became philosophers without really wishing to do so. They became, however, a quite special kind of philosophers, namely, positivistic (empirical) epistemologists. In Mach's case the transition was obvious. He began as a physicist and ended as a pure theorist of knowledge. His chief work *The Analysis of Sensation* has actually become the classical work of modern positivism. We shall return to it once more later on, but for the present we are only interested in the question of scientific hypotheses.

Mach's ideal of a scientific theory is Fourier's theory of the conduction of heat, or thermodynamics. In both of these, one or a few obvious fundamental laws are developed into a whole system of consequences, which then again appear as the expression of possible observation. The reader who is not acquainted

Ds

with the above two departments of physics may take as an example of such theories the treatment of the laws of reflection and refraction in elementary geometrical optics (theory of concave mirrors, lenses, etc.). The position in electrostatics is quite similar, where a few simple fundamentals (constancy of the potential on a conducting surface, Coulomb's law, the energy laws, etc.) likewise lead to the correct prediction of a vast number of single phenomena, both qualitatively and quantitatively.

We will call such theories 'elaborative.' Their characteristic is that they contain practically no hypothetical elements; the fundamental assumptions in them (in the case of Fourier's the law of heat conduction, in elementary optics the two laws mentioned) are themselves data of experience. In the same way, the whole of thermodynamics depends upon the two general first and second laws (energy and entropy) which without doubt may be regarded as the summary of an enormous number of individual experiences. If the word theory is understood as the logical arrangement of a large number of single laws to form a closed system of reasons and consequences, these examples from physics are without doubt patterns for such theories. Even philosophers who are not positivists will have no objection to make to this statement.

But without doubt, there is in physics, and altogether in science – a second and quite different class of theories for which the world has a distinctly different sense than for the examples just mentioned. Of this second class of theories, which we will call 'explanatory' theories, the atomic theory forms the best possible example; we have already mentioned a few others, such as the wave theory of light, the Newtonian theory of the planets, the kinetic theory of heat, etc. The characteristic of this kind of theory is that the desired logical connection and unification of the facts is only reached on the basis of a speculative assumption, which is described as the hypothesis upon which the theory is based. (This retains for the word 'theory' the former sense of connecting single elements to form a logical whole, but in this way the word theory very often also includes the sense of hypothesis, so that in everyday speech the two are confused.) The theory thus becomes, or should become, a view in the truest sense of the word.* It should give us a view of something which

* The word *theory* comes from the Greek verb *theaomai*, I look at; as in *theatre*.

really lies behind the things immediately accessible to experience, and thus enable us to understand why these things behave as they actually do. When Dalton introduced the atomic hypothesis into chemistry, he wished to make comprehensible why the elements always combine in proportions by weight which are multiples of quite definite numbers, one for each element. This fact is assured in itself, but it is an unsatisfactory state of affairs to be obliged simply to accept it without being able to explain how each element gets its characteristic proportionate number (its so called combining weight).

To take another example, it is equally a fact that in certain circumstances light meeting light results in darkness at certain points, and double illumination at points in between. This fact was already known in Newton's time to the learned Jesuit Father Grimaldi. But even a schoolboy, who sees such an experiment performed for the first time, immediately asks: How does it happen? What is the cause of it? This question is then answered by the wave theory of light as follows: Light is a wave-like process (this is the hypothesis), and waves, as is easily seen and proved, lead always and everywhere to so-called interference. Here also we have a fundamental explanatory hypothesis, and upon its basis a whole theoretical structure is then erected (in this case by Huygens, Young and Fresnel) in which all single phenomena of optics find their logical place.

Let the reader note particularly that in all the last mentioned cases, the hypothesis, and with it the whole structure of theory, contains one or several elements which are completely outside experience as so far known. No one in Dalton's time had really made atoms visible individually, or otherwise proved their existence, and no one in Huygens' time had ever seen anything wave-like in a ray of light. When Newton explained Kepler's laws by the assumption of universal gravitation, no one had ever made an experiment which demonstrated the mutual attraction of all bodies. In all three cases therefore, a real hypothesis, or in common language, an assumption was made; a supposition, which was arrived at by purely speculative methods; a product of the scientific imagination, which can no doubt be stimulated to such a production in very various ways, but in any case must perform a creative act, which is on a like plane with the productive activity of an artist.

When we are quite clear about this, we are equipped to

continue our historical and critical discussion. The actual content of the whole of Mach's and Ostwald's criticism of hypotheses may now be exactly formulated as follows: Their school of thought disputes (or disputed), the right of theories of this second kind to be regarded as a legitimate element in science. It maintains that what is useful and tenable in them, finally amounts to the same thing as theories of the first kind, but that what really distinguishes them from the latter, namely the hypothetical element found by speculation, has really no right to admission into science. These theories are therefore supposed to be of use in so far as by their help 'the phenomena occurring in nature can be described as completely and simply as possible'; but they lead to error when the belief arises that they bring us nearer to the 'nature of things,' the latter concept being a speculative notion altogether without any meaning, and conveying nothing at all to an exact investigator. All that the latter is concerned with are experimental facts and their logical connection. Anything more than this is objectionable.

It is instructive to follow the working out of this critical attitude in the particular case of the motion of the planets and gravitation. The physicist, Drude, in a well known address delivered at Leipzig, undertook to develop the programme of physical positivism, with respect to this point. Drude put the matter something as follows:—We used to speak of an 'explanation' of Kepler's laws by means of Newton's assumption of a force of gravitation between the heavenly bodies. But what did Newton really do ? He differentiated the equations of planetary motions twice with regard to time. The result was a system of equations, which state that the acceleration experienced by a planet at a given instant, is directed towards the sun, and inversely proportional to the square of its distance from the sun. If we multiply this acceleration by the mass of the planet, and call the product, as everywhere in mechanics, 'force,' we have a simple expression of the facts, which is very readily grasped on account of the anthropomorphism instinctively associated with the word 'force,' but nevertheless in reality is nothing more than the original law of Kepler. It has simply been put in another form, more economical in thought, and also more useful heuristically. For this new formulation leads to our discovering further facts (disturbances, tides, etc.).[19]

All this sounds very plausible. As a matter of fact criticisms of this kind – I could easily give a further half dozen examples[20] – brought in their time almost the whole world of science into the ranks of the positivists. If thirty years ago any one dared to put forward any other view in front of physicists or chemists, he was received with a pitying smile and regarded as suffering from incurable philosophical softening of thought, and this state of affairs continued long into the present century. The author, when he published the first edition of this book in 1913, was prepared to be ridiculed by the whole of his scientific colleagues, although individual physicists of eminence, in particular Planck, had already expressed themselves quite clearly in opposition to this excessive criticism.

The fabulous success of recent atomic physics, Planck's influence, and perhaps to some extent that of this book, which has been read by numerous young physicists, have led to some change in the meantime. Hence I had some hesitation in allowing this attack on the positivist position, which I retained in the last two editions, to be again printed, since it might easily appear like tilting against windmills. However, the very latest developments in physics, as we shall see later, have once more given such an impetus to positivism, that it would be quite wrong to regard it as finally superseded. Everything rather points to its further advance, and hence it was necessary to repeat at this point what was to be said on the subject.

The history of atomic theory in the last thirty years is a complete proof, which cannot be refuted by any sophistry,[21] that the over-critical, positivistic enmity to hypotheses in physics was a serious error. For exactly what this criticism (that of Mach, Ostwald, Stallo, Kleinpeter, Petzoldt, etc.) regarded as an illegitimate speculative element, and wished to eliminate from physics, namely the atomic theory, has turned out to be the turning point of the whole of modern physics. The atoms of which it speaks, are not, as Mach maintained, 'like all substances, imaginary things without any real existence,' they are just as real as any other object in the material world. And as events have shown, the progress of physics did not depend upon regarding atoms as a picture helping us to describe phenomena as simply as possible, but on the contrary, upon taking them as real things, the nature of which it was necessary to investigate. The whole of the physics of to-day turns about the question: What is the atom ?   No one asks any longer: in what respect do things behave 'as if' they consisted of atoms ?   It would be just as senseless for a botanist or zoologist to ask the analogous question concerning the cell, which might perhaps have been reasonable in Virchow's time.

The case is closed regarding the actual structure of matter as consisting of atoms, and there is no point in further discussing what has ceased to be a problem and has become a presumption of all further investigation.   This fact proves conclusively that there must have been a serious fundamental error in positivist criticism, since it could not otherwise have so completely failed to foresee what actually happened.   A theory of knowledge – we will turn upon positivism its own weapons – which fails to stand the test of later progress in investigation, is of just as little value as a scientific theory which is refuted by further experience.

The truth which we state in opposition to this excessive criticism of hypotheses, is this, that precisely those elements of the 'explanatory theories' found by hypothetical and speculative means form the truest and most valuable contributions to knowledge made by science.   Far from being mere aids to work, pictures, models or the like, they are in truth the most important matter, and contain exactly that around which the whole of investigation turns.   The knowledge that matter consists of atoms is a hundred times more important than all the 'consequences of the picture' which are comprised in it (such for example as the law of Dulong and Petit, that of multiple proportions, or of the existence of 32 classes of crystals).   All these are really only steps on the way to that fundamental truth; naturally they have also a value as knowledge in themselves, but from the standpoint of a criticism of knowledge which takes a view of the whole, their value is finally determined by the amount by which they bring us nearer to the root of the matter.

This is also true, to set right the example discussed above, of gravitation.   It is simply wrong to say that Newton's law of gravitation contained nothing more than another mathematical expression for the facts stated by Kepler's laws.   What is new and decisive in Newton's discovery is overlooked in Drude's description of it; namely the recognition (at first only hypothetical) of the presence of a general attraction between all bodies, of which no one, including Kepler, had thought before Newton, although Kepler had also performed the transformation pointed out by Drude. In this case history chances to make it perfectly obvious that it is not a matter of this mere simplified statement – for otherwise Kepler and not Newton would be the discoverer of the law of gravitation – but that the scientific imagination had first to produce something entirely new, this new thing being an hypothesis. But this hypothesis, as in the case of the atomic theory, has later turned out to be a fact, though in one case only a hundred years was required for this, while in the other, almost 2000 years elapsed before the demonstration took place. The presence of universal attraction is an indisputable fact, since Cavendish, and after him numerous other physicists such as Richarz, Jolly, Krigar-Menzel, proved its presence by means of the torsion balance. Eötvös recently proved it and measured it with the greatest exactness by means of other very highly sensitive apparatus.[22] This originally hypothetical, that is to say, merely suppositional, discovery of an entirely new set of facts, of which no one had thought before Newton, is the essential point, and as compared with it, 'thought economy' or heuristic value are purely by the way. It is obvious that these are always present in full degree in the case of an hypothesis which fits the facts.

Although we have thus stated in strong terms our point of view as regards 'hypotheseophobia' (Ed. v. Hartmann) we must also be fair to the other side, and not dispute that there are certain elements of truth in the positivist criticism of hypotheses, which we must now bring forward. In the first place there can be no doubt that a supposition is naturally not the same thing as the discovery of a fact. When Leverrier supposed that an eighth planet, Neptune, existed outside Uranus, but had never been seen by man, or when Mendeléeff prophesied the existence of three undiscovered elements, these were for the moment only suppositions which did not prove to be facts until afterwards. The same is true of gravitation, of atoms, and many other things. All that we are disputing is that we ought to ascribe to such suppositional judgments an entirely different kind of content than is ascribed to judgments of fact. The supposition means exactly the same as the statement of fact, but the degree of certainty is different. The critics of hypothesis believe, on the other hand, that hypothetical statements are totally different in content. According to them, the statement concerning gravitation meant something entirely different from what it actually said, namely not actual reality, but only a model. It is this that we deny.

But on the other side, it is obviously impossible to deny that man is capable of erring in the case of such suppositions as in all other cases. The notorious failure which has been the fate of so many hypotheses, defended in their time with the greatest conviction (we need only mention Stahl's theory of phlogiston, or Berzelius' electro-chemical theory) warns the scientist to be careful and modest. It would be easy to mention a couple of dozen such false hypotheses in the history of science. But much more erroneous than all these historical blind alleys is the statement made by an eminent scientist of our day, Uexküll, and believed by so many: 'a scientific truth is nothing but the sum of the errors of to-day.' It is the main object of this book to refute this false conclusion by simply pointing to what science is able to show in the way of real addition to knowledge, and most particularly by means of the formation of fruitful hypotheses. A truth that has once been really found out remains a truth, at the most it can be extended or deepened by further insight, or brought into new logical connection with other truths, but it never becomes on that account an 'error of to-day.' It is just as easy to find an equal number of examples, in the history of science, of truths which, once discovered, have never been abandoned again, as to find examples of misconstructed hypotheses. But since our whole discussion will prove this sufficiently we need not say more at this point.

But there is one further respect in which we must make an honest admission as regards positivist criticism. In numerous scientific, and in particular physical hypotheses, there is a great deal of the picture or model, indeed we may say, that we find in practice every imaginable stage between a purely pictorial hypothesis and one – according to our view – almost purely real. As a typical example of a near-picture, we may take the well known representation of the relationships in the magnetic field by means of an analogy with a current. We speak of the 'flux' of magnetic force, of the 'permeability,' of 'sources,' 'sinks,' and 'eddies' of the magnetic flux, and so on. No physicist means anything else by all this than simple pictures, which allow of a pictorial, and hence easily understood, form of expression; the whole depends upon a purely formal analogy which chances to exist between the mathematical formulation of the laws of a magnetic field on the one hand, and those of the flow of liquid on the other. Here therefore we have a true fiction in Vaihinger's sense.

In the case of electric current, for example, matters are already considerably different. In the first place, the analogy itself goes much further, inasmuch as it includes the conversion of energy in the wire and in the current of liquid respectively, a true process taking place in time therefore being present in both cases. But while until a short time ago it was still possible to maintain that this case also was nevertheless only a formal analogy, the aspect is changed since the electron theory (page 124) compels us to assume the actual existence of very small 'corpuscles' moving through the wire in one direction. Here the analogy is obviously changed into identity.

It would be easy to set up an almost completely continuous series of such ideas, from pure models on the one hand to entirely realistic hypotheses almost free from the picture element; and this series of steps, for each of which some present-day scientific notion affords an example, has often been traversed by a single hypothesis, in the course of scientific development, from one end to the other, as, for example, the kinetic theory of gases. This will be clear enough in individual cases when we come to discuss them. When positivist critics insist that the pictorial accessories of a theory are not to be erroneously given out as real objects of knowledge, we must admit that it is doing real service.

This has frequently been proved to be the case in the history of science, as we shall see particularly in the theory of light. For it thus forces science to consider what it is that is the really essential feature of the hypothesis in question, and thus protects it from a danger always present, of seeing in such pictorial components problems which in reality do not belong to the matter at all. This is also true of atomic theory, which has often been represented as dealing exclusively with single corpuscles, separated from one another by empty space. In actual fact, the sole essential notion of a 'grained structure' of matter may very well be compatible with a fundamentally continuous idea, only that the notion of separated corpuscles is simpler, and somewhat closer to common everyday talk, than a wave theory or the like.

If therefore we readily recognise that criticism in this respect has been of real service, we must not forget that, as history shows, it has always been in danger of completely denying the existence of problems which in reality form precisely the essential point of the whole matter. For only too often it has appeared, that an hypothesis originally thought of as a picture, or as a purely provisional construction 'without prejudice,' has turned out to have actual reality, and that its assumptions have later been shown to exist. Only by keeping both sides of this matter in mind, are we able to steer the middle course between dogmatism and hyper-criticism.[23]

If we now summarise the whole matter in asking what, according to what has been said, is the actual nature of a physical, or more generally, a scientific hypothesis, we may reply as follows:

A physical hypothesis is the presumption of the existence of a general state of affairs, lying at the back of certain phenomena which are matters of experience, and allowing the phenomena in the field of facts in question to be deduced qualitatively and quantitatively (mathematically) from the presence of the said state of affairs and its assumed laws. The assumption made will then also evidently allow other consequences to be drawn, and if the hypothesis is correct, these conclusions will be confirmed, the confirmation again increasing the probability of the truth of the hypothesis. *The indirect confirmation of an hypothesis obtained in this way may be so convincing that its probability practically amounts to certainty.* But the task remains of also confirming it directly, and this can in many cases be accomplished, with the result that the hypothesis is promoted to the rank of true fact (atoms, gravitation, light waves, etc.). The solution of these problems forms the most important part of the progress of science. As a rule the process is accompanied by a criticism which removes from the hypothesis in question unnecessary pictorial elements.

From this point of view we are also able to make a clear decision on the old disputed point as to what is the difference, if any, between description and explanation. One of the consequences of the critical attitude towards hypotheses which we have been considering is to deny the existence of such a difference, and hence to see in all so-called explanations nothing but a description of natural processes, perhaps simplified and adapted to economy of thought (compare for example Kirchoff's words quoted above). For it is said, we cannot speak of a real explanation when a group of facts which are strange to us is made intelligible by means of a model or picture constructed from known or familiar ideas (such as magnetism, and the picture of a current). We simply imagine that a matter is clearer to us when we find in it a number of features long known to us, and hence not exciting our surprise. Even if we admit this, it is by no means a necessary conclusion that there can be no fundamental difference between description and explanation of natural phenomena.

According to what we have said, the difference is to be found in the fact that *the explanation contains the logical subordination of the particular under the more general,* for example that of the planetary motions under the general law of gravitation, while description is the simple statement of facts as they appear. This however results in explanation becoming a relative concept, for every general set of facts of this kind (for example gravitation or the grained structure of matter) is again the object of a descriptive judgment. Every explanation therefore gives as a matter of course, if it really is an explanation and not a mere picture, a new and more comprehensive description – and this is entirely our contention to those who deny to hypotheses all real value for knowledge – but this in no way abolishes the logical relationship of subordination which exists between the formulation of the general truth as a 'fundamental law' and the particular group of facts that fall under it, and are regarded as deductions or consequences of it.

The laws of the planetary motions appear to us logically as deductions from Newton's law of gravitation and not conversely, although apparently – in this case – one is completely equivalent to the other. The logical relation of subordination comes from the fact that the law of gravitation is a quite valid general law. No one would dare to draw such a conclusion as this : ' If the Kepler laws of planetary motion are true, then all bodies must attract one another according to Newton's laws.' But the converse: 'If Newton's law is true, then the planets, among other bodies, must also act in accordance with it, which gives Kepler's laws.' Hence we can say: Explaining consists in arranging individual facts which we meet with, and first of all simply observe, in more general connections. We shall see later what consequences flow from this point of view. But we may here point out that every explanation itself calls for a further explanation, that is, a search for still more comprehensive connections, since any explanation must contain the description of a simple set of facts. This is particularly easily seen in connection with the example of gravitation, Newton himself having felt it quite clearly. He says, after he had spoken of universal gravitation:

Rationem vero harum gravitatis proprietatum ex phaenomenis nondum potui deducere et hypotheses non fingo. . . . (here follows the well known statement that hypotheses, whether metaphysical or physical, have no place in physics).

We will not go into the question here as to what Newton was really thinking of under the term hypotheses when he condemned them. But I do not think that he meant the same thing as do the positivist critics who so willingly make use of these words in the fight against the atomic and molecular hypotheses. For it would then be impossible to understand what Newton could have meant by the first word of the above quotation – the *ratio*. He is here seeking a reason for the peculiarities of universal gravitation as determined by him. Are we to take this to mean that his description of planetary motion was not simple enough from the point of view of thought-economy? Could there be anything simpler than Newton's law? I believe rather, that his attitude towards gravitation was the same as that of every physicist to-day, whether the latter be clear about it not.

Newton also had a feeling that there must be an answer to such questions as these: Why do we find in the denominator on the right hand side r^2, and not r^3, or r, or some other function of r? How is it possible that such an interaction between two bodies can exist through space without apparent connection? There must be reasons for all these things, in other words, it must be possible to deduce the facts we are discussing from a still more general and comprehensive statement of facts. We shall see later to what extent we may regard this as having been successfully accomplished to-day. Newton can have had no objection to such attempts, as is best proved by his emission theory of light (see page 104), which he certainly thought out as a reason for the properties of light.[24] We shall return several times again to this fundamentally important question of the theory of our knowledge of nature. But we will continue our description of the results of science, taking it up at another point, namely physics.

IV

THE ELEMENTS OF MECHANICS

We begin our survey with this branch of physics, since it is not merely the historical starting point, but also belongs at the beginning by its content. For the fundamental notions here developed are further applied in every department of physics. As a preliminary to actual mechanics, text books of physics usually take dynamics or kinematics, a pure doctrine of motion, which constructs its objects quite arbitrarily in a mathematical sense, without asking whether the forms of motion considered actually occur in nature, and how they are brought about. Mechanics proper then begins with the latter question, that is, the causes of real motion.

It is well known that the foundations of both branches of the science were laid by Galileo. In kinematics, his main achievement was the creation of the concepts of momentary velocity and momentary acceleration. Both of these assume conceptual work with so-called infinitely small magnitudes, that is to say, the introduction of a *limit*. In this way Galileo became, it is true without himself having any idea of it, one of the founders of a new branch of mathematical thought, the *calculus of infinitesimals*, which shortly after his time was recognised in its true importance by Newton and Leibniz, and brought by the latter into its present day form.

Galileo laid the foundations of mechanics proper by setting up the law of inertia, at first however in a special form, but his contemporary Baliani soon afterwards gave it the general form, in which it still forms the foundation of mechanics. It states that a body exposed to no external influences whatever retains unchanged its existing motion, that is to say, moves in a straight line with uniform velocity. If it does not do so, something must be acting upon it, and this something, that is to say, everything that is capable of producing a departure from motion according to the law of inertia, is called in physics *force*. *The concept of force is therefore, as it were, the other side of the law of inertia.* It then immediately appears that a further fundamental concept is very closely connected with it, the concept of *mass*. Experience teaches us that the same external effect (for example the same charge of gunpowder) produces in different bodies different rates of acceleration. Hence a different 'inertia resistance' is ascribed to them, and the physical measure of this is called the mass of the body in question. We shall deal with this in greater detail below.

In these fundamental concepts of mechanics which we have sketched quite shortly – the reader may be referred to any text-book of physics, but they are not all free from objection on this point[25] – there is an abundance of epistemological problems which for the most part are extremely difficult, and have given rise since Newton's time to innumerable discussions both among physicists and mathematicians, and among philosophers. We will mention the most important of these, but we can only discuss a few of them in detail. A proper discussion of the whole subject would require not one but several special volumes.

First a few words concerning the method of infinitesimals and the part played by it in scientific investigation. It is this method that first rendered possible the mathematical mastery of natural processes. For these appear to be – at least at first – continuous or unbroken changes which cannot be dealt with by any other method than this mathematical weapon in question. Spengler, as is well known, has seen in the development of this new method the characteristic expression of the modern 'Faustian' or 'dynamic' cultural soul, which he contrasts with the 'static' soul of Graeco-Roman antiquity. He also maintained in this sense that the beginning of an infinitesimal treatment, already observable in the case of Archimedes, in reality was something totally different from this method, which was recognised by Leibniz with such wonderful clearness. In my opinion he has not proved this thesis. In spite of Spengler, it is much more natural and simple to suppose that the Greeks had first to master the static part of mathematics, that is to work with magnitudes regarded as unchangeable, before they could begin work with variables. Every schoolboy to-day must go through the same process, although he is living in the midst of 'Faustian' civilisation. For the simpler must necessarily precede the more complicated.

More important for us than such sidelights on the philosophy of civilisation are the problems in the theory of knowledge which lie behind the calculus of infinitesimals. We may make the fundamental point clear by means of an example, since it is unfortunately not permissible to assume that the reader, unless he has studied mathematics, will be sufficiently acquainted with the matter. Let us consider the function $y = \dfrac{x^2 - a^2}{x - a}$ (dependence of two variables upon one another). y here represents the dependent variable, x the independent variable. The function gives, when represented graphically, the straight line shown in figure 2. But in this diagram the point for which $x = a$ is

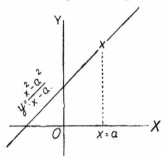

FIG. 2
Graphical representation of the function
$$y = \frac{x^2 - a^2}{x - a}$$

missing. The value of the function for this value of a is not obtainable, since division loses its meaning when the denominator becomes zero. In older text-books of mathematics we find some such argument as the following: The quotient $\dfrac{x^2 - a^2}{x - a}$ assumes for $x = a$ the indeterminate value $0/0$. This may in itself be anything at all, since conversely any quantity whatever when multiplied by 0 gives 0. In order to find the true value of the quotient in the present case we must first carry out the division generally and then put $x = a$. If we do this we get from $y = x + a$, the result $y = 2a$, when $x = a$; and this point actually lies on the line representing the function.

Considerations such as, or similar to, this were also sufficient in the beginning for the first workers with the calculus of infinitesimals and its application to geometry and physics. But in the course of the past century, particularly since the time of

Weierstrass, this point of view was subjected to strong criticism, and it is quite general to-day to proceed in a fundamentally different manner. The characteristic point of the old method was to raise, and attempt to solve, the problems as to what are the 'true values' of the quotient in the case in question (and similar expressions).

Modern mathematics, on the other hand, takes the view that there is here no problem at all, but that it is simply a question of making a new definition. Since, namely, in the explanation of division, the case that the divisor is zero was expressly and of necessity excluded, it is also necessary to fix for the function $y = \dfrac{x^2 - a^2}{x - a}$ a sense for the case $x = a$. Hence there can be no question of there being anything to find out here which was already determined. Instead we simply define that for $x = a$, the dependent variable y is to assume the value $2a$, in order to make the course of the function completely continuous. Any other definition would not be economical in thought, since it would give us some isolated point off the line, and this would be loaded with a discontinuity. In exactly the same way, it is now a general practice, even in elementary mathematics, to replace a problem by a definition, for example in introducing the rule 'minus times minus equals plus,' or the powers with negative exponents $(a^{-n} = 1/a^n)$.

Now it is easy to see that this procedure is extremely convenient and advantageous for mathematics as such, since it gets rid of all such inconvenient questions at a single blow; but this is no more, if it means regarding the problem as settled from the point of view of the theory of knowledge, than positivist and conventionalist epistemology. For the epistemologist the problem begins at the point where mathematics has left it. He is obliged, if he is not already completely lost in positivistic ideas, to consider the question of how it is that only those definitions hitherto regarded as the results of a logical process, lead to reasonable results; whereas every other definition would be a completely senseless assumption, which might be made with equal justification in an infinity of different ways. It is here that the true problem arises for the theory of knowledge; 'economy of thought' merely hides it instead of making it clear. It is not possible to go more deeply into this problem within the limits of this book,[26] since we are here only dealing with science. It is obvious that the problem of mathematical infinity is closely connected with it.

We must still however glance at a scientific side of the question. The calculus of infinitesimals, as we have said, is the method by which the continuously variable function, and hence the continuous flow of events in space and time, is rendered accessible to mathematical analysis. The question however arises as to how far nature herself actually fulfils this assumption of the continuity of her processes and connections. The question is not a new one; it was eagerly discussed at the time of the foundation of the calculus, since the atomic theory was already quite generally discussed as a hypothesis of scientific philosophy.[27]

It is easy to see that, if matter really consists of completely separated particles, all processes taking place in it, which are apparently continuous, such as sound waves, heat conduction, or the like, are in fact, in sub-microscopic dimensions, by no means so continuous as they appear. It is therefore plain that the formulae of the infinitesimal calculus can only offer a first approximation, since we know that atoms are not a mere assumption, but a real fact. It is nevertheless possible to retain the idea of a continuous process in nature by adding the further hypothesis that the atoms themselves depend upon a process taking place in a continuous medium, in a manner similar to waves or vortex rings, and this supposition has hitherto satisfied those philosophers who, like Kant, hold the idea of the continuity in time and space of nature to be given *a priori* (to lie in the nature of our mode of perception or thought); others however – and in my opinion rightly – have taken up the position that this question can ultimately be decided only by the investigation of nature, and hence have quite seriously considered the possibility that the world may after all finally consist in a discontinuous, and not a continuous manifold (Boscovich and later Boltzmann).[28] It would thus be, to use a modern analogy, a cinematograph, in which a discontinuous succession of pictures produced the false impression of continuous succession of change.

Quite recently, this supposition has suddenly acquired unexpected importance through the development of the quantum theory; we shall have to discuss this more in detail later. If this
Es

should finally turn out to be right, the calculus would not then be the method best suited to physics, and in its place, in dealing with the most fundamental matters, we should calculate with very small, but finite magnitudes; in other words arithmetic in place of analysis, an arithmetic which however would merge into statistics, the results of which could then be obtained with the greatest accuracy by infinitesimal methods (see below).

Space does not allow us to go further into the purely mathematical aspects of the question. We now come to the second complex of questions, the geometric-kinetic. The two cannot be separated, although epistemology has attempted often enough to confine itself to geometry. The unquestioned view not only of philosophers but of every physicist, was until recently, that geometry and kinetics, or in Kant's language the doctrine of the pure forms of perception of space and their combinations with time, were, from the standpoint of physics, given *a priori* or, to use Weyl's words,[29] that space and time were 'empty apartment houses,' into which physical experience had first of all to move, the laws of it (geometry and kinetics) being for this very reason given beforehand. As a matter of fact, no physicist has felt it necessary for him to carry out geometrical experiments, such for example, to attempt to prove by experiment that the sum of the angles of a triangle is equal to two right angles. Only one mathematician, Gauss, attempted to carry out an heretical experiment of this kind (a measurement of the triangle, Brocken –Inselsberg–Hoher Hagen). Kant was supposed to have proved the *a priori* truth of geometry in his 'transcendental aesthetics.'

The position is somewhat different with regard to kinetics. Here experiments have been very largely made use of, mainly for the purpose of teaching, and without seriously disputing the fact that the results obtained might have been found *a priori*. This whole point of view has been completely changed recently as a result of the relativity theory. The whole structure of this historical apriorism, is now shaken, and we shall see later what the position of affairs is at present. For the moment we must be satisfied with raising a few fundamental questions, which show what troublesome problems for the theory of knowledge are everywhere behind the apparently simple foundations of mechanics, in this field also.

Take first of all the concept of the straight line, which is made use of in the law of inertia. Whence does this concept come? Is the straight line an empirical concept? Or the idealisation of empirical ideas? A logical construction? A necessity of thought (assuming that we want and need to form geometrical concepts)? Or what else is it? How do we know whether a line is straight, and – a question immediately connected with it – whether a surface is plane? Do we judge straightness or planeness according to the rules of Euclidian geometry? Or have we on the contrary deduced the latter from our experience with real bodies, and have we therefore to judge the validity of Euclidian geometry according to the exactness with which real bodies fulfil its laws? Or finally, is the whole matter really altogether different from what it would appear from these frequently stated alternatives, and hence are the latter put quite incorrectly?[30]

Further, the law of inertia speaks of uniform motion in a straight line. Uniform motion is defined as a motion in which equal distances are passed over in equal (indefinitely small) times. A further chain of problems. First: What are equal distances? How are we to determine whether two distances are equal? We say: by using a measuring rod – but who is to guarantee that this does not change its length in the course of the operation? More correctly: What is length, which can only be defined by means of this measuring operation? But if we regard this question as solved (the further development of the theory of relativity by Weyl, Hilbert and others has shown that this is far from being so; see below), further questions remain. What are equal times? How are we to determine whether two times are equal? In actual fact, our final measurement of time is by means of clocks, which are regulated by the rotation of the earth, and we thus assume that the latter motion is uniform (in accordance with the law of inertia and its consequences). Are we not here arguing in a circle, by first defining the measurement of time by uniform motion, and then finding the latter by means of the former?

It is possible to escape from this circle, as earlier discussion of these questions (among others by Maxwell[31]) showed. For the measurement of time can be referred to the so-called principle of identical changes, which says that one and the same natural change always takes the same time for its repetition, and hence every such repetition can serve as a clock, as is actually the case in all clocks used in practice. Fundamentally speaking, a measurement of time would be possible in this way without making use of the rotation of the earth; indeed we might even think, if our clocks were accurate enough, of testing, conversely, the uniformity of the earth's rotation by means of the clock, a matter which is being seriously undertaken to-day.   But is there in nature such a thing as two exactly similar processes ?   And, if not, how about the measurement of time ?   Is the concept thus tainted with inaccuracy, or does the inevitable inaccuracy only apply with Kant's 'empirical time,' while the absolute time of Newton *aequabiliter fluit,* without taking any notice of such empirical and technical defects ?   What, altogether is the concept time and the measurement of time ?   We shall refer to further problems connected with the concept of time later on.

Finally the complex of questions into which, as we shall see later, the two mentioned above flow: the law of inertia speaks of a body 'not exposed to any external influences' but this can only be the case when no other bodies whatever are present, as all physics teaches us that all bodies are continually acting upon all other bodies.   But if no other bodies are present, there is no body there for reference, from which the movement could be judged, and then the whole statement concerning uniform straight line motion is left completely in the air.   It is simply impossible to describe a motion in any other way than by referring it to some other body, or in the ideal case, to a system of co-ordinates.   Whence is this to be derived, if the condition of the law of inertia, that of a body being left to itself, is really to be fulfilled ?   Do we not again find ourselves in a vicious circle ?   L. Lange showed, already fifty years ago, how this is to be avoided, in a very profound study on the *Historical Development of the Concept of Motion.* We can only refer here to this essay, which even to-day is of value.[32]

Another version of the same problem has fundamental importance for us.   It is possible to say, and it has been said, that a purely kinetic statement of the above question is wrong from the start, since it is not a matter of determining the presence of motion by purely optical means with reference to other bodies against which we can observe, but of determining whether motion and rest are characteristic of a body in itself, the examination of which is possible by some physical means, or whether this is not the case.   In other words: even if it is undoubtedly true that from a purely kinetic point of view, a motion cannot be described without reference to a system of co-ordinates, this by no means settles that this kinetic relativity must also be a physical relativity. It might after all be the case that within a completely enclosed space, by means of observation of certain apparatus inside it, and without any communication with the outside world, it would be possible to determine whether this space is in motion and in what way. Mechanics hitherto has actually led to a curious result, that this is possible in one respect, but not possible in another. It leads namely to the inevitable conclusion, that a uniform straight line motion of this closed 'box' cannot be observed by means of mechanical means in its interior, but that every departure from such motion, that is acceleration, retardation, or change of direction, immediately shows itself, also in the interior, as forces which appear. This state of affairs is known to everybody as an empirical fact through travelling by train. If we leave out of account the noise and shaking, which are caused by technical imperfection, it is actually impossible for anyone in a train moving at uniform speed to determine its motion. Every mechanical experiment which we perform in the compartment, such as the oscillation of a pendulum, experiments with falling bodies, or the like, would yield exactly the same result as in a stationary laboratory. Indeed, we know that this very laboratory is moving through space at an extremely high speed (19 miles per second), and that nevertheless everything takes place according to the simple laws of mechanics. But the moment the train slows down suddenly or goes round a curve, we notice the fact, often to our considerable discomfort, by the resulting effects of inertia (centrifugal force, etc.), a hanging pendulum would suddenly be deflected, a bowl of water would be spilled, etc. In short, we can thus formulate the result, that *uniform* motion can only be detected *relatively*, but *change* of motion *absolutely*. The well-known experiments which proved the rotation of the earth are a further example of this fact.

This latter law is called to-day the relativity principle of classical mechanics. Newton had already formulated it quite definitely as a result of imaginary experiments, and innumerable philosophers and physicists after him have puzzled over it.[32] Let us carefully note that this principle, if it is to remain valid, already contains a refutation of the idea, that physical relativity can be directly deduced from the purely kinetic relativity of motion (as for example Mach and others actually concluded). Kinetically, acceleration can no more be stated without reference to a system of co-ordinates, than can velocity (uniform motion). If the physical impossibility of determining the latter absolutely can be deduced from this fact, the same must equally apply to acceleration, which however, is by no means the case according to the results of Newtonian mechanics.

This conclusion is so inevitable, that Mach, in order to avoid it, could find no other method than the assertion, that in reality acceleration also can only be physically determined relatively, and not absolutely. He supposed that, in fact, it relates to the system of co-ordinates of the fixed stars. Whether Newton's real and imaginary experiments (Foucault's pendulum, etc.) would give the same result if no fixed stars existed, it is impossible for anyone to say. 'The mathematician has in this case (according to Mach) made too free a use of the method of imaginary experiment.' Mach, as we see to-day, anticipated an essential idea of the theory of relativity when he conceived this reference to the fixed stars. It is true that in his time the idea could not be accepted, since it was impossible to see what the fixed stars could have to do with experiments, which can be made in the same way in any laboratory on earth, and which do not need the fixed stars to permit their insertion in the system of theoretical mechanics. However, the unsatisfactory dilemma naturally remained, and hence the discussion continued without any tangible result. The philosophers mostly got out of the difficulty by reference to the principle of causality. Their argument ran something as follows: where nothing happens, no cause is needed. Hence as long as a motion suffers no change, we have no reason for seeking a cause for it. According to this, the law of inertia is only a special form of the principle of causality. But then it is also clear that only acceleration can be determined, since only in that case does something really happen. Where nothing really happens, nothing can be determined.

This argument sounds very plausible. Nevertheless, as appeared from a detailed discussion of the whole question which took place some decades ago between Mach, Höfler, Poske, and other eminent physicists,[33] this *a priori* deduction of the law of inertia from causality (which is also found in principle in Kant) is in reality not conclusive. It cannot really be proved *a priori* that forces can only effect *changes* of motion.   The law of inertia actually depends upon experience, not however in the sense that it can actually be proved by experience – this indeed is never the case, but that it is abstracted from certain experiences as an ideal limiting case, and then serves as the foundation for further explanation of experience.   It is a 'postulate,' but a postulate that could never have been propounded without experience.   We have already sufficient proof of this in that the ancients, who were certainly not inferior to ourselves in logical deduction, never hit upon it.   For Aristotle, circular motion was the 'natural' motion, and he would therefore have said of it, that it needed no further causal explanation.   On the other hand, he could only imagine continued straight line motion as the effect of a force, as of course everyday experience seems to prove, for example, in rowing.

But if the direct reference of the law of inertia to causality fails, the argument concerning the relativity of velocity as compared with the absoluteness of acceleration also fails.   The dilemma thus remains.   If it is to disappear, it is either necessary that velocity should after all be determinable absolutely, even if not by mechanical, than by other means; or conversely, acceleration must be relativised as opposed to the position of classical mechanics.   We shall see later how this problem is solved to-day, but we now leave for a time the discussion of the geometric-kinetic group of problems, and consider a third group of questions which relate to the fundamental physical concepts of mechanics, *force* and *mass*.

According to the definition given above, force in physics means everything which is able to change the state of motion of a body. Every such change is described, by an extended sense of the word, as an acceleration, change of direction being acceleration in this wider sense, since it can be regarded as the addition to an already existing velocity of one directed sideways.   Retardation is taken as negative acceleration.   A definition of the concept of force thus amounts to the following: *force produces acceleration.*

On the other hand, the mass of a body, as has been shortly explained above, means its inertial resistance.   One body has twice, three times . . . as great a mass as another, when the same effective force only produces a half, a third . . . as great an acceleration as with the first body.[34]   If we take any body whatever as a standard, and its inertial resistance as a standard of mass, the

other mass of any body can be compared in this way with
the standard.   We choose as our standard the mass of a cubic
centimetre of water in its condition of greatest density, that is at
4.1°C, and call this standard of mass or inertia a *gram*.   The
statement that a body has a mass of m grams, therefore means
that it is m times as difficult to set in motion as the cubic centimetre of water, or more exactly, that the same force produces an
m times smaller acceleration in it, or that a force m times as great
is required to produce an equal acceleration.   We measure force
itself by the product of mass and acceleration.

Force is equal to mass times acceleration ($F = m\,a$).

This equation, which was already given by Newton, was called
the fundamental equation of mechanics.   The unit of force is
accordingly the force which produces in the standard mass of one
gram the acceleration of one centimetre per second.   This so-
called absolute unit of force is called one *dyne*.   Since its own
weight, that is the attraction of the earth, produces in the standard
of mass, or in any other body, an acceleration of 981 units
(acceleration of gravity), the weight of a gram is therefore equal
to 981 dynes, but this quantity changes with the position on the
earth's surface, whereas mass is everywhere the same.   Galileo
showed that all bodies fall equally quickly, and hence that the
acceleration of gravity is the same for all bodies at the same place,
and from this it follows that the weight, that is the attraction of
the earth, for all bodies must be exactly proportional to their
inertia (for $a = F/m$ only remains unchanged when both F and
m change in the same proportion).   Hence the weight and inertia
of all bodies are proportional to one another, and this relationship, which was first clearly recognised by Newton, has been found
to be perfectly fulfilled, according to the extremely accurate
experiments recently made by Eötvös.[35]   Its existence is the
reason why, in common speech, weight and inertia are constantly
confused.

It required the genius of a Newton to arrive at the necessary
distinction between two concepts which, in the word 'heavy,'
are constantly mixed together in daily life.   Even Galileo did
not succeed in making this distinction; he still used the word
pondus for both.   Anyone who will understand physics must be
perfectly clear about the distinction.   Mass or inertia is resistance to being moved, weight is the force of gravitation between

a body and the earth. Hence the latter naturally changes with the relative position of the two, whereas the first has nothing whatever to do with the earth. The exact proportionality between weight and inertia forms the content of the first part of Newton's law of gravitation (the second part, the inverse proportionality of the force of attraction to the square of the distance, does not interest us here).

We know by experience that the layman (and every student of physics) finds considerable difficulty in following these fundamental conceptions, and the difficulty becomes very much greater when we attempt to get to the bottom of these concepts by a thorough analysis. This latter task is so complicated, that we find even to-day, in innumerable text-books of physics, statements which are very doubtful, or even quite wrong, when seen from the point of view of such close and critical analysis. A number of these obscurities and circular arguments arise out of the fact that even the genius of Newton was not able to find, for his fundamental definitions and laws, a mode of formulation free from objection in every respect – such a task is really beyond the power of one man. Unfortunately, his successors, instead of working to improve them, have taken his formulation for granted, since it very soon became authoritative; indeed, they have even made it worse in some respects.

We will first discuss the concept of force. Difficulties already arise from the fact that two systems of the measurement of force are actually in use in physics alongside one another, in many cases thus producing the appearance of two fundamentally different definitions of force. Here we are again brought to a standstill. Is any physical definition whatever anything other than a statement as to how a certain magnitude is to be measured? Since we shall discuss this question more in detail later with reference to a more easily understood example (see page 223), we only refer to it here without further discussion. Likewise, we shall not go further into the two methods of measuring (? defining) force, which are used in physics, the so-called statical-technical and the so-called dynamic, scientific, or absolute, since this is more a question of pure physics, and as regards the theory of knowledge would only once more lead us to the problem already mentioned.[36]

But in another respect, the concept of force takes us right into the deepest problems of epistemology. It is obvious that it stands in a direct relation to the concept of cause; many philosophers, indeed, regard it as the exact physical formulation of the latter; this is the view of Kant and most Kantians. For this very reason, other epistemologists of the empirical school, have believed that the concept of force has no permanent right to existence in a physics purified from metaphysical additions. In this connection the saying of Tylor is often quoted, that the concept of force is 'the last remnant of fetichism in modern physics.'

It is undeniable that a certain anthropomorphic element is found in its origin, since, from the historical and psychological point of view, the concept of force is derived from the feeling of muscular exertion or power. The question is, however, whether the proof of such a psychological source can have anything to do with the justification or otherwise of the use of a concept of force purified from all anthropomorphism. The concept of force shares this fate with a large number of other physical concepts. We cannot deny the concept of light or sound the right of existence in physics, because they likewise are founded first of all on simple sense impressions. The hostility of positivism is also less directed towards such concepts generally, than towards the concept of force in particular, since it is hoped that with it, the concept of cause can also be eliminated from physics; for in it one of the objectionable metaphysical additions to physics is suspected. We shall shortly deal more fully with the problem of causality, but will for the present remain by the concept of force, behind which innumerable other problems are hidden.

One of the most discussed questions of the classical epoch of philosophy (Descartes, Leibniz, etc.), is whether all forces which we distinguish in physics are not fundamentally of the same nature. We immediately recognise the analogy between this question and the corresponding one regarding substance (see above, page 2). In those days the chief subject of discussion was whether the single type of force thus assumed was a *vis a tergo* or a *vis a fronte* (a push or a pull).[37] The latter type is first plainly seen in the gravitation of Newton. The former was exemplified in the processes of elastic and inelastic collision, the laws of which had been deduced at that time by Huygens. Descartes and many others insisted that only *vis a tergo* could really be conceived, and hence that all effects of physical forces must be finally referred to such push actions, as in the case of collision.

Seen from our present point of view, the discussion of such questions at that time appears to us altogether premature. We know for example that the force exerted by the steam in the cylinder of a steam engine upon the piston really results from the collisions of the molecules with the piston, whereas on the other hand, we have not the shadow of a reason for supposing anything similar to be the case with magnetic or electric attractions. The concept of force is, like most physical concepts, to a certain extent provisional. It remains to be found what is really behind it, and until this has been done, it is of little use to indulge in premature and much too general discussion, which our present state of knowledge does not really allow of. We can, however, answer the question to a great extent to-day, but not completely; the reason for this will be clear to the reader at the end of the first part of this book.

We may say already at this point, that the concept of force does not turn out to be a simple fundamental concept to which all others can be referred. It rather appears in the physics of to-day as a necessary derived concept, but one which does not always play the same part. There can therefore be no question any longer of its *a priori* validity. It can hardly be dispensed with in teaching, since it appears to the novice as simple on account of its direct relation to sense experience, although it is in reality no more simple than is white light, which also corresponds to one of the simplest sense impressions, but according to physics is a mixture of an infinite number of different wave motions.

In this connection we may refer to a matter which is of importance for our later discussions. By acceleration we understand in physics, in the quite general sense discussed above, the increase in velocity in unit time. If we designate, as is usually done, increases of variable magnitudes by the symbol Δ, then the quotient $\Delta v/\Delta t$, is the expression for the amount of an acceleration, and the fundamental equation of mechanics given above may then be written

$$F = m\, \Delta v/\Delta t.$$

If now we multiply both sides by the time interval Δt, we get

$$F\Delta t = m\, \Delta v$$

which equation states that the product of force and time during which it acts is equal to the product of the mass of the body set in motion and the increase in its velocity. This form of the equation is called the law of impulse, since the product of the left side ($F\Delta t$) is described as the impulse of the force. On the other hand, the product of mass and velocity ($m\, v$) is called in physics momentum, and hence the law of impulse states that 'the impulse of a force is equal to the increase in the momentum of the body moved.' (For this reason the product $m\,v$ is also often, somewhat inexactly, called the impulse.)

It is noteworthy that Newton gave the fundamental equation this second form, and it was only later that the form usual in text-books used to-day was adopted: 'Force is equal to mass times acceleration.' Newton actually exhibited an almost clairvoyant prescience, since the concept of impulse must actually be regarded as the primary concept, and the concept of force as one derived from it; the law of impulse remains in the new form of mechanics required by the theory of relativity, whereas the first form becomes useless. Only the change in momentum in the impulse law must then be written, $\Delta\,(m\,v)$ since mass can no longer be regarded as constant. The true fundamental equation of mechanics is therefore

$$F\Delta t = \Delta\,(m\,v)$$

We shall return to this several times later on.

We will now turn to the concept of mass and its problems. Here the effect of the imperfections in Newton's statement have been particularly fatal, and a whole series of obscurities, circular arguments, and fallacies with regard to this point disfigure books on physics, particularly those of an elementary character. Newton's definition of mass is not the one of 'inertial resistance' given above, which, as a result of Mach's sharp criticism, has displaced Newton's fairly completely.[38] Newton called this magnitude *quantitas materiae* (quantity of matter) and defined it as: *mensura eiusdem, orta ex illius densitate et magnitudine coniunctim*. But since the density (*densitas*) cannot be defined in any other way than as the quotient of mass by volume (mass per unit volume) a most beautiful circle is the result, unless we assume that Newton meant something else by the term density.

We can only guess what that may have been. We must remember that in his time the fundamental concepts of present-day chemistry (see above) had just been set up by his countryman Boyle, but that Newton himself in all probability still adhered to the old Aristotelian assumption of a single matter, which only assumed different forms in different substances. Density might then, in principle at least, be defined by the number of the particles of this single matter in a given volume. In actual fact, such a definition might, in principle, be made use of to-day, and we might even say that it is the only really correct one. We might formulate Newton's fundamental law of mechanics to-day by the statement that the acceleration is inversely proportional to the number of protons (hydrogen nuclei, see page 133) contained in the body in question.

But even to-day, this could only be practically determined by a measurement of the weight, or of the inertia proportional to it, and in Newton's time this insight into the true unit of matter naturally lay in the far distance. Thus Mach's sharp criticism necessarily appeared justified, although even from another point of view, the concept of the quantity of matter is by no means so vague and useless as Mach represented it to be.[39] In view of a discussion which we shall enter into later on, we must deal with this fundamental point now. We hope that the reader will not find it wearisome to direct attention to a matter which appears somewhat abstract and unimportant. But if he finds it too tiresome, he may omit the next paragraph.

According to Mach, the scientific concept of mass depends exclusively upon the experience, that the relationship between the acceleration produced in two bodies I and II under the same external conditions is found to be independent of these circumstances (that is the forces acting). By reference to pre-scientific but unclear notions, the reciprocal value of the relationship a_1, a_2, is given the new name of the *relative mass* of the two bodies; and by then introducing an arbitrary body (in practice 1cc. of water) as a standard of mass, we get for every other body a number which gives its mass. This formation of a new concept only arises from the constancy determined by experience of this relationship; it gets a particular name, since it turns out to be particularly important. Now this is a quite characteristic example of that method of definition in physics, which I have described in another connection[40] as the 'formalistic' and have contrasted with another, the 'realistic.'

In general, the formalistic method consists in giving certain constant values, which occur in the relations determined (equations), a special name, and thus raising them to the level of scientific concepts. Thus for example the product of force and distance which plays a large part in the consideration of simple machines in mechanics is given the name 'work' and precisely the same is true in the case of the concepts resistance, capacity, etc., in electricity. But as has already been said elsewhere, this formalistic method is not the only one possible; in other cases the realistic method is of equal value, or even the only natural one. We will prove this to be the case as regards the concept of mass.

According to Mach, Newton's law that acceleration is inversely proportional to quantity of matter has not the importance of a true synthetic judgment *a posteriori* (judgment of experience) which it claims to be, but really contains, in a hidden form, a definition of the concept of mass (only, however, on the basis of the above mentioned experience of the constancy of the acceleration ratio). But it is easy to show that Newton's law nevertheless has a useful meaning, even when taken as a law of experience. For if for example in using Atwood's machine or a similar apparatus, we simply add to the body to be set in motion another body of any shape and size whatever, we can obviously describe this as an 'increase in the quantity of matter' without otherwise in any way stating how this quantity of matter is to be measured. In any case it is certain that there is more matter there, but experience then shows that the acceleration is now smaller, and this is therefore not a definition, but a true synthetic judgment.

But the experiment yields even more, namely, the quantitative inverse proportionality between acceleration and quantity of matter, as long as we limit ourselves to bodies of the same material, when we naturally take the volume as a measure of the quantity of matter. For doubling the volume does actually result in halving the resulting acceleration. In so far, Newton's law therefore has a perfectly intelligible meaning without any new condition or convention. Only when we now go on to deal with two bodies of different material do we need a new definition, but this also is by no means necessarily based on the constancy of the acceleration relationship, but can be made practically to depend upon another experience, likewise prescientific, namely that the weight, that is the attraction of the earth, is directly proportional to the volume of a body (as long as the material of which it is made remains the same). By reason of this generally known fact, it is a matter of everyday practice to replace the inconvenience of the measurement of volume, which would really represent 'quantity of matter' by the much more convenient measurement of weight. When the housewife buys a pound of cheese or butter, she obviously wants neither the inertia nor the gravitation of these bodies, but simply a definite amount of them, and measures it by means of the weights.

In this way it became the custom, long before the existence of physics as a science, to regard weight, as measuring quantity of matter. And when our experience with Atwood's machine shows that acceleration is inversely proportional to weight in the case of two bodies of different material, the layman usually sees nothing remarkable about this, since it has already been shown in the case of two bodies of the same material, that this acceleration was inversely proportional to quantity of matter. For weight and quantity of matter are for him, without any question, identical notions. But for an exact scientific treatment, an unfilled gap obviously still remains, which however, as is now clearly recognised, can be filled in very different ways. It is not in the least necessary to go back to Mach's formalistic definition.

On the contrary, we might just as well state by definition, in a way which would be much closer to common sense, that two 'quantities of matter' are to be regarded as equal when they are of equal weight. Whereupon experience would give Newton's law, which so regarded, naturally says nothing more than that inertia and gravitation are proportional. That both of these are again proportional to the volume in the case of the same substance, is another matter.

We shall return again to the question of the formation of physical concepts in another connection, and we shall then see how far the possibility, here apparently opened up, of reaching their formation by quite different routes, finally holds good. Here we are only describing shortly the problems connected with the concept of mass, which has played an historically important part in the development of physical knowledge, and we must now consider some other problems which are at the back of the traditional concept of mass.

It has already been mentioned in the introduction that philosophy, from the earliest times, has set up the concept of substance as a permanent substratum of phenomena, the latter consisting in a change in its states (accidents ; see above, page 2). Now the great discoveries of classical mechanics permitted this general notion to be formulated much more precisely. When it was seen how the law of gravitation and the simple principles of mechanics allowed all celestial phenomena then accessible to exact observation, to be deduced, when at the same time a large number of other mechanical processes were mastered by this system of concepts, it was only natural to believe that the key to all natural phenomena had been found. *The idea arose, that the world was a vast system of mass points, that is, points endowed with inertia but with no other property, between which attractive or repulsive forces of quite definite values were to be assumed,* and thus it must be possible in principle to reduce the whole explanation of the universe to the differential equations of mechanics.

The grandest and clearest expression of this purely mechanical world-picture is found in Laplace's fiction of a World Spirit – later used by Dubois-Reymond as the foundation of his 'Riddle of the Universe' – who knew at a given moment the position of every mass point in the universe, together with its momentary velocity, and was further in possession of an enormous system of differential equations, according to which the velocities were connected with the accelerations. This spirit would, so Laplace concluded, be in a position to calculate all events in the past, present and future with an accuracy greater because absolute, than that with which the astronomer is able to calculate eclipses or positions of the planets for thousands of years forwards or backwards.

Here we have the sum total of all mass points regarded as completely unalterable; they simply change their positions and motions relatively to one another, and in this way scientific philosophy arrives at the law of the unalterability of all mass. Mechanical mass thus appeared to be the strictly mathematical concept which was to replace the indefinite concept of substance or matter, and the 'law of the conservation of mass' coincided with the ancient law which maintained the invariability of the total amount of all substance in the universe.

Now it is actually the case that, not only in all physical processes, but even in chemical processes (as Lavoisier was the first to show by exact experiment), the sum of the weights of all substances taking part is unchanged; and since weight and inertia are proportional, it is actually true that in all processes hitherto considered, the sum of all masses remains constant. If to this we add the general habit of regarding weight as representing quantity of matter, we can see clearly why, until very recently, weight and mass were regarded as identical with quantity of matter, although the later term as applied to two different substances (for example iron and copper) is purely a matter of convention. It is of fundamental importance for our insight into more recent developments, to realise quite clearly that the experiments of Lavoisier and others finally proved nothing more than that, whilst in all physical and chemical processes hitherto investigated many other properties of bodies change, inertia and gravitation do not; or in other words, that these as far as we have hitherto followed up the matter, are always determined simply by addition of the parts (are additive constants, as we say).

But naturally this in no way proves that the same must be true for all natural processes.[41] Inertia and gravitation are properties of bodies like any others, and why should there not be processes of a deeper nature than chemical processes, which finally result in a change even in these properties? If there really be such a thing as a general law of conservation (see below), it still would not necessarily imply absolutely the conservation of just these two properties. It is true that matters only appear thus from our present-day, more comprehensive, point of view. As long as it was believed possible to reduce the whole of physics to the movement of mass points, or in other words as long as the mechanical picture of the universe held good, mass was the simple and evident measure of substance.[42]

Two aspects of this mechanistic world-picture of classical

physics may receive especial emphasis. In the first place, it is materialistic, inasmuch as the mass points have ascribed to them, once and for all, a fixed property (inertia), and are thus made 'rigid lumps of reality,' which afterwards can hardly be got rid of again. It is not a matter of chance that the first great historical advance of materialism in recent times (de la Mettrie, Holbach) was directly based on this world picture. Secondly, it must be noted that this world picture is strictly deterministic, resting upon the assumption of inevitable and unambiguous causality of all physical events, and thus finally, as it appears, of all events in nature. This latter point brings us to the problem which we must now consider at greater length though we shall only be able to deal with it finally later, the problem of causality.

V

THE PROBLEM OF CAUSALITY

This is one of the most debated problems of the philosophy of science; we may even assert that, at least so far, it has been regarded as the chief problem of all.[43]   The general opinion is, that science means the same thing as the investigation of causes. In other words, science is not only to find out the *what* and *how* of phenomena, but above all the *why*, and it has already been said that, in this sense, the concept of force has been regarded as the physical formulation of the concept of cause.   In the example discussed above of the cause of the motion of the planets in gravitation, the force of gravity discovered by Newton is thus usually regarded as the cause of the continual deviation of the paths of the planets from the straight path.   In this connection, many philosophers have also identified the law of conservation of energy with the demand for complete causal connection in phenomena.

While all this was regarded by scientific philosophers of the materialistic school in the sense of objective discovery concerning the nature of the world, these same laws were taken by apriorists of the Kantian school as categorical principles of form, which the intellect prescribes to nature, that is to say with which it approaches the mass, in itself formless, of sense impressions, and thus 'makes possible' experience (in the sense of 'ordered knowledge': Kant).

But on the other side also, the fear of every kind of speculation, characteristic of modern positivist epistemologists, has resulted in an unconquerable mistrust not only of the concept of force, but also of the concept of causality which is at the base of it; many, such as Mach, Verworn, and Pearson, have proposed to remove this concept entirely from physics, and indeed from the whole of science.   If with Mach, we describe the simplest data of sense, such as colours, tones, etc. (Mach also adds spaces and times, desires, and other things) as 'elements' which are here taken in the psychological sense) the content of so-called causal judgments is exhausted, according to Mach and Verworn, by certain functional relationships between these elements.   'The laws of nature are equations between the measureable elements, $a, \beta, \gamma \ldots \lambda, \mu,$ $\nu \ldots$ of phenomena. 'If we are in possession of all the values of $a, \beta, \gamma \ldots$ by which the values of $\lambda, \mu, \nu \ldots$ are given, then we can call the group $a, \beta, \gamma \ldots$ the cause, the group $\lambda, \mu, \nu \ldots$ the effect' (Mach, *Mechanics*, fourth edition, page 536).

But according to Verworn, it is much more correct to make no further use of these deceptive expressions, but to speak only of the conditions of a certain phenomenon (conditionalism). These suggestions have found much favour among physicists. The more mathematical physics becomes, the more appropriate it appears to physicists to regard the mathematical function concept as the real essential also of physical causality. We must now look into this matter more closely.

It is first of all necessary to be clear as to what is really meant by the usual expression 'causal relationship.' A little considera-tion shows us that this is by no means always the same, and that for this reason alone the whole problem is therefore from the start not a single problem. Hence it is wrong to regard what has been settled by the examination of one, or several similar examples, straight away as the whole solution of the problem of causality. The commonest concept of causality is found in such sentences as the following: the shot through the heart was the cause of death; the treatment of this meadow with ammonium sulphate is the cause of the improved yield of grass. Such causal judgments as these are already made by children and savages and even by animals, though not in words. The general nature of such causal judgments consists in the fact, that we bring a certain complex of phenomena B into a peculiar inner relation-ship with a complex A, which preceded the other in time. We are convinced that B would not have occurred without the previous A, but that conversely when once A has happened, B had to happen, since like things happen in like circumstances.

In a second kind of causal judgments, this succession in time is missing, as for example in the judgments: the heat in the boiler is the cause of the pressure; or: the current in this wire is the cause of the magnetic field that we observe. Here cause and effect are present simultaneously. Nevertheless – and this is of fundamental importance for what follows – the relationship between the two retains its direction. Cause and effect are, here also, not interchangeable; it is obviously absurd to describe the pressure in the boiler as the cause of the high temperature. In the same way, no one ever thinks as a rule that the magnetic field could be described as the cause of the current. While in the former case we have a succession according to law, in the latter case we have a co-existence, also determined by law, but directed in a certain sense, of certain phenomena or groups of phenomena.

A third kind of causal judgments is found in such sentences as the statement that the cause of the splitting of the heredity factors is to be sought in the processes affecting the chromosomes in the so-called reduction division, or the cause of the thunder storm in certain phenomena of atmospheric electricity. Here we obviously intend to express by these sentences what is really at the back of the phenomena in question, in what they really consist.

A fourth kind of causal judgments is reached in the following way: the investigator of heredity, let us say, shows that the highly gifted geniuses of a country are related to one another on the average in a much higher degree than other pairs of people in the same country. He concludes from this that talent is mainly determined or caused by inheritance and everyone accustomed to scientific thinking will agree with this, if this accumulation of relationship between the talented cannot be explained in other ways. In the same way, the meteorologist shows that certain meteorological phenomena change periodically with the number of sunspots, which shows maxima and minima at periods of about eleven years. He concludes from this that there must be a causal connection' between the two, although at present we are not able to understand its nature. There are whole branches of science which are mainly dependent upon such statistical methods, for example, the investigation of human heredity. In all these cases it is a matter of determining whether the 'coefficient of correlation' between two or more phenomena is greater than is to be expected from general probability. If this is the case, a 'causal connection' between the phenomena in question is probable. This, however, does not necessarily mean that A is the cause of B, but it will generally be the case that both are effects of one and the same complex C.

At this point we will not go further into other distinctions between causal judgments. They are naturally indispensable for a more complete theory of causality.[43] For example, we have the distinction between so called 'effective causes,' and 'occasional causes' (the first, for example, in the case of an electric bell is the current, the second is the pressure on the button). If we now summarise what has been said, we reach the following result: Causal judgments are always statements concerning the co-existence or succession of phenomena or groups of phenomena which are regarded, not as matter of chance but of necessity (either determined by law, or at least regular); whereby either the one, A, is described as the cause, the other, B, as the effect, or several A's (A_1, A_2 ......) are described as the causes of the effect B, or both A and B are regarded as effects of the same third complex C (C_1, C_2......). In both cases the direction, cause – effect, is not reversible. A theory of causality must of necessity fulfil these two main conditions, firstly that of necessity, and secondly that of direction. Now there are almost as many theories of causality as there are philosophical systems, and a complete discussion of them would require a whole book. We are therefore obliged to confine ourselves to the most important, and will regard as such the two most widely accepted, namely Kantian apriorism, and Mach-Verwornian conditionalism (functional).

According to Kant and the great majority of later epistemologists, the invariable direction of the causal relationship depends upon the direction of time. The statement of this point in Kant's *Critique of Pure Reason* runs as follows: 'Everything that happens (comes into being), presumes something upon which it follows according to a rule.' Kant further interprets this sentence as follows: 'The reason transfers the order of time to phenomena and their existence, by ascribing to each of them a position in time determined *a priori* with regard to preceding phenomena. . . . The determination of the position cannot be made from the relation of phenomena to absolute time (since this is not an object of experience), but conversely, the phenomena must determine one another's position in time, and make this necessary in the time order. . . . Hence the law of sufficient cause is the foundation of possible experience, namely the objective knowledge of phenomena with reference to the relationship of the same in temporal succession (*Critique of Pure Reason*, Müller's Translation, Part II, pages 166, 175).

According to Kant therefore, causality is a 'category,' a form in which our reason orders the mass of phenomena 'with regard to their temporal succession.' The direction of time is then naturally also that of the causal relationship. The relationship of succession thus becomes a relationship of dependence, but Kant is at pains to show in his discussion, why, as opposed to Hume's purely empirical theory (*post hoc, ergo propter hoc*), the two are nevertheless not fundamentally one and the same. Here we need only consider the point we have quoted, but since in the case of causal judgments of the second kind mentioned above time obviously plays no part, Kant is compelled to introduce a further new category, that of 'interaction,' which he relates to spatial proximity in the same way as true causality is related to temporal succession. This is a decided weakness in his theory, for we have no direct sense whatever of any essential difference between causal judgments of the first and second kind, while here an explanation of direction fails, since no directional sense can be deduced from spatial co-existence. [44]

As however Kant's doctrine has also been adopted by many philosophers of other schools, among these the advocates of 'critical realism,' such as Ed. von Hartmann and Erich Becher, some of them have attempted to fill the gap mentioned. According to Becher, the sense of direction is actually produced in the case of heat-pressure, or current-magnetic field, by the fact that we imagine, at least in thought, the processes leading to the cause A as having already concluded or being in process of conclusion, when the process B, the effect, begins. 'Causality is a connection between two objects of reality (processes, states, situations), in which a stage of production of the one object (the cause) precedes the appearance of the second object (the effect), although the object which first comes into existence immediately brings the second into existence by law as soon as it is itself ready.'

He also points out in this connection that, for example, in the production of heat by the electric current (Joule's law), the movement of the electrons in the wire, which is the current, actually precedes in time the molecular movement, constituting heat, which it causes. Only, it is a pity that we have only known this for the last thirty years, whereas the causal relationship current-heat, had direction ascribed to it long before. On the other hand, the best will in the world does not allow a temporal direction to be ascribed in this way to the causal relationship between current and magnetic field, since both are always present together or not. It thus appears to me that the problem of the direction of causality cannot be solved in this way.

Matters are still worse in the case of the positivistic resolution, described above, of the causal relationship, into two simple functional connections. It is a necessary property of all mathematical relationships that they are reversible. The equation $y = x + a$ and the equation $x = y - a$ are mathematically completely equivalent. The physical formulae which give, say, the connection between temperature and pressure in the steam boiler or that of current strength and magnetic field, are likewise mathematically reversible. But this will lead no one to regard the logical relationship of the concepts as also reversible without further question. In both cases, only the equations $p = F(T)$, and H (magnetic field) $= f(I)$ have a causal sense. The reversed equation $T = \phi(p)$ and $I = \phi(H)$ (tangent galvanometer) have quite a different meaning, namely, that we can deduce from the observed magnitude of p, or H, the magnitude of T or I: in other words, we can measure the latter by means of the former. In this sense, the manometer on a steam boiler can actually be used to measure the temperature, but no one will assert that the pressure which it indicates is the cause of the temperature we determine.

The positivistic explanation affords no indication whatever of the complete logical difference between the two forms of the functional connection. It is true that Mach has made an attempt to answer this question. He thinks (*Mechanics*, page 513) that the causal relationship is only a provisional way of regarding phenomena. As soon as we are accustomed to the connection of the elements in question, we always immediately regard them as belonging together, and no longer speak of cause and effect. 'Heat is the cause of the pressure of steam. When this relationship has become familiar to us, we always imagine the steam as having the pressure which belongs to its temperature. The acid is the cause of litmus turning red. But later this turning litmus red is regarded as a property of acids.' This obviously contains the view that the directional sense of causal judgment is that from the already known to the unknown.[45] We already know that the substance in question tastes acid. We then learn that it also turns litmus red, and hence we call the acid the cause and the reddening the effect.

In the same way we should have to explain the directional sense of current-magnetic field as follows: When Oersted, in 1820, made the fundamental experiment, he already knew that in his arrangement of metals, liquids, wires, etc., that process was taking place which we call electric current.   But he knew nothing of the magnetic field, and for this reason the latter was called the effect and the former the cause.

This explanation is so plausible that in earlier editions of this book I followed it to a considerable extent.   But I have since convinced myself that there must be an error in it.   For one only needs to consider the following question: How would it be if the historical course of events had by chance been reversed?   If long before Galvani and Volta, it had already been noticed that in the neighbourhood of certain arrangements of metals, liquids, wires, etc., a magnetic field is present?   Should we then, when the simultaneous electric phenomenon, the current, had afterwards been discovered, have regarded the latter as the effect of the existing magnetism?   The above positivist explanation requires this question to be answered in the affirmative.

But the actual course of physical research raises a serious doubt as to whether this is the case.   For there are enough cases of this kind, in which the conclusion has by no means been drawn that what is already known is the cause of what has been newly discovered: on the contrary, the newly discovered has been regarded with instinctive certainty as the cause of what is already known. Thus for example, the cause of radio-active radiations was recognised in the decomposition of the atom, which was discovered later.   Both cases are equally common, and hence it is not possible to assert that if history had taken the course in question, the current would not have been regarded as the cause of the magnetic field, but the magnetic field as the cause of the current.

It would be worth while to take this course in teaching physics by way of experiment, and see what conclusion the pupils would draw.   I would almost be ready to wager that they would immediately come upon the idea of seeking the cause of the magnetic field in that mysterious something, which makes known its presence in the wire in so many different phenomena.   But if this is true, then the psychological and historical explanation of the sense of direction fails, and we are obliged to admit that what we describe as cause and what as effect is settled by the nature of the matter itself and not the chance history of discoveries. We are thus again confronted with the question as to the real essence of direction in the causal relationship.

It appears to me that the solution is to be found in the statement by Mach which we have given above, only we must think it out correctly. 'If we are in possession of all the values of α, β, γ . . . by which the values of λ, μ, ν . . . are given, we can call the group α, β, γ . . . the cause, the group λ, μ, ν, . . . the effect.' We only need to lay the emphasis upon the words 'are in possession of' and to recognise clearly that this does not mean stating mentally the values in question in the sense of the mathematician, but that they are already laid down by the whole of the circumstances necessary for realisation of the case in question. The rise in temperature in the steam boiler is given by the whole of the manipulations which lead to it, even when we leave the steam valve open so that no excess of pressure can result. But if we also assume its closing as a necessary condition, then the excess pressure is produced; it is then formed with the same necessity as that with which a certain conclusion follows from certain premises. The sense of direction depends upon this logical relationship and not upon a succession in time. The logical 'prius' is here the whole complex of those circumstances which we describe as the cause, and the logical 'posterius,' the conclusion, is the effect.

The clearest case is seen in a causal relationship mentioned by him in chapter 3, in which the pressure exerted by a stone upon its support, is regarded as caused by the gravitation of the earth, or also in the reference of the planetary motions to the same force. The latter in this case is regarded as the cause of the effects in question, because it forms the object of a general law, from which the particular cases can be deduced as logical consequences. Let me be quite clear: I am naturally not maintaining that cause and effect are identical with reason and consequence,[46] though they are often used interchangeably in every day language. Cause and effect are concepts which relate to real things and their relationships (at least it is in this sense that we are at present using them, whether correctly or incorrectly is a further question that cannot be discussed here); reason and consequence, on the other hand, are concepts which relate to a logical relationship between judgments.

I maintain, however, that the former two concepts are at the last resort to be referred back to the two latter, and in particular, that the directional sense of the causal relationship A→B always depends fundamentally upon the fact that, upon the determination of A (in the logical sense) and upon general laws already settled for other reasons, the simultaneous determination of B follows logically.   If we assume universal gravitation between all bodies as existing, and make use besides of the general fundamental laws of mechanics, it follows that the elliptical form of the planetary orbits is a logical necessity.   For this reason and only for this reason, is gravitation called the cause, and the elliptical form the effect.

Also, it is by no means always necessary here that the deduction in question should be really possible.   It need only be imagined as possible.   In the case of the steam boiler, for example, we can only carry out the deduction very approximately, since the kinetic theory of heat has not yet been able to give a satisfactory solution of the saturation curve, but in spite of this defect in the theory we are fully convinced that such a logical connection exists, since without any doubt both heat and pressure are only different sides of one and the same thing, namely molecular motion.   Indeed, it is not a causal explanation in the sense of physics, if when asked why the pressure increases in the boiler I simply answer: because the temperature rises.   A real explanation, an answer to the question as to how the thing happens, is only obtained when the further question can be answered: why, when the temperature rises does the pressure also increase? This question is answered by the kinetic theory of heat, though not quite completely, but even if that theory did not exist at present, we should still be convinced that such a connection is logically necessary for some reason or another, and it is in this, and not in anything else, that the 'causal connection' required by science actually consists.

We have now only to inquire as to the explanation of the fact that a reference to order in time has repeatedly been brought into the causal problem by philosophy, and still is, even to-day.[47] The answer is not difficult.   It arises from the fact that in far the greater number of cases, and just in those of greatest practical importance, the general laws which enter into the question are of the kind which appear to connect the momentary values of certain magnitudes with their changes with respect to time, a point mostly clearly recognised with regard to the fundamental laws of mechanics. It is in this way that time actually comes into the content of almost all, or at least of the most important and frequent, causal judgments, and this has led to the error of including it in the form, the structure, of the causal relationship itself, instead of in its content.

By far the most important for us are those causal judgments which allow us to predict in advance a certain course of events; much less frequently do we need those which allow us to conclude from a state of affairs at one point or another, what is happening simultaneously elsewhere. But in principle, these two cases of 'dynamic' and 'static' causality, as we might call them, have no priority one over the other. The essential feature, the conclusion from a determined A to a co-determined B, is in both cases exactly the same, only that in one case time occurs among the necessary variables, and not in the other case. In the first case the direction of the causal relationship coincides with that of time; whereby, it is true, the further question immediately arises as to how time itself gets its direction. The 'Laws of Nature' allow us, as is particularly clearly seen in astronomy, to calculate the course of events backwards just as easily as forwards. Hence it is obviously necessary to find some other principle, if the invariable direction of time is itself to be understood. We shall return to this question in another connection (page 154). In the opinion of the writer, a really satisfactory answer to this question cannot yet be given. The little that we can say with regard to it to-day, requires us to take account of matters which will be discussed later. The same is true of the further question as to how far nature can really be captured by laws of the form here considered, namely of differential equations capable of unambiguous integration. Just this, which, to classical epistemology (Kant), was the self-evident foundation of all science has become to-day a problem, as we shall see later. But it is useless to go further into this matter until we have examined the new knowledge itself which has led to a new problem.

We therefore will now lay aside the discussion of the problem of causality for the present and turn again to physics, continuing our statement of the fundamental concepts of mechanics.

VI

THE ENERGY LAW AND THE DIVISIONS
OF PHYSICS

What we call to-day the classical or mechanical world-picture, shortly sketched in the last chapter (it found its most exact expression in Laplace's fiction), was without doubt a bold attempt to formulate a unity, which at that time was not even approximately comprehended. No one could, at the time of Laplace and Kant, have given any kind of exact idea as to how optical, or the few known magnetic and electric phenomena, were really to be brought under the common general concept of the 'motions of mass points under the influence of forces.' This was postulated because people were convinced that a science (mechanics) which had shown itself to be so fertile in explaining celestial phenomena, would be capable of still further achievement, and finally comprehend all natural phenomena. Kant's theory of knowledge is, when seen from this point of view, finally nothing but the attempt to give this conviction, held by scientists themselves more for instinctive reasons, a clear expression and a general philosophical foundation. The concepts of mass and force appear, as already indicated in the beginning, in this view only as the two categories of substance and causality transformed into measurable physical concepts. If these two were really only *a priori* forms of thought, it was of course not remarkable that the whole of science had finally to do only with the motion of mass points under the influence of forces.

But in reality, direct physical experience presents, as is generally known, a much more complicated picture. If we describe the various forms in which natural events confront us, for example the fall of raindrops, lightning, thunder, the rainbow, etc., as manifestations of various 'forces of nature,' this amounts at first to not much more than giving them a common name. For each of them appears in itself, for the moment, as something quite different qualitatively from all the others. And we do not need to seek far for the origin of these qualitative differences between heat, light, sound, etc.   They are obviously only the differences, unconsciously projected into the outer world, between our various sense impressions (Locke's 'secondary qualities').

But at this point, epistemological reflection commences.   The rays of the sun, for example, produce on the hand the sensation of heat, in the eye, the sensation of light.   Now do they therefore contain two 'forces of nature,' or does one and the same cause, operating upon two different sense organs, or through them upon two different parts of the brain, excite different impressions ? This example shows us how difficult it has been, even for scientific physical investigation, to get completely free from 'naïve realism.' For only in our time has the conviction become general that the second answer alone is correct, and that the sun's radiation is of one kind only, the difference first occurring in the body or the mind.   We have no other means of deciding these questions, than the activity of the reason in combining, analysing and comparing.

All physics commences therefore, not only historically, but even to-day as regards teaching, with a separate investigation of all these single departments.   The usual division of physics into mechanics, sound, light, heat, electricity, and magnetism, is generally known.   The second stage of knowledge then consists in finding the threads which connect the various divisions.   In reality, these are all connected together, since there is no such thing as a purely mechanical, purely thermal, purely electrical process.   But these connections can only be clearly recognised, when the single division itself has been cleared up as far as possible.   This investigation is at first mainly concerned with setting up quantitative measures of the phenomena to be dealt with. Extraordinarily instructive examples of this are given by the development of the measurement of temperature, and of electrical magnitudes.   We may here devote a few words to the first of these, which will also be useful in a later connection.[48]

Direct observation by the senses gives us a continuous succession in the difference between 'hotter' and 'colder.'   But more exact observation very soon leads us to find cases in which this simplest thermometer, our skin, misleads us.   Thus we can find lukewarm water to be cold, if we previously dip the hand in hot water.   (We will not go further here into the idea of the deception of our senses, which thus arises.)   In order to reach a more trustworthy measurement of temperature we use a principle which rules the whole of measurement in physics, that of placing side by side two series of changes. We find, namely, that the change of temperature first of all observed by means of the nerves of our skin, is accompanied by simultaneous changes in innumerable other properties of the body in question, among others, its volume. We choose one of these properties, the most easily and exactly measureable, as a measure of the temperature. Here we cannot proceed without taking certain arbitrary steps. In ordinary thermometry, it is in the first place arbitrary to choose the volume of a body as a measure of the temperature. Secondly, the choice of the body itself (the thermometric substance) is arbitrary. Thirdly, we are free to choose in what way, that is according to what mathematical laws, we are to connect the magnitude to be determined, the temperature t, with the volume measured, v. It is by no means a matter of course that this should be a linear law (direct proportionality between the changes in t and v), as is generally assumed as self-evident in the case of the mercury thermometer.

In the further development of thermometry, hydrogen or helium has been finally chosen as the thermometric substance, and it is further advantageous not to measure the volume of the gas at a constant pressure, but conversely the pressure of the same at constant volume, and to use this as the measure of the temperature. On these assumptions, Amonton's law then holds: The pressure in any gas rises or falls on heating or cooling between two fixed temperatures to the same extent for all gases, rising between the freezing and boiling points of water by $100/273$ of its value at the freezing point. If this range of temperature is divided, as is usual, into 100 equal parts, so that each part represents an increase in pressure of $1/273$, this scale agrees very closely with the ordinary centigrade scale; that is to say, the increase in volume of mercury is nearly proportional to the increase in pressure of a gas.

If we imagine this scale continued downwards in the same way, the pressure of the gas would become zero at 273°C. below the freezing point. This (purely imaginary) point of temperature is called 'absolute zero,' and the temperature measured from it is called the 'absolute temperature' ($T = t + 273$). In reality this is a fiction, if only for the reason that all gases are already liquid above this temperature, and also that the simple law of pressure no longer holds as we approach the point of liquefaction. Nernst's so called Heat Theorem tells us that absolute zero is unattainable. Hence it has recently been proposed to connect pressure with temperature in a different manner,[48] in such a way, namely, that at absolute zero we have the temperature $-\infty$°. The temperature scale itself then includes the fact that this point can only be reached 'asymptotically.'

A further arbitrary matter is the choice of the size of a degree. If it is desired to find for this a 'natural scale,' that is to say one directly connected with other physical measurements, it would be necessary to go back to the transformation of heat into work which we are about to discuss. This would then take us away from the science of heat strictly so called, and bring us into other matters, and the same would be true if we should enter into the question as to what the absolute zero really is.

The problems of the theory of knowledge lying behind this theory of measurement will be dealt with later on; we must first add a few words here concerning what is called calorimetry, or the measurement of quantity of heat.

The concept of quantity of heat results in physics from something like the following considerations and experimental results.   The conduction of heat between two equally warm bodies, the process by which their temperature becomes equal, produces a strong impression that something passes from the hot body to the cold body.   Hence it is said also in daily life that, for instance, the heat of a hot oven passes into a vessel set upon it. Upon this follows easily the quantitative notion that the amount passing is the same in equal times for equal temperature of the oven, and when we observe that it takes a large quantity of water longer than a small quantity to reach the same temperature on the same oven, we are immediately led to the notion that a greater quantity of heat is required to warm the greater quantity of water.

We now define as unit quantity of heat, or one calorie (cal), that quantity of heat which is necessary to raise one gram of water in temperature by one degree.   In order to heat, say, 20 grams of water one degree, 20 calories are necessary.   But on the other hand experience teaches us that in similar circumstances we may also have four grams of water warmed five degrees, or ten grams two degrees, or one gram twenty degrees (approximately).   Hence, generally speaking, the quantity of heat necessary to heat m grams of water t degrees is mt calories.   One gram of another substance requires in general a different amount, usually less; one gram of copper for example 0.1 calorie.   This number, which is therefore almost always a fraction, is called the specific heat of the substance in question, and is designated by c.   In order to warm m grams of any substance by t degrees mct calories are necessary.

If we now mix various bodies at different temperatures in a suitable vessel insulated against transference of heat, we have the law of mixture of Richmann and Black: the whole quantity of heat present in the bodies remains unchanged.   Hence the quantity of heat given up by the warmer bodies is equal to that gained by the colder bodies.   According to this rule, which is easily stated in a mathematical form, we are able to calculate the final temperature if we know the masses, the specific heats, and the original temperatures before mixture.   We have only stated the rule here because it shows most clearly that there is a certain magnitude, the quantity of heat, which remains constant in all purely thermal processes.

Now such processes, as we have said above, never exist strictly speaking, and we are now interested in the reciprocal relations between thermal and mechanical processes. Let us consider say the manufacture of fire by savages, or the steam engine. In the first case heat is obviously produced by mechanical exertion, in the second case, heat conversely produces mechanical work. It is our next task to consider these relationships.

In order to be able to do this, we must first say a few preliminary words concerning the development of some other mechanical concepts, which the reader acquainted with the matter may omit.

For the working of the simple machines (lever, roller, set of pulleys, inclined plane, etc.), which were already known to the ancients, the so-called 'Golden Rule' of mechanics holds: What is saved in power is lost in distance; or more exactly: The smaller the force the greater is the distance through which it must move. The product of the two (F. s) is therefore an unchangeable quantity, and it is only possible to change one factor if the other changes as well. This product therefore represents a magnitude which is given once and for all in a given case (as, for example, the raising of a heavy body to a certain height) and this magnitude is given a special name.[49] It is called *work*, and is measured by the product of force and distance, F. s. If one dyne is taken as the unit of force (the force which gives a mass of one gram an acceleration of one centimetre per second per second) the unit of work is accordingly one dyne times one centimetre. This is given the name of *erg* (from *ergon*, work) and since it is too small for practical purposes, a unit ten million times as great is used: one joule. (Using the value of the acceleration of gravity, 981 cm. per second per second, it is easy to calculate that in order to raise a weight of one kilogram a distance of one metre an amount of work 9.81 joules is necessary. We add this merely as illustration.)

FIG. 3
Motion of
pendulum

While we have hitherto assumed that the force necessary for doing work is given by our muscles, or by some other force in the narrower sense, we must now take note of the fact that a moving body is also able to do work. We can, for example, by shooting a bullet at another body, not merely burst it apart, thus doing work against the forces of cohesion, but also bring it about by suitable means, that parts of the body, or the whole of it, fly into the air, and hence do work against gravity (the weight multiplied by the distance raised).

This fact is shown us still more simply by the pendulum (see figure 3). By moving this from its position of rest at A, work is done, since the bob of the pendulum is raised to a higher level. If we now let it go, say from the point B, it swings back to A, which it reaches possessing so much 'kinetic energy,' that it rises on the other side just as high as before. This kinetic energy is therefore able to do the same amount of work which we ourselves did previously. Now in physics, every magnitude which is equivalent to work is called energy, and hence we speak of *kinetic energy*, or energy of motion.   The energy of motion of a moving body is obviously the greater, the greater its velocity and also the greater its mass.   Calculation shows, as may be seen from any text-book of physics, that it is $\frac{1}{2} mv^2$.

A further fundamental notion of the doctrine of energy is also to be seen by considering the pendulum.   When the latter has again reached its highest point on the other side, its kinetic energy is used up, but to make up for this it is again at a higher level and is able in falling to do the same amount of work as before.   It is therefore said that the pendulum (or more correctly the system formed by the pendulum and the earth) possesses at the point B energy of position, or *potential energy*, which is transformed during falling into kinetic energy, while conversely, the kinetic energy is transformed back into potential energy on the upward path.   The measure of the energy of position or state is given by the amount of work originally put into the pendulum, that is the product of the weight of the pendulum bob and the height by which it was raised.

Thus the pendulum continually transforms one form of energy into the other, and it is easy to prove by calculation, that not only are the two final amounts of energy (kinetic at A, potential at B) equal, but also that, at every intermediate point of the path, the potential energy has always increased by the same amount that the kinetic has decreased, and conversely.   *In other words, the sum of the kinetic and potential energies remains constant during the whole process.*

This law is called the law of the conservation of mechanical energy.   It is naturally only approximately true, for in reality we are not able to allow the processes to take place without friction or hindrance.   But it is true for all purely mechanical processes, no matter how complicated, apart from such opposing forces as friction, resistance of the air, and the like.   This knowledge, which was founded by Huygens and Leibniz, remained, curiously enough, without application for over a hundred years, before its fundamental importance was recognised, although Leibniz had already seen in it the universal cosmic law of the 'conservation of force,' as was the term used in those days.[50]

Recognition of the true state of affairs was hindered by the doctrine founded by Richmann and Black, of the conservation of the total quantity of heat in purely thermal processes.   This law, which was in itself undoubtedly a great advance in our knowledge, formed the foundation of Black's 'substance' or 'fluid' theory of heat, according to which heat was a peculiar substance having no weight, which could be poured from one body to another in thermal processes without changing in amount. The possibility that this substance might be fundamentally identical with mechanical energy did not occur to the scientists of that period, since for the time they were held fast in naïve

Gs

realism, according to which we are able to experience heat directly as a peculiar quality, and hence regard it as self-evident that what appears to us to be so different from motion, must necessarily be so absolutely.

It was the great historical achievement of Robert Mayer (1842) to break away and reach the correct view. He recognised that in the heat produced in frictional processes, for example, we have the equivalent of the mechanical energy apparently lost in these processes, and conversely, when heat is transformed into work, as, for example, in the steam engine, the mechanical energy obtained ($J.s$) is the equivalent of an amount of heat (a certain number of calories) which actually disappears. *Hence quantity of heat, and mechanical energy or work, are equivalent to one another, or transformable one into the other.* According to the latest determination, one kilogram-calorie or 1000 cals, is equivalent to a work of 426 metre-kilograms, that is the work that is done when 426 kilograms are raised a height of one metre.

But the matter does not end here. Mayer already recognised, and Helmholtz was the first to state clearly, that this principle of equivalence is a fundamental law of the whole of physics, and that not only heat and mechanical energy obey it, but that in every department of physics there is a magnitude, which is equivalent to mechanical work, and heat, and all other quantities of this kind. In electricity, for example, it is the product of quantity of electricity and voltage, which stands in this relation of equivalence to mechanical work. All these quantities are called by the common name of energy (electrical, magnetic, chemical, radiant, etc.), and hence the general law may be thus stated.[51]

In every physico-chemical process the total energy, that is, the sum of all single kinds of energy, remains unchanged. This is the law of the conservation of energy, known shortly as the energy law. It states that whenever any kind of energy apparently disappears, some other kind of energy appears in corresponding amount. Every natural process is a transformation of energy, but energy is never produced or destroyed without equivalent.

This discovery brought to exact physical expression a philosophical idea many centuries old, which along with the 'law of the conservation of substance' (see above, pages 2, 64), has been the concern of scientific philosophers of all periods; the notion, namely, of the conservation of the total sum of all happening (all motion, force, etc.). This idea, it is true, only came to a very few particularly penetrating minds in antiquity (among them probably Democritus), for the ancients thought statically and not dynamically, for which reason also they were denied insight into the law of inertia. At the beginning of modern times however (Descartes, Bruno and others), we find appearing everywhere, the idea that motion once present in the world never really ceases completely to exist, but is only passed on to other bodies (see also above, page 58). Hence in the classical period, the philosophers (Leibniz and Kant) already placed quite definitely alongside the law of the conservation of substance, that of the conservation of force, or of motion, only they did not succeed in raising this law from an indefinite general notion to the level of a clear and certain piece of quantitative physical knowledge. This was reserved for Mayer, Joule and Helmholtz.

Conversely, this discovery has exerted a considerable effect upon scientific philosophy. A whole school of philosophy based upon it arose, commonly called that of the energists, and is chiefly represented by Wilhelm Ostwald in Germany, and W. J. M. Rankine in England. Although its one-sided view is abandoned to-day, it is worth while devoting a few words to considering it. The energist regarded the law of energy not only as a general law of nature, but as *the* law of nature, to which all others could finally be reduced, and thus made energy, to use Lipps' neat expression, the 'horse in the world machine' or in other words, the only substance in the world process. What we buy in a pound of dynamite, is, so Ostwald argued, not the quantity of substance (carbon, etc.) contained in it, but the energy contained in it. What the consumer pays the electricity supply station for, is electrical energy; what I buy when I buy food, is the chemical energy of it, which is transformed in my body into heat for mechanical work.

Indeed, we can say that everything that we experience is energy, for the light waves and sound waves, and so on, which give us the sensations of things, are also nothing but certain forms of energy; matter is thus nothing but a certain constant connection of forms of energy, and strictly speaking nothing exists but energy, since nothing else can be perceived (here Ostwald linked his energetics, which in itself is undoubtedly a system of metaphysics, with the anti-metaphysical epistemology of Mach, which was only prepared to recognise the 'elements,' that is, the simple data of sense, as real).

The following must be said by way of criticism of energetics. It is quite possible in itself that all matter may be regarded as a certain relatively constant connection of definite forms of energy, and we shall even see later that this fact may almost be taken as proved from our present point of view. But the energists were making matters too easy for themselves, when they supposed that so comprehensive a doctrine could be founded only upon the doctrine of the transformability of the various forms of energy, which was their basis available at the time. The law of energy only states the equivalent of the various forms of energy. It is therefore convenient to use the same measure (erg or joule) for all the different forms; but we must not imagine that this proves the actual identity of the different forms.

This identity is perhaps a postulate resulting from the law of energy, but not a result; it is a task which is given us, but not a solution. The unity of all the different forms of energy must first be sought for, that is, we must first really perceive how far and in what sense heat and work, for example, are actually one and the same, before we can assert that they are identical. As long as we only have the equivalence of heat and work the statement of their identity is merely on paper; it remains our task to give it real content. We shall see in the next chapter but one that this problem has actually been solved to-day as regards heat and work. But as regards a whole number of other kinds of energy it remains in the form merely of a postulate. For every kind of transformation of energy is a simple fact in the sense already discussed, which requires further explanation. It is not sufficient for us, for example, to observe that electrical energy is transformed in the electric lamp into heat and light; we must know what really happens in the course of this transformation, and that is only possible if we are able to develop such an idea of the electrical processes on the one hand, and of those of heat and light on the other, that from the two ideas taken together, both the phenomena in question and all sorts of other relationship between the three types of phenomena follow.

The positivism so strongly supported also by Ostwald[52] obscures here, as everywhere, the problems actually present, or simply declares them abolished by putting forward the purely superficial and mathematical relationships as the final result of our wisdom. At this point also, he does not realise that the actual problem only begins with the determination of this functional relationship, the real question being how just this particular relationship comes to exist.

The law of energy is only the most striking example of such relationships, first discovered as pure matters of fact, and then requiring further explanation. We may examine physics at any point, we always find the same state of affairs. This does not however prevent our recognising that the law of energy is actually the most important of such relationships, by reason of the fact that it is all-inclusive. This statement already implies that all physical processes must be finally of one kind, for the equivalence of the various kinds of energy is only comprehensible, if they are fundamentally only different forms of one and the same thing. In so far, the law of energy also affords from our realistic point of view an extremely important instance for our judgment concerning the whole system of physics. Only we must be clear that this judgment cannot be founded upon it alone.

This is also clearly seen from another point of view, namely when we consider the relationship of the law of energy to the particular laws which hold for single departments of physics, such as mechanics or electro-magnetism. In the language of theoretical physics, the law of energy appears as a general integral of the differential equations in question. This means that it represents a consequence of them, which holds in every special case where they apply, but that conversely, it is not possible to deduce from it the special laws in question, at least not without the help of further assumptions.

The layman not acquainted with theoretical physics can also make this fundamental idea clear to himself as follows: the law of energy is a necessary but not a sufficient condition of all physical happening. Nothing happens which conflicts with it, but it is not possible to deduce from it what really happens. For in every special case there are usually an infinite number of possible processes which fulfil the condition that energy is neither lost nor produced. If for example a ball is lying on an inclined plane, the law of energy would hold whether it ran down by the shortest path (as steeply as possible) or by any other; indeed, even if it ran upwards by itself. The law of energy does not even allow us to conclude that it runs down at all, let alone that it follows the shortest path. This alone shows quite clearly that the energists are wrong when they state the law of energy to be '*the law of nature.*' *In all cases something else must be added, which really determines the actual course of events among the innumerable courses which are compatible with the energy law* (compare also page 264).

The latter consideration gives us an opportunity for referring to a complex of problems which played an important part in scientific philosophy 150 years ago, and is also of great importance for pure physics. We refer to the so-called *minimum principles of mechanics*, in particular the principle of least action. The essentials of the matter may be stated as follows:

When a number of mass points or rigid bodies are connected by arrangements such as rods with ball joints, threads or the like, the single mass point is no longer completely free to move, but it is on the other hand by no means confined by such arrangements to a quite definite path (such for example as is the case when a hinge is used). Those motions which are consistent with the conditions of the system, are called *virtual* motions. If the mass system in question is now exposed to the influence of external forces, its parts will carry out quite definite motions, which on the one hand are determined by the external forces, and on the other by the linkage forces resulting from the existing connections. It then becomes a problem of general mechanics, to find what conditions govern the appearance of the actual motions as compared with the numerous possible or virtual motions. This question was first answered by the famous French mathematician d'Alembert in a very elegant rule, the so-called 'principle of virtual motions.' After him, his countryman Maupertius, and his contemporary Euler, somewhat later Gauss, and in the middle of last century Hamilton, all stated similar general rules. Common to all these is the fact that they state certain functions of the positions, velocities, masses etc., of the mass points in question, functions which assume for the actual motion, as compared with all other possible motion, a greatest or least value.

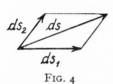

FIG. 4

As example we may take Gauss' 'principle of least compulsion.' In a very small interval of time dt the mass point moves as a result of the connections through a real small distance ds, which can be regarded as the resultant (according to the parallelogram of forces) of the small distance ds_1 through which it would move freely in the same time as a consequence of inertia, and a small distance ds_2 determined by the forces (external together with system forces) (figure 4). The product of the mass of the point and this small distance ds_2 was defined by Gauss as the 'compulsion' which the point in question suffers during the motion considered, and his principle states, that the sum of all these products is smaller for the actual motion than for any other virtual motion. Similarly, the Maupertuis-Euler 'principle of least action,' that was later given by Hamilton in a more general form, states that in the case of the real motion a quantity described as 'action' is a minimum, this quantity being defined in the older writers as the product of mass, velocity and distance ($m.\ v.\ s$), but by later authors generally as the product of energy and time ($E.\ t$). We mention this here because at a later point we must return and examine in detail this concept of action, which to-day stands in the forefront of physics. Hamilton's principle is a principle of integration. It states that in the real motion, the sum of all products $E.\ dt$ taken over a certain interval of time (that is the time integral of the energy $\int_{t_1}^{t_2} E.\ dt$) is a minimum.

In the latter form we are able to see with particular distinctness the peculiar character of this statement, which has caused it to play so important a part in scientific philosophy. For all these rules have a certain teleological character; it looks as if nature sought, out of many possible motions, those which allow the greatest possible effect being produced by the least means, or those in which the exertion (compulsion) is as small as possible. In actual fact, the discoverers of those principles, above all Maupertuis, already looked at them from this point of view, and Maupertuis actually bases a physical proof of the existence of God upon his principle, which, however, he stated in a very unclear form. But Euler also, who was the first to give the principle a form free from all objection, was not indisposed to regard it as a direct proof of Leibniz' idea of 'the best of all possible worlds.'

Right up to the present day, these principles have been called in to support arguments of this kind, and there is really something surprising in seeing a process take place, in accordance with Hamilton's principle, in such a way as if, so to speak, nature already considered at the commencement of the time interval $t_2 - t_1$, what it must do to make the value of the integral $\int_{t_1}^{t_2} E.\ dt$ as small as possible at the end of the time. The truth is, however, as Mach and after him Petzoldt showed, that the principles in question contain nothing more than the statement that under certain definite circumstances something definite happens; or in other words, the principle of causality in a somewhat unusual form.

In reality, what happens is determined by the differential

equations of mechanics (see above). But mathematically speaking, every minimum or maximum condition amounts to this, that the so-called differential co-efficient of the function in question, which is to be a maximum or minimum, must become equal to zero; in other words, it amounts to a differential equation. Hence conversely, every differential equation may be regarded as the condition for a certain function becoming a maximum or minimum; and we even have a considerable liberty of choice as regards the functions. There are thus several functions which have the property of becoming a maximum or minimum when the differential equations of mechanics are assumed to hold, and hence it is possible to find many other such functions besides those generally put forward. The statement that certain differential equations hold is completely equivalent to the statement that this or that function becomes a maximum or minimum.

The teleological character only arises from the fact that such functions as Gauss' sum of the products $m.\ ds$, or the Hamilton integral in which the real motion becomes a minimum, have been given names adapted from human purposive activities such as 'compulsion,' 'action,' etc. This brings teleology into a purely objective set of facts. If, not this integral but another were a minimum, the other would have been given a name of this kind, and nature would have been credited with the intention of making this other a minimum. We could only speak of a real foundation of teleology if the quantities in question had already been recognised by quite different means as the natural criterion of purposiveness of the whole process, but there is no question of this.

Although we must recognise that attempts, which occur even to-day,[53] to give a teleological character to these minimum principles are not tenable, the fact still remains that the principles in question, and particularly that of least action, are again playing a very important part in the system of physics. The reader may be referred for further details to Planck's excellent essay in the volume on physics in *Kultur der Gegenwart*, but we should add that meanwhile, and as a result of Planck's own investigations (see below), the quantity called 'action' has acquired much more fundamental importance than before. We shall return to this at length later, and shall then see that in a certain sense, a kind of teleological character must be ascribed to some physical processes, namely those in atomic dimensions.[54] In the present connection the only point which interests us in the whole problem is what distinguishes real natural processes from the many which would be compatible with the law of energy.   This distinction is made, as is now clear, by an addition to the general law of energy of at least one further law, which actually settles the direction of the course of events.   This results in showing that a purely energetic physics is inadequate from the beginning (this is particularly important in view of a later biological discussion, see page 393).

While we have thus reduced the claims of Ostwald's ideas (which have lost their importance to-day) to their correct position, we must nevertheless defend them against an objection which has not seldom been raised, as in a frequently quoted essay by Lipps.   It is said that we are dealing in the case of energy with the product of a long and difficult conceptual development in physics, and that for this reason alone it should not be regarded as the beginning of physics.   Argument of this kind is again based on Mach's positivist theory of knowledge, according to which the simple data of sense are the only real things, while all physical 'elements of reality' such as atoms, light waves, forces, etc., and even, strictly speaking, all 'things' whatever, are simply 'thought-symbols for sensation-complexes of relative stability.'

In the second part, we shall consider the criticism of this solution of the problem of reality of knowledge, called to-day the conscientialist solution.   The fact that it is not a sufficient basis for physics becomes obvious when once we picture to ourselves what form the simplest physical law, such as the acceleration of gravity being 981 cm.-seconds, would take when translated into this language.   Precisely the main object of physics is to free us from the limitations and errors of direct sense experience (see above), and to provide us with a system of judgments firmly founded in itself, which, it is true, are of a purely quantitative character, and contain only spatial, temporal, material, and causal relationships.   The doctrine of the subjectivity of so-called secondary qualities (sensation qualities such as colours, sound, temperatures, pressures, etc.) is, since Locke's time, almost the only law about which all philosophers are agreed. For the refutation of naïve realism is too striking to fail to convince anyone who really thinks about it.

As soon, however, as we grasp this fact, the system of physical realism follows of necessity.   In place of sound and colours we have frequencies, in the place of the direct sensation of heat, we have molecular motion, and it is the task of this first part of our book to develop this system as a logical whole. This does not amount to the assertion that the world of physical realism is already the world of things-in-themselves, which in philosophical terminology is described as the world of noumena. The relation which exists between the two must form the subject of a further investigation of an epistemological character, into which we shall enter at the close of this discussion. But it is clear that in the world of physics which we are considering, concepts such as energy must be placed systematically at the beginning, for the very reason that they were only discovered late. The road to knowledge of this world leads actually backwards from the phenomena directly given to the causes which lie at the bottom of them, hence the axioms are necessarily found at the end and not at the beginning of knowledge. But since in this book, as was already said in the introduction, we are not following the plan of a systematic philosophical investigation, but are allowing philosophical insight to develop naturally from consideration of purely scientific matters, we shall now proceed with our statement of the system, and not again interrupt it, but wait until the conclusion before stating the epistemological result, apart from a few remarks here and there more or less by the way.

The law of energy leads, as we said above, to the necessary postulate of a final unity lying at the base of all the different varieties of physical phenomena. In actual fact, the whole development of physics shows such a tendency towards unification, a fact which we shall now have to exhibit in detail.

Present-day physics has now only two chief divisions, the physics of matter, and the physics of ether. To the first division belong, besides mechanics, sound as the science of oscillatory motions (for the most part perceptible to the ear) and heat, as we shall immediately proceed to show. The physics of ether, on the other hand, comprises electricity, magnetism and optics, as we shall set forth in the next chapter but one. Numerous threads already connect these two departments, so that there can hardly be any doubt that the final result will be complete unification, a view which we shall later on support by detailed argument. Our next task is to show how the science of heat may be brought under the general concepts of mechanics.

VII

THE KINETIC THEORY OF HEAT

This was already propounded in 1738 by Daniel Bernouilli, but then abandoned again. Its actual foundation took place in the years 1856 and 1857 by Krönig and Clausius, and it was then further developed especially by Maxwell, van der Waals, and Boltzmann and more recently by Planck, Einstein, Debye and others. Since almost every part of it is full of difficult mathematical demonstrations, we must here confine ourselves to a fairly superficial sketch, in which unfortunately the most surprising and beautiful successes of the theory cannot be properly exhibited. The reader must bear this in mind in what follows.

The fundamental idea of the kinetic theory of heat is the hypothesis shortly referred to on page 11, that what we call heat, and experience as a sensation of the skin, is nothing but the kinetic energy of the molecules of bodies, which molecules are in continual motion. More exactly, our skin experiences warmth, when the average energy of the external molecules (say the air) is greater than that of the skin molecules, so that during their collisions with one another, the external molecules give up on the average more energy to the skin molecules than vice-versa.

From the standpoint of the theory of knowledge, this hypothesis is no more remarkable in itself, than is the fact, which no one doubts, that the oscillatory motion of the air is what excites the sensation of sound by stimulating our ears.

Generally speaking, the molecules of a body have, according to the hypothesis, unequal velocities; and what we call the temperature is decided by the average velocity of all molecules in the region considered; or more exactly the average of the squares of the velocities, since we must regard the absolute temperature as proportional to the square of velocity. The absolute zero of temperature (see page 78) is the point at which the molecules are completely motionless.

A 'dynamic' equilibrium of temperature exists between two different bodies or two parts of the same body when, at the boundary surface, as much kinetic energy is transferred in one direction as in the other. If one part of a body is heated this means that kinetic energy is given to the molecules of the body at that part, and this is then spread over the body by the continual motion. This is the equalisation of heat by conduction. But by no means the whole of the energy added will be found again in the increased kinetic energy of the molecules, that is, in the rise of temperature. On the contrary, a greater or less fraction of the energy added is always employed in 'internal work,' work that consists firstly in the overcoming of the forces exerted by the molecules upon one another, and secondly in internal molecular changes.

The impossibility of determining this internal work more accurately and introducing it into calculation, was the reason for many failures in the theory. It becomes particularly troublesome in dealing with solid bodies, where the internal molecular forces are strongest. Hence solid bodies, until quite recently, could hardly have the theory applied to them. Recently, however, considerable progress has been made in regard to them by Planck's theory.[55]

Matters are simplest in the cases of gases, where the molecule moves in almost complete freedom, and of the gases, the monatomic are particularly simple (helium and others, see page 24); hence it is in the case of these that we may expect the best agreement between theory and experience. In actual fact, this expectation is fulfilled in the fullest degree, as, for example, is seen by the value for the Loschmidt number found in the case of helium. Helium may be regarded as an almost 'perfect' gas, but it is also possible in the case of the other gases to deduce mathematically from the theory almost all the laws of heat conduction, diffusion, internal friction, etc., in good agreement with experiment, and even many of the phenomena appearing in the transition between the gaseous and liquid state can be satisfactorily explained by the kinetic hypothesis.

We cannot here enter into the many details, in part highly remarkable, of this agreement between theory and experiment, nor into the disagreements, which are also not absent. Taking it all round, the theory gives results correct as regards essential points, and the departures from the theory, which occur in complicated cases, can be explained quite naturally as due to our ignorance concerning the 'internal work,' which makes it necessary to introduce somewhat arbitrary suppositions or simplifications, in order to calculate a result. It is not the hypothesis as such, but the special assumptions, which may fairly be regarded as responsible for such failure as occurs.[55]

Nevertheless, there would be no question of even approximate certainty regarding the truth of the fundamental assumption, and the kinetic hypothesis would rather have to take its place as a moderately probable supposition, or even as a mere model or picture, were we not acquainted with a phenomenon which, when taken in conjunction with the results of theory, can be regarded

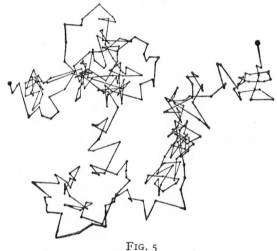

FIG. 5
Brownian motion (from Perrin)

as a direct experimental proof of the kinetic hypothesis. This phenomenon, the fundamental importance of which was strangely long in finding recognition, is the so-called Brownian motion. It was discovered in 1827 by an Englishman, Brown, but has only recently been more closely investigated.

Anyone in possession of a good microscope can easily observe it. If a turbid liquid is brought under the microscope, say a suspension or emulsion of the kind described above (page 20), the particles of which are particularly small but still plainly visible (diluted milk, gutta-percha emulsion, colloidal gold solutions and others), we see that the suspended particles are in a state of continuous motion, giving the impression of perpetual flickering and trembling. With very strong magnification, especially when the ultramicroscope is used, we can clearly recognise the path of a single particle as a complicated and quite irregular zigzag, which is best imagined by scribbling on paper with a pencil as children do. These observations have recently been photographed and even cinematographed (figure 5). There can thus be no doubt concerning their reality. For the comparison between observation and mathematical theory, it is better to use, instead of the liquid emulsions referred to, suspensions of fine dust particles or liquid drops in gases, which exhibit Brownian motion equally clearly. Thus recent investigators, who have examined this phenomenon (Perrin, Svedberg, Seddig and others) have frequently used suspensions of cinnabar or gold dust in air.

The theory of the Brownian motion has been worked out from the kinetic theory of gases, chiefly by Einstein and von Smoluchowski. According to the latter, the suspended particles are set in motion by the molecules which strike them in much the same way as if small balls of cork were hung up in a swarm of gnats (assuming that the gnats would not notice them and collide with them blindly). This theoretical deduction has been most completely confirmed by the painstaking experiments of the investigators we have mentioned, particularly Perrin, and the result has been a new method for determining the Loschmidt number (see page 22). It may be said that, nevertheless, the true existence of molecules is not thereby proven; that it might be so, but that it is still not impossible for another explanation of Brownian motion to be found. The suspended particles are still gigantic in size as compared with the molecule.

That is true. But we must recollect what was said on pages 20 ff. with regard to the results of modern colloid chemistry. We must remember that, starting from suspensions and emulsions and colloidal solutions, the particles of which are still visible in the microscope and ultramicroscope, there is a perfectly continuous transition to so-called 'true solutions,' in which the dissolved substance is distributed in the form of true molecules. It is quite impossible to suppose that Brownian motion suddenly ceases as the particles become smaller and smaller. On the contrary, as far as it can be followed, the motion increases as the particles get smaller, in entire agreement with theory. The principle of continuity thus requires that we assume the motion to exist, even when on account of the smallness of the particles, we can no longer recognise it, and this brings us to the actual motion of the molecule. Hence Seddig says quite rightly at the close of his account of his beautiful investigations[56]: 'These quantitative experiments may surely be regarded as a direct experimental confirmation of the correctness of the kinetic theory of heat, which is now almost generally accepted. Brownian motion gives us a coarse picture, directly visible to the eye, of the heat motion of the molecules, of which motion it is the consequence.'

At this point we will introduce a small digression by referring to a law which we shall need in another connection. This is the so-called *law of entropy, or second law of thermodynamics*, the first law being the energy law. We may conveniently connect it with Brownian motion. Many careful experiments have shown that Brownian motion continues with quite undiminished intensity as long as the temperature of the suspension observed remains constant. It thus represents true perpetual motion. As we know, only such inventors as have had no education in physics attempt nowadays to realise this century-old will-o'-the-wisp. For the fact that it is impossible to construct an apparatus which perpetually produces work out of nothing, is the substance of the law of energy, which is the most certain of all physical laws.

But we must here make a clear distinction. The law of energy denies the possibility of an apparatus which produces work without using up energy from another source. But it actually incites us to seek for an apparatus which allows a complete closed circle of energy transformation, a transformation of one kind of energy into another, and converse transformation to the point of complete restoration to the original conditions. Such an apparatus would fulfil completely the demands of the law of energy, for the sum of the various energies would remain constant during the whole process. No work could be obtained from it, but it would itself continue in motion to all eternity. Every periodic motion, such for example as that of a pendulum, gives an approximation to such a process, but it is always only an approximation.

The assumption for its construction would be that we should succeed in preserving undiminished the whole effective energy taking part in a process. In the case of Brownian motion, and also in that of true molecular motion, this assumption is fulfilled but *the statement that a corresponding apparatus cannot be constructed of greater dimensions than a molecular,* in *molar* dimensions as we say, *is the essential content of the second law of thermodynamics.*[57] If that were possible, we might for example suppose that a ship could take from the ocean, which represents an enormous reservoir of heat, any desired quantity of heat, transform this into work, use it to drive the screws, and thus again heat the water by means of friction. The heat would then have simply passed through a circular course water-engine-water, but the ship would have crossed the ocean for nothing. Work would be neither gained nor lost, and hence the law of energy would be fulfilled. Nevertheless, everyone has an instinctive feeling that this kind of perpetual motion is just as Utopian as the first. Why is this? In order to understand the answer, at least approximately, we must enter a little into thermodynamics.

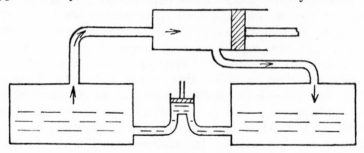

FIG. 6

Scheme of a heat engine; boiler on the left, condenser on the right

Fig. 6 shows the plan of a heat engine. We may assume that it is a steam engine. In the boiler, the working substance (water) is heated and turned into vapour; in the condenser it is again liquefied by cooling. The excess of pressure on the side of the boiler drives the piston forward in the cylinder, and enables it to do work. Now the fundamental question, which was asked and first answered by Carnot in 1824, is: how much work can the engine do in the most favourable case, when we exclude all loss by friction, conduction of heat, and radiation? This maximum work is by no means equal to the total heat Q (calories turned into metre-kilograms) given to the boiler. For it is certain that the whole process necessarily requires a part of the heat to be given up to the cooling water in the condenser, for the process of cooling consists in taking heat from the working substance and thus liquefying it. Without this cooling the machine could not work at all, since the pressure on both sides would be equally great. When we calculate the course of the process, we find, as Carnot and Clausius showed in 1828 and 1850, that the amount of heat which can be transformed into work has the maximum value $Q \dfrac{T_2 - T_1}{T_2}$ in which Q represents the total amount of heat stored in the boiler, and T_1 and T_2 represent the absolute temperatures (see page 78) of the condenser and the boiler respectively. The fraction $\dfrac{T_2 - T_1}{T_2}$ is therefore called the theoretical efficiency of the engine.

For example, if the temperature in the boiler is 160° C., in the condenser 40° C., which gives as absolute temperatures $(t + 273)$ the values $T_2 = 433$ and $T_1 = 313$, we have $\dfrac{T_2 - T_1}{T_2} = \dfrac{120}{433} \approx 0.28$; that is, about 28% of the heat could be transformed into work. The remainder, 68%, goes to waste in the cooling water.

Hence heat can only be transformed to a limited extent into work or other forms of energy; the fraction capable of transformation is proportional to the difference of temperature present $(T_2 - T_1)$, *and inversely proportional to the absolute value of the higher of the two temperatures.*

This law is, in the narrower sense, the 'second law of thermodynamics.' We furthermore have the fact that heat is in any case dissipated by the tendency to natural equalisation of temperature, without work being done. The practical result of this is, that the efficiency of every heat engine is considerably less than the theoretical; thus in a steam engine under the above mentioned conditions, it is only about one-half of theory. Hence a perpetual motion of the second kind is impossible for two reasons. In the first place, the heat losses last mentioned are irrevocable; but in the second place, even if they were not present, the periodic transformation of heat into work, and conversely, is rendered impossible by the second law. For if, for example, we stored the maximum work delivered by our heat engine (28%), say by raising a weight, this could, it is true, be made to give back to the boiler the same amount of heat that had been previously transformed into work $(Q \dfrac{T_2 - T_1}{T_2})$; but of this heat, only the fraction $\dfrac{T_2 - T_1}{T_2}$ would be again transformed into work the next time, and so on. The resulting amounts of work would thus become smaller and smaller (according to a geometrical series, e.g. 0.28; 0.28^2; 0.28^3 . . . .)

The deeper reason for this law, which at first sight appears so curious, lies, as Boltzman showed, in the nature of heat as an irregular molecular motion. The transference of heat from one body or part of a body to another, the equalisation of temperature, really consists in the molecules furnished with a higher average energy of motion transferring their energy to the slower molecules of the other part by continual collision. Now Boltzmann showed

Hs

that this equalisation means nothing else than the transition of the whole system from a less probable to a more probable condition. The second law, or the law of entropy as it is usually called,* can thus be deduced from the kinetic theory by simple application of the statistical rules of the calculus of probability. The only assumption that must be made is the 'hypothesis of elementary irregular motion,' that is, the assumption that all possible molecular velocities in all possible directions are in themselves equally probable (just for example, as in the application of probability to dice, the assumption must be made that the dice are not loaded on one side).

This supposition is very closely true for ordinary temperatures, whereas later investigations (Planck, Nernst, Debye and others) have shown that it can no longer be assumed to hold in the neighbourhood of absolute zero. Hence in this case, new investigations are necessary, and these have given a whole series of quite surprising results. But for ordinary conditions, the hypothesis of the equal probability of all molecular velocities, and hence of the law of entropy, may be taken as unquestionably correct. Its application has been even more fruitful in many directions than the energy law itself. In theoretical chemistry in particular, it has produced magical results since it was introduced into this science by van t'Hoff, Horstmann and Gibbs.

From what we have said, the continuation unchanged of Brownian motion, which itself is, as it were, nothing but a coarser molecular motion, will require no further discussion. The law of entropy is no more valid for it, than is the law of probability for the single case. It is only subservient to the energy law, which holds for molecular processes.

After this short digression, which we need with a view to later applications, we return to our main line of thought, the meaning of the kinetic theory of heat for our physical world-picture. In view of the results we have mentioned, it may be regarded as settled to-day that the molecular motions assumed by the kinetic theory of heat are real in the same sense as the existence of the molecules themselves is real (see page 31). We have thus again a confirmation of what was said above concerning the meaning of physical hypotheses. What a few men of genius imagined

---

* *Entropy* means *transformation*; the name was introduced by Clausius for a quite definite mathematical quantity.

some two hundred years ago, as an explanation of the behaviour of gases in respect of changes of pressure and temperature, can be taken to-day as an experimentally ascertained fact. A supposition, and one very daring at that, has turned out to be correct, and not merely a useful model.

But the critic will not regard this as a sufficient reason for yielding his position. He will rather say: Let us assume it to be the case, although there are a good many objections still to be made; but even if it be taken as a fact that molecules exist and that they move as calculated by the kinetic theory, the latter is still far from being proved. It has merely been shown that molecular motions take place proportionately to what we call temperature, and measure by means of the hydrogen or air thermometer, and also perceive somewhat coarsely by means of the nerves of our skin. But who can tell us that heat is identical with these motions ? Would that not mean, that all phenomena of heat could be referred to these motions, and must thus be capable of deduction from the kinetic theory, and would this not confront us with the demand, impossible to fulfil, for a complete induction, that is, infinite experience ?

What are we to reply to this criticism ? Something very simple, something so simple that it is easily overlooked. Namely, that *heat is now and for the future identical by definition, as far as we are concerned, with molecular motion.* We actually always measure temperature by means of the hydrogen or air thermometer, whereby we measure nothing other than the molecular motion, since this produces the pressure of the enclosed gas. Every other thermometer is in practice graduated according to the air thermometer. But if we should in any way come to the conclusion – as is actually the case – that in certain regions of temperature, say very low temperatures, the readings of the air thermometer no longer correspond to the actual molecular energy, there is nothing to prevent us from declaring that nevertheless, the latter is still to be regarded as temperature, and that the thermometer we are using does not give this temperature sufficiently accurately. This is a pure matter of convention, and is in no way remarkable, when once we have clearly realised that in any case we have long ago freed, and been obliged to free, our physical concept of temperature from our direct sense experience of heat.

The question whether all phenomena of heat can be reduced to molecular motion, is pointless, when we simply define heat as molecular motion. It would only have a meaning, if we continued to understand by the term heat our original feeling of heat. It would then be a matter of whether this feeling of heat always stood in the same relation to other phenomena as does molecular motion. That this is not the case we were already aware before (see page 77), for it was for this reason that we made use of the thermometer instead of the skin since our reason led us to say that our sense of heat deceived us. The same is also true, for example, when we 'burn' our skin with acid or alcohol. We are obliged to conclude from the readings of the thermometer that the sensation of heat is here produced without increase in molecular velocity, and we regard ourselves as justified in declaring the feeling to be an illusion of sense. Here we mean no more than that the condition which, as experience teaches us, normally produces it, is not present.

This condition is the increase of molecular energy. *Everything else also capable of producing a sensation of heat, is simply excluded by definition from the physical concept of heat.* Here we are not acting more arbitrarily, or with less justification, than when we understand in acoustics under the terms musical note and sound, only the oscillatory motion of substances, and speak of an illusion in the case of everything else also capable of producing the sensation of sound in the ear. The singing in the ears of a fever patient is just as little an object of physical acoustics, as are fever heat (which does not arise only from the slight rise in temperature), and the above mentioned case of a burning pain, objects for the physical science of heat. Otherwise we might regard a dream concert as belonging to acoustics, and the common dream of flying as belonging to mechanics. The whole object of physics is to use the sense impressions to check one another, and, with the aid of the intellect, to form a picture of reality as independent as possible of the sense impressions; a picture carrying the guarantee of its correctness in itself (see page 27). For this purpose, such exact definitions as the one we have been considering, definitions which are only apparently arbitrary, form an indispensable means.

It is true that the objection can be made and often has been made, that all this comparison, combination, and analysis, nevertheless entirely fails to take us out of the realm of sense experience, and so can only, fundamentally speaking, reduce an unknown X to another unknown, Y. When, for example, the kinetic theory of heat resolves heat into motion, it does not thereby give any fundamental explanation, since the 'motion' goes back to certain sense-impressions just as does the heat. That is, in a certain sense, true. But it overlooks the most important point. For it is by no means a matter of indifference whether I have in heat and in motion two separate sense experiences, which being isolated, are doubly incomprehensible to me, or whether I reduce both to a single unknown. *Only by way of the reduction of groups of phenomena to one another, or to a third common and fundamental phenomenon, is it possible to find an answer to the question of the nature of reality.* If finally this should be only a relative answer from the point of view of the theory of knowledge, if the unknowable 'thing in itself' behind this reality is to continue a riddle for ever insoluble for us, physics can be indifferent. From its point of view, there is no longer any riddle, as soon as everything has been reduced to one single thing. This sounds like a contradiction, but is not so. For it is easy to see that what is mysterious about a certain group of phenomena is really only its isolation. What appears to us as a 'brutal fact' without any connection with others, appears for this very reason to us as mysterious, and demands that we find an explanation for it, that is to say, bring it into subservience to a more general statement. The same is true, for example, as regards gravitation. We should regard it as explained the moment we succeeded in proving that the law of gravitation was a consequence of the fundamental electro-magnetic laws.*

If now all phenomena were thus reduced to one single fundamental statement, it is evident that this would have to be regarded as itself inexplicable, since it would explain everything existing. Only two questions could then be stated with regard to it:
1. *What is its origin, and why is it just as it is, and not otherwise?* That would be the problem, pursued to the last position accessible to physics, of why the world is there, and why it is as it is and not otherwise (the cosmological problem of the contingency of the

---

* The above sentence from the first edition has been allowed to stand intentionally. Einstein's general theory of relativity has fully justified it (see page 147), only that we must put something else in place of the fundamental electro-magnetic laws.

world). This problem goes beyond the limits of physics, although a discussion concerning it is, in a certain sense, possible. We shall return to it later (see page 214). 2. But it might also be asked: What has the hypothetical physical constitution of the world, in which all physico-chemical phenomena are to be regarded as included, and hence also those processes which take place in the brain of man when he feels, *what has this physical state of affairs to do with the psychical, the sensation,* etc.? This is the psycho-physical problem reduced to the last imaginable physical position, and we shall return to it in another connection. As regards these two questions, we may certainly talk of 'world riddles,' but to seek other impossible mysteries behind physics, is merely making mystery for its own sake.

The kinetic theory of heat is one of the most important steps on the road of modern physics, the road which leads to the ordering of everything particular under something more general, to the search for the fundamental laws, of which all single natural laws are consequences and special cases. As a result of the kinetic theory of heat, the apparent qualitative difference between two departments, those of mechanics and of heat phenomena, disappears and becomes simply the quantitative difference between molar and molecular emotion. This tendency towards unification becomes much clearer, when we now turn to the second chief department of physics, the physics of the ether. We shall begin this with a short account of the theory of light.

VIII

THE DEVELOPMENT OF THE THEORY OF LIGHT

Scientific optics (the science of light) begins with the investigations of Newton and his contemporaries, Hooke, Grimaldi, Huygens and others. One of the first important discoveries was that of the law of refraction by Snellius (1626), after many other investigators, including Kepler, had sought for it in vain.

The law tells us that when a ray of light passes from one medium to another (for example from air to water) its direction is changed in such a way, that the ratio of the sines of the two angles which the ray makes with the normal to the surface between the two media (angles of incidence and refraction, see figure 7) always has the same value. This is called the refractive index of the particular pair of substances. When we speak of the refractive index of a single material, we mean that against air, or more accurately speaking, vacuum.

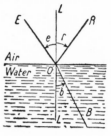

Fig. 7

Reflexion and refraction of a ray of light. *e*, angle of incidence; *r*, of reflexion; *b*, of refraction

Whenever refraction takes place, reflexion also takes place at the boundary of the two substances. The law of reflexion, which says that the angle of incidence is equal to the angle of reflexion, was already known to the ancients. Since the same law is also true for the reflexion of elastic balls (for example billiard balls) at plane surfaces, the supposition was natural that light consists of small 'corpuscles,' which are reflected in a similar way at a boundary. Newton actually succeeded in deducing the law of refraction in this way. For if we assume that the atoms of the medium exert an attractive force on the light corpuscles, but that this only acts over a very short distance, it can only take effect in the very close neighbourhood of such a boundary. If we assume that in the figure, the denser, and hence more attractive medium is below, there is a resulting force downwards on the corpuscle while it is passing through the boundary layer. Assuming the corpuscle to possess inertia, and to be moving very rapidly, the effect of the force is to divert it in the direction towards the perpendicular, its velocity also being increased. Calculation shows[58] that the law $\sin \alpha / \sin \beta = c_2/c_1$ must hold. According to this *corpuscular theory of light* of Newton, the index of refraction (the constant ratio of the two sines) must be equal to the inverse ratio of the velocities of light in the two media.

Soon after the publication of Newton's theory in 1672, Huygens' work appeared, in which he proposed to regard light as a form of wave motion, the theory of which was founded by him. His argument showed that the wave theory also gives both laws, that of reflexion and refraction, correctly. Only according to it, the index of refraction (the ratio $\sin \alpha / \sin \beta$) is equal to the direct ratio of the rate of propagation of the waves in the two media, or $n = c_1/c_2$ whereas it should be the inverse of this according to Newton. Since in the case of air to water the index of refraction is about 4/3, light should travel in water 4/3 times faster than in air according to Newton, whereas the converse should be the case according to Huygens. At that time a decision was impossible, since it appeared impracticable to determine the rates of propagation of light in various substances; indeed, doubt was still felt regarding Römer's discovery of the finite velocity of light (1675), which was made by astronomical means, using the moons of Jupiter. In view of Newton's great authority, however, and also no doubt on account of its greater concreteness, the corpuscular or emission theory first gained the day, although even at that time some of those phenomena were known which form a direct proof of the wave theory, namely interference and diffraction. On the other hand, neither Newton's nor Huygens' theory was able to explain the fact of the separation of colours which takes place at every refraction, with the production of a spectrum; this shows us that the index of refraction is not actually a single constant, but slightly different for each colour.

When towards the end of the 18th century, the English oculist Thomas Young, and somewhat later the French engineer Fresnel (1820) showed how all known interference and diffraction phenomena (colours of thin plates, grating spectra, etc.) can be explained by the wave theory, whereas the emission theory completely fails, the position was changed. From then onwards the wave theory reigned undisputed in physics, and hence the direct refutation of Newton's theory by Fizeau and Foucault's determination of the velocity of light in water (1850) was no longer needed, though it actually showed that, as Huygens' theory requires, the velocity of light was smaller in water. Interference phenomena cannot possibly be explained without the help of a wave conception.

An experiment carried out in 1895 by the German physicist Wiener gave a direct demonstration of the wave nature of light. Wiener allowed light to be reflected at a silver mirror (figure 8, CD). We then get so-called 'stationary waves,' with nodes and loops which lie parallel to the plane of the mirror. They are shown in the figure in section. Such waves are easily illustrated by means of any stretched rope or string, only that in the present case we are dealing with waves in space and not on a straight line. Hence the nodes and loops are not points but planes. Wiener set a photographic plate (shown in section in the figure by AB) at a very slight inclination to these planes, so that it cut them successively. Where it cut a node plane, it was not blackened; where it cut a loop plane, the blackening was a maximum. The nearer to parallel we make the inclination of the plate to the planes, the further the distance between successive blackenings; this distance, and hence the wave-length, could thus be measured directly on the plate with a divided rule. A particularly original and curious experiment is that of placing the plate exactly in a node plane, when it is not blackened, though light goes through it from both sides. It would be impossible to demonstrate the oscillations even of a string or rope more directly than this.

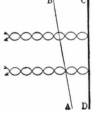

FIG. 8

Weiner's experiment; photography of stationary light waves

While from all we have said, there can be no doubt as to the correctness – not merely the utility (see above, pages 38 ff.) – of the wave theory of light, it nevertheless had great difficulties to meet at first, since it was attempted in vain to give the waves a concrete or mechanical sense. As we know, physics distinguishes between transverse and longitudinal waves. In the case of the first, such as the waves on a stretched rope or vibrating string, the oscillation of the single particles takes place at right angles to the direction of propagation of the wave. In the second case, as with air waves (sound waves, for example), it lies, on the contrary, in the direction of propagation. It is easy to see that transverse waves are only possible in media having elasticity of shape, and hence not in liquids and gases. In these we can only have longitudinal waves.[59] But light, as is proved by the phenomena of polarisation, consists of transverse waves.

Huygens already knew of such phenomena, and had even investigated them more closely (calc spar); but he could not make up his mind to conclude from them the transverse nature of light waves, since light comes to us through empty space, and this cannot well be given the properties of an elastic solid, which would further have to have an enormous modulus of elasticity, on account of the very high velocity of light. It is a very great

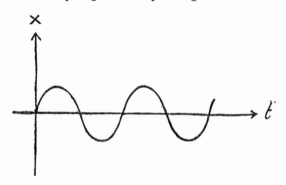

FIG. 9

Sine curve as graphical representation of an oscillation

service to science of Fresnel's that he nevertheless ventured to make this assumption. He told himself quite correctly that the quantitative working out of a theory is much more important than the ease with which it may be imagined. If it works out correctly, the other question will solve itself, often in ways of which we have no idea. This also happened in the present case, but only, it is true, after new data had arrived from quite another direction, about which we shall speak immediately.

It was first of all attempted once more to settle the matter by simply ruling it out positivistically, and here we have an example of the greatest interest in relation to the theory of knowledge, and one which bears a close resemblance to the situation occurring to-day. The argument was something as follows: The physical concept of an oscillation means in the first place a to-and-fro movement, and we usually think of this as a quite definite form of movement, namely so-called simple harmonic, or sine motion, in which the momentary distance of the oscillating point from its middle position is shown graphically, as a function of time, by the sine curve of figure 9. (The figure will no doubt also be intelligible to the layman). Now this same curve can also be used as the representation of any other kind of periodic change of state, taking place according to the same law. For example, the temperature in a room might be changed in such a way by alternate heating and cooling, that a recording thermometer would draw this curve. We may then speak of an oscillation of the temperature or other magnitudes, thus using the word oscillation in a broader sense, and meaning the same thing as a periodic change of state.

The concept of a wave may also be enlarged in the same way. A wave, in the ordinary mechanical sense, is produced when an oscillation (in this case in the ordinary sense of the word), produced at any point, is transferred to neighbouring points on account of elastic connections, or as in the case of air, by the transference of pressure, so that each point commences to oscillate a little later than the point behind it. If we now make use of the more general concept of the 'periodic change of state,' in place of that of mechanical oscillation, we obtain the corresponding generalised concept of a wave. Hence *a wave, in the physical sense, means any kind of periodic change of state which is propagated in space with a finite velocity.* Any kind of physical magnitude such as temperature or the like, is caused to change periodically at one point, and these changes affect the neighbourhood, since the latter is connected with the point of the original change in some way. It is a matter of indifference as to what is the nature of the physical magnitude in question which changes at every point in space, but at each point somewhat later than at the previous point, nor does it matter what is the nature of the connection.

The distance between two points in space which are separated by a distance of one complete period, is also called, in this transferred sense, a wave length. To pass over this distance, the wave obviously requires the time occupied by a complete oscillation T. Hence we have for all wave processes (not only the mechanical) the equation $c = \lambda / T$ or if we use in place of the time of oscillation T the number of oscillations per unit of time $n = 1/T$, we have $c = n\lambda$. This equation, which we shall often make use of, could hardly be omitted here.

So far we have been dealing with an enlargement of the two concepts of oscillation and wave, to which no one can reasonably object. Physics is free, if it should appear convenient, to give these a larger sense than they have hitherto carried. We only leave physics for epistemology when we declare with regard to the theory of light, as was customary about the middle of last century, that there is no further sense in asking, in view of this enlarged concept of oscillation and wave, what it really is that oscillates in light. One should rather be content to possess in the equations in question a formal statement which applies to all oscillations and waves in the enlarged sense, and to show that the properties of light necessarily result from this formal statement. For physics has no right whatever to claim more than a suitable formal description of the fact. Here we have the positivist position in its purest form. We have a perfect example of its value and also of its dangers.

Its value consists in the fact that this formal theory of light does actually set us free from the difficulties of forming a mental picture, which were peculiar to the older, mechanical or elastic, theory. If we leave unanswered the question of what really oscillates in light – and it was in fact simply abolished from physics as unanswerable – we naturally avoid all the difficulties which arose in the mechanical theory from the necessity of ascribing to empty space the properties of a perfectly solid body, in order to render transverse waves possible in it. Huygens had already imagined space in this sense, in view of its functions as a carrier of light waves, as filled with an imponderable and invisible, but truly existing substance, the luminiferous ether, which however, since he only imagined longitudinal waves, he regarded as a kind of extremely rare gas. This something which could not be perceived, and hence fell under the suspicion of being merely a metaphysical hypothesis, was now simply abolished. There was no further use for the ether; only formulae were required to describe the way in which light behaves. It is an undeniable service performed by positivist criticism that in such cases as the present, it frees theory from superfluous special notions, which were here quite incapable of being carried out. But its error is to regard a problem as settled, which it simply refuses to state;

whereas the problem still exists and is merely set aside for the moment.

This problem in the present case is the following: What really is the quantity, the periodic changes of which form light? Is physics not to be allowed even to raise this question? As soon as positivism lays down a negative requirement of this kind – which it has done more than once in the present case – it exceeds permissible limits and becomes negative dogmatism (it would be more correct to call all positivism 'negativism,' since it lives solely upon negation). In this case it quite arbitrarily ruled out a question, which, however we look at it, was really the fundamental point of the whole matter, as compared with which all pure formalism was simply the external clothing. It is obvious that this was the case as regards the theory of light, since the investigations of Maxwell and Hertz have actually been able to answer the question of the nature of the magnitude which changes periodically in light. According to the purely formalistic (that is positivistic) theory of light, this question would not even have had any meaning, whereas it actually forms the central point of later insight into the problems. In order to explain this more fully, we must first enter a little into the modern theory of electricity.

THE ELECTRO-MAGNETIC THEORY OF LIGHT

The science of electricity and magnetism received a powerful new impulse at the end of the 18th and beginning of the 19th centuries by the discoveries of Galvani and Volta, while about the same time, Coulomb first discovered a quantitative law of electrical action. The attractive or repulsive effect of two magnetic poles or electric charges (of unlike or like sign respectively) is, according to his investigations, firstly directly proportional to a quantity which we call pole strength or charge respectively, and secondly, inversely proportional to the square of the distance.

$$\left(\sim \frac{m_1 m_2}{r^2} \text{ or } \frac{e_1 e_2}{r^2} \right)$$

The analogy of this law to Newton's law of gravitation is quite obvious, and in consequence of this agreement as regards the fundamental law, the laws of magnetism and static electricity are to a large extent formally similar to those of gravitation. The development of the mathematical theory of these actions is mainly due to Laplace, Poisson and Gauss. Among the useful 'fiction' we have the concepts of the 'field,' of 'lines of force,' of 'potential' and others. On account of their fundamental importance for the whole of our physical world picture, we must shortly discuss the most important of these.

We call the space around a heavy mass, an electrified body, or a magnet, in which the effects of gravitation, etc. can be shown to exist, its *field* (gravitational, electrical, magnetic). If we bring into the neighbourhood of a magnet say, a small north pole, which we will suppose to be a single point, this moves along a certain curved line through the field (in practice the pole can be caused to follow this line by allowing it to float on water). It also describes a line of this kind starting from any other position. We call this line a line of force, and imagine the whole space to be permeated by such lines of force, which give at every point the direction of the magnetic force (figure 10). Most people know how they are made visible by means of iron filings. Apart from the latter statement, all that we have said is also true of the electrical and the gravitational field. The direction of the lines of force of a magnetic field is given by the direction of motion of a north pole, of an electric field by the motion of a positively charged small body, and of a gravitational field by that of a mass. The force which is exerted upon the body thus introduced appears in this way to be divided into two factors, on the one hand the 'field strength,' on the other the amount of the charge possessed by the body introduced (pole strength, mass). Speaking mathematically, the expression $\frac{m_1 m_2}{r^2}$ is divided into the two factors, m_1/r^2 and m_2, the first being the strength of the field of the mass m_1, at the point where m_2 is (at a distance r), while m_2 only depends upon the mass itself (the test body or point of reference, as the technical expression goes). But this is all for the present, naturally, a purely mathematical way of looking at the matter; a picture or model, or

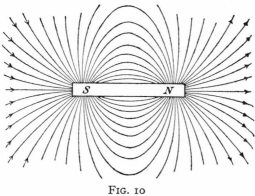

FIG. 10

Lines of force of a bar magnet

better, a scheme, which turns out to be extremely convenient for further mathematical development. Poisson and Gauss, who were the chief founders of this point of view, had for the moment no other intention. We can save ourselves any further discussion of the mathematical development, and will now consider the transformation of this purely mathematical theory into a physical theory by Faraday and Maxwell.

Faraday was probably led to this concept of a field of force, which has fundamentally changed our whole world-picture, by a fact which he was the first to investigate, namely that the magnitude of the electrical and magnetic action between two bodies is essentially dependent on the substance which is between them. In Coulomb's law this is expressed by the fact that the products $\frac{e_1 e_2}{r^2}$ and $\frac{m_1 m_2}{r^2}$ are multiplied by a factor $1/D$ or $1/\mu$ respectively, which depend upon the substance forming the medium. D and μ are called the *dielectric constant* and *magnetic permeability*

respectively of the substance in question. Faraday rightly concluded from this fact that these effects must be transmitted through
space through some intermediary, and that we are by no means
dealing with simple 'action at a distance,' as we had generally believed since Newton's time. Newton himself, it is true, had sought for
an idea which would make the transference of gravitation through
space comprehensible (see page 44). But he did not succeed in
finding one, while the supposition of a simple action at a distance
excellently explained all astronomical phenomena. Some success
was also obtained in attempting to explain actions such as elastic
forces, hitherto regarded as direct contact actions from point to
point, by means of distant forces between atoms separated in
space. The whole of physics thus appeared to be resolving itself
into a sum of forces acting at a distance between mass points
separated in space, and the idea of a medium of transference
through space completely disappeared; in particular, magnetic
and electric forces were regarded entirely from this point of view.

Faraday's view was the direct opposite of this. *It replaced
action at a distance by direct action via the medium: and since the
effects are, as we know by experiment, transmitted through empty
space, this itself now appeared as an electro-magnetic medium.*
The ether of the Huygens-Fresnel light theory thus gained a new
importance. It now became the carrier of an electric and
magnetic field, which represented real states in space (ether) or
in the medium, the lines of force being the mathematical expression of this actually existing state, and the field strength (m/r^2)
the measure of it. It is characteristic of the striving of the human
mind after *mechanical* explanations, that Faraday himself tried
to picture this whole state of affairs by comparison with the state
of tension which exists in a sheet of rubber which has been
stretched in one direction. He thus again returned to a
mechanical picture of phenomena, as is still attempted to-day,
particularly in England. But this is by no means necessary, as
we shall see later. The point is whether the lines of force, field
strength, etc., actually correspond to something real at a given
point in space, which is also present when no material body is
there to indicate its existence.

This idea, which at first was practically neglected, only exhibited its full fertility when Faraday's great fellow-countryman,
Maxwell, succeeded, in 1860, in putting it into a mathematical

form, at the same time enlarging it considerably. The well-known fact that a magnetic needle is deflected by an electric current (Oersted, 1820), means that an electric current produces a magnetic field in its neighbourhood (figure 11). Maxwell first recognised that this is only a particular case of a more general fact. An electric current is the equalisation of electric tension, and hence always amounts to the change of an electric field (disappearance of a field). Maxwell therefore enlarged the relation discovered by Oersted, by stating that *every change in an electric field produces a magnetic field in the neighbourhood.*

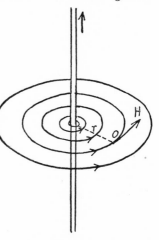

Fig. 11

Magnetic field of a current. The current is flowing upwards

Now Faraday had discovered the exact converse of this in the fundamental fact of so-called induction; the change in a magnetic field produces in a conductor placed in it an induced electric current. If we imagine this conductor removed, no current can be produced, but nevertheless an electric tension is produced, in other words, an electric field. This gives us Maxwell's second law: *A change in a magnetic field produces an electric field in its neighbourhood.* The two laws were formulated by Maxwell in two sets of three mathematical equations, which also give the quantitative relationships. These equations contain the change with time, and the spatial distribution of the electric and magnetic field strengths, the dielectric constant and the magnetic permeability of the medium, and a constant c, which defines numerically the magnetic field strength produced by a certain change in electric field strength or current and conversely.

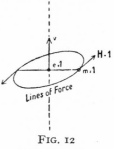

Fig. 12

Definition of c

The constant c, practically speaking, gives the speed at which an unit charge of electricity must be moved in order that the resulting current may act upon a magnetic pole of unit strength at a distance of 1 cm. with a force of one

Is

dyne (figure 12). According to the measurement of Kohlrausch and Weber (1847), the value of the constant is 300,000 kilometres per second.

Maxwell's equations lead directly to the consequence that any change in the condition of an electric or magnetic field must be propagated in all directions from the point of change with a finite speed. The speed with which this propagation takes place is, according to the equations, $= \dfrac{c}{\sqrt{D\mu}}$; and since D and μ are both $= 1$ in empty space, this velocity is there c itself. The constant c therefore also signifies the rate of propagation in empty

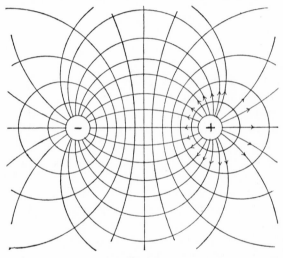

FIG. 13
Field of force of two electrically charged balls

space of electro-magnetic changes of state. If we now imagine this state to change periodically at any point, this periodic change will likewise be propagated with velocity c. But this simply means that we have electro-magnetic waves propagated with the velocity c. In order to form an accurate picture of this, let us imagine two electrically charged balls $+$ and $-$ (figure 13), the equal and opposite charges of which are continually reversed, let us say a million times a second, so that around them the electrical field oscillates to and fro in the general sense discussed above. Figure 13 shows a single instant; after the elapse of half the time of a whole oscillation, the lines of force will be exactly reversed in sign. All lines of force in the whole field change everywhere with the same frequency, but those farther away are always later in their change as compared with those nearer, so that at a certain distance λ there are points which just commence their first change when the second is beginning at the balls, and so on. Simultaneously with these electric oscillations, magnetic oscillations take place everywhere according to Maxwell's equation, their lines of force being at right angles to the lines of electric force. Both lie at right angles to the straight lines radiating from the balls, the direction of propagation of the wave, and we are therefore dealing with transverse waves.

Now the constant c of Maxwell's equation, found in the experiments of Kohlrausch and Weber, is exactly the velocity of light in empty space. It is evident that Maxwell was obliged to conclude from this agreement that the light waves are nothing but electro-magnetic waves. In other words the physical magnitude which changes periodically in a light wave, and is left undetermined by the formal light theory, is the electric and magnetic field strength. In accordance with Maxwell's laws the two changes always take place simultaneously in directions at right angles to one another, and to the direction of propagation.[60]

It is well known how Hertz succeeded twenty years after Maxwell in producing experimentally these electro-magnetic waves which had been predicted on purely theoretical grounds and in showing that they possess all the known properties of light (reflexion, refraction, interference, diffraction, polarisation). These waves are now known to everyone through broadcasting and wireless telegraphy. But very few of us have a clear picture of what the word electric wave really means. Popular description in which ether oscillations and the like are still mentioned, simply confuses the matter instead of making it clearer, since the illustration of mechanical oscillations is always made use of instead of attention being directed to the essential point, namely the periodic change of the electric and magnetic field strength. The latter are concepts which can be made just as comprehensible as a mechanical oscillation.

The well known experiment with iron filings is surely to be found in every course of instruction in physics. It is difficult to understand why ether waves should still be continually brought into the matter.   It is very necessary that the reader of this book should get away from the much too narrow mechanical notion. Naturally there remain the questions of what electric and magnetic field strength really are, why they are connected by Maxwell's laws, and how a change of them at one point can act at a point in the neighbourhood, even in empty space.   But we are not dealing with these questions for the moment, but only with the theory of light.   To the fundamental question: what is light ? we have a perfect answer: light consists of electro-magnetic waves, that is, periodic changes of the two field strengths, of exactly the same kind as in wireless telegraphy, these changes being propagated through space.   Only, in the case of visible light, the frequencies are much greater, and the wave lengths much smaller, than in the case of wireless waves.

Since we are here dealing with one of the chief supports of theoretical physics, it is necessary at this point to carry the matter a little further.   We will make a short survey of the whole range of electro-magnetic waves, since without it all later discussion would be difficult.

We must first remember that the colours of the spectrum from red to violet, as was already shown by the older theory of light (Fresnel), differ from one another as regards their wave lengths. The red end of the spectrum has a wave length of about 760 $\mu\mu$, and hence according to the equation $c = n.\lambda$, a frequency of about 400 billions per second, violet light having about double this frequency (about 800 billion), and hence half this wave length, 380 $\mu\mu$.   In terms of sound, the visible spectrum thus covers fairly exactly one octave.   Since the beginning of the last century, we know that the spectrum is continued at both ends by radiation invisible to the eye, but recognisable by suitable physical apparatus, the so-called ultra-red and ultra-violet (Ritter, Herrschel).   For ultra-violet, the photographic plate, which is very sensitive to it, can be used; for the ultra-red other apparatus is needed, which is sensitive to radiation of every wave length, since it simply turns the energy of the radiation into heat, and this again into electric currents, which can be measured by means of highly sensitive galvanometers.   The sensitivity of these apparatus (bolometer, thermopile) is so great that the radiation from an ordinary candle at a hundred metres distance upon one square centimetre of surface can be measured; indeed even the radiation arriving from the larger fixed stars such as Sirius or Vega can be measured.

With the help of such apparatus on the one hand, and of the photographic plate on the other, the invisible spectrum has of recent years been conquered octave by octave. For a long time the lowest ultra-red known was the wave length 342 μ, that is 0.342 mm., reached by Rubens and his collaborators. From this to the shortest electric waves which had at that time been produced, about 2 mm. in length, there was long a gap of from two to three octaves, but this has since been bridged. In the over-tones of a so-called Lebedev radiator (electro-magnetic), Möbius found waves down to 0.1 mm., these being smaller than the longest ultra-red waves of Rubens. Indeed, waves as short as 30 μ have recently been obtained by means of electrical apparatus, and these could be measured in exactly the same way as Rubens measured the ultra-red rays of the spectrum.

If any proof was necessary of the identity of the ultra-red waves produced by these two different means, it has been given by a recent experiment of Rupp,[61] in which a direct interference between ultra-red waves and oscillations was exhibited. A further similar experiment was made recently by Lewitsky.[61]

There thus cannot be the slightest doubt concerning the completely continuous connection between Hertzian electro-magnetic waves, such as are used for broadcasting in lengths of three kilometres down to a few centimetres, and the visible red rays of a wave length of 760 $\mu\mu$. The accompanying figure 14 makes the whole matter quite plain.

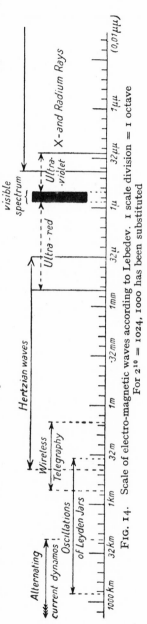

FIG. 14. Scale of electro-magnetic waves according to Lebedev. 1 scale division = 1 octave

For $2^{10} = 1024$, 1000 has been substituted

But at the other end, the diagram carries us far beyond the visible violet. First come the ultra-violet rays of the sun and of our 'artificial sunshine' lamps, the physiologically active rays of which lie between 340 and 280 $\mu\mu$. This region has been followed by Lyman down to about 20 $\mu\mu$. From this point, there was until recently, another small gap before the X-rays were reached, but this is now also filled. The longest X-rays known to-day, as far as the author is aware, are about 37.5 $\mu\mu$. At the other end the waves produced by the X-ray tube join without a break the γ-rays of the radio-active substances; and these again are followed, with a small gap however, by the cosmic rays which have quite recently become famous, and which, by the way, were not first discovered by Millikan, but by the two German investigators, Hess and Kolhörster. Regener has quite recently found a part of this radiation to have a wave length estimated at about 0.000001 $\mu\mu$ or 1 $\mu\mu\mu\mu$.[62] These rays are so penetrating that a thickness of lead of $1\frac{1}{2}$ metres is required to reduce their intensity to one half.

We thus now have a complete picture of the whole of electromagnetic radiation, and it would be idle to doubt that all these rays represent qualitatively the same process, and differ only quantitatively as regards wave length. It should therefore be frankly admitted that the original Maxwell hypothesis, from the very beginning, could have had no other meaning than an assertion of this identity in nature between visible light and electric rays in the narrower sense. This state of affairs was obvious in 1860, and became still more obvious when Hertz actually produced electric waves. What is the sense of talking in this connection of 'thought-economy,' and the 'heuristic value' of hypothesis?

In view of the case in question, which was just as clear 40 years ago as to-day, it has always been a mystery to me how so many physicists, with such an obvious example before them, could support a theory of knowledge which attempted to ignore the main point, that of identity in nature.[63] Surely at that time it was not simply a matter of whether the continuous transition from electric to light waves had been proved, or whether it was merely part of the experimental programme. Epistemology cannot make its fundamental position depend upon the momentary state of physical discovery. A statement such as that of Maxwell, that light consists of electro-magnetic waves is either right or wrong; there is no third possibility. If it is right, it is also fruitful from the heuristic point of view, and of that of economy of thought. But any one who sees therein its only value, appears to me to be like a person who sees in singing only a healthy exercise for the lungs and the diaphragm. If there is anything in the history of science which can show us plainly the nature of physical hypothesis, it is the theory of light.

While clearly recognising this, we must not on the other hand, overlook the fact that the solution of the question of the nature of light accomplishes a part only, and as we see to-day, only the smaller part of the task. Maxwell's genius brought light under the concept of the electro-magnetic wave. But when we examine the matter more closely we see that this deals with only one side of the question: the propagation of light, and strictly speaking only with propagation in empty space. Maxwell's theory does not tell us, firstly, how the electric waves in our ordinary sources of light acquire their enormous frequencies (for we cannot recognise a candle as an electrical apparatus); and secondly, how it comes about that the propagation of these waves is essentially modified as soon as matter is present in the space through which they pass; and thirdly, how the energy of the electric waves is changed into heat again when light is re-absorbed by matter. The first and last questions, as we shall see, are very closely connected, but they form the most difficult chapter of the whole theory of light.

But a number of single problems are also contained in the second question, among the chief of these being that of *dispersion*, or the variation of refractive index with wave length. How does matter, such as glass, occupy a certain space, refract different waves differently according to their wave length? Maxwell's theory gives us only one single refractive index for all wave lengths; it should be equal to the square root of the dielectric constant, which is a purely electrical magnitude (see above). This relationship is fairly true for a number of gases and liquids which are moderately good insulators, but is even then not exact; while in the case of many substances such as water, it does not agree at all, the refractive index of water against vacuum being about $4/3$, while its dielectric constant is about 80. Furthermore, the refractive index is in any case not a single constant, but different for every wave length. But the dispersion problem is only one of many which remained unsolved by Maxwell's theory.

We may say quite generally that it was only able to explain correctly the connection between light and electro-magnetism in general, and the propagation of light in empty space. But it fails as soon as the interactions of light and matter come into play. This is obviously due to the atomic structure of the latter which is not taken into account in Maxwell's original equations, since they are strictly continuous in character. But it appeared for many decades almost impossible to find a way out of the difficulty, and optics had therefore no other resource, in spite of Maxwell, than to treat such constants as the refractive index, or absorption co-efficient of a given substance for a given colour, simply as experimental values, and likewise simply to register all the facts of spectroscopy (see below) without being able to explain them.

It is useful before continuing our description of present-day physics, in which these problems have come much nearer solution, first to make a short survey of the position of physics at the end of the 19th century. Physics was divided, as already mentioned, into two chief departments, the physics of matter and the physics of the ether. The first included mechanics, and sound, heat, the second electro-magnetism and optics, or more generally radiation, to which we may add so-called heat radiation, such as is sent out by a stove below red heat, and is simply ultra-red radiation of the kind already described. We already see from this remark that lines of connection between these two main departments run to and fro, since this dark radiant heat (that is, ultra-red light), results from the stove being heated, and hence is a result of increased molecular energy. But the question as to how this came about was at that time (before 1900) quite insoluble. In each separate department there were certain fundamental facts, which had simply to be taken as such. Thus mechanics devoted much attention to movement under the influence of gravity. But there was no answer to the question why this exists and follows Newton's law that weight is proportional to inertia. Mechanics was also much interested in elastic forces. But here again we were confronted with a fundamental fact, and could give no answer to the question of how it comes about, for example, that steel or ivory strive to restore their form, and why lead or glass do so not at all, or much less effectively. The 'modulus of elasticity' is here simply the 'material constant' found by experience, like many others throughout physics.

The same was true everywhere. Wherever we look at physics at that period, we meet a few great general laws of the department in question (the fundamental equations of mechanics; Maxwell's equations), and a number of special laws, with constants derived from experience; while many other such special laws, once likewise matters of experience, such as the law of refraction or the propagation of waves, could now be derived from the general law. At the same time, the constants, first of all found purely empirically, afterwards often show surprising connections. Thus, for example, in the year 1819, the two French scientists Dulong and Petit found that the specific heat of the elements is inversely proportional to their atomic weight, and in the year 1853, the two Germans Wiedemann and Franz found that electrical and thermal conductivity are nearly directly proportional. The first relation, as Richarz showed in the eighties, can be derived from the kinetic theory of heat, which even gives numerically the constant value 6 for the product of atomic weight and specific heat.[64] The second fact has only become comprehensible to-day, on the basis of the latest theory concerning the conduction of electricity by metals (see below).

If we had at that time (1890–1900) formed such a picture by sober criticism, it would have been natural and sensible to say: we are not yet at the end of our labours, and what we cannot do to-day, our children and grandchildren will in all probability succeed in doing in great part. If we therefore cast our minds forward in time for one or two hundred years, we perhaps may imagine a state of physics in which the unification of our explanation has been completed and when all special laws can really be deduced as the necessary consequences of one or a few fundamental laws. Such a position might have been taken up. That this was not done is again the fault of positivism, which declared the main problems non-existent, and we now see quite clearly the true psychological reason for this type of thought. Positivism is always and everywhere the epistemology of premature resignation, of the resignation which comes from disappointment. The mechanistic solution of the whole of physics, formerly regarded as so simple, was seen to disappear more and more beneath the mass of special phenomena, for no possibility could be seen of reducing electro-magnetism to mechanics. It was therefore believed that the attempt at a unified explanation of all phenomena must be regarded as a misleading ideal, and the question of essence was therefore fundamentally rejected in favour of a purely formal quantitative description of phenomena by means of pictures or models useful from the point of view of thought-economy.   We have now seen how foolish this attitude was.

This fact will become still more evident when we now turn to the further development of physics, which clearly shows that the goal in question was no delusion, but rather that we are very close indeed to it.   This new phase in the development of physics was brought about at the turn of the century by the discovery of X-rays, radio-activity, and all that is connected with them; above all by the investigation of the electric discharge through gases. Anyone who, like the author, passed through that time as a student, will clearly remember how astonishing and revolutionary was the effect on that generation of these discoveries.   At first the impression was that the goal of a unified explanation for the whole of physics had now receded into the farthest distance, and that after such unheard of discoveries, others still more remarkable were possible in unlimited numbers, and that it was therefore hopeless even to think of such a thing as unified physics, since every day might bring something quite new.   This was how most people thought and wrote in those days.   But they were wrong, and the very opposite was true.

These new discoveries turned out to be the entrance to a region in which all the roads of the physics of the past met. Indeed physics had, without knowing it, the thread of Ariadne already in its hand, which enabled it to pass with assurance through this labyrinth of apparently limitless new regions of fact. All that was needed was the genius who would recognise this and go forward boldly.   This genius was Hendrick Anton Lorentz, and he was followed at the beginning of the new century by a whole series of men of equal genius, so that in this respect, our own time puts the classical epoch of Copernicus, Kepler, Newton, Leibniz and Huygens in the shade.   Lorentz, Planck, Einstein, Rutherford, Bohr and all the others are names – we can say this to-day with complete certainty – that will be named as long as men live who pursue science, and know something of the achievements of their forefathers.   The decisive step was the development of an 'atomic theory' of electricity.

X

THE ATOMIC THEORY OF ELECTRICITY
(ELECTRON THEORY)

This was effected almost simultaneously by Lorentz and Wiechert in 1896. As we have already remarked, they did not really produce anything new, but they showed what was contained in the already known. The concept of an atomic division of the electric charge, of an elementary unit of electricity, an indivisible unit of all electrical charges, arises from consideration of the laws of electrolysis, which were found by the great electrical experimentalist, Faraday, in the thirties. It is necessary to recall to the reader's mind the simplest fundamental points. Certain liquids, for example, the aqueous solutions of acids, bases and salts, are separated by an electric current into two parts which fall out at the anode (+) and cathode (–). The two parts were called by Faraday *ions* (meaning wanderers) and each separately *anion* and *kation*. If, for example, the current is passed through hydrochloric acid, H Cl (figure 15), H$^+$ is the kation, Cl$^-$ the anion, the former separating at the negative, the latter at the positive pole.

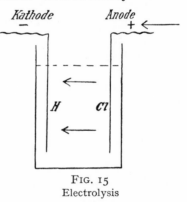

FIG. 15
Electrolysis

For many reasons which we cannot possibly go into here,[65] the conviction was already general about 1860 that the two parts are not actually separated from chemical combination by the current, but are already present as free atoms (the substance being dissociated, as it is called); though not as ordinary atoms, but as such with opposite electric charges. In the present case, the hydrogen atom has a positive charge, H$^+$, the chlorine atom a negative charge, Cl$^-$. The effect of introducing the electric voltage is merely to cause these charged atoms to move towards the corresponding oppositely charged electrode, where they are discharged and so transformed into ordinary atoms, which can then be recognised by their properties (chlorine, for example, by its green colour and smell).

This theory, founded in 1857 by Clausius, and in 1884 by Arrhenius, assumes that the charged atoms, which are now called ions in a narrower sense, have essentially different properties from uncharged atoms. Thus, for example, the ion Cl^- is, in contrast to free chlorine, colourless; for hydrochloric acid, in which it is contained, has no colour when pure. According to the quantitative laws of electrolysis discovered by Faraday, we must conclude that every monovalent (page 17) atom, or atomic group, possesses one and the same charge, and carries this to the electrode, while every bivalent atom possesses double this charge and so on. All atomic or ionic charges are thus multiples of a certain smallest quantity of electricity, which was called by Helmholtz the elementary electric charge, and is designated by e. Its absolute magnitude can easily be determined from the Loschmidt number N (see above, page 22); and conversely, every determination of e leads to one of N. The most exact value of e is, according to Millikan's recent determinations, 4.772×10^{10} electrostatic units, or 0.159 trillionth of a coulomb.[66]

These theories were, as already mentioned, generally accepted about 1880.[67] The essential further step taken by Lorentz and Wiechert was to assume that ions of this kind do not exist only in electrolysis, but also in all substances, either actually or potentially; and that above all, gas molecules and atoms may be 'ionised' under certain conditions, and thus become electrically conducting. It was then possible by this means, as the sequel showed, to arrive at a comprehension of almost the whole of the complicated phenomena of the discharge of electricity through gases (spark, arc light, Geissler tubes, etc.) which hitherto had been quite inexplicable. The most important and fruitful discovery was, that the negative electrical elementary charge is not necessarily bound to material atoms, as in electrolysis, but can exist free as such, and that the cathode rays, which had already been investigated by Hittorf in the sixties, were none other than a cloud of freely moving negative elementary charges (figure 16). This made the elementary electric charge itself a kind of material atom, and the ordinary chemical atoms now appeared as complexes of such negative charges, which are called *electrons*, together with a main positive mass, which is to-day usually called, following Rutherford, the *atomic nucleus*. This must naturally contain as many positive units of charge as there are electrons present in the atom, if the atom is to be electrically neutral.

According to Rutherford, who developed this idea further about 1910, we are to imagine the atom as a kind of planetary system, inasmuch as the electrons rotate about the positive nucleus in the same way as do the planets about the sun. Other investigators have proposed, in place of such dynamic models,

FIG. 16
Straight-line propagation of cathode rays

static models with stationary electrons, but a further hypothesis is then needed to explain why these electrons do not plunge into the oppositely charged nucleus. We shall see later that these questions are probably meaningless, and why this is the case. In electrolytic dissociation, for example, of hydrochloric acid, the chlorine atom robs the hydrogen atom of one electron, so that the latter contains one positive elementary charge in excess, while the former carries one electron over its normal amount, and hence has a one-unit negative charge. Further details must be left till later.

We have now to add to this picture a few experimental details. The conviction that we are dealing in the case of the cathode rays with a negative 'corpuscular radiation' (as opposed to the radiation of light waves) depends upon the following grounds. Cathode rays are propagated strictly in a straight line (figure 16). They can be diverted by an electric and by a magnetic field exactly as would be expected from the general laws of electro-magnetism for a cloud of negatively charged particles (figures 17 and 18). Further, they actually carry a negative charge along with them, for example, they charge an electroscope placed in their path negatively, and finally – and this is the decisive point – their so-called specific charge, that is the ratio of charge to mass (e/m),

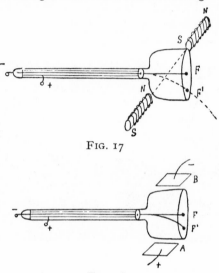

Fig. 17

Fig. 18

Magnetic deflection of cathode rays (schematic)

which can be exactly determined from the amount of their deflection by electric and magnetic fields, is always found to have the same value, whatever the exact conditions of their production, or whatever the material from which they originate. It follows of necessity from this constancy of e/m that we are dealing in every case with the same kind of corpuscles. The velocities of the cathode rays are found to be, according to their mode of production, about $1/400$th to $\frac{1}{4}$ to the velocity of light. The fastest electron rays are the β-rays of radium, which reach a velocity up to 99% that of light.[68]

Since other experiments[69] allow and compel the conclusion that the charge of an electron is identical with the elementary charge, whereas the ratio e/m is found to be much larger than that for electrolytic ions – it exceeds the specific charge of the lightest of these, hydrogen, by 1800 times – no other conclusion is possible than that the mass of these particles is correspondingly smaller. Since, on the other hand, they are produced from every kind of cathode material, they must necessarily be interpreted as component parts of the atom, as we have already stated above. Besides the negative electron rays, we also know rays consisting of positive corpuscles, the so-called *anode* and *canal* rays. For these, values of e/m different from those of electrolytic ions have never been found, and hence we are likewise dealing with ions such as H^+, Na^+, etc. *The smallest known positive particle is the H^+ ion, or what is the same, the hydrogen nucleus.* For, as we shall explain more exactly later, we must imagine the hydrogen atom as consisting of only one electron and a positive nucleus carrying an equal and opposite charge. For the present let us keep in mind these two facts: there are free negative electrons, but positive charges only exist in conjunction with atoms of normal weight.

When cathode rays, that is moving electrons, fall upon substances such for example as platinum, these send out X-rays. But these are not corpuscular in nature; they are wave rays, as is shown by the facts that (1) they are propagated with the velocity of light; (2) are not deviated by electric and magnetic fields; (3) are capable of interference and polarisation (see below). X-rays are ultra-violet of very short wave length, as we have already remarked in connection with the spectrum.

All three kinds of rays; moving positive (also negative) ions, moving electrons, and X-rays, are found again in the phenomena of radio-activity, which, by a fortunate historical coincidence – not entirely a coincidence, since at that time particular attention was being paid to new kinds of radiation – were discovered at almost the same time as the X-rays (Becquerel, 1897). The three types of rays, which can be emitted by radio-active substances, are called α- β- and γ-rays. The α-rays correspond to the canal or anode rays. They are positive corpuscular rays, namely helium atoms with two positive charges (or more correctly helium nuclei). The β-rays are, as we have already mentioned above, very fast cathode rays, and the γ-rays are X-rays of very short wave length. The energy of these radiations is enormous: a gram of radium develops, all told, about three thousand million calories. In order to explain it, all kinds of external causes were first thought of, such as the reception of cosmic energy, and even the invalidity of the law of energy, but the conviction finally became stronger and stronger, that it was really a matter of energy which was already potentially present in the radio-active substance. According to *Rutherford's hypothesis of atomic disintegration* (1902), which is generally admitted to-day, a radium atom explodes as it were, with the result that positive and negative corpuscles are hurled out while short-wave radiation is also produced. More exact investigation shows that one of two things happens: either the projection of positive α-particles, or the projection of β-particles (electrons) with simultaneous production of γ-rays. Only in rare cases do both processes appear to take place with the same atom (see the table).

In this explosion it is not the planetary system of the atoms which is destroyed, but the nucleus itself. *Radio-active processes are changes in the nucleus*, and hence when they occur, a new element is always produced, the process being one of atomic degeneration. Elements of lower atomic weight are formed from elements of higher atomic weight. Thus from radium, via a whole series of intermediate stages, we get finally radium G, which is identical with lead. On the basis of certain investigations concerning the passage of X- and corpuscular rays through thin layers of metal, Rutherford concluded in 1912, that the nuclear charge, measured in multiples of the elementary charge, must be equal to about half the atomic weight, excepting that in the case of hydrogen, it is equal to the whole atomic weight 1. Van den Broeck then recognised in 1913, that what came into play was not the atomic weight, but the number of the element in the periodic system, the so-called *atomic number* of the element, which in fact closely approximates to half of the atomic weight. Since on the other hand, the positive nuclear charge must be equal to the number of electrons revolving round the atomic nucleus, we get the following fundamental rule of van der Broeck:

Nuclear charge = number of electrons = atomic number
which has been confirmed by all later investigations.

From this law follows a further important consequence. Since the loss of an α-particle (helium nucleus, charge 2, mass 4) lowers the positive nuclear charge by 2, the element passes up the periodic system by two columns when an 'α-transformation' occurs. This is, for example, the case with radium itself, which changes by loss of an α-particle into radium emanation, a noble gas with the atomic weight 222 (see table at the end). If the nucleus loses an electron, its atomic weight is not sensibly altered, but the nuclear charge becomes greater by one unit, and the element then moves up the periodic system by one column without change in the atomic weight. (The external system of electrons adjusts itself automatically to the nuclear charge.) These two rules form the content of the Soddy-Fajans displacement law. From it there further follows, that after one α-and two β-transformations, the elements must again reappear at the same point in the periodic system although it has in the meantime lost four units of atomic weight. Accordingly, therefore, there exist both elements of like chemical properties but different atomic weights, so called isotopes, as well as elements of unlike chemical properties with like atomic weights, so called isobars (the latter being such as differ only by a β-particle).

These consequences of Rutherford's disintegration theory combined with van der Broeck's hypothesis at first appear somewhat fantastic, but have been completely confirmed by certain famous investigations, one of the best known of these being Hönigschmidt's determination of the atomic weight of uranium lead. Radium itself, in the sense of the disintegration theory, is only one step in a series of transformations, the ancestor of which is uranium. After a total of five α-transformations, and four β-transformations, radium is finally transformed into radium G, which is identical with lead. Hence this lead, which originates from uranium, must have an atomic weight of 226 (radium) – 20 ($=5 \times 4$), that is 206, since the β-transformations have no effect on the weight; whereas the chemical atomic weight of lead is 207.2. The difference is much greater than the experimental error, and hence this was for long regarded as a decided obstacle to the acceptance of Rutherford's disintegration theory.

It will ever be regarded as one of the most remarkable triumphs of scientific prediction, recalling the discovery of Neptune and Mendeléeff's prediction of unknown chemical elements, that an atomic weight determination of lead obtained from uranium minerals by Hönigschmidt actually gave the required value of 206. Quite recently atomic weight determinations have been developed to such a degree of accuracy especially by the last

Ks

named experimenter, that even the second decimal place may now be regarded as exact, and here small departures from the calculated value have been found, so that further undiscovered causes of a still finer description must be at work; these are being sought for with great energy.[70] But this fact in no way detracts from the fundamental importance of the result. The achievement of Leverrier and Galle was not diminished by the fact that Neptune was found one half a degree distant from the position calculated by Leverrier, nor that of Mendeléeff by the fact that the elements gallium, germanium, and scandium did not fulfil his predictions in every single particular. We have here a particularly characteristic example of the way in which physics progresses step by step with continually increasing accuracy. At another point we shall discuss this question more fully from the point of view of the theory of knowledge (see below, page 228 ff.).

Before we proceed, it is necessary to mention a few discoveries of a more experimental character, which are of importance for matters to be dealt with later, and also complete what we have said earlier, and give it a further foundation.   J. J. Thomson and Townsend were the first to determine the value of the elementary electric charge, and hence of the Loschmidt number, by purely experimental methods.   The method is somewhat as follows.   It has long been known that air can be cooled considerably below the dew-point without condensation (to fine drops of mist) of the water vapour contained in it, the condition being that the air should be completely free from dust, since the dust particles act as nuclei for condensation.   But condensation occurs when the air is ionised; and, as Thomson first showed by special experiments, it occurs first on the negative ions, and then on the positive when the degree of cooling is greater.   Thomson invented an extremely ingenious method of counting the drops in a cloud of mist, and in doing so, he could likewise count the ions which served as condensation nuclei; combining this with a direct measurement of the total charge, he was able to find the charge on a single ion, e, by simple division.

This method of Thomson's was later frequently repeated with improved apparatus, and gave values of N in excellent agreement with values otherwise found (27.2×10^{18}).

Two further determinations of e (and hence N) were made by other methods.   Regener, and also Rutherford and Geiger, were able to make a direct determination of the number of α-particles thrown off by a small quantity of radio-active material in a given time, by allowing a very small, but exactly measurable, fraction of the total radiation to pass through a very small hole in a block of lead, the effect produced by every single α-particle passing through the hole being observed (the two methods differ only as regards the method of reception).   But more remarkable

PLATE I
Wilson's photographs of α- and β-ray tracks

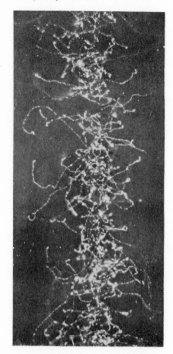

FIG. 19   FIG. 20

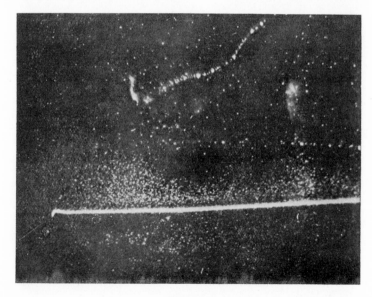

FIG. 21

than all these was the method devised by C. T. R. Wilson, who succeeded in actually photographing the paths of single α- and β-particles. For this purpose he again made use of the condensation of water vapour on the ions of the air. Every single α- or β- particle passing through the air ionizes the air molecules, probably by shattering them, and the water vapour then condenses on the ions then produced. Wilson's method consisted in producing the cooling and the illumination at the same moment, and thus retaining a picture of the momentary state of the air. The accompanying figure 19 (opp. this page) shows a fragment of radium, which is sending out α-rays in all directions. Figure 21 shows a single ray, in which the peculiar hook at the end is noticeable, above this is the trace of a β-particle with a characteristic bead-like arrangement of the water particles. Figure 20 shows the track of a narrow bundle of X-rays through air (from bottom to top). We see clearly how the electrons (β-particles), immediately recognisable by the characteristic form of the path, fly out of the air molecules in all directions. In view of these wonderful results, there can hardly be any further doubt as to the real existence of these particles, which previously was hypothetical.

We may also add, that Millikan and Ehrenhaft also arrived at new determinations of e and N by analogous methods. The best result found by Millikan was $e = 4.772 \times 10^{-10}$ E.U., that is, $N = 27.2 \times 10^{18}$. Later, Ehrenhaft observed charges on particles suspended in air, which appeared to be considerably smaller than e, his result being one thousand times smaller. This was in contradiction to all previous results, but strong objections were raised by other investigators such as Regener and Baer = ;[17] the question has not been quite cleared up so far. In view of the large number of arguments otherwise telling in favour of the constancy of the elementary charge, it becomes very difficult to accept Ehrenhaft's results.

We will now proceed with our development of modern physical knowledge. As we have stated, Hönigschmidt's investigations proved the existence of isotope elements. Ordinary lead must therefore be regarded as very probably a mixture of different atomic isotopes (RaB, RaD, RaG, and others) and the obvious question arose whether the ordinary chemical elements which are not radio-active might still be such mixtures of kinds of atoms of similar chemical character but different atomic weight. Strange though this assumption appeared at first, it has been fully confirmed by the investigations, which quickly became famous, of Aston, a pupil of J. J Thomson.[72] The latter had already determined the ratio e/m by magnetic and electric deviation experiments for canal rays (positive corpuscles). Aston succeeded in so perfecting the method, that he was able to determine from the position of the deviated rays, the mass of the particles to the third and even the fourth figure exactly, the charge being regarded as given.

It now appeared, the first case being that of the noble gas neon, that the position of these points does not correspond, in the case of a large number of elements, to the chemical atomic weight, but that in place of this we generally have several points for each single element, the masses being whole numbers closely approximating to the chemical weight. Neon, for example, has the chemical atomic weight 20.2; but in Aston's 'mass spectrograms' (figure 22, opp. page 133) it gives a point at $m = 20$ and one at $m = 22$ (the points seen on the figure in the first column come from other gases). In the case of chlorine, likewise, Aston found no point corresponding to 35.4, the chemical atomic weight. But he found points at 35 and 37, and also two at 36 and 38, which he ascribed to the monovalent ion HCl. In this way, by far the greater number of the elements of chemistry were found to be mixtures, only a few (carbon, nitrogen, fluorine, sodium, aluminium, phosphorus and arsenic, perhaps also iodine) appeared to be pure elements, that is to contain atoms of only one kind. Many contain a quite surprising number of isotopes (thus, for example, 8 of mercury are known); generally speaking the number appears to increase with increase of atomic weight.

We may further say a word here concerning the remarkable fact, that in spite of so many elements being undoubtedly mixtures of isotopes, the chemical atomic weights have hitherto always been found to be constant. This fact had always been taken, since the time of Dalton and Berzelius, as a proof of the identity of all atoms of the same element. Specimens of elements of entirely different mineralogical and geological origin have been investigated with great care to discover whether slight differences in the atomic weight might not be present, which would indicate that the elements in question had been formed under somewhat different conditions, and hence were perhaps made up of isotopes in somewhat different proportions. The results so far, apart from the case of lead referred to above, have always been negative, or at least not positive with any certainty. This result leads almost necessarily to the conclusion that our chemical elements found to-day on the earth were formed together at a certain time, and have only separated subsequently. But it still appears possible that sooner or later a case will be found where the same element shows a different weight according to its origin, as is actually the case with ordinary and uranium lead.

The physical separation of isotopes has recently been accomplished

PLATE II

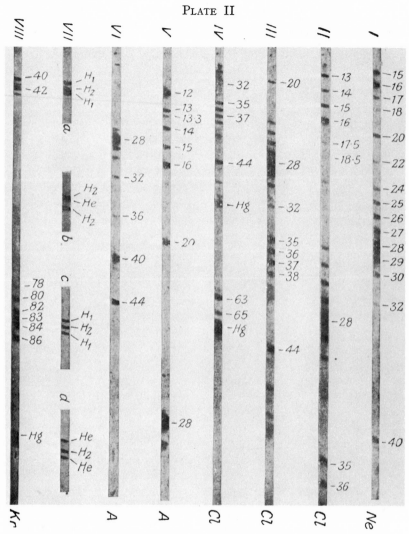

FIG. 22. Mass spectrograms of Neon, Chlorine, Argon, Hydrogen, Helium, and Krypton. (Aston)

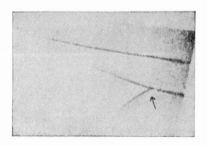

FIG. 23. Shattering of Atoms. (Rutherford)

successfully in a few cases. The chief method is that of the diffusion of gases, since in them the velocity of the atoms depends upon the atomic weight. For example, chlorine gas has been separated into two parts of different atomic weights. By similar means it has been shown that the weak radio-activity of potassium is due to a small quantity of an isotope of atomic weight 41, while the isotope of atomic weight 39 which forms the chief constituent is not radio-active. The first isotope can be enriched by distillation, whereupon the radio-activity increases.[73]

The new knowledge which has thus been won, opens the way to a physics of the atomic nucleus, although we have to-day, it is true, very little satisfactory material for it. The fact of isotopy proves at least this much, that the same resultant nuclear charge, upon which alone the position of an element in the periodic system depends, may be obtained in different ways, and hence with different internal structure of the nucleus. But what the elements of this structure really are, is a matter about which we know very little. The fact that so many radio-active elements throw off α-particles (helium nuclei) suggests that such particles are already present in the nuclei of the heavier atoms. On the other hand, the single atomic weights as determined by Aston appear to be almost exactly whole numbers. This gives a new weight to the hypothesis first put forward in 1815 by the English physician Prout, but later given up on account of the apparent deviation of atomic weights from whole numbers, the hypothesis that hydrogen (H = 1) is the fundamental element, out of which all others are built. In this sense, Rutherford calls the hydrogen nucleus a *proton*, and regards all larger atomic nuclei as built up of it and electrons, so that we should now be reduced to two original elements.

This hypothesis Rutherford supported by experiments, which have become famous under the name of the 'disruption of atoms.' He 'bombarded' substances such as paraffin, which contains along with hydrogen atoms carbon, nitrogen, and other atoms, with the α-rays of RaC, which have the greatest energy of all known corpuscular rays. He then observed, using Wilson's method, that in the rare cases of the collision of such α-particles with the other atoms, the α-particles were strongly deflected, and at the same time secondary rays proceeded from the point of collision (figure 23, opp. this page), which upon investigation by means of electrical and magnetic deviation, turned out to be hydrogen. These particles appeared very plainly in nitrogen, but were absent in the case of elements the atomic weight of which was divisible by four (carbon, oxygen and others). Rutherford concluded from this, that in the latter the protons, if they are present at all, are present in groups of four, that is, in the form of helium nuclei (α-particles); which are so stable that they can even resist the energy of the RaC particle, whereas, in the nitrogen atom, for example, one or two single protons are present, which can, so to speak, be shot out of it by means of the RaC particle.

Later investigations (Kirsch and Petterson) appear to show that the energy of these atomic fragments is greater than the difference between the energy of the RaC particles before and after the collision, so that it must be derived at least in part from the nucleus itself. These experiments would therefore be interpreted as artificial radio-activity, and the blow would only be the cause of release. Among atoms which could be shattered in this sense were those of boron, nitrogen, sodium and aluminium. The central blows necessary for this happen very rarely. Ten million RaC particles produce in the case of boron only about three, of nitrogen about fifteen, of sodium about three, and of aluminium about twenty destructive collisions.

Apart from this method, the attempt is being made to-day to discover the secret of the nucleus by careful statistical investigation concerning the distribution of the isotopes in the different parts of the periodic system, by finding certain empirical rules concerning the numerical values of the atomic weights of isotopes, and by other methods. But taking it altogether, it must be said that the physics (which would better be called the chemistry) of the nucleus is still only just beginning to be developed.[74] For this reason, we cannot assert yet with any finality that the assumption of two original elements, proton and electron, will suffice to develop the system of the nuclei, when the principle of construction has been actually found. But we can nevertheless regard it as very probable, and assert with assurance, that some kind of common units will be found at the basis of all kinds of atoms, since without this supposition the whole body of facts represented by radio-activity and isotopes, and nuclear structure, would be incomprehensible. For our further discussion we shall make use of the hypothesis, which is exceedingly probable, that

protons and electrons are the only fundamental elements.   If
the further progress of nuclear physics should make necessary
some other statement of these fundamental assumptions, it would
not in any way change our conclusions.

We may describe the last statement as the highest point which
has been reached by chemical science in the analysis of the nature
of matter.   It is as good as certain that it will carry us a few
steps further, but the reader will have already noticed that these
steps cannot be many in number, after the three-quarters of a
million of known chemical individuals have been reduced to 92
different kinds of atoms, these again to mixtures of isotopes, the
single isotope finally to structures of proton and electrons, leaving
us with only two fundamental kinds of matter.   As regards the
latter, we can further make the following statement.   The nucleus
determines, as we have explained, the place of the element in the
periodic system and thus the chemical nature of the element.
Since the single isotopes only differ as regards their mass, being
chemically and also spectroscopically identical (see below), it
follows that the chemical and spectroscopic properties can only be
functions of the electron system surrounding the nucleus, since
this is always the same for all isotopes on account of the equal
charge of the nucleus.   As we shall see later, it can be asserted
with great probability that only the external regions of this
electron system determine the properties in question.   In all
likelihood, the electrons belonging to the system are arranged in
several rings, or 'shells,' or levels, and only the most external of
these is concerned in chemical processes, and in the ordinary
emission of light.[75]

This point of view suddenly throws surprising light on the old
chemical paradox of which we have been aware since the time
of Boyle, that nothing is seen of the properties of the elements
when these are present in compounds, although they must be
assumed to be existing.   In actual fact, only the external regions
of the atom change during the chemical processes of combination
or decomposition, although this change is a very thorough one,
while the interior of the atom, that is the internal shells, to say
nothing of the nucleus, remain quite unchanged.   We thus have
an explanation both of the complete change of chemical and
physical properties and of the retention of chemical nature.
We are now able to say to what extent the atom remains the

same, and to what extent it is changed, when it combines chemically with another atom.

When we now finally view the whole matter from this standpoint, we find that modern chemistry has approached very considerably nearer to the old Aristotelian doctrine of substance and
form which so long deluded the alchemists of the Middle Ages.
If all elementary atoms consist of the same two fundamental
components proton and electron, it is in fact only differences of
form, namely the arrangement of these components in the nucleus
and the external rings of the nuclei which finally determine all
differences in matter. The number of qualities to be ascribed
to matter as such (the two fundamental substances) is already
reduced to a minimum. Charge $+ e$ or $- e$, mass of the proton
(hydrogen nucleus) $M = 1.63$ quadrillionths of a gram, mass
m of the electron $= 1/1800\ M$; this is all that we need in the way
of substantial qualities for the structure of the physico-chemical
world.

To these, however, we must add the field laws, that is the
relationships which connect substances together and decide their
interactions. In the theory of Bohr, which is still usually made
use of, it is assumed that the two elementary charges act upon
one another with the electrostatic forces of attraction and repulsion, according to Coulomb's law; and that when in motion,
they obey Maxwell's laws of electro-magnetism on the one hand,
and the fundamental mechanical laws on the other. To what
extent these are really appropriate ideas is more than doubtful.
Whatever may be the true state of affairs, it is certain that some
such mutual relations must be present between protons and
electrons, which explain on the one hand the structure of atomic
nuclei and electron shells, and on the other hand the macroscopic
electric, magnetic, gravitational, and other forces. At this point
the fundamental problem of chemistry is linked with that of
physics, as we already foreshadowed in the introduction. We
must now therefore turn to the further development of the theory
of fields of force in physics, and then conclude by considering
the relations between the field and atomically divided matter
as far as they are clear to-day.

XI

THE RELATIVITY THEORY AND THE GENERAL
THEORY OF FIELDS OF FORCE

It is necessary to go into these matters, since we are now embarked upon discussions, in which the boundary between physics and philosophy ceases to be clear, and which also carry us far into the old question of our view of the world. In order to realise the necessity by which physics was led to these problems, we must first recollect that Maxwell's theory had made of the field something really existing; and since such fields could without doubt also exist in empty space, it was either necessary to regard the latter as something real, or to regard it as filled with a real something, which carried the fields. The ether of the light theory therefore returned, as we have already stated, in the form of an electro-magnetic medium. This immediately raised the further question as to how processes of an electro-magnetic or optical nature (now one and the same thing) take place, when the media in which they are occurring move relatively to the ether, as must certainly be the case, for example, on account of the earth's motion round the sun. This appeared to present the possibility, so long sought for in vain, of determining the absolute velocity of the earth's motion (see page 52 ff.). In view of the fundamental importance of this matter, we must deal with it in somewhat greater detail, although this cannot be done entirely without the use of mathematics.[76]

Imagine a light ray sent from A to B (figure 24) and reflected back to A by means of a mirror at B. If the length of $AB = a$, the velocity of light c, we have for the time required for the passage of the light to and fro $t = \dfrac{2a}{c}$. Let the whole path AB (figure 24a) move in the direction AB with the velocity v, so that the mirror B runs away, as it were, from the ray moving towards it, while the returning ray is met by the observer A at A'. A' and B' represent the positions of A and B after motion. It is then easily shown that the total time required for the to-and-fro passage is not the same as before, but

$$t' = \frac{2a}{c} \cdot \frac{c^2}{c^2 - v^2} = t \frac{1}{1 - \beta^2}$$

in which the fraction v/c, that is the ratio of the velocity of motion to the velocity of light, is designated by β. If on the other hand, the whole path AB moves at right angles to itself with a velocity v (figure 24b), the light has to take a transverse

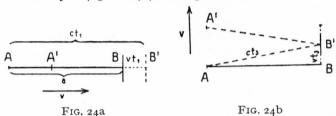

FIG. 24a   FIG. 24b

course, in order to return to the observer at A, we then get for the total time the expression*

$$t'' = t \cdot \frac{1}{\sqrt{1 - \beta^2}}.$$

The two values t' and t'' are thus not equal, but different from one another and from the original time t. With the help of a sufficiently accurate measurement of time, it should be possible to detect this difference. In practice the experiment is carried out by dividing a ray of light into two parts, which are arranged to follow two exactly equal paths at right angles to one another, and on their return along these paths, to interfere with one another. If the whole apparatus is first set up so that one of

* Let the light take time t_1 in going, and t_2 in returning.
1. In time t_1, B moves through a distance vt_1 to B', so that the ray has to travel a distance $a + vt_1$. On the other hand, the distance to be travelled is ct_1. Hence $ct_1 = a + vt_1$, from which it follows that $t_1 = \dfrac{a}{c - v}$. In the same way, we get for the return time $t_2 = \dfrac{a}{c + v}$, and for the total time $t' = t_1 + t_2 = \dfrac{a}{c - v} + \dfrac{a}{c + v}$

$$= \frac{2ac}{c^2 - v^2} = \frac{2a}{c} \cdot \frac{c^2}{c^2 - v^2}$$

2. If t_3 is the time for the one path, this $(AB') = c t_3$. This is the hypotenuse of the right-angled triangle ABB', the sides of which $AB = a$ and $BB' = vt_3$. Hence $(ct_3)^2 = a^2 + (vt_3)^2$. Thus $t_3{}^2 (c^2 - v^2) = a^2$, and therefore $t_3 = \dfrac{a}{\sqrt{c^2 - v^2}}$, and $t'' = 2t_3 = \dfrac{2a}{\sqrt{c^2 - v^2}} = \dfrac{2a}{c} \cdot \dfrac{c}{\sqrt{c^2 - v^2}} = t \dfrac{1}{\sqrt{1 - \beta^2}}$

these two directions coincides with the earth's motion, and is then turned round through 90°, the interference image must be displaced. The experiment was first carried out by the American scientists Michelson and Morley (1887), the arrangement being sufficiently sensitive to show an effect ten times smaller than that which we should calculate theoretically from the earth's motion. The result was completely negative. The difference of time sought for does not exist, and here again we are disappointed in our attempts to measure velocity of the earth by measurements made upon the earth itself. Repeated experiments of this kind with the most refined means of modern technique have always given the same negative result.[77]

We might at first suppose that this could be quite simply explained by supposing the earth to move as a whole relatively to the ether, but to carry along with it the ether in its immediate neighbourhood, just as a sphere moving through a stationary liquid carries with it a layer of liquid close to itself. But this idea unfortunately contradicts another experiment which was made in 1851 by Fizeau. Fizeau investigated the velocity of light in a moving liquid. Contrary to his expectation, he found neither that the velocity of motion of the liquid was simply added to that of the light moving in the same direction, nor that it had any influence on the velocity of the light. The light was found to be carried along by the flowing liquid, but only with a fraction of the velocity of flow. This fraction is given by the expression

$$v \left(1 - \frac{1}{n^2} \right)$$ where v is the velocity of flow and n the index of

refraction of the liquid against a vacuum, or what is the same thing, the ratio of the velocity of light in empty space to that in the stationary liquid. Now n for air is only slightly greater than one (1.000294), and the value of the quantity in the bracket amounts to about 0.0006. If we thus apply the result of Fizeau to the atmosphere of the earth, we find that the motion of the air through the ether of space cannot carry the latter with it to any considerable degree, or what is the same thing, in the Michelson experiment we can neglect a drag on the ether. But it should then give a positive result.

If in spite of this we are not willing, as some modern opponents of the theory of relativity are, nevertheless to retain the idea of ether drag (with the assumption that the great mass of the earth has an altogether different effect upon it than the small mass of water which circulates in Fizeau's experiment), the next resource is offered by the idea that light, like a projectile, may take with it the velocity of the body emitting the light. This hypothesis, worked out by the Swiss physicist Ritz, unfortunately comes to grief on other difficulties, chiefly that of explaining the Doppler effect (see page 276). It could not be maintained, and has been generally abandoned to-day.

In this difficult position, which was strongly felt by physicists round about 1900, a brilliant idea of H. A. Lorentz appeared to offer salvation. Lorentz assumed that the motion of matter through stationary ether may have the effect of shortening all lengths, including the diameter of molecules and even electrons, in the direction of motion by the fraction $\sqrt{1-\beta^2}$. We immediately see that if we put instead of a in the formula for t', the fraction $a\sqrt{1-\beta^2}$, then $t' = t''$ (Lorentz's contraction hypothesis). This hypothesis is not in itself so fantastic, when we consider that all substances are supposed ultimately to consist of electric charges, which are surrounded by an electric field, and therefore motion of these charges relatively to the field-medium (the ether) might very well result in forces which would effect this contraction. But it had naturally the objectionable character of being a hypothesis produced *ad hoc*, and one which, furthermore, could never be proved, since all measuring rods suffer the same contraction.

For this reason, Einstein, in 1905, gave another solution, which then developed in a very short time into an astonishing structure of theory. According to Einstein, the error lies in our conception of time. We ordinarily imagine all processes to take place in the world according to one single time. If there were a great clock in the universe, we suppose that every event would take place at a quite definite time as given by it. The fact that we are only able to compare our determinations of time made at different points in space by sending signals, does not alter this conception. For the time which is required to send the signal can be brought into these calculations if we know the distance and the velocity of propagation of the signal. If an observer, for example, wishes to compare his clock with that of another observer by means of light signals, he only needs to divide the distance by the velocity of light c, in order to know the time taken by the signal. If he deducts this time, he can compare his clock directly with the other one.

Einstein abolishes the whole of this assumption, which appears self-evident to the layman, and had hitherto appeared the same to the physicist. According to him, the comparison of two clocks cannot simply be made by taking into account the distance existing between the two at the moment. The relative state of motion of the two clocks enters into the matter, that is to say, it is a question of whether one possesses a velocity relatively to the other. There is no general world time, but only times for each observer. These can, however, all be related to one another, but only by taking into account the relative motion of the observers. Only when this assumption is made, as Einstein shows mathematically, can the requirement of fundamental relativity of motion be carried out electro-dynamically and the negative results of the Michelson experiment be explained. If it is impossible to determine absolute motion also by means of optical (electrical) processes – and Einstein makes this the first and fundamental postulate – it becomes necessary to develop another theory of kinematics (see page 45). The final reason for this is that the constant c is actually a peculiarity of empty space, which according to Einstein is really empty. If this does not represent any *physical* property of an ether, it becomes necessary to make it into a *kinematic* property, that is to say belonging to our space-time scheme in itself.

In mathematical language the requirement is as follows: If x, y, z, is a three-dimensional co-ordinate system of an observer, t being the time measured by him on a clock, and x', y', z', t' are the corresponding magnitudes for a second observer, and if the two systems and observers are moving relatively to one another with the velocity v, all laws of nature must retain their mathematical form when we pass from one system to the other. But this is only possible on the basis of Maxwell's electro-magnetic equations, when the transition from one system to another is expressed by formulae, in which the ratio of the relative velocity to the velocity of light plays a part, and in which t' is not equal to t.[78]

The most striking point is, first of all, that the time is also transformed (t' is not equal to t). That is to say, every observer has his own time, and therefore two events at different places which are simultaneous for one observer, are not so in general for the other. It is extremely difficult to grasp this idea, since the notion of a universal world time (*tempus absolutum, quod aequabiliter fluit*, says Newton) is in our blood. The kind of mental revolution is necessary which must have been required of the contemporaries of Copernicus. The Lorentz contraction (see above) follows as a consequence, but only with the sense that a rod, which retains for an observer moving with it the length l, experiences an apparent shortening (to the extent $l\sqrt{1-\beta^2}$) for another one, moving relatively to him. This is not brought about by a physical cause through forces exerted by ether on matter, but appears thus for the purely kinematic reason of relative motion.

I cannot here omit a remark concerning what I feel to be the effects of much popular literature concerning the relativity theory. It is a general custom, even on the part of scientists – for Einstein's own popular statement[76] is not free from it – to lay too great a stress on these paradoxes which result from the theory of relativity, with the result that the layman, who naturally at first cannot take them in, believes that the whole theory of relativity is incomprehensible to him. This does a serious injury to the theory of relativity itself.

The emphasis ought rather to be laid upon the alternative expressed in the question whether c is a physical or a kinematic constant. The first means that space, in the sense of Maxwell's theory, is capable of exhibiting physical states (fields), which are then propagated with the velocity c. The development of this conception necessarily leads to the absolute theory, and hence to the invalidity of the classical principle of relativity (page 53) for the phenomena of electro-dynamics and optics. The actual impossibility of determining absolute motion relatively to the ether must then be explained by accessory hypotheses (drag of the ether by the earth, but not by objects moved on the earth; or by the Lorentz contraction). On the other hand, we have Einstein's solution, which makes c an inherent property of the space-time system as such. But this obviously means not only that c, strictly speaking, enters into the formulation of all physical laws, including those of mechanics, but the conclusion must also be drawn that, since c is without any doubt the fundamental constant of electro-magnetism, the latter belongs directly to the fundamentals of reality (see below, pages 194 ff.).

The theory we have just described is called the *special theory of relativity*. As we have seen, a number of interesting consequences follow from it, but these are put in the shade by those which follow from the extension of his theory made by Einstein in 1915, the so-called *general theory of relativity*. The transition to this extended form was given by the mathematical restatement by Minkowski of Einstein's original theory. In his famous lecture on 'Space and Time' which he gave at Cologne in the year 1908, Minkowski showed that the new kinematics of Einstein

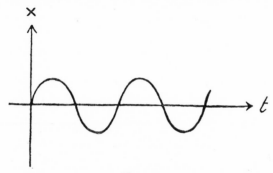

FIG. 25
Sine curve as graphical representation of oscillation

could be represented in a very clear and elegant form as a kind of four-dimensional geometry, by simply treating time as a fourth co-ordinate along with the three co-ordinates of space *x*, *y*, and *z*. This world-geometry is, however, only a pseudo-euclidian system.[79]

In order to make these ideas intelligible to readers who are not familiar with them, we may look upon the matter as follows. A motion may be one-, two-, or three-dimensional, that is, it may be limited to a line or surface, or it may have the whole of space at its disposal. An example of the first case is the motion of a railway carriage on the rails, or that of an ordinary pendulum which is confined to the arc of a circle, for the second case we have a skater upon the ice, a stone rolling down a hillside, or a conical pendulum (which has the surface of a sphere). As example of motion with 'three degrees of freedom' we may take a bird or an aeroplane. In order to determine the position of a moving point, we require in the first place one, in the second case two, in the third case three co-ordinates (x, y, z). Let us now take the simplest case, a linear motion, for example the usual motion of a pendulum (figure 5, page 80). If the distance of the pendulum from the middle position is x, then x is represented graphically, as a function of the time, by a sine curve (figure 25). In this figure, we see at a single moment, adjacent in space, what in reality is a succession in time. This has been effected by taking time as the second co-ordinate, the displacement x being the first. Motion in a single dimension thus appears as a two-dimensional

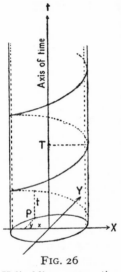

FIG. 26

Helical line representing circular motion

picture. But exactly the same is also true for higher dimensions. Motion in a plane, for example, taking place uniformly in a circle could be represented as a curve in space, by setting up the axis of t vertically at the middle point of the circle. The graphical picture of this circular motion would then be a screw line (figure 26). In general, n-dimensional motion gives us an $(n + 1)$ – dimensional picture.

In exactly the same way, the Einstein-Minkowski theory regards all three-dimensional motion as a 'world-line' in four-dimensional space. This however is not simply the four-dimensional analogy of our ordinary three-dimensional space, since the fourth co-ordinate, time, must be regarded as imaginary, if the matter is to be treated in this way. Treatment of it on complete equality with the space co-ordinates, such as is often given in popular statements of the theory of relativity, does not take place. Rather, the magnitudes dealt with can be sharply divided into 'space-like' and 'time-like.' But this does not prevent space and time being fundamentally inseparable, which fact is made still more evident, when we give up the condition that the four-dimensional space is to be euclidian. For then we actually get time as a real fourth co-ordinate quite analogous to x, y, and z. But space ceases to be euclidian. What is the meaning of this?

In order to understand this notion better[79] let us imagine for a moment a being having only two dimensions (length and breadth) but possessing intelligence. If such beings lived on a plane, they would discover our geometrical theorems, such as that of Pythagoras, for plane geometry.   They would further be acquainted with the concept of a straight line, and if they were accustomed to working with systems of co-ordinates, they would, like ourselves, represent a straight line by an equation of the first degree $(ax + by = c)$.   A particularly clever specimen of these 'flat-heads' might then arrive at the idea of imagining three-dimensional space and operating in it with a three-fold system of co-ordinates (x, y, z).   The 'space' of his world, that is to say what we call a plane, would then be represented by an equation of the first degree in x, y, and z, $(ax + by + cz = d)$, just as in our own analytical geometry of space.

If such two-dimensional beings, on the other hand, lived on the surface of a sphere, their geometry would be entirely different. They would define a straight line as the shortest way between two points of the sphere, that is, as an arc of a great circle, and would then be able to work with triangles, etc. (our spherical triangles) but naturally they would find quite other fundamental rules.   For example, the theorem of Pythagoras would not hold (there could be no such thing as a square) nor would the sum of the angles in a triangle be equal to two right angles.   But if we imagine their sphere to be very large, that is to say, very slightly curved, they might after all arrive approximately at the laws of euclidian plane geometry, as long as they did not move over too great distances.   They would then only discover the curvature of their space when their measurements became sufficiently exact and extensive, and then they would be able to determine it numerically.   Naturally, however, the mathematician who lived on the plane could invent spherical geometry, while the corresponding spherical genius could invent euclidian plane geometry.

This whole argument may now easily be transferred to higher dimensions.   The two-dimensional mathematician can construct our three-dimensional space, and represent his own plane as the trace of an equation of the first degree of three variables, and the spherical surface of his colleagues on the sphere as the graph of the equation $x^2 + y^2 + z^2 = r^2$.   In the same way, we can rise in thought above our own three dimensions, and imagine ourselves in a four-dimensional space with the co-ordinates, x, y, z, and u, and we see that our euclidian space would again be represented by an equation of the first degree of these four variables

Ls

$(ax + by + cz + du = e)$, but that other three-dimensional figures may be placed alongside it, which correspond to other equations (for example, $x^2 + y^2 + z^2 + u^2 = r^2$, the four-dimensional hypersphere). These other non-euclidian spaces are then curved, as contrasted with our plane space, and the geometry which would hold therein can naturally be fully worked out, although we are just as little able to form a mental picture of it, as the 'plane' mathematician could of the spherical surface, or conversely, either of them of our three-dimensional bodies.

The problem whether our space may not even be curved in reality, and euclidian geometry be only an approximation for us, because the distances which we measure are too small, is often discussed. Gauss, in order to decide this question experimentally, measured as exactly as possible, and taking into account all possible sources of error, as large a triangle as possible (Brocken, Inselberg, Hoher Hagen), but could not find a difference between the sum of the angles of the triangle, and two right angles, which exceeded the limits of error. We shall return almost immediately to the question as to how far experience can decide a question of this kind. In any case, we must first consider the possibility that our geometry may not be, as generally assumed by epistemology since Kant's time, certain *a priori*, but rather a kind of 'physics of space,' which has always been maintained by empiricism as against Kant's view. It is true that, even if regarded in this way, this physics of space would always occupy a position completely separate from all other physics, since the physical laws would still do no more than fit into the frames of the geometrical, without having any inner connection with them (see page 53).

The genius of the chief creator of 'metageometry,' Riemann, conceived the possibility, that the last mentioned, widely held idea might perhaps be wrong, and that the 'basis of the metrical relationships of space' might rather be found outside the latter itself, in 'binding forces acting upon it.'[80] But it was not until seventy years later that this idea was carried into practical effect by Einstein. In order to understand this achievement, we must first be clear about the way in which Riemann enlarged the concept of space. We supposed just now the mathematicians living on the plane or spherical surface to be capable not only of working out the geometry of their two-dimensional 'metric field,' as we will now term it, but also of constructing other such geometries than their own, by calling to their aid a three-dimensional field (our space) and interpreting any equation whatever between the three co-ordinates of it, x, y, z, as a two-dimensional field, which in general would be curved. In every such case, however, ordinary (euclidian) planimetry is valid on an infinitely small scale on such a surface, no matter how strongly, and according to what law, it is curved; in other words, it is valid to any degree of approximation, if we only limit ourselves to a sufficiently small region. (A small part, for example, of the surface of a sphere can be regarded as plane.)

Now this argument, as we have already indicated, may be applied to our space of three dimensions, and it is upon this fact that Riemann's achievement depends. According to him, the concept 'three-dimensional space' is to be enlarged so as to include all possible spaces, but so that nevertheless ordinary euclidian geometry holds in every case upon an infinitely small scale. This has been rightly regarded as a parallel achievement to that of Maxwell and Faraday in physics. Just as these replace Newton's action at a distance by direct action (or an integral law by a differential law), so Riemann requires euclidian laws in geometry only as differential laws (for the infinitely small scale) and thus obtains for the integral law, that is for regions of finite size, the freedom of every imaginable kind of three-dimensional geometry.

From this point to the summit, only two further steps are required. The first is the inclusion of time according to the method given by Minkowski. The relativity theory really deals not with space, that is with geometry alone, but with kinematics. In the special theory of relativity, as we have shortly indicated, we have already arrived at a semi-euclidian system when the time is introduced as a real co-ordinate. The three-dimensional sub-space, x, y, z, that is, space alone, is not curved, but euclidian or plane, and the whole region becomes euclidian, as we have said, when an imaginary time-co-ordinate is introduced. The matter becomes altogether different from the standpoint of the general theory of relativity. Here neither the kinematic totality nor space alone has a definite pre-determined character. Just as little as Riemann postulates only one non-euclidian geometry, does Einstein postulate only the one (non-galilean) kinematics of the special theory of relativity. On the contrary, we have an infinite number of such non-galilean systems. And when Riemann inquired after the 'binding forces' which determine the metrics of space (the metrical field), Einstein answers that these binding forces are the gravitational forces. The kinematics which hold here in my neighbourhood depend upon the distribution of matter around me, and only where the gravitational field is 'homogeneous' do we have the kinematics of the special relativity theory.

The time-space relationships are therefore a function of matter, and thus geometry, or rather kinematics, has become for the first time a physics of space, and indeed a part of physics itself. There is therefore no such thing as a generally valid kinematics (and geometry) but we have everywhere a particular kinematics which is determined by the existing distribution of matter. In other words, the metrical field is a function of matter in exactly the same sense as an electric or magnetic field is a function of neighbouring electric charges or magnetic poles. Just as without the latter no electric or magnetic field exists, so without matter no metric field exists, that is, no space or time. These latter, far from being empty, *a priori* forms of cognition, are actually assimilated to the physical 'thing.' The world becomes space, time, and matter, as a single inseparable unity.

This brings us to the highest point, and we will now proceed to look around, seeing much that is known in a new light, but other things that are quite new. In the first place, we have the whole complex of questions, which may be summed up by the names of Kant and Einstein. If nothing else should permanently survive of the relativity theory, it has at least to its credit the historical achievement of compelling us at the present day once more to investigate thoroughly the problem of space and time, instead of relying, as was the custom in very wide circles, upon Kant's authority. The relativity theory has, in my opinion, also brought us a step forward in matters of fact beyond Kant. The purely speculative 'metageometry' which was developed after the time of Bolyai, Lobachevski and Riemann was restricted to pure mathematics, and hence was not capable of the achievement of the general theory of relativity, which regarded as a physical reality what metageometry had merely put forward as a possible construction of the imagination. We are obliged to conclude from it that our view of space and time corresponds objectively to some properties of physical reality, exactly as our sensation of light corresponds to appropriate waves, of our sense of heat to molecular motion, and that we are no longer in a position to assert the sole validity of euclidian geometry merely because our view of things is bound up with it. It is simply a matter of using our minds to free ourselves from this prejudice derived from experience, as from so many other similar prejudices.

Kant saw quite correctly that space and time, as 'forms' of sensation, are just as much of an ideal and subjective nature, as colours, sounds, heat, etc. (as opposed to Locke's distinction between primary and secondary qualities). His mistake was to believe himself obliged to link this view (whether as reason or consequence) with the doctrine of the 'apodictic validity' of the geometrical axioms, which for him were self-evidently the euclidian. About this there can, in my opinion, be no doubt, and the present-day disciples of Kant should not give themselves so much trouble to avoid the recognition of the fact that Kant's demonstrations fall to the ground for the most part, as soon as the general relativity theory leads us to recognise that euclidian geometry, in certain cases, is not even one among many possible geometries, but actually false.

Not only did Kant never suspect this, he started his whole argument from the simple validity of euclidian geometry. This validity did not mean for him merely validity in immediate experience – against which there would be as little to be said as against the assertion that colours appear to us subjectively as arranged in a circle, while objectively or rather physically, they form a linear succession. On the contrary, Kant, without any question, simply deduces from the assumption of the (subjective) pure form of perception, the validity of euclidian geometry for the whole extent of our knowledge of empirical reality.

This can certainly not be made to agree with the general theory of relativity. For according to it, euclidian geometry is only valid approximately, because we are limited to a comparatively small region of the objective, non-euclidian world. For this law, which undoubtedly makes good sense, there is simply no place whatever in Kant's theory – at least if we take Kant's own theory, and not its form as given by his modern supporters – because in this department, as in every other, it joins together inseparably the reality of the object of knowledge and the certainty of judgments of knowledge.[81] This finally depends upon the fundamental error of all idealism, using the word only in the epistemological sense, that it neglects too much the part played by the object in the bringing of knowledge into being. However that may be, for the theory of relativity the question as to what space-time scheme really holds in individual cases can only be decided on the basis of physical experience. Though immediate space-time experience, like the qualities of sense, may only exist as such in the subject, the world is itself determined in some way, possessed of some kind of manifold order, which, though it allows our space-time experience to be euclidian in normal cases, nevertheless compels us to apply to the world as a whole the super-euclidian scheme, and thus to raise ourselves above the narrowness of our immediate experience, in a similar manner to that in which we do so, as regards the circle of colours, by means of the scale of wave lengths (see pages 117 ff.).

These few short remarks do not, as I am well aware, by any means exhaust the problem. Whoever wishes to study the matter more closely must read a small library of special literature. In doing so we observe the very curious fact, that both opponents and friends of the relativity theory are to be found in all philosophical camps. Even in the positivist camp of Mach's followers, in which direction Einstein himself has a strong tendency, we find along with such enthusiastic followers of relativity as Petzoldt, some decided opponents, who condemn it as useless mathematical speculation exactly from the point of view of the pure empiricism which they uphold. Among these opponents was to be found, as appears from posthumous publications, Mach himself, who, by the way, was also while he lived a sharp opponent of all metageometry. Opponents of this kind are, it is true, exceptions on this side. Generally speaking, the conviction is held that the relativity theory is nothing but Mach's philosophy turned into physics. This is all the more natural because Mach himself actually anticipated a very important notion in the theory of relativity on the basis of his fundamentally empirical point of view (see page 53).

But I fully agree with Geiger when he points out that the positivistic friends of the relativity theory only see its destructive side. They usually lose sight of the fact that the theory, by relativising observation, proceeds to determine a set of facts entirely independent of any subjective point of view. Much the same must have been the experience of those who were the first to discover the spherical shape of the earth. Here it is the notions 'above' and 'below' which lose their absolute significance, simply because a generally valid fact, the direction of gravity towards the centre of the earth, is brought to light.

On the other side, Neo-Kantianism (Idealism, Phenomenalism) is fairly divided in opinion. While a number of its supporters approach the relativity theory for the present with more than critical caution, and in any case reject it without question in so far as it contradicts Kant's doctrine, a number of others (Sellien, Cassirer) have taken great trouble to show that Einstein finally brings us nothing but a new development of Kant's programme. I have already indicated that I do not agree with this assertion, but cannot go further into the matter here, and must refer the reader to the statements by Winternitz and Reichenbach.

Between apriorism based on Kant, and pure empiricism based on Mach, we have a group of later epistemologists, which have been suitably termed 'conventionalists.' According to them, physical knowledge only comes into being on the basis of certain definitions and conventions, which are arbitrary within certain limits, their choice being mainly determined by the principle of economy of thought. Thus according to Poincaré, who has recently been joined to a certain extent by Dingler (see, however, below), euclidian geometry is neither *a priori* in Kant's sense, nor *a posteriori* in the sense of naked empiricism, but it is simply the most convenient geometry, and hence the basis which our thoughts involuntarily make use of, in forming judgments concerning the spatial qualities of bodies. In particular, a body as 'rigid' body – and thus as foundation for a cartesian system of co-ordinates – is judged according to the way in which it fulfils the laws of euclidian geometry, but these are not conversely, as empiricism maintains, read off from it purely receptively.

In this camp also there are both friends and opponents of relativity. While Dingler himself rejects it as a superfluous speculation which only complicates matters unnecessarily, it is regarded by others, more or less, as the simpler system, and hence as that depending upon more useful conventions. This view is taken by Carnap in a recent paper in the *Kantstudien*, to which I

shall have to return later.   Here we draw particular attention
to his dissertation on space, in which the whole problem of space
is developed with altogether extraordinary clarity.   Carnap has
shown particularly plainly, that the whole question deals with
three concepts of space which must be carefully kept distinct:
first, the purely formal space of mathematics, this being nothing
but an arrangement of manifolds, which can be just as well
applied to colours or equations as to points, lines, and surfaces.
Secondly, the space of direct experience, which on the one hand
is a special case of general mathematical space (namely a three-
dimensional, plane, or euclidian continuum), but on the other
hand contains, in our direct experience of lines, planes, surfaces,
something that cannot be grasped by any mathematical definition,
however exact, but must be experienced, like colour or sound.
Thirdly and finally, physical space, which – here I cannot agree
with Carnap – represents the real order of things which can only
be determined by the analysis of experience.   This order likewise falls under the general doctrine of order, that is, geometry
in the first sense, but here again only represents a special case
which appears to be, unlike the euclidian space of direct experience, a non-euclidian continuum (or also perhaps a non-con-
tinuum; see below).

The further discussion of the problem will probably turn
mainly about the question as to how far the combining activity
of the intellect is already operative with respect to the space of
immediate experience, and what reasons cause it to set up a
euclidian order.   Whether the principle of economy here appealed
to by the empiricist and conventionalist explains the matter,
appears to be very questionable.   A very peculiar solution has
recently been attempted by Dingler.[82] He points out that a plane
surface is produced in practice in the workshop by grinding three
surfaces upon one another and until all three fit together.   According to him, this gives us a definition of the plane which precedes
all theory, as a surface which is firstly alike at all points and
secondly alike on both sides.   The straight line is then
immediately defined as the intersection of two planes, and the
rigid body is obtained by a further definition, which likewise
simply depends upon 'indistinguishability.'

The author is inclined to think that these ideas are worthy of
more careful consideration than they have usually received.   In

the ordinary epistemological argument for non-euclidian geometry too little attention is paid to the fact, that while a space (nth order) with zero curvature is simply, from the mathematical point of view, a special case of an infinite number of imaginable spaces of all possible curvatures, nevertheless, logically speaking, the absence of all curvature is a singular, peculiar space contrasted with all other spaces, which are curved and hence distinguished by individual constants. Euclidian space (the straight line, plane, etc.) is actually the only one which possesses no special constants in itself, and hence can be reproduced anywhere and everywhere with certainty. This logical peculiarity disappears, in the usual statement, behind the purely formal position of the figure zero in the series of all numbers, and it is in this way that we are led to seek in thought economy or other considerations, the reason why euclidian metrics are chosen among an infinite number of possibilities.

I do not venture to decide whether Dingler's 'philosophy of the act' gives the correct solution. However the matter may appear after ten or twenty years, one thing may be asserted, that it is no longer possible to do as some pure Kantians have done, and assert that Einstein must be wrong in so far as he contradicts Kant. Such a proceeding is pure dogmatism, and will never gain a footing in science. On the other hand, all sides must agree that the whole epistemological discussion raised by relativity is to a great extent independent of the acceptance or non-acceptance of relativity by physics. For philosophy, the simple possibility of such a theory is sufficient to cause the abandonment of epistemological views which make it impossible, since philosophy must have room for every possibility.

A further very important question is that of the relationship of the four-dimensional space-time order assumed by Einstein and Minkowski to the form of immediate experience, which completely separates space and time. How does it come about that we experience this fourth co-ordinate of the objective world-order, which, it is true, also plays a special part in the relativity theory itself, as something quite different from space? Hitherto, as far as I am aware, this problem has only been worked upon very little even by supporters of relativity, while their opponents have been satisfied with the simple remark, that the spatialisation of the time co-ordinate is simply a mathematical picture without any real value. The following remarks may be taken as a small contribution to this question.

In the classical view of physics, as expressed by Laplace's picture of the World-Spirit (see page 63), an initial physical condition of any kind is, as we have already several times stated, set in a space in itself devoid of qualities, and the course of events in this physical state can then be calculated both forwards and backwards by known laws of nature. More exactly speaking, the motions of a system of points are completely determined, when the initial state, that is the initial values of the three co-ordinates, x, y, z, and the three velocity components x', y', z', are given.[83] The laws of nature then appear mathematically in the forms of differential equations of the second order, which connect the changes with time of the magnitudes in question with the magnitudes themselves.

In the Maxwell-Hertz-Lorentz electro-magnetic field theory, we have, in place of a part of these initial conditions, the 'boundary conditions.' If namely, we isolate a certain portion of space by a closed surface, the events in this portion of space, since we are now abolishing action at distance, are only dependent upon the initial state and the conditions which are ascribed to the boundary surface during the period of time considered. The latter form what are called the boundary conditions; for example, if the whole system is enclosed in a metallic wall, the electrical field strength at this is permanently zero, and the 'law of nature' is here given by the Maxwell-Lorentz equations.

In order to apply this to the Minkowski world of relativity, we have first of all to ask what the law of nature which governs the process is, in the language of this theory. If time is made a fourth co-ordinate of space (see page 147), the kinetic law of motion of a point is obviously transformed into the purely geometrical law of a line, and the statement that the whole motion is given by the law and the initial state $(x_0, y_0, z_0; x'_0, y'_0, z'_0)$, now means that this line is completely determined by a point and the direction (tangent) at this point (this, by the way, is only unconditionally true, when the equations in question give 'analytical functions').[84] If we now take any number of points and assume a 'dynamic' law which connects these with one another, for example, the Newtonian law of gravitation, the meaning of this, when translated into four-dimensional geometry, is as follows. We are given any number of points on the 'base' of a four-dimensional cylinder (this base being a limited three-dimensional part of space), and further, a 'time-like' direction through each of these points, that is, a small part in the direction of the axis of the cylinder.[85] A law then exists (the 'dynamic' law) by which the equation for each one of these single lines is predetermined; which law, in other words, defines the whole contents of the cylinder in 'world lines' up to any height.

If we pass from Newton and Lagrange to Maxwell and Lorentz, all that is changed is simply this, that the dynamic law no longer contains forces acting at a distance, but only direct actions, and that we must therefore fill the whole space part (the base of the cylinder) with a continuous field, generally containing electrons. Thus the whole cylinder is to be imagined as filled with an infinite number of world-lines, the law of which is, however, again given completely by the fundamental dynamical law (the Lorentz equations), and the boundary conditions, which represent the external surface of the cylinder. So far everything seems to be fairly simple. (The reader not accustomed to multi-dimensional geometry may, for the sake of understanding it more clearly, first translate it all into one less number of dimensions, thus taking the collection of n points, not in a space within a closed surface, but in a plane within a closed curve.)

But as soon as we pass to the point of view of the general relativity theory,[86] the matter becomes essentially different. For in all previous arguments, we have still been assuming an arbitrary four-dimensional space, in which we place some kind of initial physical conditions. From these it is possible to calculate the state in the immediately following moment according to distant or near laws which, in themselves, may likewise be arbitrarily constructed. But all this can no longer be regarded as permissible according to the general theory of relativity. Space itself (the metric field) is here a function of the matter assumed, and the law of this dependence is at the same time the fundamental dynamic law, from which all special natural laws must follow.

This appears to me to result in very far reaching philosophical consequences. For in the first place, it is consequently possible only in a sense needing quite special definition, to assume an 'isolated system' (within the world'); it is fundamentally only possible to take the world as a whole as the object for consideration. For example, the practical consequence of this is seen in the fact that the most elementary of natural laws which we here consider valid, the law of inertia, is a consequence of the material distribution of the fixed stars. We can, however, speak of isolation in space in the sense that a system of material parts is separated from others by great distances (this word being now used in the ordinary narrow sense), and therefore, in the terminology of relativity, that the world lines crowd together in certain narrow regions of the Minkowski world, which are widely separated spatially from one another.

It is obvious that the time co-ordinate does thus exhibit the peculiarity that most world lines run by preference in its direction, and that in this direction everything appears to be filled up continuously and uniformly. It is perhaps not too daring to formulate the hypothesis, that this is one of the objective reasons for the distinction between time and space, and further, for the preferential occurrence of causality in time succession. We should thus, in other words, think in terms of time, because we are spatially isolated. The close connection between the measurement of time and the law of inertia (see above, page 51) thus appears in a new and peculiar light.

Very detailed and fundamental investigations of the time problem have recently been published, principally by Reichenbach, to which I should like to direct the reader's special attention.[87] One of his chief results is the connection of the unidirectional character of time with the circumstance, that causal conclusions as regards the past are always definitely determined, but as regards the future always only more or less probable, since as regards the latter we are never able to perceive exactly what circumstances of the present are to be regarded as taking part in the determination of a certain event in the future. According to Reichenbach therefore, the present is that section of the world at which the transition between causal definiteness and mere probability occurs. We shall return to these investigations, and to the whole problem of time, once more in a quite different connection, but we will first again deal with questions of a more physical nature, which are connected with the relativity theory.

In this connection, we must first once more refer to mass. The founders of the electron theory (J. J. Thomson, Lorentz, and others) already recognised that mass, that is inertia, may appear in certain circumstances as a pure phenomenon of the field, and that the very small mass of the electron (1/1800 of the H atom) may perhaps be purely electro-magnetic. For if an electrically charged body – for the sake of simplicity we may imagine a ball charged on its surface – is set in motion, this motion represents an electric current, and this current generates a magnetic field in its neighbourhood. But this magnetic field contains a certain store of energy, and this amount of energy must naturally be provided by the cause which sets the ball in motion. In other words, in order to set the charged ball in motion, more energy is necessary than if the same ball were to be set in motion in an uncharged condition.

This has the same effect as an increase in its mass or inertia, and it is easy to calculate,[88] that for slow movement (v small as compared with c), the amount of this apparent increase in mass is equal to $\dfrac{2e^2}{3ac^2}$ (e being the charge, and a the radius of the ball). If the velocity is greater, the calculation becomes considerably more complicated.[89] The electro-magnetic mass then becomes dependent on the velocity, dependent, moreover, in different ways, according as the charged body is to be accelerated in the direction of the previous motion, or at right angles thereto (longitudinal and transverse mass).

Such rapid motion then allows us to conclude, according to the law governing the increase of mass with velocity, what the relation between the electro-magnetic and mechanical mass actually is, and experiments made about 1800, first by Abraham and Kaufmann, and later by Bucherer, Hupka, and others, showed that the whole mass of cathode ray particles (electrons) appears to be electro-magnetic according to this formula. In other words, an electron would actually have no mechanical mass, and its whole inertia would simply be the effect of its field. On the basis of the conception of a charged ball, these measurements also allow the radius a of an electron to be calculated, but calculations of this kind have now been given up as useless, since, as we shall see below, the simple assumption as to the nature of an electron cannot possibly be correct. But even the consequence of the old theory which we have just discussed, that the whole mass of an electron is electro-magnetic, could not be maintained on the basis of the relativity theory. According to the latter, it may be true, but does not need to be so.

From the point of view of the relativity theory, all mass, also mechanical mass, varies with the velocity; this follows upon a purely kinetic basis. A mass which has the value m_0 for an observer at rest relatively to it (moving with it), has for an observer relatively to whom it is moving, the mass $m = \dfrac{m_0}{\sqrt{1 - \beta^2}}$. The change of mass with velocity, actually observed in the case of rapidly moving electrons (β-rays), is thus a self-evident consequence of kinematic fundamentals, though the amount of the change is slightly different from that required by the older theory of Abraham. This affords a possibility of testing the relative correctness of the two theories, and a repetition of Kaufmann's experiments made later by Bucherer and Hupka, by refined methods, has actually decided in favour of the relativity formula.[90] This gave at least one direct confirmation of the relativity theory, for everything else, the explanation of the negative result of the Michelson experiment, etc., was essentially only indirect evidence in its favour.

The theory appeared to be the one which brought together, in the simplest manner and greatest unity, all these phenomena.[91] But its greatest interest naturally appears at the point where it proceeds beyond classical mechanics and electro-dynamics, and predicts new and hitherto unobserved phenomena. These are by no means easy to find, for, as our short description has already indicated, the departure of the Einstein theory from the older view only becomes perceptible at velocities which are comparable with the velocity of light (for all velocities v which are small as compared with c, $\beta = v/c$ is almost zero, and hence $\sqrt{1 - \beta^2}$ approximately unity, whereupon Einstein kinematics pass over into those of Galileo and Newton). But all velocities which are at our disposal in practice, are small – even the greatest, that of the earth, amounting hardly to $1/10000$th of the velocity of light. Only in the motions of electrons and ions in cathode and canal rays, in the Bohr atomic model, and above all in the α- and β-rays of radium, do we find motion at a velocity approximately comparable with c, and it is therefore only in such cases that we can hope for direct confirmation of the relativity theory. And as a matter of fact, as Sommerfeld has shown, the relativity theory is well confirmed by the theory of the 'kine-structure' of spectrum lines, whereas the absolute theory gives incorrect results.[92]

We must not omit to mention, that these successes of the relativity theory have recently been questioned on several sides, for reasons having considerable weight. Above all, Lenard has been at great pains to show that the results given by relativity can also be obtained by suitable modifications of the ether theory, without any necessity for the objectionable relativisation of time. It is therefore too early to regard the matter as completely settled, as is done by many over-enthusiastic relativists. Only the further development of physics itself can give a final decision. We may however feel sure that many results of Einstein's will survive even a possible overthrow of his theory. One of these is of very great importance, and we must make a short mention of it. Einstein concluded from the special theory of relativity that the inertial mass of a body must be quite generally regarded as its total content of energy, divided by the constant factor c;

thus $M = \dfrac{E}{c^2}$ or $E = Mc^2$ (for one gram of mass, this gives

us 9×10^{20} ergs, or approximately 9.2 billion metre-kilogrammes; converted into heat, about 22 billion calories).

It appears for the moment incredible that such enormous quantities of energy should be contained in bodies. But we only need to consider the quite enormous energies concerned in radio-active transformations, and also the Rutherford-Bohr atomic model, to find it less incomprehensible (the complete transformation of one gram of radium into emanation produces approximately 3,000,000,000 gram-calories, or about 1,000,000,000 metre-kilograms; though even this, it is true, is only about 1/7000th part of the amount of energy assumed by Einstein to exist in one gram of mass). This would result in a body becoming lighter by loss of energy, for example by radiation; while conversely, the acquisition of energy would increase its mass; both of these effects being of course far too small in ordinary processes to be directly measurable. At the most, it would only be possible to test the theory experimentally in the case of the great loss of energy which takes place in radio-active transformation. But in any case, the law of the conservation of mass becomes one, from the point of view of this theory, with the law of the conservation of energy.

This in itself is so fundamental a result, that the scientific philosopher cannot overlook it. But the general theory of relativity enlarges this result in an essential respect. Let us recall what was discussed on page 53. We saw there, that even if we were enclosed in an opaque box, we could detect an acceleration of this box and its whole contents by means of effects which would appear in its interior, for example, by a bowl of water, and we saw that the effects of so-called centrifugal forces can in the same way be used to make an absolute determination of rotation. Both of these phenomena are explained by classical mechanics as due to inertia, which causes apparent forces to come into play. But it is now clear that our box would exhibit precisely the same phenomena, if instead of being accelerated, it were placed in a stationary gravitational field of suitable strength, and that we in its interior, therefore, could not decide whether acceleration and hence the effects of inertia caused say the rise of the water at the edge of the bowl, or whether this was due to attraction by a body of sufficient size. For inertia and gravitation are exactly proportional to one another for all bodies, so that the phenomena would be exactly the same in the two cases.

Einstein's theory turns this proportionality into identity, and at the same time takes up again the idea of Mach to which we then referred, and which at first appears so monstrous, namely that the law of inertia does actually depend upon the fixed stars. For Einstein's theory leads to the conclusion that, in the metric field of sufficiently distant masses, sufficiently great in number and distributed quite at random, the ordinary kinematics holds for velocities which are not too great. But instead of being a fundamental law, this simple case is only the limiting case of a state of affairs in general much more complicated, which can only be mastered by a mathematician familiar with all the methods necessary. In the same way, Newton's gravitation law also appears in the general theory of relativity as a limiting case, which is true as a first approximation for masses sufficiently far apart.

The differences between the result of the more exact Einstein theory and the classical theory afford a number of possibilities for experimental tests of the new theory, three of which have become especially famous: the rotation of the perihelion of the inner planets, the deviation of the light from the fixed stars in passing by the sun, and the displacement of the spectral lines towards the red end under the influence of the gravitational field of the stars. Einstein's theory gives us a second approximation, that the planetary orbits are not exact Kepler ellipses, but ellipses which slowly rotate in the course of long periods of time. Now this so-called rotation of the perihelion has been long known to astronomers in the case of the planet Mercury.[93] Newcomb found its amount to be 43" per century. This is exactly the amount calculated from Einstein's theory. But this success, which was at first so astonishing, has lately become doubtful, since Newcomb's value appears to be somewhat large. A final decision will only be possible in some years time, when the amount of the rotation has been redetermined with better appliances than Newcomb had at his command.

The second consequence, the deviation of light rays when passing the sun, which should amount to 1.7" for rays passing close to the sun's edge, was determined with certainty by the two English eclipse expeditions of 1919 and 1922.[94] The opponents of the relativity theory, however, here point to the fact that older absolute theories also lead to this consequence, if we ascribe to light, as to all energy, inertia, and hence weight.[95] A formula for this deviation was actually given by Soldner in 1809, but here the numerical factor but only half that given by Einstein. The first eclipse observations after Einstein's theory was published appeared to confirm his value very closely, but the most recent result is 50% too large. This is still unexplained.

Finally, the red displacement forms the most direct proof for the Einstein theory; for the opponents of the relativity theory have hitherto been unable to find any other explanation of this effect, as opposed to the other two. Unfortunately the results hitherto obtained here are not completely certain. The results of Freundlich, Grebe, Bachem, and others,[96] which depend upon statistical investigations, have been met by objections hitherto not completely refuted, and a remarkable observation made some time ago on the companion star of Sirius, which had at first appeared to be a brilliant confirmation of the Einstein effect, later received, at least it would so appear to-day, a quite different explanation.[97] We are thus still without an absolutely unquestionable experimental proof of the general relativity theory, and it is more the enormous daring of its construction, and its wide range, which attract the majority of theoretical physicists,

Ms

than the existence of experimental results which only admit of another interpretation with difficulty, or not at all.

But in this connection we should not omit to mention, that the opposition to the whole of the relativity theory has quite recently reappeared several times. The American physicist Miller, who some time ago repeated, with Michelson, the latter's famous experiment, and by obtaining a negative result, supplied one of the strongest supports for relativity, has recently repeated the experiment on higher mountains, and believes that after all a positive result is obtained, in other words, the proof of an 'ether wind'.[98] He has therefore ceased to be a supporter of the Einstein theory and become an opponent of it. These experiments of Miller have not, however, remained unchallenged. A repetition of the experiment by Piccard and Stahel in a balloon, and also other ether wind experiments by Tomaschek on the Jungfrau saddle, again gave negative results as before, so that we cannot reject the supposition that Miller's results, which are not in good quantitative agreement with the requirements of an absolute theory, may have been caused by sources of error still unknown.

On the other hand, we must not overlook in this connection another recent investigation carried out by the Berlin astronomer, Courvoisier, which had for its object nothing less than a direct proof of the Lorentz contraction of the earth resulting from its motion relatively to the ether.[99] Courvoisier argued that, if such a contraction really exists, the earth must be shortened by it in the direction of the momentary motion of the earth in space (relatively to the fixed stars and systems of stars) and hence gravitation must be subject to periodic changes. He now believes that he has actually proved the existence of such a half-day period of gravitation in various different ways, and calculates from it, as well as from other data following from the Lorentz contraction of the earth, a motion of the earth (after deducting the known motion) of about 600 to 700 kilometres per second, approximately in the direction of Capella. He was led to quite similar results by the exact measurement of the angles of incidence and reflection of light reflected at a very carefully arranged mercury mirror. According to the absolute theory, small differences should appear in this case, and Courvoisier believes that he has actually found them. If his results should be confirmed, it would be necessary to regard the whole relativity theory as a castle in the air, since its most important foundation, the experiment of Michelson, would likewise then be referred to the Lorentz contraction, and the necessity of making use of more far-reaching hypotheses would cease to exist. For the present, it is true, very strong doubt is felt by the most qualified physicists regarding the reliability of Courvoisier's methods.

While on the one hand, the controversy regarding the theory itself continues, numerous eminent physicists have, on the other, attempted to extend it further, and above all to solve the problem left open by Einstein, of including the theory of the electro-magnetic field in general 'world geometry.' A first attempt of this kind was made shortly after 1915 by Weyl, by casting doubt on a further assumption which had hitherto been made implicitly in physics; the possibility of transferring a length from one position to another. He showed that, on this basis, a system of very general field equations can be set up, from which the mechanical fundamental equations of Einstein on the one hand, and Maxwell's laws on the other, are deducible as special cases. The objection raised to this theory, which was further extended by Hilbert, had not yet been overcome, but Einstein himself has quite recently developed a theory of the same kind on the basis of a similar generalisation of a fundamental assumption, and this theory appears, as far as we know at present, to have better prospects of success.

We have every reason to believe that, sooner or later, a single theory will be found capable of including all field action, that is, the transference of all effects between the parts of the world, both in space and time. Indeed, the possibility of such a theory is a necessary consequence of the fact that the electro-magnetic constant c (velocity of light) also enters into the equations of Einsteinian mechanics. On this basis, electro-magnetic mass will naturally become identical with mechanical inertia (and so again with weight). These two properties, inertia and weight, which are fundamental qualities of all material things, will then both appear as field phenomena, and we shall then have to take up the task of deducing the other properties of matter from the general field theory which we are assuming. Such attempts have already been made several times, that of Mie having been most discussed.

None of these theories has yet succeeded in achieving the main point, namely a deduction of the fact that matter only occurs in the form of those quite definite smallest quantities which we designate as atoms, or electrons and protons. We shall see later that this fundamental fact is unquestionably connected with another fundamental fact, namely that energy itself, even that of the field, is divided into 'quanta.' But so far neither the one fact nor the other has been deduced from Einstein's or any other extended field law. Let us bear in mind that here a very deep problem still awaits solution, and proceed to consider how far physics has gone in the investigation of this quantum structure of the world. A great part of the latter problem has already been dealt with by what was said in earlier chapters concerning the investigation of atoms. But all that was said there only referred to the material atoms themselves, whether they appear to us as chemical atoms, or as charged ions, or electrons and protons.

We must now turn to the phenomena which depend upon the interaction of these material corpuscles and the field, in particular the electro-magnetic wave-field, light radiation. As already remarked on page 119, the first question is that of the generation of light in the interior of matter, and also of its transformation there into other material forms of energy there; that is, its emission from, and absorption by, matter.

XII

THE PHENOMENA OF
LIGHT EMISSION AND ABSORPTION

The word 'light' will now be taken in the widest sense, and will include invisible heat radiation (ultra-red), ultra-violet, X-rays, and 'cosmic' radiation. In other words, we are to deal quite generally with the transference of energy from matter, in which it is present as heat (molecular motion) or electric current, to the field, where it appears as radiation, and conversely.

The light emitted by solid or liquid incandescent bodies gives us, as is well known, a continuous spectrum, that is light of all possible wave lengths, when decomposed by a prism. As opposed to this, an incandescent gas (see figure 28) gives what is called a line spectrum, that is, it only emits light of single, quite definite wave lengths and not of all possible wave lengths, although the number of lines (as, for example, in the case of the spectrum of iron) may be very great indeed. This distribution of lines is a peculiar characteristic of chemical elements, which they also usually retain in all their compounds. Along with the characteristic emission spectra, we have the absorption spectra. If light of all colours, for example from an arc lamp, is passed through a substance, this also extinguishes certain parts of the spectrum, either quite broad parts or single definite colours, so that we then see, in the continuous spectrum of the original source of light, broader or narrower black strips. This is true for all substances, even those which are colourless and transparent, their absorption regions chancing to lie only in the invisible part of the spectrum, and not in the visible part. Incandescent gases, again, also exhibit the characteristic property that their absorption lines coincide exactly with their emission lines, that is, that they absorb most strongly precisely the light which they themselves emit (Kirchoff and Bunsen).

The explanation of this fact, which at first sight appears somewhat paradoxical, is quite simple. We need only assume that structures capable of vibration are present in the atom, which, just like a tuning fork, send out on the one hand waves of a quite definite length, and are able on the other hand to absorb energy of the same frequency from waves falling on them. This is the well known phenomenon of resonance in acoustics. The behaviour of incandescent gases thus proves that the oscillatory mechanisms in their atoms (oscillators, resonators) are tuned to a quite definite frequency, while we must assume the existence in solid and liquid bodies of resonators of all possible frequencies. If we also keep in mind what was said above concerning the enlargement of the concepts of oscillation and wave, we shall be on our guard against a hasty, and all too mechanical, interpretation of this concept.

These quite general assumptions are already sufficient, as appears when they are worked out exactly and quantitatively, to render comprehensible part of the interactions between light and matter; above all, the phenomenon of the separation of colours (dispersion), and its connection with absorption, can be more or less explained.[100] But this is obviously not sufficient, for we must in the first place explain why quite definite wave lengths are emitted by glowing gases, and why on the other hand solid and liquid bodies give a continuous spectrum, and why also in the latter, the energy is distributed among the wave lengths according to a definite law; and above all, how the emission of electromagnetic waves comes about. For the fact that radiation is of this nature is now proved, and was already settled hypothetically in Maxwell's time.

The arrival of the electron hypothesis appeared to solve the latter question straight away. If the atom consists of electrically charged sub-divisions in relative motion, say the positive nucleus surrounded by negative electrons in the manner of a planetary system, as Rutherford's hypothesis suggested, the emission of electro-magnetic waves appears to be given at once. For a periodically rotating electron naturally means a periodical change of field, and this would then be propagated in space, according to Maxwell, as a wave of corresponding frequency, with the velocity c. But this would mean the gradual diminution of the energy of the electron if it were not continually replaced from outside, and the path of the electron would then become smaller and smaller, the frequency continually increase, and the electron finally fall into the nucleus. At this point our picture obviously fails, for the frequency in fact remains exactly constant, and this is a serious objection, which was raised against the theory from the start.

Nevertheless, it appeared for a time as if it might be made to solve the riddle of spectroscopy, since by its aid H. A. Lorentz, the originator of the electron theory, succeeded in predicting a completely new phenomonon with quantitative accuracy, and this was then discovered by his pupil Zeemann in 1897: one of the most memorable triumphs of theoretical physics. Lorentz said to himself, that the revolving electron must be influenced by the action of an external magnetic or electric field in its motion, in exactly the same way in which an electron moving in a cathode ray is influenced. More careful consideration showed that an influence of this kind must be easier to observe in the case of a magnetic field than of an electric field (the same being true of the cathode ray tube), and that the result would consist in a change of frequency of the spectral lines. Calculation shows[101] that, in place of the light of one colour emitted by the particular electron, light of three somewhat different wave lengths will be emitted, that of the original wave length and one somewhat shorter and one somewhat longer; and that these three parts must be at the same time polarised in a certain way.

The proof that this is really the case was given a year later by Zeemann. Figure 27 (opp. page 170) shows the normal Zeemann effect. From the characteristics of the polarised light, and the amount of the difference in the wave lengths, we can show by calculation that the particles emitting the light have a negative charge, and that the ratio e/m in their case is almost exactly the same as in the cathode rays. A more beautiful indirect confirmation is hardly imaginable, and it is not surprising that a large number of physicists soon devoted themselves to the investigation of this phenomenon, which leads us so deep into the secrets of the atomic world. It soon appeared that the simple case first calculated by Lorentz only occurs rarely. The effect of the magnetic field is usually to give division into more than three components, and as many as 19 have been observed. In spite of this want of agreement, the numerical agreement found in simple cases cannot possibly be regarded as accidental. But the explanation is obviously much more complicated in most cases than assume at first.

This became quite evident when further investigation, encouraged by this first great success, penetrated more deeply into the secrets of electro-and magneto-optics.[101]   A whole series of further discoveries were made which appeared to confirm the theory.   Thus shortly after Zeemann, Voigt observed the reversal of the effect, the magnetic splitting up of absorption lines, and a few years later, the double refraction of glowing gases in a magnetic field (for light close to a spectral line) was observed by Macaluso and Corbino, as predicted by Voigt.   Finally, Stark succeeded in 1914 in observing the long-sought-for effect of an electrostatic field upon spectral lines.   In this connection, further long known facts found an explanation, among these the so called Faraday effect, the magnetic rotation of the plane of polarisation, already observed by Faraday in 1849 (a gas or vapour becomes, like quartz, optically active, that is, rotates the plane of polarisation, when it is brought into a strong magnetic field).   But

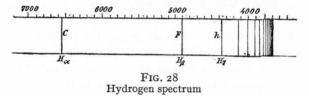

FIG. 28
Hydrogen spectrum

repeatedly, and particularly in the investigation of the Stark effect, it was found that the theory led to quantitative results which are quite different in many respects from actual observation.   Something was therefore obviously wrong with it; quite apart from the original, fundamental difficulty, that it was unable to explain the first and most fundamental fact of spectroscopy, the emission of single quite definite colours.

Altogether, spectroscopy had exhibited, for decades, the spectacle of a branch of science in which theoretical explanation did not keep pace with experimental results.   The extraordinary fineness and exactness of measurement in this field is quite on a par with the proverbial accuracy of astronomy, or even exceeds it.   Most spectroscopical data are accurate to at least six figures, and many even to seven or eight.   It is therefore not surprising that, given these innumerable exact measurements, quantitative relationships or laws were now and then determined purely empirically.   The most important of these was that relating to the spectrum of hydrogen, discovered in the year 1885 by the Swiss scientist Balmer.   The hydrogen spectrum (figure 28) is particularly simple in construction, the frequencies of its lines may be determined according to Balmer with great accuracy by means of the formula $R\left(\frac{1}{4}-\frac{1}{n^2}\right)$, where R is a constant which has the numerical value 3,291 billions, and n is to be given the successive values 3, 4, 5, 6. . . . Hence $R\left(\frac{1}{4}-\frac{1}{9}\right)$, $R\left(\frac{1}{4}-\frac{1}{16}\right)$, $R\left(\frac{1}{4}-\frac{1}{25}\right)$ etc. are the frequencies of these lines.

Two other similar formulae hold for other spectral series, and it was further found that chemically similar elements from the same column of the periodic system also give similar series, only with other values of the constants, which themselves stand in simple relationship to the above mentioned constant R. The chief credit in discovering these relationships is to be ascribed to a Swede, Rydberg, and they are therefore called 'Rydberg constants' in his honour. The investigation of the Zeemann effect then immediately gave the remarkable result that the lines of one series all show the same Zeemann effect (being therefore divided, all of them, into 5 or 7 components, for example) and that as a rule, only isolated lines belonging to no series give the simplest type of result, the so-called triplet (figure 27, opp. page 170).

This supplies us with a very simple and frequently applied means of discovering series, a task which is by no means easy in the case of spectra with many thousands of lines – iron for example. In the latter case it has only been accomplished within the last few years. In spite of all this, the main problem, the emission of single lines, remained altogether unsolved. Why hydrogen gives just one particular series, the alkali metals just another set of series; how this peculiarly complicated phenomenon of the series comes about, reminding us of the acoustic relationship of fundamental note and overtones, but still being so entirely different; all these questions could not be answered.

Quite recently, X-ray spectroscopy has been developed, in addition to the ordinary or optical spectroscopy of which we have just given a short sketch, and we will now devote a few words to it. As already shortly remarked on pages 21 and 127, the wave nature of the X-rays, and at the same time the lattice structure of crystals, was proved by the famous experiments of von Laue. The illustration (figure 29, opp. page 170) shows a diffraction phenomenon which appears when a narrow beam of X-rays is sent through a regular crystal (rock salt) parallel to the axis of a cube. The position of the spots immediately allows the quadratic symmetry to be recognised (figure 1, page 15). We owe, chiefly to the two Braggs, but also to Debye and Scherer, further methods which, on the one hand allow of very exact separation of the X-rays according to their wave length and of the determination of the latter, but on the other hand also, the elucidation of the actual structure of the crystal lattice. We shall not go further into the latter here, although results of very considerable theoretical importance have been obtained.

The spectrum analysis of the X-rays[102] by the above methods has yielded the following general results. The X-rays are produced by cathode rays of great velocity meeting an 'anti-cathode' (usually platinum or tungsten). The electrons are here slowed down, and the resulting disturbance of the electro-magnetic field is propagated into space. At the same time, however, the electrons contained in the atoms of the anti-cathode are themselves made to move, with the result that they also send out radiation. The continuous spectrum produced by the slowing of the electrons, containing all possible wave lengths, is therefore accompanied by a characteristic radiation peculiar to the anti-cathode material, and this radiation can be separated comparatively easily from the other. It forms the so-called X-ray spectrum of the substance in question (for example, platinum, aluminium, etc.). This exhibits, as contrasted with the extreme complexity of optical spectra, highly simple laws, the discovery of which we mainly owe to the English physicist Moseley, who fell at the Dardanelles in 1916.

In the first place, the X-ray spectra are extremely simple in construction as compared with the bewildering abundance of optical spectra; they only consist of two or three groups of lines. In the second place, these three groups of lines, which are described as the K-, L-, and M- groups, are the same for all elements, with the restriction, however, that in the case of the lower elements from lithium to zinc, only the K-group is known, in the middle elements the K- and L- groups, in still higher elements only the latter, the very highest giving in addition the M-group. These three groups are displaced in the direction of short wave lengths quite regularly as the atomic weight increases. In accordance with this the K-group has the shortest wave length, the M-group

PLATE III

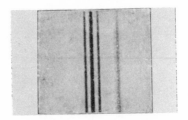

FIG. 27. Normal Zeemann Triplet (the feeble line
on the right is not part of the triplet)

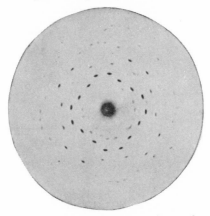

FIG. 29. Laue Diffraction Figure, using rock salt plate

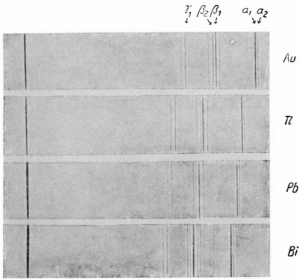

FIG. 30. L–series of the elements Gold, Thällium, Lead,
and Bismuth. (Siegbahn)

the longest.   The K-spectrum has four to six lines, the L-spectrum about fourteen, the M-spectrum six lines (figure 30, opp. page 170). The regular displacement of the lines with increasing atomic number is regulated by Moseley's law, which says that the square root of the frequency of each single line is a linear function of the atomic number Z $[\sqrt{\nu}=k(Z-a)]$.

The discovery of this extremely simple law was of first class importance for our whole knowledge of the periodic system. In the first place, every missing element is immediately indicated by a gap in the series of points determined by the law.   In figure 30, which shows the L-series for the elements gold, thallium, lead, and bismuth, we immediately recognise that an element (mercury) is missing between gold and thallium, since the amount of displacement between these two is twice as great as between the others.   The irregularities in the periodic system are cleared up completely in this way.   The arrangement required by the system, which does not quite correspond to the atomic weights, is actually that of Moseley's law (K, A; Te, I; Co, Ni).   We can now say exactly what elements are still missing.   Since the discovery of the two higher homologues of manganese (masurium, 43, and rhenium, 75) by Noddack and Tacke, and of illinium, 61, by Hopkins, only two are missing, namely No. 85, a halogen, and No. 87, an alkali metal.   These two may possibly be so strongly radio-active, that they cannot be obtained in measurable amounts, owing to their decomposing again as soon as formed.   But their discovery has already been announced from America, though not yet confirmed.

Moseley's law can be explained by the Rutherford atomic model, as soon as we assume that the emission of X-rays is due to those electrons of the system which are nearer to the nucleus. These most likely form, as we have seen, closed concentric 'shells,' of which the first probably contains two, and the second and third each eight electrons, this bringing the structure of the system as far as argon.   How it goes further is not quite certain, but the X-ray results appear to show that the addition of further electrons always takes place on the outside, so that the internal and already completed shells remain unaltered, only that the electrons belonging to them are more firmly bound with every further unit added to the positive charge of the nucleus.   If the emission of X-rays is effected by these internal electrons, it is clear that a system which remains similar will on the whole emit similar waves, and that these will have increasing frequencies in accordance with the increasing force of attraction as we go from element to element.

We have already mentioned that the external electrons are the 'valency' electrons of chemistry, the arrangement of which can easily be disturbed by external action, and that these same external electrons are also the agents of the optical spectra, which therefore not only have much longer wave lengths, on account of the weaker attraction, but also much greater complexity. Many peculiarities of spectra are thus excellently explained by this model. But the one most important point is still wanting, the explanation of the fact that single, quite definite waves are emitted.

We now come to this question, which has become fundamental for the whole of physics, and which has to-day been brought near to its final solution, though not completely solved, by two investigators of genius. These two investigators, whose names will be remembered in the history of our times alongside Rutherford and Einstein, are Planck and Bohr. The first decisive step was the foundation of Planck's quantum theory.

XIII

THE QUANTUM THEORY

This new fundamental theory of modern physics did not arise out of the problem of line spectra, but of the continuous spectrum of glowing solid and liquid bodies; in the first instance that of the so-called absolutely black body, that is, one which absorbs completely all radiation falling upon it (lamp black is approximately such a body). As already mentioned above, we are dealing with the conversion of electro-magnetic energy of radiation into molecular movement, and vice versa. We assume that oscillators (resonators) are present in the molecules or atoms, and bring about this conversion. In glowing solid or liquid matter, we have oscillators of this kind having all possible frequencies. The question is the following: how is the energy divided among the different oscillators, and how does this distribution change as the temperature rises?

It is well known that a body when gradually heated first only emits dark (ultra-red) heat rays, then red light gradually becomes visible, then yellow, green and so on are added, until we finally arrive at a white heat. If an apparatus sensitive to the reception of all wave lengths is used to measure the amount of energy at different parts of the spectrum (page 116), we obtain a curve of distribution of energy (figure 31), the highest point of which at lower temperatures lies far in the ultra-red; as the temperature rises, the whole curve also rises, while at the same time, the maximum is displaced in the direction of shorter wave lengths, and finally reaches the visible spectrum. In the case of the sun's light, it lies in the middle of this. The problem consisted in finding a means of deducing this curve theoretically. This attempt resulted in a contradiction with the theory of heat hitherto assumed. According to this, all modes of motion of the molecules, 'degrees of freedom' as they are called, were equally concerned in absorption and radiation of energy, and hence 'equipartition,' or equal distribution, of energy among all degrees of freedom was assumed to exist (pages 91 ff.).

On the basis of this assumption, as was first shown by Rayleigh in 1900, the radiation curve of energy distribution can be calculated theoretically with the help only of the laws of energy and entropy, but the result does not agree with experience. For according to it, the energy of the continuous spectrum should increase steadily towards the violet end, whereas, as we have already said, it always reaches a maximum at a certain wave length, and after that falls away very quickly. Planck's fundamental idea consisted in abolishing the assumption of the uniform distribution of energy, or what amounts to the same thing, in no longer assuming that energy is given up by the atoms to the electro-magnetic field, and vice versa, in a continuous manner, in quantities of any degree of smallness. According to Planck this transference of energy can only take place in multiples of a certain minimum 'quantum,' which however is different for every frequency, being directly proportional to the frequency. If this frequency is called v, we have in the equation

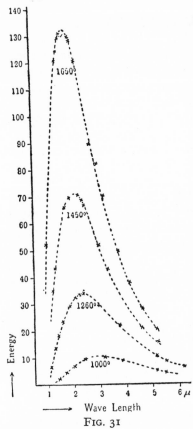

FIG. 31

Planck's radiation curve. The crosses give the observations of Lummer and Pringsheim

$$E = h\,v$$

the expression of this assumption, v being a general or universal constant, the value of which can be calculated from measurements of radiation. According to the best recent determinations, this value is 6.53×10^{-27} erg-seconds. Since v is the reciprocal of a time (the time of oscillation), E/v is a product of energy and time, which is called in mechanics, as already mentioned on page 86, 'action.' h is usually called *Planck's quantum of action.*

Furthermore, radiation measurements also give us the Loschmidt number, since the constants of Planck's radiation formula depend upon it and h (together with c), in a simple manner.[103]

While Planck in 1900 had, at first, no other object in making this new fundamental assumption than that of solving the problem of energy in the continuous spectrum, it very soon appeared that this assumption, and the equation $E = h . v$ were keys to a number of hitherto closed doors. The largest share of credit for this enlargement of the quantum theory again falls to Einstein. He first recognised that we have here the solution of certain problems in the kinetic theory of heat, which the latter cannot deal with in its classical form, though for want of space we are not able to discuss them here.[104] But more important than this was the application of the theory by Einstein to the photo-electric effect, to which we must devote a few words on account of its other and more distant consequences. This term is used to denote the fact that clean metallic surfaces, when illuminated with ultraviolet and X-rays, emit electrons, that is to say, cathode rays.

The phenomenon itself had already been observed in 1888 by Hallwachs, and more exactly investigated by Lenard, Elster, and Geitel, who constructed the most suitable apparatus for the purpose, consisting of an exhausted glass bulb, containing a liquid alloy of potassium and sodium. The quantitative investigation of the photocathode rays gave velocities up to 1/300th of the velocity of light, and for e/m the same value as usual. Now this in itself would not be further remarkable; but the curious part is the quantitative relation, namely the law, according to which the energy of motion of the electrons emitted depends upon the quality of the incident light. We should at first expect it would be intenser the more intense the light. This is not the case. Increase in the intensity of the light only increases the number of electrons emitted but not their velocity, or kinetic energy. This only depends upon the frequency of the light. It becomes the greater, the greater the frequency or the shorter the wave length of the light, the form of the law being the simplest imaginable. If E is the energy of motion of the electrons, n the frequency of the light, h the Planck constant, we have

$$E = - E_0 + h v$$

in which equation the constant E_0 represents the work of

release, that is, the work necessary to separate an electron from atomic attraction, or from the metallic surface. The energy of motion together with this work $(E + E_0)$ is therefore the total energy which the light gives up to the electron, or in other words, the total energy is here also equal to $h. v.$

This equation further holds, as Einstein showed later, for photochemical processes.* In the decomposition of bromide of silver in the photographic plate, a quantity of energy $h v$ is taken from the light for each atom of silver separating (the photochemical equivalence law). In short, it appears as far as we have been able to observe, the equation $E = h v$ holds for all interaction between matter and field (radiation). The investigations mentioned are in part of fairly recent date. We now come to the most important and fundamental of all the theoretical developments which have sprung from Planck's quantum theory, the solution of the riddle of spectroscopy by Bohr.

* This has recently been disputed by other investigators ; for instance L Cramer.

XIV

BOHR'S THEORY OF SPECTRA

In the year 1912, the Danish physicist, Nils Bohr, a pupil of Rutherford, and at that time 26 years of age, came upon the brilliant idea of connecting Rutherford's atomic model with Planck's theory of quanta, in order to find a solution of line spectra. The original electron theory assumes, as we have seen, that the frequencies of the emitted waves must obviously be identical with the periodic times of the electrons revolving in the

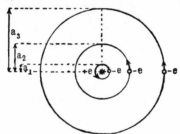

FIG. 32

Quantum orbits of hydrogen atom

atoms. After Lorentz had succeeded in predicting the Zeemann effect on this basis, no one had any doubt that, at least in this respect, the Rutherford-Lorentz model was correct. But an explanation of the effect of line spectra, could not, as we have seen, be given in this way, since the electron could rotate like a planet around its sun in any orbit whatever and thus assume any frequency.

Bohr's fundamental idea was that, among these infinitely numerous possible orbits, only those are allowable for which the Planck quantum conditions hold, the action calculated for such an orbit being a simple multiple of the universal quantum of action h. If this so called 'Bohr's first quantum condition' is applied we get – first for the simple case of circular orbits – that only certain simple orbits are possible, the radius of which increases according to the square of the quantum number (one-quantum, two-quantum . . . orbit). These are represented in figure 32 for the simplest case, the hydrogen atom, which

Ns   177

according to Rutherford consists of only one proton and one electron. The radius a of the inmost, one-quantum orbit measures 0.0532 $\mu\mu$. But since the frequencies of revolution thus calculated are thus in no way identical with the hydrogen lines, Bohr made the further assumption that the electron when revolving in such a quantum orbit does not radiate, but that when it passes from one quantum orbit to another, the energy set free is transformed into radiation again according to the formula $E = h\nu$ (Second Bohr condition, figure 33). If we now use the values of e, n, and h,

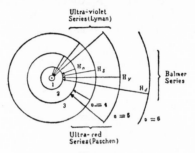

Fig. 33

Radiation of H atom according to Bohr. The radii of the circles are drawn arbitrarily

derived from cathode ray experiments and Planck's radiation formula, to calculate out these assumptions, we do not merely get Balmer's law of series, but also, very approximately, the actual value of the Rydberg constant,[105] and the agreement becomes practically perfect when, in accordance with Sommerfeld's development of the Bohr theory, we assume Kepler ellipses for the orbits, instead of circles, and apply the relativity theory in place of ordinary mechanics.[105]

This way finally leads us to the so-called fine structure of the spectrum lines, which could not be deduced from the simple fundamental assumptions of Bohr. (Most spectrum lines turn out to be multiple when greatly magnified.) The achievements based upon Bohr's discovery and developed in quite a short time (ten years) are truly remarkable, and we can hardly find anything to equal it in the whole history of physics. The long needed 'Newton of spectroscopy' had appeared at last. The magic formula which he used to open the enchanted castle, was again $E = h\nu$. Anyone who cannot understand from such results that the most important part of physics consists in the hypotheses (naturally the good ones) is beyond all hope.

Nevertheless the sceptic has a certain right to make use of this case also as one of the proofs of his thesis, that the finally important matter is only the quantitative results which we determine and not the pictures or models of which we make use to help us to find them. The Bohr model has not escaped the fate of suffering many and fundamental subsequent corrections. This was to be expected from the start, since it had certain unacceptable peculiarities. In the first place, the assumption that the revolving electron does not radiate is in direct contradiction to Maxwell's electro-dynamics, as we have shown above on page 166. In the second place, we cannot see by any manner of means how the wave emitted really gets its frequency as calculated by the formula $E = h\nu$. This frequency must surely be first produced in the interior of the atom. If the electron is not to produce it by its revolution, what does produce it? Or how otherwise does the surrounding 'ether' or field come to pick out just this particular frequency which suits the formula, since the electron merely delivers the necessary energy?

It has been said, not without justice, that this second assumption of Bohr's was still more of a 'pistol shot' than the first, and if this bold hypothesis had not been justified by fabulous success, no one would have taken it seriously. Furthermore, this new basis makes the explanation of the Zeemann effect considerably more difficult, and entirely without any mental picture. And finally – and this is the main point – although the riddle of the line spectra is referred by this theory to the quantisation of the electron orbits in the atom, the question at the back still remains: What really causes this quantisation? Why must the action of the revolving electron be an exact multiple of h?

These and still other questions immediately arose as soon as Bohr's model became known, but found for the time no solution. Nevertheless, research naturally proceeded on the basis of the Bohr model, since it proved to be fruitful in the discovery of new knowledge; and Bohr in particular attempted to throw new light on the periodic system by means of his theory. He also succeeded in this endeavour to a certain extent. The increase in size of the periods from eight to eighteen elements at higher atomic weights, by the introduction of the elements scandium to

niobium, and yttrium to palladium, and the further introduction
of the 'rare earths,' numbers 58 to 70 (see above, page 7) may,
as Bohr showed in 1921,[106] be explained fairly plausibly by the
assumption of a gradual building up of higher atoms according
to the quantum laws.

Bohr was able to put the crown on this theoretical structure by
using it to determine the element No. 72, hitherto regarded as one
of the rare earths, as a higher homologue of zirconium, whereupon it was immediately found by Bronsted and von Hevesy in
zirconium minerals. It was therefore called, in honour of Bohr's
home Copenhagen, hafnium. But even this discovery did not
make matters quite satisfactory, since a considerable number of
disagreements had been found in spectroscopy itself, while the
agreement as regards the periodic system was never just what it
should have been. Many physicists – and this again was entirely
in the sense of positivism – found a kind of relief in the fact that
the Bohr model gradually began to dissolve, as regards its more
directly picturable features, as a result of quite other considerations and experiments.

It is again Einstein's achievement to have clearly grasped the
fundamental problems, and this is perhaps the greatest of the
numerous services to science of this incomparable genius, who
is still only 50 years old. Planck's theory of quanta dealt, as we
have seen, with the transference of the energy of the field to
matter, and conversely. Planck himself, and with him most
physicists, were originally of the opinion that the inner reason
of this quantisation must be sought in the arrangement of the
interior of the atom. It is quite easy to imagine mechanisms
having the property of only taking up and retaining energy when
it exceeds a certain limiting value. This property is characteristic of a cross-bow or a spring pistol. Even if the atoms are thus
arranged internally, we could still keep the energy in the field
itself unquantised, and thus continue to retain the old continuity
of classical wave theory.

As opposed to this, Einstein took from the beginning the view,
based on his explanation of the photo-electric effect, that the
energy in radiation itself, in the field therefore, is to be regarded
as divided into quanta. This means, as it is easy to see, more or
less of a return to Newton's corpuscular theory of light. The
most curious part of the matter is, that the determination of the

velocity of light in water or other media (see page 104), previously regarded as decisive against Newton's theory, no longer carries this interpretation on the basis of Einstein's new mechanics. A light quantum has the value $h\nu$, and hence according to the Einstein mass-energy formula (see above) the mass $h\nu/c^2$, and in empty space the impulse $h\nu/c^2 \times c = h\nu/c$ (see page 59). If it passes through the boundary between two media, whereby the boundary forces do work upon it, its impulse increases, but this does not mean, as was hitherto unsuspectingly assumed (also by Newton) on the basis of the formula 'force equal to mass times acceleration,' that the velocity must also increase correspondingly (which is contradicted by experience). On the contrary, an increase of $h\nu/c$ can only take place as long as h and ν remain unaltered by decrease in the denominator, the velocity therefore decreases, varying inversely as the impulse, and so giving us the same results as the Huygens wave theory. The refractive index becomes equal to the direct ratio of the velocities (compare page 104). The paradox of the result disappears when we consider that the work done nevertheless reappears in the increase in mass.[107] Hence the Fizeau experiment is no longer opposed to the corpuscular theory.

While, regarded from this point of view, the Einstein assumption is quite possible, the photo-electric effect is explained by it in the simplest possible manner, whereas the continuous wave theory is quite powerless to explain it. In this effect, the energy $h\nu$ of the light quantum is simply transferred, according to Einstein, by a kind of collision to the electron, the energy of which consequently increases in direct proportion to the frequency, while the intensity of the light, which is again settled by the number of light quanta arriving, obviously determines only the number of electrons emitted. (According to the wave theory, on the contrary, the energy of the electrons must correspond to the intensity of the light.) In exactly the same way, certain phenomena connected with X-ray spectra are readily explained, and we even acquire in this way a number of new methods for the accurate determination of h,[108] these methods being superior to the determination by means of the radiation formula.

A further success was the deduction of the latter formula from the assumption of the existence of free light quanta without the use of material resonators, simply on the basis of the notion, that the light quanta $h\nu$, existing in otherwise empty space, move and collide with one another like gas molecules (Bose, Einstein).

In spite of all these successes of the corpuscular theory of light, most physicists could not bring themselves to accept it unreservedly, Planck being one of these.   They pointed out quite justly that, although the corpuscular theory is excellently adapted to the recently discovered phenomena we have mentioned, it fails completely to include just those phenomena most successfully explained by the old wave theory, namely interference and diffraction, and that it further threatens to destroy the connection which undoubtedly exists between light and electric waves of any length down to the fields of ordinary alternating current and mechanically moved electrostatic charges.

The problem is the more mysterious since the latest experiments[109] have rendered it certain that interference takes place at such incredibly low intensity of light, that a light quantum would occupy many cubic metres if the wave energy were distributed in space according to the old theory.   Interference has actually been found in the case of experiments in which only a single light quantum passed through the whole apparatus.[109]   How this is to be reconciled with the corpuscular theory of light, it is impossible to see at present.

But on the other hand, phenomena have been found of recent years which appear to afford a direct proof of the corpuscular theory, the most important of these being what is called the Compton effect.   The English physicist, A. H. Compton, investigated in the year 1922 the process of scattering of radiation, known to everyone in the production of ordinary 'diffused daylight.' According to the wave theory, this process is caused by the incoming waves setting in oscillation some kind of resonating structures in matter, and these structures then radiate in all directions waves having the same frequency.   But when he investigated the scattering of X-rays, Compton found the curious peculiarity that the diffused or secondary wave length is not identical with the incoming wave length.   We see on the photographs of figure 34, (opp. page 186) alongside the original lines, quite plainly another line of somewhat greater wave length, the difference between the two wave lengths depending on the angle of diffusion, but being constant for all X-rays for the same angle. This result cannot possibly be explained by means of the wave theory, but Compton himself, together with the German physicist Debye, soon succeeded in explaining it by means of Einstein's light quantum theory, and in deducing it with quantitative accuracy. According to this view, the process of scattering consists simply in a kind of elastic collision between light quanta and electrons loosely bound in the atom, a collision quite analogous to that between two billiard balls.

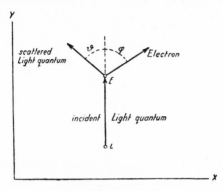

FIG. 35
Explanation of Compton effect

As figure 35 shows, the electron will then move in some oblique direction, and the scattered light quantum, the energy and impulse of which have naturally decreased, will move in another direction. If these two assumptions are calculated out by the simple laws of elastic collision, which have been known since Huygens' time (energy and impulse being given by the usual formulae), we actually arrive exactly at the result found experimentally.[110]   The wave theory is entirely unable to explain this result.

The Compton effect must be regarded as a direct proof of the corpuscular theory of light (for the present for X-rays), and we actually observe, since the year 1923, that this theory is steadily gaining ground even in the case of physicists such as Planck and Sommerfeld, who had hitherto retained the wave theory as far as possible.   The result is to make this dualism still more unendurable, and it is not surprising that the greatest efforts are being made by physicists to find a compromise between these two views, which are apparently completely opposed to one another. This movement is in full progress to-day, and it might therefore appear rash to attempt to give here the latest phases of the development, since as soon as it is printed, it will probably be out of date. We must nevertheless take up this task if our description is not to remain a torso. It is already clear that recent work on this field has again entered upon a new phase – is it the last or all but the last? – of the problem of light and matter, which we can describe, with reference to its most important new theories, as wave and quantum mechanics. To this development we shall devote the next chapter.

XV

WAVE AND QUANTUM MECHANICS

Unfortunately in this case, as in the theory of relativity, matters are so difficult from a mathematical point of view that in a book designed for the lay public, only the barest outlines of the theory can be given, and the best part of them cannot be clearly set out.

Einstein's bold hypothesis made the light quanta possess a corpuscular structure even in free space, and hence possess, along with their energy $h\nu$, also mass $h\nu/c^2$ and impulse $h\nu/c$; in other words, the two most fundamental properties of matter were also ascribed to light. It was not a long step to ask whether this idea might not be reversed, and whether the particles of matter might not be capable of being regarded as possessing the two most important properties of light, frequency and amplitude ; in other words, whether the desired union between light waves and material corpuscles might not be found by way of a resolution of the latter into the former. The credit of having first put this idea into a useful physical form is due to the French scientist, Louis de Broglie (1924). De Broglie recognised that the problem of the quantisation of the electron paths inside the atoms, which the theory of Bohr could not explain but simply postulated, is immediately solved if we put in place of a material (corpuscular) electron, revolving in a circle, a wave also moving in a circle. For it is immediately obvious, that such a wave can only exist as a permanent process if its wave length is a complete multiple of the path (see figure 36), in every other case the wave would be annihilated by self-interference. If we now calculate, according to Einstein's postulates of energy and impulse, the wave length to be ascribed to an electron having a circular orbit, we are surprised to find that this can only be a proper fraction of the circumference of the orbit, when this is one of Bohr's quantum orbits.

According to Einstein, we have for a light quantum in empty space, as stated above, the impulse $G = h\nu/c = h/\lambda$ (since in empty space $\lambda = c/\nu$).

This equation is first applied hypothetically by De Broglie to any kind of moving material particle, by ascribing fundamentally to every such particle possessing an impulse G a wave length $\lambda = h/G$. On the other hand, the first Bohr quantum condition requires the products of impulse and circumference $(G \times 2a\pi)$ to be a multiple of the Planck quantum, and conversely therefore, the circumference to be a multiple of h/G. But since this, according to De Broglie, is the wave length of the corpuscle, Bohr's condition obviously amounts to the condition that the circumference be a multiple of λ.[111]

According to De Broglie's method of attacking the problem, we can easily calculate the wave lengths for material corpuscles

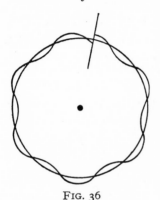

FIG. 36

De Broglie's Wave Theory
of the electron

of a definite velocity; for example, we get for moderately slow cathode rays (about one to ten per cent. of the velocity of light) an equivalent wave length of about 0.1 to 0.01 $\mu\mu$ and for the slower electrons emitted by incandescent filaments (in valve tubes) about 1 $\mu\mu$. These wave lengths lie in the region of X-rays, and the question immediately arises whether these waves of matter can be shown to exist by diffraction at crystal gratings.

The decisive experimental proof of this diffraction, found to give wave lengths exactly as required by De Broglie's theory, we owe to the American scientist, Davisson, who made experiments in 1924 on the scattering of electrons at crystal surfaces, and thereby discovered by chance (for the moment without any idea of matter waves) phenomena having the character of diffraction. After De Broglie's theory became known, he recognised the importance of the result and proceeded, with his co-worker, Germer, to repeat it in 1927 in so improved a form, that the diffraction was proved without any doubt, exactly as in the case of Laue's photographs (see figure 29, opp. page 170). Photographs quite recently obtained in the same way by Kikuchi of electron rays diffracted at thin plates of mica (figure 37, opp. this page),[112] appear at first sight so similar to Laue's diagrams, that one would take them for such, if one did not know that they were produced by corpuscular rays.

PLATE IV

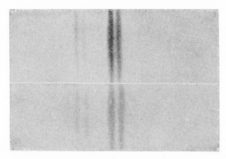

FIG. 34. Compton Effect. Photographs by Kallmann and Mark. Angle of scattering θ, upper figure 72°, lower 90°. The unchanged line on the left

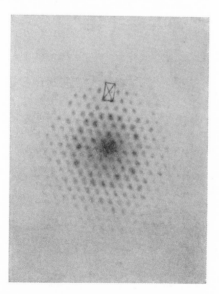

FIG. 37. Diffraction of electron rays by crystal grating. (Kikuchi)

However simple this sequence of ideas may appear, its consequences turn out, on further consideration, to be very complicated. The equation $h/\lambda = mv$ was first of all, according to Einstein, valid for light quanta in empty space. When De Broglie turns it round and uses it to give a material impulse mv a wave length, he states what is for the present an arbitrary hypothesis, the further consequences of which must be tested. In general, according to the relativity equation above discussed, the energy of the mass m has the value mc^2. But this energy, when we replace the mass by an oscillatory process, must further be equal to hv; and hence $m = hv/mc^2$, and we get as consequence of De Broglie's assumption of $\lambda = h/G$, that

$$\lambda = \frac{h}{mv} = \frac{hc^2}{hvv} = \frac{c^2}{vv}; \text{ or } \lambda v v = c^2.$$

Now according to the general theory of waves (see page 107), the product of wave length and frequency λv is always equal to the velocity of the waves. If we put this equal to u, we get that $uv = c^2$; and it follows, at first sight paradoxically, that the velocity of propagation u of matter waves is greater than the velocity of light, since v, as the velocity of a material particle, must, according to the relativity theory, always be smaller than c. In other words, the wave velocity u is in no way identical with the velocity of motion of the material particle, which according to De Broglie is to be replaced by a wave process. How is this possible?

At this point, unfortunately, the mathematical difficulties become almost insuperable for the layman not acquainted with theoretical physics. It is necessary first to introduce at this point the concept of what is called a wave group. By this we understand a collection of waves, the frequencies of which may assume all possible values within certain limits. It is then easy to see, on the basis of the general principles of wave motion, that if all these waves have random phase differences between one another, their superposition will in general always result in a value very slightly different from zero. Within this wave group, however, there always exists at a certain moment a definite spot, at which the phases of all wave components of the group are very nearly in agreement. At this point, which is called the centre of energy of the wave group, there exists a resultant amplitude of vibration which is different from zero. It is then comparatively easy to show mathematically that this spot changes its position within the whole wave complex with time. This change in position occurs with a velocity which is carefully to be distinguished from the rate of propagation of the wave. The latter is called the wave or phase velocity, and the former, that is the change of position of the centre of energy, the group velocity.

In De Broglie's formulae, v, the velocity of the corpuscle, is the group velocity, while u, the super-light velocity of the matter waves, is the phase velocity. The further development of these ideas, which were known as such in mathematical physics before De Broglie's time but had found no practical application, is due to the investigations of the German physicist, Schrödinger. It is simply impossible to go more fully into them.[113] It must therefore suffice to remark that, according to Schrödinger, the material corpuscle is resolved into a 'wave-packet' and the foundations of mechanics are correspondingly referred quite generally to the principles of wave motion, in particular to what is known as Huygens' principle. The most important success of the new method was the final explanation of the Stark effect, that is, the electrical separation of the spectrum lines, which the classical theory was unable to explain, although it had been successful in the case of the Zeemann effect.

A few months previously, a competitor of the Schrödinger theory had arisen in the theory of Heisenberg (1926), which makes still greater-demands on the power of mathematical abstraction. The fundamental physical notion of this theory is as follows: It is shown generally in theoretical physics, that any kind of periodic change (that is an oscillation of any form) may always be regarded as compounded of a number of single oscillations of strictly sine form (figure 38). In acoustics, this method is used to resolve a given sound into its single 'partials,' which may be determined both by physical means, employing resonators, and by purely mathematical means, from the curve of oscillation which represents the sound.[114] The formulae in question are known as Fourier's theorem. Now the difficulty of Bohr's atomic model lay in the fact that the frequencies (spectral lines) generated by the revolving electrons had apparently nothing to do with the frequency of revolution of the electron itself.

Heisenberg's theory reverses this connection, and that is the

happy idea at the bottom of it.   It reg rds this apparently simple
periodic motion of the electrons as a resultant of the partial
frequencies represented by the single lines of the spectrum,
whereby, however, it is not possible to operate with the simple
Fourier analysis known to us from acoustics; it is necessary to
reckon with a more complicated, two-dimensional manifold.
According to Heisenberg and his associates, Jordan, Born, and
others, the intention is to liberate the theory of spectra from the
assumption of motion which cannot be observed in any way.
These statements are determined by a decided positivist point of
view; they wish the theory to contain in the future only observable
quantities, and hence have also expressed their rejection of
Schrödinger's wave-packet,[115] since this again, in their opinion,

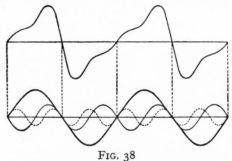

FIG. 38

Analysis of a compound oscillation (upper curve)
into three single sine curves (Fourier's theorem)

brings in an element of the unobservable.   On the other hand,
Schrödinger was able to show very soon after the appearance of
Heisenberg's work, that the latter was completely equivalent
mathematically to his own theory, and both lead to the same
results.

The further development of these ideas of Schrödinger and
Heisenberg, naturally in continual conjunction with experimental
investigation, has already led to such an enormous output of new
theories, that we can only refer here shortly to the most important
of these.   Heisenberg's idea was developed by Born and Jordan
into a method of still greater mathematical abstraction, namely
matrix mechanics.   Uhlenbeck and Goudsmit then introduced
the hypothesis of the spinning electron, according to which the
electron rotates on its axis.   This rotation makes it a small
magnet, the energy of which is naturally also quantised.   By

using this hypothesis, which ascribes to an electron altogether four quantum numbers (the full demonstration cannot be given here), Landé and Pauli finally succeeded, in 1926, in making the periodic system really comprehensible. The number of elements in the periods, 2, 8, 8, 18, 18, 32, had already shown that we were dealing with the double of squares (2×1^2; 2×2^2; 2×3^2; 2×4^2), but no explanation of this had been given; even the Bohr ideas above mentioned did not lead of necessity to this conclusion. A rule given by Pauli (the 'Exclusion Principle') states that two electrons never agree in all four quantum numbers, and from this, the sequence of numbers in the periodic system follows of necessity.[116]

Dirac was subsequently successful in extending Schrödinger's theory so that it yielded electron spin as a necessary consequence without special assumptions. Further, Sommerfeld showed, on the basis of a new method of quantum statistics founded by Fermi, that the difficulties of an electron theory of metallic conduction, which had previously appeared insoluble, can now be overcome.[117] Riecke and Drude had attempted in their day to explain metallic conduction by regarding the external electrons of the metallic atoms, which are obviously very loosely bound (as is seen by the ease of formation of positive ions), as completely free to move among the atomic nuclei; as a kind of gas in fact, which can be treated by the usual mathematical methods of the gas theory. Serious want of agreement was found however, and in particular, the curious phenomenon called hyperconductivity (Kammerlingh Onnes, 1918) is in no way predictable by the old theory. At a definite, very low temperature, certain metals (mercury, lead, and others) suddenly become hyperconducting, that is, they have practically no electrical resistance left. A current once set going in them, say by induction, lasts for days. Sommerfeld's new theory not only gives this new phenomenon, but also the law of Wiedemann and Franz (proportionality between the conductivities for heat and electricity; see above). It has been quite recently further extended by Eucken, by introducing, in the place of electrons, Schrödinger's waves.

Successful advances have also been made on the basis of the new theory into the physics of the nucleus, which was hitherto practically unknown territory. The most famous of these is an investigation by Gamov,[118] who succeeded in deducing on the basis of wave mechanics what is called the Geiger-Nuttal relation between the range of the β-rays of radio-active substances, and their life (disintegration constant, or half period). In this, as also in other investigations of the problem of disintegration and reformation of material atoms (corpuscles) from radiation quanta, the Einstein relations ($m = h\nu/c^2$, etc.) are of course generally taken as a basis. The hypothesis has also been put forward, that in certain circumstances, an electron and a proton may completely unite, the material corpuscle then disappearing, and a certain quantity of radiant energy being produced in its place. Calculations give for such a 'neutron' a wave length which lies far beyond the shortest known rays, and is probably identical with the shortest wave length hitherto measured in the cosmic rays of Kolhörster (see above). As regards these, later experiments[116] of Kolhörster and Bothe have led to the supposition that these rays are not waves, but corpuscular. It has however been rightly objected from another side,[119] that in regions of such high frequency or velocities (corpuscular energies), the boundary between wave and corpuscle disappears in all probability, and we therefore find ourselves at the point at which radiation and material corpuscle become one and the same. The profound idea of certain old mystics that the world consists of 'frozen light' acquires through these investigations a new interest.

[A new aspect is given to the matter by the actual discovery of neutrons. See note 119. *Trans.*]

But apart from consequences such as these, which lead us into philosophy, the new theories deserve the greatest interest. They have effected a revolution in all departments of physics and chemistry, and we can as yet see no end to this new development; a well-founded hope exists that still further problems, hitherto insoluble, may now find their solution. It is already possible to say that the contradiction between wave and corpuscular theory, which once seemed insoluble, is already beginning to be resolved into a higher unity, at the cost however of – at least at present – an almost complete loss of physical comprehensibility; as in the case of the relativity theory, only a highly trained specialist is able to have any real insight into the full range of the ideas. This is much to be regretted from the philosophical point of view, but we may still retain the hope that finally, when the new ideas have once been thoroughly worked out in all directions, it may be possible to put them into such a form, that the educated layman may also grasp their essentials, as is already possible within certain limits for the relativity theory. This is the more necessary in the case of quantum mechanics, since, as we shall now see, its philosophical consequences lead us far, if not further, than those of the relativity theory. To these consequences we shall devote the next chapter, regarding our survey as finished.

XVI

THE PHYSICAL WORLD-PICTURE OF TO-DAY

We shall see that, in particular, the two categories already discussed earlier in this book, those of substance and causality, and the concept of the law of nature, all appear in an essentially new light.

The development of physico-chemical investigation in the last three decades has shown quite plainly that all the lines upon which investigation takes place finally converge, and we may now feel convinced that all the various physico-chemical phenomena are finally founded upon one and the same set of facts; this conviction not being based on any supposed deductions *a priori*, but upon the whole course of three hundred years of investigation, upon historical experience; our conclusion is thus *a posteriori*. While, however, classical epistemology, as we saw above (page 61), believed itself able to give these fundamentals *a priori* (namely in the sense of the mechanical world-picture which it drew), we know to-day, that a final formulation of them will only be possible when the ultimate problems are really solved, and that we must for the present be content with parts only of the whole picture, between which only single lines of connection are more or less clearly visible.

Let us first see what elements are fundamentally required by the physicist to-day who builds up a world-picture. They are easily stated. He needs the proton and the electron with their opposite charges of 4.772×10^{-10} electrostatic units, and their masses 1.661×10^{-24} grams and 9.00×10^{-28} grams. He further needs the elementary quantum of action $h = 6.53 \times 10^{-27}$ ergs per second, and the Maxwellian or Einsteinian field laws, or perhaps a field theory including both. The characteristic constants of the field laws are c and κ (the gravitation constant), which latter has hitherto not been deduced from the field theory. The laws otherwise necessary for building up the physico-chemical system deducible from these elements, those for instance of mechanics or electro-dynamics, must be taken as consequences of the general field laws, and hence are not counted in here. As we

Os

see, the number of the fundamental assumptions which have to be made by the physics of to-day has already become quite small, this being true both of 'substances' and fundamental laws.

It is, in the first place, quite clear that physics will not cease to advance beyond these few single data hitherto necessary. It will not rest until an answer has been found to the question as to why the opposite charges of proton and electron are equal to one another and of a certain size, and why their masses are unequal. The Schrödinger mechanics raise a new problem, demanding absolutely a solution, namely why, if material corpuscles are to be interpreted as wave-packets, two essentially different packets of this kind exist, the relationship of their masses being just $1 : 1800$. Light may perhaps be thrown on this matter when the Davisson-Germer experiments, which have hitherto been carried out almost exclusively with electron rays, can be extended systematically and successfully to all kinds of corpuscular rays. Experiments also with positive (anode) rays by Sugiura have quite recently given the diffraction phenomena required by De Broglie's formula.[120]

More generally speaking, the problem is as follows: What internal connection exists between the values of the constants which, in present-day atomic physics, stand unconnected alongside one another? Many attempts have already been made to discover connections of this kind purely empirically, naturally in the hope that this would lead to the right clue for an explanation. It may be shown (Perles, 1928)[121] that the elementary quantum of action is exactly the $(\pi - 1)$th part of the product of e^2/c and M/m. But a relation like this, which might also be a matter of pure chance, only represents a real advance when we are able to say why it is thus and not otherwise. The most discussed attempt is one made by Eddington.[122] From the fact that the new fundamental wave-mechanical laws were most simply represented in a 'configuration space' of 16 dimensions, he concluded that the reciprocal value of the Sommerfeld fine structure constant $\dfrac{1}{a} = \dfrac{hc}{2\pi e^2}$ must have the numerical value 136 $= 16 + \dfrac{16 \times 15}{2}$ which is actually very nearly true. By extending this theory of Eddington, R. Fürth of Prague recently succeeded[123] in also deducing the theoretical value of the relationship

M/m with very close approximation, the theory giving 1838.2 while the best observed value is 1838.1.   The basis of this theory is the assumption of a possible transformation of matter into radiation (light quanta) and conversely.

Since all these questions are still in a state of flux, it would be premature to put these hypotheses forward as final truth.   They are only given in order to make more clear what was said above. This much is already quite evident: some kind of connection exists between the various fundamental constants of present-day atomic physics.   The present position is quite remarkably like that of the periodic system about 50 years ago.   At that time also, physics stood before internal relationships which could be guessed at with fair certainty; as a physicist of those days said, 'Like children at Christmas in front of the closed door.'   Sommerfeld has quite rightly said, in a report[124] on Eddington's investigation: 'If Eddington is right, the conclusion would be unavoidable that the electron charge could be constructed from the quantum theory (h) and the relativity theory (c).   We believe the quantum theory to be capable of every kind of marvel; we have long been convinced that Planck's h plays a part in all elementary processes of non-living nature.   But the construction of e would perhaps be its greatest triumph, and would open up vast prospects as regards the simplification of the physical world-picture.   It is significant . . . that we regard this triumph, if not as certain, at least as possible.'   To these words of Sommerfeld, we will add the following.

The general field theory, which we will assume to comprise in one unity electro-magnetism and gravitation (though this is not yet finally established) affords us in itself, as we saw above, no sufficient reason[125] for the fact that the something, which fills the field and may be described by the so-called energy impulse tensor of the relativity theory (or other quantities derivable from it), exists in just those separate smallest quanta which we know on the one hand as light quanta, on the other as material corpuscles. But at this point, however, it becomes very highly probable that the elementary quantum of action h is the real fundamental constant, to which we shall finally be able to refer the other atomic constants, that is, e, M, and m.   In other words, it is highly probable that matter and electric charge are atomistically divided because action is divided into quanta.

This depends upon the following considerations. In Maxwell's classical field theory every point is characterised by the energy density there pertaining, by which we understand the energy per unit volume. (The value of this in an electro-static field is $E^2/8\pi$, in a magnetic field $H^2/8\pi$, where E and H are the respective field strengths.) If the value of this quantity varies continuously from point to point, we get the energy contained in a given finite volume, by multiplying each infinitely small element of volume dV by the value of the energy density belonging to it, and adding together these infinitely numerous, infinitely small amounts over the whole region considered, in other words, by integrating the energy density ρ over the space $(\int \rho \, dV)$. In the four-dimensional world of the relativity theory (x, y, z, t), the corresponding quantity would not only have to be integrated over a space, that is over elements of volume, $dx \cdot dy \cdot dz$, but over the elements of the four-dimensional Minkowski world. The result is then a quantity which represents a product of energy and time, hence in other words an action, and the fundamental importance of the quantum h of this action is undoubtedly connected with this, but likewise also the circumstance that the law which sums up the whole of mechanics is the principle of least action (see above, page 86).

Classical physics and chemistry set up, as we have already seen above, the laws of the conservation of mass and of energy, side by side. The Einstein mass relation, $E = mc^2$ suggests that the two are really one. We can now recognise the deeper reason for this. Energy and mass (or impulse mv) are not fundamental magnitudes at all, they are only the spatial and temporal projections or components, respectively, of action, and if they both only occur in quanta, the reason must be that action itself, which is at the bottom of both, only exists in this form. For want of a final field theory, we are not yet able to state this internal connection completely and exactly; we can only suppose that it must exist. But so much appears a legitimate further supposition, that the quantisation of action is the primary fact, that of electricity and mass, as well as of wave energy, the secondary.

However this problem may be solved in the future, we will assume for the present that we shall succeed in referring the other atomic constants, M, m, and e, to which κ will also probably be added, to c and h. We should thus be left with only the two fundamental constants c and h, one of which decides the character of space-time order, the other – what are we to say? The answer to this final question may be, I should suggest with great caution, a half physical and half philosophical hypothesis. The philosophy of the classical period (Kant) speaks of two 'forms of perception,' space and time, and two fundamental 'categories,' substance and causality. Both pairs were for it empty frameworks without qualities ; we have already quoted Weyl's remark concerning 'empty apartment houses,' referring to time and space; but exactly the same thing is true of the other pair. Kant was entirely of the opinion that the categories also are unquestionably devoid of any definite content in themselves, but are simply an empty frame, into which all special physical and chemical experience fits for the very reason that the frame is in itself without properties, and hence in no way prejudices special experience. Now the relativity theory shows that in the first place, the two forms of perception are indissolubly connected; and secondly, that this frame or scheme of order is in itself by no means devoid of qualitative definiteness, for the constant c is peculiar to it.

My supposition is, that the constant h plays the corresponding part for the two categories of substance and causality, which are now fused together in the concept of action. Its existence proves that these two categories, like the space-time scheme, are certainly not characterised by complete absence of qualities, for the constant h already gives them a definite content. This content at present seems to be simply a matter of experience, in the same way as c is. The meaning of this will be clearer to us subsequently. We will first draw the general epistemological conclusion that apriorism proves to be inadequate in this region also, in view of the advance of physics. The concepts of substance and causality, which it gives us, and more than which it cannot give, may be an inevitable plan of thinking, in the same sense as colours and sound are to be regarded as secondary qualities, Euclidian geometry and Galilean kinetics as laws of *our* space-time *perception*. But for this very reason, they can be transcended, and are not, as Kant and his successors supposed, binding on all physical knowledge to all eternity. What is behind them is rather to be determined more exactly by physics in its further progress, and only at the end shall we be able to say what the objective content of these forms really is,[126] just as it was only possible to say after the introduction of the prism and experiments in interference, what was really and objectively at the bottom of colour. This is the standpoint of critical realism, to which we also subscribe in this connection.

We shall now consider these fundamental ideas somewhat more in detail, first investigating a little more closely the concept of substance in present-day physics. From what we have said, it is already clear whither we are tending: Physics is on the point of finally and completely throwing the concept of substance, in the old sense, overboard. What remains of it is in any case something entirely different from what was originally meant by it. The 'law of the conservation of substance' runs in Kant (second edition of the *Critique of Pure Reason*) as follows:— 'In all changes of phenomena substance remains unaltered, and the quantity of the same is neither increased nor diminished in nature.' Kant regarded this law as certain *a priori*, and his famous example of the weight of smoke,[127] which he gives two pages later, proves without question that he did not mean it as a vague generality, but in a very special sense. (It is naturally based on the mechanistic world-picture of classical physics, see above.) The objection that this world-picture has already been superseded by the field physics of Maxwell and Hertz, could hitherto be answered by a defender of Kantian apriorism by admitting this fact, but asserting that the Kantian concept of substance was still valid in a wider sense. The ether, with which Maxwell's theory reckoned, is also, he would say, a substance, and the electrons of the later electron theory also. This rejoinder is undoubtedly a good one, as long as we retain the view of Maxwell and Lorentz, which to-day is also described as classical.

But Einstein's extension of the field theory on the one hand and the resolution of the electrons into wave packets on the other, have changed the situation essentially. The ether has become superfluous in the relativity theory, as we saw above. If Einstein has, on occasion, subscribed to the re-introduction of this word, he wished to designate by it something quite different from what it has hitherto meant. It was now no longer to stand for the carrier of the field, but to describe the essence of the fields themselves. On the other hand, it might naturally again be said: the Schrödinger waves themselves also need a carrier, for a wave can only be imagined when we first have something there which moves in this wave form. But this would be an undoubted slip back into the purely mechanistic conception of a wave, which was already superseded by the formal theory of light (see above, page 107). We there saw that wave and oscillation in themselves have no other meaning in modern physics than periodic change of some quantity, no matter of what kind. When now everything that can be stated regarding the processes in question depends only on the form of these processes, while the nature of the quantity itself which changes according to the wave formula has become a matter of complete indifference, then all that plays any part in physics is the process itself, and it does not in the least matter what the something is in which the process takes place. This last sentence itself then appears as a mode of expression which only leads to superfluous and insoluble questions, a mode of expression which arises out of our primitive experience (in the same way as do the metrics of euclidian geometry) but which loses its sense as soon as we are dealing with the final fundamentals.

The correctness of this view is proved by the reinterpretation of Schrödinger's theory by Born and Jordan, which has already been indicated above. According to Schrödinger, the quantity occurring in these formulae, the so-called field scalar S or ψ, the periodic changes of which are given by the wave formulae in question, is the density of the electric charge. According to Born on the other hand, it can also have ascribed to it the purely formal meaning of a statistical probability that a light quantum takes a certain direction.[128] In this sense it has therefore no real physical meaning whatever. I do not by any means intend – for reasons which will be immediately given – to maintain that only this view of Born can be final; I merely make use of it here to prove the thesis, that for the theories of atomic processes we are discussing, it is to a great extent a matter of indifference what the magnitudes themselves used in the formulae really mean. In other words, in these theories only the form of the process is the essential matter, and no longer in what it takes place. This fundamental idea needs only to be carried to its logical conclusion for us to see that a final state of physics is not merely imaginable but already very nearly reached, in which the notion of a substance is no longer made use of, since everything which can be stated physically is founded only upon the law of the processes.

In this sense therefore, the world-picture of present day physics is distinctly a dynamistic one. Materialism in the narrower sense, that is the belief in eternal indestructible matter or in atoms as 'rigid lumps of reality,' is thus finally abandoned.[129] What we call matter is a sum of certain processes. An hydrogen atom or electron does not simply exist, it happens. We do not first need a substance, so that, secondly, something can happen to it, as Kant's statements above quoted assume to be self-evident; we simply need a something which distinguishes the world from nothing, from an empty four-dimensional scheme of order. The formal laws which this something obeys are the physical laws of all being and happening whatever. There is no real, but only a provisional or practical sense in making a distinction between laws of being, that is the properties of so-called matter, and laws of process, that is the properties of the forces, energies and so on (the field). Both are only different sides or projections of the same fact, as obviously follows from the fact of the internal relationship between the electrons and the light quantum. The layman may find it at first a little difficult to accustom himself to such abstraction. He may try to make the matter a little clearer by a further simple example. If we write on paper the formulae, say, of a so-called stationary wave, or even merely imagine them, these formulae state for the expert a whole number of properties of this kind, for example, that nodes and loops exist, how two such processes act together, and many other matters. In all these statements it is a matter of complete indifference what the nature of the quantities may be, the periodic change of which is described by the formulae. The formal properties of a stationary wave hold in every case. If this is also the case with formulae which describe the behaviour of the final 'material elements,' that is, the light quanta and electrons, we obviously have no further use for a 'substance.'

This whole point of view has incontestably a certain grandeur. In place of cold, dead, rigid matter we have the restless flow, the crossing and recrossing of billions upon billions of 'actions,' which together, however, do not result in chaos, but in 'a cosmos of a symmetry, order, and regularity, which appear to us actually improbable' (Titius).[130] It is, further, especially noteworthy that the line of development of the world is in the direction of complication.[131] If any one were told: 'Here you have the quantum of action h and the field laws' (possibly also the two masses M and m of proton and electron and the elementary charge e): 'build a world out of them,' he would presumably suppose that not very much could be done about it. We know in reality, that out of these few fundamentals is built first the world of the atoms (the periodic system); out of these again a million and a half of molecules (chemical compounds) and out of these again, by mixture, the innumerable substances of nature in ever increasing complication of form, and finally the cosmos of the fixed stars, and possibly higher units.* The unity and simplicity of these fundamentals do not therefore exclude a continually increasing individualisation and production of new forms; it includes them. Without question, we have here a very strong stimulus to teleological lines of thought, and further to such of a more tenable character than those with which we became acquainted in connection with the minimum principles of mechanics, or such as have often been linked with single peculiar facts, such as the abnormal properties of water, or the like. The teleological interpretation of the latter often reminds us of the providential interpretation of the presence of great rivers in large cities. On the other hand, the teleological conception of the physical cosmos as a whole impresses itself almost irresistibly upon even the soberest physicist, just as it has always done upon the astronomer, and as it is expressed in Kant's well-known words regarding the starry heavens.

But we are digressing. We are not dealing here with teleology and the production of form, but with the being and becoming of matter and processes. The physical world-picture of to-day corresponds, as we saw, to the old aphorism of Heraclites, πάντα ρεῖ. The whole of simple existence is resolved into perpetual becoming or happening. But we can now – and this is the most important fact from the point of view of scientific philosophy – equally well regard the matter from the opposite point of view, and say that modern physics has also cleared the way for the equally ancient view of the opponents of Heraclites, the Eleatics, according to which all happening is in reality timeless existence. We need only recollect to this end the formulae which we just took as an example of a process such as that of stationary waves. When a physicist has these formulae in front

* How far we should include here biological forms will be discussed in Part III.

of him or merely thinks of them, they describe the process in question for him over its whole time range. As human being, he naturally needs a certain finite, although very small time to grasp their contents. But we can obviously neglect this time without in any way altering the contents of the formulae ; we can therefore imagine the act of grasping this content as timeless.

But on the other hand, the physicist in question can regard the time entering into the formulae as limited in any arbitrary fashion forwards or backwards, or also as unlimited; he can indeed, if he imagines the whole affair in Einsteinian instead of in ordinary kinematics, suppose the time as well as the space to return upon itself – and all this without his being in any way bound to time or space. These formulae have for him only a timeless and spaceless 'validity.'[132] Furthermore, he can, by connecting together several such process formulae, draw conclusions concerning the connection between many such single processes, that is to say, reconstruct in his mind the casual connection of nature, for which purpose, likewise, theoretically speaking, he needs no time for thought. If we further imagine that he has not merely innumerable such formulae in his head, but that he is able merely by thinking them to make their content real for other spiritual beings similar to himself, we have, when we eliminate as far as possible all anthropomorphism, a new and enlarged edition of Laplace's World-Spirit, whereupon we are naturally confronted with the problem as to how single limited systems within the compass of this thought-complex, namely man and animals, come to possess an inward psychical grasp of the whole system.

This is the psycho-physical problem, seen, as it were, from above, and we shall return to it in another connection. But for the present we will lay it aside and simply draw one conclusion only from the whole discussion, namely that we may very well regard the whole flow of events in time from the opposite point of view, that is to say, as the timeless effect of something simply 'laid down.' If philosophy has long known this – the old Scholastics taught that God sees the world *uno aspectu* – it is only through modern physics that this notion has been made possible as a direct perception, in particular as a result of the relativity theory. This opens up the prospect of our finally getting beyond the contrast between being and happening (Eleatics, Heraclites). When we think out both of them to the end, we finally discover that both are one and the same world, which we regard at one time *sub specie temporis*, at another *sub specie aeternitatis*. The something which distinguishes the world from nothing is at once being and happening, matter and energy. A distinction between 'substance' and 'accident' has no further sense when we get down to the last fundamentals.

At this point, a positivist opponent might make a fairly obvious counter-attack to what we said against his view earlier (pages 36 and 122). He might perhaps ask what right we had previously to make so great a to-do about the question of being, since we now ourselves admit at this point that, after all, the whole thing depends upon the formulae which describe the process, and that the question concerning the underlying 'essence' (the substance) is a matter of complete indifference.

The answer to this counter-attack is as follows: There is a fundamental difference between refusing to admit the question of essence at the beginning and recognising at the end that it ceases to be. For in the first case, the way is closed to the most important statements of problem, as our discussion of the kinetic theory of heat, the electro-magnetic theory of light, and other questions, must have sufficiently demonstrated. As long as physical phenomena, different in essence, appear to exist, the question of essence has every right to exist, for, as we showed in another connection, the unity behind these various phenomena is then, for the time, only a postulate, but not a result. Only when, in the sense of the immediately preceding chapters, all varieties of phenomena have been reduced to one and the same in essence, are we able to take into account the possibility that we no longer need any 'substances.'

Positivism everywhere makes the fundamental error of too early refusing to admit problems, which in their place still have sense. Furthermore, we should now discuss the question whether at the point at which physics may finally abolish the question of substance, this may not perhaps still retain a valid meaning from another point of view. For, after all, it is not unthinkable that physics may be far from including the whole of human knowledge; we still have biology and psychology and these might perhaps give the starting point for further problems. We shall return to this at a suitable point in the third and fourth parts of this book, but we cannot omit drawing the reader's attention meanwhile to the excellent statement of the present problems given by K. Riezler (*Naturwissenschaften*, 1928, p. 705). Riezler quotes a picture of which Eddington makes use. Many thousands of years hence, an archaeologist discovers a book containing numerous games of chess, given in the usual sign language of chess players. After long investigation, the rules of play are discovered, but the true nature of the chess men and the chess board, cannot be discovered. While Eddington uses this analogy, to illustrate the relationship between the system of order, of the relativity theory, and possible physical realities (electrons, etc.), Riezler now interprets it as follows: Supposing that the archaeologist never arrived at the notion that the signs he had discovered might stand for a game played by living players according to their humour and power, but had got it into his head that these signs were a part of a single self-consistent system of laws, he would be successful with this hypothesis up to a certain point, in so far as like often follows like in the case of these signs. . . . 'Now there comes along a humorist devoid of all sense of shame and respect, and explains: Do you not see that these signs are very far from belonging to a system of unambiguous order ? Order at certain points proves nothing at all. Observe more closely, get right down to the smallest of the small, and the order disappears. Suppose that the whole were a game . . . since you know neither the idea of the game, nor the players, you will never learn why one move is followed by another in one case, and by a different one in another case. The true reality is the idea of the game and the exertion of the players. If you wish to learn something about it, ask your own soul, its thought, struggle, and striving – perhaps it is condemned to take a part in the game, and hence knows something of its meaning. (For a similar remark by Bohr see page 411). And when you have then guessed something of the meaning of the game, try treating this meaning as the true invariant of the world and the key to the interpretation of the signs. Look upon the regularities which you already know, merely as a surface, the order of which is derived in part from the interplay of the many players and the interconnection of their intentions and mistakes, in part also from the rules of the game (that is to say, the so-called laws of nature).'

We have thus been led to touch upon the most important problem arising out of recent development, the problem of causality, and we must now go into this a little more fully.   The conclusions which we shall arrive at are of decisive importance as regards our whole world-picture.   As far as I know, Schottky[133] and Nernst were among the first expressly to cast doubt, on the basis of the new light quantum theory, upon the ideas of causality hitherto generally accepted.   The latter's address as Rector of Berlin University[133] was the first means of bringing them to the notice of a wider circle.   In the previous edition of this book, I made especial reference to this address, and then expressed dissent, in the same sense as Planck[134] and other physicists, from radical scepticism concerning the strict rule of law in nature.   In view of recent experimental results, I must somewhat modify this point of view, although I am not yet convinced, as are Born[135] and others, that the final abandonment of strict causality of all happening is really the last word.

It is useful at this point first of all to recollect something which was shortly pointed out further back (pages 48, 63 ff.).   The world-picture of classical physics is characterised by three fundamental assumptions.   The first is continuity in space and time, the second the unlimited mutual interpenetration of all actions, and the third strict determinism.   All these three postulates are most obviously expressed by Laplace's fiction of the World-Spirit, which we have already discussed.   It is true that here matter itself was given a discontinuous structure (it was imagined as made up of single mass points), but the forces exerted by these points upon one another are extended in space according to the strict law $1/r^2$ or a similar law, and thus fulfil the requirement of continuity.   This is true, in a still more ideal form, of Maxwell's field equations, which present themselves to us, both as regards space and time, in the form of strictly differential laws.   According to both fundamental laws, all partial influences arriving from all points interpenetrate at each point of the universe (in Laplace's case timelessly, in Maxwell's case with a phase difference proportional to the distance).   This is also, by the way, the starting point of Leibniz's doctrine, according to which the Monad is the mirror of the universe.[136]   This universal mutual interpenetration is most clearly perceived in the wave theory, while broadcasting now brings it home in the most striking manner to everyone.   We have already said all that is necessary on pages 65, 75 concerning the strict determinism of classical physics.

It was naturally observed quite early that, in many departments of physics, these requirements are only apparently fulfilled, and that in reality we are dealing with discontinuous processes, the laws of which cannot be strictly described by differential equations and hence are not strictly valid. For example, the propagation of a wave of sound in air, or the conduction of heat in a solid body such as a crystal, is described by certain differential equations, in which the pressure and the temperature gradient are given respectively by the differential quotients dp/dx and dT/dx. That is to say, the pressure or temperature change is calculated over an infinitely small distance (that is, imagined to be as small as we like.) This is certainly not an exactly appropriate form of expression, if only because in the region of molecular dimensions (about 1 to 0.1 $\mu\mu$), there can no longer be any question of a truly continuous change in the quantities in question.

In the region of these small dimensions, neither pressure nor temperature exists, for these are only average statistical values of the energy of motion of very large numbers of molecules. A single molecule has no temperature, but only the definite value at a definite moment of its energy of motion, while shortly after collision with another molecule, this value may be quite different ; and in the empty space between two molecules the concept of temperature loses all meaning whatever, unless we ascribe to each of the light quanta there a definite amount of energy, but one again capable of sudden change. These formulae are therefore without doubt merely approximations, which nevertheless lead to very exact results, since the distances over which we afterwards integrate (sum up) are in practice very large as compared with molecular distances. Exactly the same is true of the propagation of sound, of the transmission of elastic forces, and many other matters. In all these cases, therefore, the assumption of continuity is a conscious fiction, which is only made use of because the method proves extremely effective.

Laws set up in this way by physics obviously only have the character of average statistical rules. This has already been pointed out with regard to the law of entropy, but the same is true for the laws of heat. We may take as an example of this the so-called gas law (the Boyle-Amonton-Gay-Lussac law). According to this, the pressure of a gas is proportional to the number of molecules present in a given space and to their temperature, that is the average value of their energy of motion. If we imagine a moderately large surface, say the top of a table, or the surface of a brick, the whole pressure of the air on this surface is the sum of all the blows given to it by the air molecules, and a statistical calculation naturally gives this average pressure, the number of molecules being so enormous, to be the same from above as from below.

But this is only a statistical rule, and not the law in the strict sense. If we take the surface much smaller, say only a square micron, and inquire as to the probability of the difference in pressure between the upper and lower sides assuming a sensible value during a ten-thousandth of a second, the Brownian motion (page 93) proves to us that this probability is not by any means very small; that on the contrary, the consequence of this difference of pressure, namely the movement of the small surface in one direction, occurs within a short time. In other words, it is only a question of the amount of movement to be expected, its duration, and the size of the space considered, for the probability in question to become perceptible. Perrin once calculated that for a brick to jump from the ground into the hand of a bricklayer working on the third storey, he would have to wait on the average for $10^{10^{10}}$ years. This number is 1 followed by ten milliard noughts. If written, it would go half round the earth if each figure were two millimetres wide. The meaning of it cannot possibly be imagined with the best of will. If the whole of humanity, since the appearance of man on earth, let us say for two hundred thousand years, had done nothing but count this number, it would not, up to to-day, have counted any appreciable part of it. An improbability of this kind is a practical impossibility. But as soon as we take the dimensions in time and space sufficiently small, the possibility of variation acquires more and more perceptible values.

This and similar examples have therefore long ago accustomed physics to ascribe to a great part of its laws merely a statistical character. But nevertheless, no one in earlier times seriously cast doubt upon the existence of fundamental causal determination. Although, to take the example of gas pressure, we are only able to calculate it by means of a statistical method, this in no way prevents our being able, and indeed obliged (so at least it was said at one time), to regard the path of each single molecule as strictly determined by the laws of mechanics. While we are not practically in the position to calculate the position of such paths exactly, since our mathematical equipment is not sufficient (it already fails with regard to the famous three-body problem of astronomy), this practical difficulty is naturally something totally different from the indeterminism in principle of natural events. In other words, while the statistical rules are true in coarse dimensions, in the so-called macroscopic region, the exact law is still at the bottom of it in small directions, as regards the single atom or molecule.   It is as when playing dice: a single throw results without question from causes which strictly pre-determine it.   But for what interests us in respect to the dice, we have no need to know these causes, the supposition being sufficient that they make the cube fall with equal frequency upon any one of its faces, so that the probability for each single throw is $1/6$.   This view has hitherto been always found satisfactory.

But what has now come into sight in physical investigation is nothing more or less than a fundamental doubt concerning the whole causal substructure of statistical physics.   This attack took place in two stages.   First of all, the light quantum hypothesis, as such, dealt a severe blow to the ideas of continuity and mutual interpenetration of all actions, which ideas lie at the bottom of classical physics.   The idea of regarding the world as a cinematograph film (see above, page 49) no longer appeared so preposterous, as it would have done in Boltzmann's time.   In the classical theory of light, the light wave, like any other electro-magnetic wave, was a process which was propagated continuously and uniformly in space with decreasing amplitude. Every electron in its path took up just as much of the energy of the wave as corresponded to the value at that time and place.   But now light is to consist of real corpuscles, which (in the photo-electric and Compton effects) either hit or do not hit the electron like a billiard ball.

This has made it necessary to ascribe, as it were, to the light quantum, the knowledge in advance of what electron will again absorb it.   The whole process of light emission and absorption thus acquires a kind of teleological character.[137]   This might, however, still have been compatible with a strictly causal view, since the same state of affairs is present in the minimum principles of mechanics, in particular the principle of least action, and we have seen above that a causal foundation is nevertheless not thereby excluded.

Now came the second attack, and this time the kernel of the idea of causality was struck. It started from Heisenberg's formulation of quantum mechanics. Since the matter is difficult to explain without mathematics, we must limit ourselves here to a sketch of the fundamental idea. Action is, as we saw above, a product of energy and time, or otherwise a product of impulse and distance. Now the task of physics obviously consists in measuring as exactly as possible the phenomena observed, and these measurements will always have for their object on the one hand distances or intervals of time, on the other hand, quantities of energy or impulse (see above, page 59). The method of measurement consists in continual approximation, in which we try to eliminate unavoidable errors by taking their supposed value into account in a given case from other results of measurement. If, for example, a temperature is to be measured, the use of the thermometer necessarily causes a small change in the temperature itself which is to be measured, since it takes away a little heat from the body concerned. But this amount can be taken into account beforehand by using the known data of the thermometer with the aid of the material constants concerned (specific heat of glass and mercury, etc.), and thus the error can be corrected more or less perfectly. This example shows that every measurement necessarily changes a little that which is to be measured.

It has hitherto been classically assumed, that by means of this kind of correction of measurements, the refinement of our sense organs (Wiener) could in principle be carried to any extent, or that, as Reichenbach[138] puts it, every existing degree of accuracy for any measurement may still be superseded by greater accuracy, if only we have the patience and good fortune to try. But Heisenberg's theory leads to the consequence – and this is its fundamentally new result – that this belief in the convergence of accuracy of measurement to absolute certainty was an error, and that, on the contrary, a finite, though very small, lower limit to accuracy of measurement is fixed by nature itself. According to what is called the Heisenberg 'Uncertainty Principle'[139] (a certain fundamental equation of quantum mechanics), action can never be determined more accurately than to the order of magnitude of the Planck quantum. The consequence of this is, that in the region of the extremely small, the accuracy of the measurement of distance can only finally be increased at the cost of the accuracy

Ps

of measurement of impulse, and that of the measurement of time only at the cost of the accuracy of the energy measurement. If, for example, we wished to measure the distance absolutely, that is with infinite accuracy or zero error, the error of the other factor, the impulse, would automatically become infinitely great ; that is, it would remain quite undetermined.

It must be carefully noted that this uncertainty is not merely temporary and capable of being overcome later by further refinement in measurement. There is no more delicate means of measurement than light. Physics naturally strives to eliminate subjective elements in measurement more and more, in order to arrive at the objective facts, independent of all points of observation. We saw above that this, and not a general relativism, is the true meaning of the relativity theory. But this endeavour is brought immovably to a halt by the Planck quantum. Within its order of magnitude, no real independence between observer and observed can exist, since the process taking place between the object on the one hand and the sense organs of the observer on the other itself consists of such quanta of action.

The consequence of this has been formulated by Haas in the following words:[140] 'If a precise description of atomic events in the classical sense is impossible in itself, the causal principle naturally loses its meaning for physics. For this principle, according to which the exact knowledge of the present allows an exact calculation of the future, ceases to have any object when an exact knowledge of the present cannot be acquired. According to quantum mechanics causality must be denied for the elementary processes of physics, and can only be confirmed for the probabilities which are to be ascribed to these processes for statistical reasons.'

The latter remark of Haas refers to the above-mentioned restatement of Schrödinger's theory by Born and Jordan.[141] Schrödinger's differential equation is naturally of the type of an exact law, precisely like the classical equations of mechanics of electrodynamics.[142] The phase waves assumed by Schrödinger are propagated by the same kind of strict law as applies to the classical wave theory of light. Here, therefore, we should have the causal sub-structure of statistics, and one entirely corresponding – though, so to speak, a storey lower – to the strictly causal substructure of statistical heat theory afforded by classical pointmechanics.

But the unfortunate thing for the concept of causality is this, that this original interpretation of Schrödinger is not necessary. The 'field scalar,' which he interpreted as electrical charge density, may, according to Born, be regarded as a pure probability, which naturally has a numerical value, and may be thought of as con- tinuously variable and hence comply with a differential equation. Schrödinger himself appears of late inclined towards this inter- pretation.[143] At any rate, he has expressly assented to the radical doubt concerning the traditional concepts of causality. Born, who together with Jordan was probably the first to draw these conclusions resolutely, asserts directly in a recent publication[144] that those who oppose this sceptical position should experiment and not argue if they wish to refute it, since it is a necessary conse- quence of the Heisenberg quantum mechanics, which itself is confirmed by experience.

It is difficult at the moment to stand sufficiently far off from these matters to form a satisfactory judgment. The development is in full swing, and has recently been enriched by an extraordin- arily valuable contribution from Reichenbach,[145] who shows how it is possible to work quite exactly with 'continuous successions of probabilities,' in place of continuous causal chains, using for the purpose a mathematical formula language invented with this special object. It would be premature to take up the position at present, perhaps based merely on a prejudice in favour of a classical concept of cause, that a causal substructure for statistics must be required in every case, and that physics is therefore com- pelled to seek it until it finds it.

The epistemological problem[146] is exactly similar to that of euclidian geometry in the relativity theory. There also, apriorism tried to persuade us that we were bound once and for all to euclidian geometry as the form of our perception of space, whereas in reality, the development of physics was in gradual process of freeing us from this form, as well as from the secondary qualities (sounds, colours, etc.). It might also be that the road finally leads to our recognising, in our usual concept of causality, another such limit to our power of thought; which nevertheless, as soon as we see it as a limit becomes passable for us, and requires to be replaced by the more accurate concepts of quantum mechanics. Episte- mology must not exclude such possibilities; it is another question whether physics will in the future finally remain at this point.

On the other hand, epistemology must also issue a clear protest, when positivism at this point once more believes it possible to abolish a problem for all future time simply because for the moment no solution is in sight.[147] It did this once in respect to the atom. When it was certain that the atoms in any case must be so small, if they existed at all, that they would for ever remain invisible even for the ultra-microscope, positivism declared categorically that atoms therefore had no place in physics, since they were for ever excluded from direct perception by the theory of physics itself, and what could not be perceived, did not exist for physics. How false this conclusion was, has been sufficiently shown by the whole previous discussion in this book, and quantum mechanics itself makes it clearer than ever, since its central concept is the corpuscle. It must therefore be careful to avoid falling into such negative dogmatism, the absurdity of which has already been so strikingly exposed. It is somewhat amusing to find that one of the most valiant champions of the old positivism of Mach, Petzoldt, feels called upon to stand for the maintenance of natural causality against Jordan and others,[148] although those whom he attacks use, almost word for word, the same arguments against this belief which Mach used thirty years ago against atoms.

Positivism, as we saw in discussing the theory of light, has unquestionably performed the service of repeatedly freeing physics from the discussion of superfluous questions. But, just as often, it has led to an intentional deafness and blindness towards the most important and progressive statements of the problem. If it had had its way, the theory of light, for example, might, and ought to, have remained at the formalistic stage (see above), and Maxwell's electro-magnetic theory would have been quite superfluous. Who can guarantee that sooner or later we may not be led to further problems, which may finally result in further light being thrown on the Schrödinger waves. The interpretation of Schrödinger's field scalar as a simple probability is suspiciously similar to the purely abstract and formal view of the light vector in the formalistic light theory, while real progress at that time lay in the direction of the question as to the true nature of this vector. As we saw above, the problem of the fundamental constants has hitherto not been really solved in any way, and no one, for example, can say with certainty at the present moment, why the wave packets, or corpuscular numerical values of quantum mechanics occur in just the two different forms of proton and electron; nor in what way a positive differs from a negative corpuscle; why the first is 1800 times heavier than the second, etc. It must, therefore, be premature to try to arrive at a final decision to-day as regards causality. We are neither willing to admit the right of a dogmatic apriorism to tie us for ever to the classical view of the concept of causality, nor of a dogmatic positivism (which here as usually would better be called negativism), to cut us off from further questions as to causes, on the basis of the present state of knowledge.

But if we assume for the moment provisionally, that a modification or enlargement of our earlier concept of causality may really become necessary, corresponding to the enlargement of the euclidean geometry into that of Riemann in the relativity theory, it may in any case be said that such an enlargement will not affect what we described above as the most important element of the causal concept, namely, the construction of a logical connection between the various parts or partial processes of world events. Statistical physics actually gives us, for example in the kinetic theory of heat, as good logical foundations as the classical and deterministic. When we have perceived, that the increase in entropy (equivalent to decrease of free energy) happens because there is a transition from a less probable to an enormously more probable condition, this is just as good a satisfaction of our logical instinct as when we deduce on strictly causal lines, the form of the planetary orbits, by means of Newton's law of gravitation. In both cases, our question as to the reason is answered as far as such an answer is possible within the frame of a definite small region of world events.

It is perhaps a prejudice to suppose that this question as to the reasons of phenomena can only be satisfactorily answered by a system of differential equations, capable of definite integration, such as are given by the classical dynamics and electro-dynamics, this prejudice being on the same lines as that which assumes the metrics of the world to be euclidean and galilean, or the weight of the smoke plus the ash to be equal to the weight of wood plus oxygen. Heisenberg's quantum statistics actually give us as full an explanation of the problem of spectroscopy as that given by the classical and strictly determinist theory; indeed a better, as closer examination proves. There is thus no question of the cosmos hitherto known to us having to be replaced by a chaos of arbitrariness or miracles.[149] From the new point of view also, the world remains susceptible of rational explanation to a certain and very large extent, only that the relative extent of the rational and of that simply given irrationally, becomes slightly different than before. Something more must be said concerning this. We are now confronted with a problem which has already been mentioned in passing, and now needs more detailed discussion, the problem of contingency.

By the words 'problem of contingency,' the older philosophical literature understood the question as to why the world exists. This problem is complex in nature, and contains various subsidiary problems, which we must consider singly if we are to arrive at a clear answer. Here also we may start from the old classical and mechanistic view (the Laplace fiction), according to which the world consists of a system of mass-points (by analogy with astronomical systems), which act upon one another according to certain laws, that is, produce accelerations in one another; for example, according to Newton's laws of gravitation. For the course of events to be determined without ambiguity, these laws must first of all be given, and secondly the exact distribution of the single mass-points, and their momentary velocities, must be settled for a given moment. This latter datum is usually described in mathematical physics as the 'initial condition,' since the time, the zero of which must of course be arbitrarily fixed, is usually reckoned from that moment for which the distribution of positions and velocities in question is given. By integration of the differential equations forwards or backwards, we then arrive at the state of the world at any other required point of time in the past or future (compare page 64).

In this classical mechanistic view, the world is obviously contingent (a matter of chance), on the one hand with respect to the general laws in question, on the other hand in respect to the initial condition. It is evident that the latter is incapable of any rational explanation, as soon as we clearly perceive that another world would result if two atoms only were to change places. This is in no way altered by our attempting, for example, to refer the present condition, say of our planetary system, or even of the whole system of fixed stars, to an earlier condition, in accordance with general laws and on the basis of cosmogonical hypotheses.

This merely means referring the problem of contingency to the earlier condition, which again must have a single quite definite constitution, in order that exactly the present state of affairs with distribution of matter and energy may result from it. If we make use of Dingler's happily chosen expression,[150] and call the essence of all that we know concerning the general laws of the world, the 'theoretical prime structure' (*Urbau*) and all that we have discovered concerning the special states and systems actually realised, the 'historical prime structure,' we may say that the latter unquestionably possesses a contingency which is insoluble for human thought. The case is different as regards the contingency of the theoretical prime structure, that is, of the general laws of the universe. The question as to whether this might be otherwise, has been frequently discussed of recent times.

First of all, the well-known Russian physicist, Chwolson, has given the question the form: May we apply the laws of physics to the universe?[151]   The equally well-known French mathematician and philosopher, Poincaré,[151] then propounded the analogous question regarding a possible change with time of natural laws. Poincaré shows that the question is fundamentally meaningless, since it would always be possible to refer the laws regarded to-day as natural laws to a more general law, which included this change. The author followed out the analogous process of reasoning in earlier editions of this book[151] with respect to Chwolson's question, but it was also shown in the same connection that the question has after all another and deeper sense, namely the more general problem of 'thusness' contingency: Is it also possible to imagine a world with other physical laws, or must these be as they are? If we describe non-euclidean and super-euclidean geometry as 'meta-geometry' we are presented here with the prospect of a corresponding 'meta-physics,' as was described in greater detail in earlier editions of this book.

The question, however, is whether this 'thusness' contingency is really a final one.   When we cast our eyes over the way hitherto travelled (compare page 192), the thought is forced upon us that we must be prepared for a still further simplification of fundamentals.[152]   Let us suppose for a moment, that in the first place, only c and h remain as fundamental constants, in other words, that everything can be referred to the general (Einstein) field law and the quantum of action.   We should then again be confronted with the question whether these two values were the last word of wisdom, and could only be accepted as such, or whether, on the other hand, it might also be possible to give them some kind of a foundation, or at least find a functional connection between them, which would allow us to perceive that this particular value of c should only be associated with this h and no other. It would then be possible to imagine other worlds or parts of worlds having other pairs of values of these constants, and their realisation somewhere and somewhen would therefore also be imaginable. But since this latter question then belongs to the sphere of the contingency of being (dealt with above), there would actually be no room left for the question of 'thusness' contingency. But this would then mean nothing less than that the nature of the world in general (its thusness in respect of its laws) would really after all be deducible *a priori*.

This may appear to a thorough-going empiricist as absurd; but it obviously lies, after what we have said, entirely within the bounds of epistemological possibility, although we cannot at present assert that it must really be so. But if it should finally turn out to be true, it would mean the victory of rationalism over empiricism. Physics would then be, in other words, completely arithmeticised (Haas). Its fundamental constants would then all be reducible to pure numbers, such as π or e or Eddington's 136 (see above). Since one could accordingly deduce the general nature of the world *a priori*, in other words, show why the world, if it exists at all, must, in a general way, be as it is and not otherwise, we should finally be brought back to Plato's doctrine, and the sentence 'In the beginning was the Word', the *Logos*. I remark explicitly that these are by no means vague speculations of a philosophical dreamer, but that scientists of the first rank, for example, Eddington, Weyl, and others, have pursued similar lines of thought. They are in the air to-day, since it is actually overwhelming how strongly the tendency to unity in physics has already become effective.

We are obviously only a short distance away from the goal of a final unified summary of all physical knowledge, and the further question as to whether this goal will be the expression of a necessary, or a chance, 'thusness,' is obvious and inevitable. This question also we cannot answer at present, since so far we do not possess the necessary knowledge. But the time will come when it will be answered. It is not superfluous to notice how apriorism also fails in this respect. The whole problem here discussed simply does not exist for it any more than for its antipole, logical empiricism. Only critical realism, for which the whole of knowledge is a process of gradual approach to an objective set of facts already existing before all knowledge (see below), deals properly with this deepest of all problems.

From this point of view, we now return to the earlier question, as to how the physical world-picture appears, if the strictly deterministic view-point turns out to be untenable, or at least superfluous. We can now answer this question somewhat as follows.[153] For the sake of concreteness we will suppose for a moment that the world is only two-dimensional (as upon a photograph) reserving for the time, the third dimension, which we are capable of depicting, as the t axis of a system of co-ordinates. In the sense of the relativity theory, then, every movement in our three-dimensional field (in the same way as the movements on the screen in a moving picture) will be represented by a 'world line' in the three dimensional region (compare pages 143 ff.). In other words, we can imagine a number of transparent pictures of the field at successive instants of time superimposed to form a pack, and connect the positions of a corresponding point on each picture right through the pack of films, this connection then representing a world-line.

Our whole classical conception of natural law means, in the case of this mental picture, that every one of these world lines in a closed physical system is determined by the 'laws of nature,' as soon as two of the sections (single pictures) are given. (The ordinary requirement of a given initial state, namely that the position and momentary velocities must be given, is a special case of this more general one, for the requirement of given momentary velocities amounts to the new position being given for a second, infinitely near, point of time, dt seconds later. But it is not absolutely necessary that the two points of time should be infinitely near to one another.) This initial state is now, as we saw, contingent in any case, and it is in itself a peculiar assumption from this point of view, that in direct contrast thereto, the structure of the world (that of the pack of films) should be of such a nature that the whole structure of the world-lines may be calculated with logical necessity from two such completely contingent states. *The new conception of quantum mechanics simply amounts to this strict and complete distinction between contingent initial state, and non-contingent structure, being untenable, the contingency being instead distributed uniformly, so to speak, over the whole pack, that is to say, the whole process in time of the world.*

This latter in no way possesses the "thread-like" structure of the classical picture, for such individualised points of substance as were originally assumed as in the case of the Minkowski world-lines, do not exist. On the contrary, only quanta of action (in all probability) exist, units which extend over a certain small region of the whole four-dimensional world. The structure of the world would thus be rather that of a network of very small meshes or rings, which it is impossible to cut through without thereby destroying such single units (which in our case, however, is impossible); or of a mosaic picture, the single stones of which would no longer be divisible. The macroscopic laws of nature are nothing but statistical rules concerning these quanta and their distribution, which on account of the extreme smallness of them, hold with extraordinary accuracy in the orders of magnitude which we usually deal with. It is, however, an error to suppose that this macroscopic regularity must be founded upon a similar one in dimensions no matter how small. In these small dimensions there exists – within the limits of the Heisenberg relation – possibly or probably a certain freedom, so that every calculation of a future state of the world based upon the present state has in it an element of uncertainty, which becomes greater, the greater the time interval. In place of the old continuous causal chain, we should then have Reichenbach's continuous succession of probabilities, with the condition that every new determination of the path actually taken by the electron from among the various possible paths, would give a new basis for a calculation of the next following state by means of probability.

Here a further prospect is opened out for us. We quoted above the remark of Haas, according to which 'the old principle of causality ceases to have any object, if the exact knowledge of the present which it assumes, is not attainable' (according to Heisenberg's relation). This statement of Haas is, in my opinion, still too positivistic and subjectivistic. According to our realistic view, the assertion of positivism, that only what is observable can ever exist for physics, is wrong, and hence the circumstance, that according to Heisenberg's relation, impulse and co-ordinate, or energy and time, cannot be determined with absolute exactness, is in itself by no means a sufficient reason for banishing these ideas from physics. Why should not physics work also with elements which are fundamentally not directly observable, assuming only that these stand in a logically necessary relationship to those which can be observed? The true reason for accepting or rejecting a theoretical idea can never lie in the purely formalistic conditions of the process of knowledge, but only in the definite system of the content of knowledge, or in other words, always in ontological points of view, and never in epistemological. We shall adhere to this unconditionally. But it is exactly from ontological considerations that we can now arrive at quite a similar result to that of Haas.

If the world is really arranged as we have just indicated, the question of the absolutely exact and simultaneous knowledge of impulse and co-ordinate (position) is in itself without meaning. It is just as meaningless as, according to the kinetic theory of gases, it would be to seek to know the temperature at every single point whatever of a gas. This question is not unanswerable because we are not able to measure it for some reason or another, but because the theory itself, according to which the temperature is only an average value, causes this concept to lose its meaning as soon as we consider atomic magnitudes. If only quanta of action really exist, all single determinations of classical mechanics concerning times, spaces, substances (masses) and forces, are in reality statistical results having only a macroscopic sense, and for this very reason cannot be applied to the quanta themselves. The connection assumed by us between c and h, and also the dependence of world-metrics on mass distribution already existing in the general theory of relativity (see above, page 147), point rather to the four fundamental determinations of physics (time, space, substance, and causality) being based on a common physical reality. This cannot yet, it is true, be stated quite exactly, but we already have good reason for assuming that, in any case, substance and energy are not two fundamentally different things, but one and the same thing regarded from different points of view, namely in time-like and space-like projection respectively.

But times and spaces themselves are probably, from the objective physical point of view, only forms under which certain principles of order[154] of quanta appear to us, just as molecular collisions, taken as a whole, appear to us as temperature. Physically speaking, only quanta of action in a four-dimensional (four times infinitely manifold) order, or something closely allied to this, exist. From this point of view, it is entirely comprehensible when the question of the simultaneous absolute determination of co-ordinate and impulse, or time-point and energy in the quantum theory, proves to be a question in appearance only, and one just as little soluble as the question of the temperature of a single molecule. Bohr[155] proposed for the relationship existing between the two problems, the name 'complementarity.' The formation of quite new concepts becomes necessary here; we cannot reasonably expect that the 'forms of perception' and 'categories' familiar to us from the macro-world, will be directly applicable in the region of the sub-atomic micro-world.

One question still remains to be answered: What will be the influence of this new conception upon our whole view of the world? In the rectorial address of Nernst to which we have already referred, we find the remark that 'A certain similarity between the new view of the law of causality and theological lines of thought, cannot be overlooked.' But Born, on the other hand, has quite rightly pointed out[149] that it would be entirely wrong to proceed to introduce again a belief in miracles into our view of nature. Anyone who should put his trust in the Perrin brick that we mentioned jumping into his hand by itself, or in a stone falling through the air being diverted from the head of a passer-by by a suitable unequal distribution of molecular velocities on two sides of it, would be a lunatic, since improbabilities of this kind are practically completely identical with impossibilities.

For all macroscopic conditions, the deterministic predictability of events continues to remain as good as perfectly strict, and the astronomers, for example, will continue to calculate eclipses to the fraction of a second, and will continue to prove correct in their calculations. In this sense what Goethe expressed in the well-known verse concerning 'Great, eternal, iron laws,' remains true. But nevertheless, these laws have now acquired an essentially different complexion. They lose, so to speak, their rigidity, and hence the cold and dead aspect, the element of meaningless mechanism, which has been a cause of offence to so many spirits full of life and faith. The repulsive idea[156] that the world as a whole is a gigantic clockwork, which plays its tune mechanically, an idea which is an inevitable consequence of the mechanistic picture of the world, now falls to the ground. This idea, as we know, found its theological expression in the Deism of the age of enlightenment, according to which God determined the initial state of the world once and for all, a very long time ago, the world then going its own way according to laws laid upon it, without God having any further need to bother with it.

Religious thought has always protested against such deistic notions, which are only half concealed atheism, and has always insisted that if an idea of God is justified at all, it could only be a theistic one, that is to say, that God is to be regarded as perpetually active, and creation therefore not as a single act, but as *creatio continua*, as uninterrupted action. The new physics gives this requirement an overwhelmingly clear background. Not only does it actually allow us to regard the course of the world *uno aspectu* (see above, page 202); it now distributes, as we saw, contingency justly and uniformly over this whole process in space and time. In theological language, it maintains complete freedom of the divine will together with full practical validity of the "laws of nature" (as statistical rules) for mankind. While man, as a macroscopic being, does not escape from the rule of these laws, the divine act of creation itself is not bound by them, an act which, of course, is not in any case to be imagined as taking place in time, but as one of timeless determination (compare page 202).

And the really admirable feature of this whole creation is, as it appears, that by the sole means of the logic immanent in statistics, without the aid of further special and contingent general laws, the whole multitude of the world of phenomena: electron and proton, atom and molecule, crystal and rock, perhaps also cell and organism, may be deduced by successive synthetic steps, each new step forming a fresh and more complete synthetic unit, a *Gestalt* (see below, pages 401 and 433), so that a cosmos of individualities is formed, the higher of which always includes the lower and absorbs it.

Goethe's lively and organic view of nature, as expressed in the lines concerning the Earth-Spirit (*Faust I*), thus receives its justification also from physics, without the latter thereby making any concessions. And, perhaps in this way we may even approach nearer to the solution of a problem that has mocked mankind from the earliest times: the problem of the freedom of the will.[157] But we can only deal with this in a later part of this work (see below, pages 532 ff.).

It now only remains for us at the close of this first part, to enter into a short discussion of the process of physical knowledge as a whole. For we take the view as realists, that a discussion of this kind finds its proper place at the end, and not at the beginning. Epistemology exhibited false arrogance in supposing that it should precede ontology. In truth, it is first necessary to discover what the content of knowledge really is before it is profitable to deal with the process which has led to it, and the principles upon which it depends. As a matter of historical fact, a new theory of knowledge always appeared when new and great extensions of knowledge took place, and not conversely. The whole of classical epistemology from Locke to Kant is obviously orientated towards the great achievements of the time between Copernicus and Newton, as was the Greek epistemology towards the achievements of geometry, as Plato's well-known remark already expresses. In the same way, the great new wave of scientific knowledge will now bring with it new effects for epistemology, and a glance at the present literature proves that this has actually been the case to an enormous extent.

XVII

THE PROCESS OF KNOWLEDGE IN PHYSICS

All physical knowledge is acquired in three stages. It is first determined that this or that happens under certain conditions, or is simultaneously present at another point; in this way we arrive at *qualitative* laws. Secondly, the concepts connected together by such laws, such for example as length of pendulum and time of oscillation, or electric charge and repulsive force, are made into measurable quantities, and the *quantitative* law is then discovered; it is determined, for example, that the time of oscillation is proportional to the square root of the length of the pendulum. The functional equations, tables, or graphical representations thus obtained then form the basis for the third and last stage, the question of why; that is to say, the question of the explanation of the phenomena observed, and their quantitative laws.

The above facts result in a number of epistemological questions. First of all: What is a physical concept, such as temperature, quantity of electricity, density, capacity, or the like, and how is this concept connected with the method of measurement? Secondly: Upon what is the formulation of the laws founded, and what is the relationship of the law to the single case? Thirdly: What is the real meaning of 'explanation'? And what do the hypotheses which appear in it signify? Since we have already anticipated the last question, it only remains for us to discuss the first two; that is to say, on the one hand the formation of physical concepts; and on the other, the problem of induction, which term describes the process of setting up a general law from single cases. We will first consider shortly:

(*a*) *The Problem of the formation of Physical Concepts*, and will here take up again what was said on pages 77 ff. in connection with the concept of temperature. For the ordinary realistic view of physics, the thermometer is in the first place an apparatus introduced for the purpose of finding the 'real' temperature of a body, as opposed to our easily deceived feeling of heat (see above). But a large number of recent epistemological theorists, namely the so called conventionalists, now object to this view. For them there are not three different things: sensation, physical state of heat, and thermometer reading, but only two: sensation and thermometer reading. The latter is, for the physicist – so say the conventionalists – the temperature. This, they say, can already be seen from the fact that a mercury thermometer gives in reality different readings from a spirit thermometer. Which temperature is then 'real' ? Even when we use gases as thermometric substance, which at first apparently gets rid of the arbitrary element, since all gases have equal expansion with heat, it nevertheless again proves necessary on closer investigation to introduce a convention as to which gas is to be the thermometric substance, since in reality, individual gases again show small differences in their co-efficients of expansion according to Regnault's exact investigations. As a matter of historical fact, the air thermometer was followed by the hydrogen thermometer, and then by the helium thermometer. Does the choice of the latter really depend finally upon pure convention ?

The answer that we give from the realistic point of view is: No ! For in truth, we choose helium because this gas proves to be the most 'ideal,' that is to say, because in its case there is the most complete fulfilment of the requirement that the whole energy given to it in the form of heat appears again in the form of increase in molecular velocity. The real or 'absolute' scale of temperature is the thermo-dynamic: Steps of equal increase of temperature are such, as a result of which, the average energy of motion of the molecules increases by an equal amount. An absolute measure of the temperature would be obtained, for example, by deciding to regard as one degree that rise of temperature which is required for the energy content of a mol of an ideal gas to take up one unit of energy, or if we required the mechanical equivalent of heat to have the value of 1. Helium is chosen as the thermometric substance because its readings approach most closely to the absolute scale.

It is therefore clearly recognisable from this example that the road to the formation of physical concepts is paved with conventions, or, in other words, that the structure of physics cannot be built up without a scaffolding of conventions, but that this scaffolding can and is taken down when the building is finished. This is only possible, as we see most plainly from the example of therm-ometry, when the concept in question (here that of temperature) is itself made part of a larger and more comprehensive set of re-lations (here the kinetic theory of heat).   In the realistic sense, all formation of physical concepts contains a 'prospective potency'. it always points beyond itself towards the aim of know-ledge.   Only from the point of view of the latter do all physical concepts receive their final logical position.   In this connection, we may once more refer in particular to the concepts of mass and velocity on the one hand, and impulse on the other (compare pages 66, 181).   In this, as in numerous other cases, the final result exhibits the historically and psychologically later concept as, generally speaking, the prius from the logical and systematical point of view; this is particularly true of 'action,' as we have pointed out in greater detail above.

Leaving out some other sides of the problem, more of interest for the specialist,[158] we now turn to a further question, which has often been discussed in epistemological works; quite recently in a valuable study by R. Carnap:[159] namely the question of relation-ship between definition of concept and the method of measure-ment.   Carnap said in the study referred to:

'It has sometimes been supposed that a physical magnitude, such as time, for example, may have a meaning in itself, without any reference to the way in which it is to be measured, and that the method of measurement is a second question.   As opposed to this view, it must be strongly emphasised that the meaning of every physical magnitude consists in certain numbers being ascribed to certain physical objects.   Until it has been settled how this ascription is to take place, the magnitude itself is not settled and statements concerning it have no meaning.' Carnap then dis-tinguishes, for every physical definition of a concept, five stages, two 'topological' and three 'metric.'   It is first of all necessary to determine what we mean when we say that two magnitudes of the kind in question (for example, two masses, weights, forces, quantities of electricity, etc.) are equal; secondly, what it means when one is greater than the other, and these two relationships must be 'transitive' (that is to say, when $A \geq B$ and $B \geq C$, then must also $A \geq C$).   Then follows the settlement of the metrics; the 'scale form' must first be determined, that is to say, it must be determined under what conditions the differences of the two

Qs

magnitudes of the kind considered (two parts of the scale) are to be called equal; and then a zero point, and finally a unit, must be settled. It is easy to see that these five points actually occur in every definition of the measurement of a physical quantity, and this is shown by Carnap from a few examples. The realist will, however, always raise the further objection to the whole deduction that all this refers only to the measurement, but not to the explanation of the concept itself.

The precise statement of the question thus becomes as follows. Is it correct that the concept of time, or energy, or light intensity, or suchlike, only has a meaning in so far as 'certain numbers are to be ascribed to certain physical objects'? Or do these concepts, or at least many of them, already exist before the ascription of numbers? To the author the latter answer appears the more correct. It is not easy to see how those concepts, which physics holds in common with the daily experience of the most primitive of mankind, for example, time, temperature, intensities of light and sound, etc., are already related to numbers in this most primitive experience. The difference between two lengths of time, or two degrees of loudness of sound, or a high and low musical note, already forces itself upon the child and the savage, before they have any notion of counting and measuring, and pairs of concepts such as loud and soft, long and short, bright and dark, are therefore found at every stage of human thought. Are we then to say of a savage or a child, that they have no concept of loudness of sound or intensity of light, because they are not able, as is the trained physicist, to ascribe numbers to the scale of their sensation?

This is surely going too far. It seems more natural to the author, as to numerous other recent and also older philosophers (for example, Hofler, Poske, Volkmann, Frost), to distinguish two, or if necessary several stages in the formation of concepts. Even the savage has a concept of time. But the measurement of time is only known to civilised man, and this is developed out of the former concept quite gradually and with ever increasing precision. The same is also true of temperature. In this case, as we saw above, we can clearly distinguish two successive stages in the formation of the physical concept; so that, including the pre-scientific stage, we have to distinguish three stages. The first is formed by the simple concept of temperature based upon simple sense impression common to everyone, the second by scientific thermometry, in so far as it attains to the arrangement on a definite scale of the sensational series, on the basis of certain conventions.   The third begins with the question as to what is really at the bottom of heat, and the final definition in physics then runs: temperature is the average energy of motion of the molecules.

Conventionalism, in accordance with its origin in pure mathematics, is inclined to ignore both the earlier and the later stages. In the first respect it can take advantage of the fact that physics actually contains a number of new conceptual formations, which, arising directly out of measurement and not supported upon direct sense impressions, appear to rest upon pure convention. In this class belong, for example, almost all electrical concepts, since we have no direct sensation corresponding to these.   Nevertheless, even in the case of these, it is not true that formation of concept and measurement are one and the same, as Carnap maintained.   The concept of an 'electric charge' as a peculiar something, which somehow adheres to, or dwells in bodies that are rubbed, is formed without there being any question of any kind of connection, no matter how primitive, with a scale of numbers.

But the third stage also must not be left out of consideration, as is likewise done in Carnap's explanation.   Here also, electricity affords an excellent example in the strength of current.   According to the conventionalist view, physics could, and ought to, have been content with the fact that Ampère showed us from the magnetic effects, Faraday from the electrolytic effects, and Joule from the thermal effects of a current, how to ascribe to it scales of numbers, which scales were also easy to relate one with another.   But, in actual fact, current strength is not the deflection of a galvanometer, nor is it an amount of gas or silver, nor the increase in length of a hot wire.   All three are only effects of the current, which can be used to measure its magnitude, since they are functions of I. I itself is something quite different, namely the quantity of electricity flowing per second through the circuit.   This alone is the thing which corresponds to the physical definition of current strength.

If we wished, we could naturally begin the whole science of electricity, in a course of instruction or in a book, with the tangent galvanometer or the ammeter, and then define quantity of electricity as the product of current strength and time, as is actually often done in technical work.[160] But the conventionalist is mistaken in believing that this purely formal possibility, of beginning the logical structure at this end, guarantees the epistemological equivalence of this method with the correct definition of current strength which we have just given as quantity of electricity per second. Although the two equations $I=Q/t$ and $Q=I.t$ are mathematically identical, they are far from being so epistemologically, for the reason namely that quantity of electricity plays a fundamental part in the whole method of physics, whereas current strength is only a derived quantity. This example also shows that conventionalism, as well as apriorism, places obstacles in the way of understanding physics, by trying to measure it by the standards of mathematics. Only realism, which teaches us to base our understanding of the whole process of knowledge upon its goal, is everywhere able without difficulty to judge fairly the actual processes of investigation.

We now turn to the second problem referred to above, that of induction, with which, as we shall immediately see, an age-old problem of epistemology is closely connected, a problem that once in the Middle Ages played a great part, the problem of so-called 'Universals.'

(b) The Induction Problem

All 'laws of nature' depend upon the observation of single cases, however great in number these may be, whereas they themselves purport to be general, and hence to include an unlimited number of cases. When Galileo first made his experiments on falling bodies, he observed associated values of time, velocity and distance. These values, which we may imagine as set out in a table, then led him to the general laws $v=a.t$ and $s=\frac{1}{2}a.t^2$. The same thing happens to-day in every corresponding case. We have then, first of all, the question as to our right to undertake this generalisation. It represents mathematically an interpolation, and eventually also extrapolation, of any number of intermediate or external values on the basis of a curve or formula, which is to-day generally carried out in practice by plotting the observed values in a system of co-ordinates, drawing a curve through them as nearly as possible, and then deducing, either from the curve itself or by means of a formula arrived at from the form of the curve, any desired intermediate or external values. What right has the investigator in a case like this to assert something that he has not directly observed, and indeed in the case of extrapolation, which is much more important than interpolation, is not able to measure ?

We call the drawing of conclusions from the particular to the general the inductive method. The method just described forms the first stage of this process, which prevails throughout the whole of science; in it we are always dealing with the generalisation of single cases into a law. The second and still more important stage, the generalisation of special laws to construct more general laws, was already considered at length in chapter three.

We find the question above stated answered in epistemological literature in very many different ways. The rationalist of the Kantian type, and also the positivist, bid us seek the explanation in the fact that man reads the order of the 'law' into phenomena; according to the rationalist because he is obliged to do so upon the basis of an *a priori* category, according to the positivist and conventionalist, because it makes the facts easier to grasp as a whole, and of more use in practice. The observations, so say both, show us to all eternity, however numerous we may make them, after all only the single points of the curve we draw – and this is, as we know beforehand, furthermore affected with error. The curve, that is to say the general law, is an addition made by man on his own account, and in this respect he has a certain degree of freedom, since a curve which is to pass through fixed points, can always be varied within certain limits. If we nevertheless chose a quite definite law, for example, for gravitational attraction, the simple law $1/r^2$, and not $1/r^{2.0000007}$ which is equally possible, our reason is simply economy of thought. It is more convenient to calculate with 2 than with 2.0000007, and since the observations agree equally well with both formulae, we take the simplest.

Instinctive feeling surely warns everyone, who has these considerations put before him for the first time in so crass an example (it is, by the way, by no means fictitious),[161] that something after all must be out of order. Can it really be a matter of indifference, that we find in the exponent of the law of gravitation such an odd fraction, instead of the whole number 2 ? A cosmos appears to collapse, and give place to an arbitrary chaos. For this reason also, the representatives of a religious view of the world, who of late are glad to make use of positivistic theories of knowledge in order to evade the objectivity of natural laws, should give the matter further consideration.

Let us see whether we cannot look at the matter in another way. Let us take the laws of falling bodies as the simplest example for this purpose. It is first of all clear, that the first two simple laws found by Galileo are only arrived at on the basis of a certain stylisation or idealisation of the problem. They neglect the effects of air resistance, friction, and other disturbances. According to conventionalism or rationalism, this idealisation is simply the means of which the understanding makes use, so to speak, to handle phenomena, and nothing further. This would be true if it could be shown that one could have gone about the matter quite differently, but exactly the opposite is true. The resistance of the air, etc., are not thought of in order to simplify the problem of falling bodies, but are facts on their own account, which can be, and are studied on their own account (in aerodynamic laboratories). We may imagine its laws determined empirically for the present (whether they can be referred back further theoretically is another matter) and we are then able, on the basis of our knowledge of these laws, to describe 'falling in resistant media' by means of more complicated formulae, which are then used in place of the simple ones of Galileo, and already give a much more accurate result. In exactly the same way, we can then further consider 'sources of error': the rotational energy of the rolling ball, friction, and so on, and thus approach the actual state of affairs with ever increasing exactness, at the cost, however, of mathematical simplicity.

This whole system of successive approximations is – and this is the decisive point as regards our question – not arbitrarily constructed at any single point ; each of its parts is forced upon us unmistakably by nature again from quite different directions. We already know quite certainly from other considerations, that when a ball rolls down an inclined plane, friction must appear, a part of the energy must be transformed into rotation, and so on. All these are therefore anything but 'conventions for simplifying the problems of falling bodies'; they are rather sub-problems or preliminary questions, which are presented to us, and which we cannot alter. The 'stylisation' in Galileo's law, consists only in the fact that we quite consciously ignore for the moment these subsidiary effects, and only consider one influence, the constant gravitational field of the earth. Only when we have thereby rendered clear the mode of action of the main cause, can we then go further and add that of the subsidiary causes; and the physicist proceeds everywhere exactly in this manner. Galileo already spoke in this sense of the analysis and synthesis of phenomena (resolutive and compositive methods). Volkmann coined for the whole method the expression 'principle of isolation and super-position.'[162]

If the conventionalist were right, the whole of physics would have a totally different appearance. For it would then be much simpler, more 'economical in thought' to develop all functions needed in series, and determine the co-efficients empirically. This is actually done in numerous cases, but only *faute de mieux*, because we do not yet know the true partial causes of the phenomena in question, and in absence of this knowledge the analysis of the problem into its single parts is not possible. What use would it have been to Galileo (if the theory of series had already existed in his time) to have used, in order to reproduce the actual process of the rolling of a ball in a groove, a series $s = a_1 t + a_2 t^2 + a_3 t^3 + \ldots$ instead of $s = \frac{1}{2} at^2$, which could naturally have been made to represent all observations with any required accuracy. He would, it is true, in Kirchoff's words (often quoted by the positivists), 'have described the phenomena happening in nature as completely as possible and in the simplest manner,' but the very essence of the matter would then have been overlooked, namely the knowledge, based on his laws, that constant force produces constant acceleration.

The objection of course, still remains, that observation can never give the exact theoretical value, on account of inevitable error, even in a case where the phenomena are determined by only a single cause; and conversely therefore, the theoretical curve can never be concluded from observation with absolute necessity, but always allows of shifts or variations within certain limits. If we nevertheless choose the simplest law, this, it is said, is only to be explained by 'thought economy.' At this point we will leave the laws of falling bodies and take an example from the most recent physics, Moseley's law already mentioned (see page 170). Does anyone seriously believe that Moseley, when he first saw this extraordinary and simple relation between the frequencies of an

X-ray line and the atomic number of the corresponding element appearing from his series of observations, believed that thought economy required the linear form? He must at that moment rather have had the feeling – one could wager a hundred to one – that comes over every fortunate investigator at such a moment: the feeling of satisfaction of having once again torn from Nature one of her secrets, of having discovered an important fact yet unknown, and not of having read it into Nature. The conviction must have forced itself upon him, as in every such case, with irresistible power, that something was behind it. This behaviour of the X-ray lines, simple beyond all expectation, points unerringly to a deeper reason for which this linear connexion must exist. And this expectation has been confirmed by the whole later development of the atomic theory, as we have seen.

Neither positivist conventionalism or apriorism has a satisfactory explanation of all this. The most important psychological moment, that of complete surprise, is entirely ignored by both, for both maintain·that it is really the human mind itself that first reads the law into nature. Only realism, which actually sees in such a 'discovery' the uncovering of a state of affairs already existing apart from the activity of our understanding, gives full value quite naturally to this experience, which is decisive from a psychological point of view. But, it will be objected, here also the simple linear law is actually only a first approximation. In reality, the curve of Moseley's law is not a straight line at all, it is slightly curved. This is undoubtedly true, but no physicist in Moseley's position would allow himself to be blinded by a more exact determination to the fact that, all the same, the approximation to a straight line must also have a special reason, but that secondary causes of departure from exact linearity are also present.

As long as humanity has been engaged in physics, its procedure has never been otherwise; and as far as I know, this manner of drawing conclusions has never deceived it. This method exhibits what Volkmann has so neatly described as the principle of isolation and superposition, and what we can also describe equally well as the principle of permissible neglect. 'We live in a world,' says Zilsel in his valuable study on the *Problem of Application*,[163] 'in which one may neglect circumstances and still know something. . . . Though the world always remains a variable thing, a thing full of partial indistinctness, these variations still compensate one another, these indistinctnesses are so fortunately distributed, that man is able to determine quite definite relationships in spite of all vagueness, although, it is true, we have to keep on adding to these determinations without end.' It is only necessary to add to these accurate statements, that both the matter and manner of neglect is completely dictated to us by the object, at least in by far the most numerous and important cases.

It follows from this that the whole principle relates only to the process of knowledge, and that the content of knowledge finally remains untouched by it. The old epistemological controversy as to whether the world is rational in itself, or whether we first make it so (by means of categories or conventions) is therefore answered by saying that it is 'rationalisable'; from which we can surely draw the further (metaphysical) conclusion that it is itself finally rational. All that is further to be said regarding this conclusion has already been given in the discussion of the problem of contingency (page 214). For epistemology, we may take at this point of our First Part as sufficient the statement: *The world is so constructed, that it can be known, by means of rational concepts, in rational judgments (laws) of increasing approximation, and it prescribes the path of these approximations to our understanding.*

We cannot leave this subject without devoting a few words to the old problem already mentioned above, which in reality lies behind this whole discussion, the problem of 'universals' of the Middle Ages. In those days the controversy turned about the question whether concepts as such – this word being taken in the sense of the classical Greek philosophy – are to have reality ascribed to them, or whether they are mere names (*flatus vocis*). The representatives of the latter view were then called nominalists, those of the doctrine, derived from Plato, of the reality of concepts, realists. A certain compromise between these opposite views was given by the doctrine of Abélard, according to which 'universals' (general concepts) are for human knowledge *post rem* (in the sense of nominalism), in nature *in re*, and for God *ante rem* (in Plato's sense). We of to-day are inclined to smile at this controversy, in which it was discussed in all seriousness whether the concept 'table' or 'horse' exists as such in a metaphysical sphere; nominalism appears to most people to-day as self-evident. But Windelband is right, when he says (*Geschichte der Philosophie*, page 245) : *Mutato nomine de te fabula narratur*; that is, as long as present-day science cannot state, with full certainty and clarity, in what consists the metaphysical reality and effectiveness of what we call, for example, a law of nature.

Let us consider as an example a simple empirical law, such as Snellius's law of refraction or Newton's law of gravitation. Present-day nominalism (that is conventionalism) says: There is in nature in reality no such thing as a law of refraction, but only light waves, water, air, glass, and so on; and likewise no law of attraction, but only bodies, acceleration, and so on. The law exists only in the mind of man, who uses it to arrange phenomena on this thread. The view of aprioristic rationalism is similar. Very well, but the nominalist leaves one question unanswered: if it is a matter for us to form concepts such as angle, sine, light ray, and laws such as the laws of refraction, how does it come about that from a hundred observations of pairs of angles α and β, we get exactly the law $sin\ \alpha / sin\ \beta = n$, and not $tan\ \alpha / tan\ \beta = n$, or $\alpha / \beta = n$, or any other equation ? These facts can be turned around as we please: it is impossible to eliminate the fact that, in the totality of actual cases, there lies a something which is simply forced upon us, an 'objective' in Meinong's phrase,[164] and that it is just this given thing that is of importance to us; and not all that we ourselves do in connexion with it. But the corresponding statement is already true of the simplest formations of concepts. Even concepts such as plant and animal, or oak or beech, have not been put into nature by us or invented in order to make it more easy for us to grasp; they are forced upon us by an objective something that was there long before men with power of conceptual thought existed.

An epistemology which blurs or ignores such obvious facts, to emphasise by contrast always only the little that man for his part does in such formation of concepts or induction of laws, is insufficient, since it overlooks the main point. Realism is obliged to-day, to protest against such subjectivisation of knowledge, as it did in the time of Anselm and Wilhelm von Champeaux. In our opinion, therefore, we should recognise that to-day too, the existence at least of *universalia in re* (possibly also *universalia ante rem*) is a well-founded thesis. This becomes quite clear when we consider the fact referred to above, that every determination of one of the general laws in question immediately opens up the problem as to why this law reads just so and not otherwise. Consider, for example, the part played in theoretical optics by the law of refraction (see page 103). This question becomes meaningless, or must be entirely avoided, if, as conventionalism maintains, we are only to see in the formulation of empirical laws guides for finding our way most conveniently in the world of phenomena. We saw above (page 43) how in reality, every special law demands an explanation from more general laws, and how there thus finally results a total system of knowledge in which every single thing finds its quite definite place. This leads us to a final and concluding thought, the insight into:

(c) *The Convergency of Scientific Investigation*

If the part played by the subject of knowledge in arriving at the knowledge of nature were as great as conventionalist and aprioristic epistemologies would make it, one thing would remain quite incomprehensible, namely, the agreement between results found by the most different means, an agreement which repeatedly exceeds our boldest expectation, and only appears subsequently; an agreement not only of single numerical values, but also of theoretical ideas. Better than anything words can give, is the impression of two such pictures as the two annexed figures taken from Auerbach's *History of the Development of Modern Physics*. Both represent the historical development of the determination of an important physical constant: the first the determination of the mechanical equivalent of heat, the second the specific charge of the electron (page 126). They really hardly need any comment.

Conventionalist positivism is obliged, in order to explain such obvious convergence to a quite definite limiting value, to take refuge in the meaningless excuse, that the conventions on which the method of measurement is based are changed until agreement is attained.[165] This may pass as long as we are dealing with a few methods concerning which it is just possible to show that, in each case, the method has been changed in its fundamental assumptions until it agreed in its results with the old. But it becomes absurd when, quite unexpectedly and without any further convention or any change in earlier conventions, half a dozen different

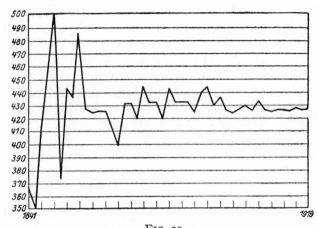

FIG. 39

Determination of the mechanical equivalent of heat (in
metre-kilograms per kilogram-calorie)

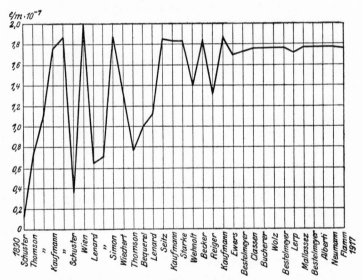

FIG: 40

Determination of the specific charge of electrons

methods also lead to the same result, as is beautifully illustrated
by the determinations of the Loschmidt number (see above). No
unprejudiced physicist will, in view of such cases, allow himself
to be talked out of his direct feeling that these agreements have

been enforced, not by his subjective methods, but purely by the object itself, which is only one, and hence can obviously only give a single value for the constant in question, without regard for the side from which we attempt to approach it.

This assumption of the existence of truth is the fundamental assumption of all investigation whatever, and I am glad to be able to state that this also is the opinion of the three greatest living German physicists, Planck, Einstein and Sommerfeld, in express opposition to conventionalist and pragmatic views recently expressed by certain representatives of the Viennese School of Mach.[166] The view of Planck is that longest known. But Sommerfeld has recently, at the 1929 congress of physicists at Prague, openly opposed a previous speaker[165] who expressed pragmatic and conventionalistic ideas in a crude form; he said that he had discussed these questions with Einstein, and could declare, also in the latter's name, that mankind should be thankful from the bottom of its heart, that Nature has equipped it with the bodily and mental powers for the gradual discovery of some of her secrets.

This is a clear and unambiguous renunciation of the doctrine according to which man does not discover the truth, but so to speak, invents it for the practical purpose of finding his way about the world (pragmatism).   For example, Frank maintained at that Congress, that we really mean nothing more by the 'existence' of Planck's quantum of action, than the actual agreement between the numerical values found in all sorts of different ways of certain factors in equations, (for example, in the law of radiation, or in photo-electricity).   We oppose to this purely formalist view, which overlooks the essence of the matter, the fact, which cannot be abolished by any nominalist sophistry, that this very agreement has not been brought about by the conventions in question, but – to the greatest surprise of the discoverer himself and of everyone else who hears of it for the first time – was only determined after the event and by experience.   And it is equally superficial, as we have already seen, when Frank in the same address again attempts to regard the purely formal agreement of the laws of propagation, in Mach's manner, as the only inward meaning of the assertion of identity in nature between light and electro-magnetic waves. What we are actually dealing with here has been made clear enough above.

We therefore adhere, in spite of Frank, to the 'school philosophy' which he so despises, and which according to him is wrong in starting from the opinion that the truth exists prior to our judgments concerning it. Frank and others attempt to make this opinion ridiculous by saying that one might just as well ask whether the irrational numbers of mathematics exist, which are in reality only another name for convergent infinite series without rational limiting value. This comparison fails in two respects. In the first place, because in pure mathematics, convention, that is definition, actually plays quite a different and more decisive part than in physics, where definition must always be adjusted to experience. In the second place, because even for pure mathematics, as we have already stated above (page 48), it is not by any means settled that the purely nominalist avoidance of the problem really abolishes it. Even in this subject, there may be much more truth in the old Platonic doctrine of knowledge as recollection, than nominalism, which is to-day completely dominant in mathematics, would like to admit.[167]

However that may be, in physics at any rate, there cannot be any question of such pure nominalism, since here everything of decisive importance comes not from the subject, but from the object, and this not only holds for the values of constants; it holds – and this is of much greater importance – finally also for the whole system of theoretical knowledge itself. This also obviously converges towards a single end. However often several various theoretical views first compete with one another, and however often it may also seem that two, or several such, might stand with equal authority alongside one another, the history of physical knowledge shows in reality that ambiguity of this kind does not last, but that one theory, the true one, finally wins the day over the others, which are false. If conventionalism were right, there would be several quite different sciences of physics. Carnap, for example, actually counts up, in an otherwise excellent paper on the problems of physics,[168] four different systems of physics.

If we look more closely, it is easy to recognise that the systems of physics which he puts forward as systems of equal authority, in reality only represent partial truths, and hence by no means exclude one another, but rather supplement one another's deficiencies, inasmuch as each system only expresses certain sides of the whole physical reality. For example, it is distorting the facts to put the quantum theory as a system alongside relativity. We have sufficiently shown, above, that the general theory of the field given by the latter, and the fundamental conception of the quantisation of all matter which is characteristic of the first, are for the moment not deducible one from the other, with the result that a certain dualism doubtless exists; but for this very reason there is no contradiction between the two bodies of doctrine, one being for the moment as necessary as the other, if we are to understand the totality of known physical phenomena. Hence, conventionalism is completely unable to bring forward any proof that even as many as two systems of physics exist.

The following point of view is still more important in this connexion. Even if it must be conceded to conventionalism that, in the beginning, a greater or less freedom in the formation of subsidiary ideas, concepts, measurements, and so forth, exists, and also that initially the active contribution of the subject of knowledge plays an essential part, these arbitrary elements (or 'particular determinations,' to use Lange's expression) are more and more eliminated as time goes on, as we have seen from the history of the concept of temperature. The 'degrees of freedom' therefore diminish in number, and as far as we can see, only one possibility really remains as a final result. Everything else simply converges towards this, often in an overwhelming manner. Conversely, on the other hand, we find that the number of difficulties meeting every totally or partially wrong theory grows, at first perhaps only slowly, but finally like an avalanche; it is obvious that in this case, to make use of the instructive arithmetical expression, we are confronted with a divergent series.[169]

This divergence in the one case, and convergence in the other, form, in the author's opinion, the decisive argument against every kind of theory of scientific knowledge which lays too great a stress on the part played by the human mind, for they are historical facts, which could not themselves be deduced *a priori*, but are simply recorded *a posteriori*. They prove that the tacit assumption behind the whole of scientific investigation, our trust in the possibility of knowledge concerning an objective set of facts, is not an unjustifiable prejudice, but is based upon good reason, and that we are therefore really approaching a final condition of physics, in which all part-discoveries are comprehended in a single unitary logical structure, that may naturally be approached from all sorts of different points, as in teaching, but in itself is always the same, no matter at what point we begin.

This point of view is proof against two final objections which are frequently raised in course of the discussion. The first claims to be based upon the undoubted historical fact, that several times in the history of physics, a unitary physical world-picture has been supposed very near, only to be followed by the necessity, as a result of entirely undreamed-of new discoveries, of beginning again from the beginning. We may recollect the fate of the mechanistic world-picture of the 18th century, or the electron theory round about 1900. This objection plays an important part in the essay of Chwolson's referred to above. He tries to prove that man makes himself ridiculous when he supposes himself to have exhausted the possibilities of the universe by the little that he is able to learn of natural law upon his earth. This objection depends upon a false generalisation which amounts to much the same thing as if people had believed, in Cook's and Magellan's time, that new continents might still be discovered *ad infinitum*. A limit was set to such discoveries by the finiteness of the earth's surface. A time had to come when the last continent was discovered, and, as we know, it came with the discovery of the Antarctic. From now on, geography can plan out its investigations of the few blank spots still to be found on the map, and can thus tell us exactly what still remains to be discovered. No sensible man would start off to-day without any plan to travel about the world in the hope of finding unknown treasures, kingdoms, and countries.

We have reached a similar point in physics to-day. We are now able to state very exactly where lie the problems still to be solved, a fact demonstrated by the whole of our discussion in previous chapters. It is a vain hope, still sometimes believed in by the laity, that a fundamentally new discovery could be made in the manner of Galvani or Robert Mayer, by random experiment or purely speculative theorising. All really great theoretical and experimental discoveries of the last decade have been made by trained specialists, and by far the greatest number by those who had long stood upon the highest points of science before they made the discoveries which astonished the world. The time of happy chances is long passed, a fact most clearly proved by the high point already reached in the development of theory. This first objection therefore amounts to nothing at all.

The second objection, which is often raised at this point, is of no greater potency. It is supposed to be a danger of the view we are putting forward, that the belief in our closeness to a unitary solution of all physical problems will destroy the attraction towards investigation, and result in stagnation of our mental life, at least in this department. The answer in the first place is, that if a problem is really solved, it is solved. But in the second place, it is a great mistake to suppose that the attainment of such a unitary system of physics as we have foreshadowed, would cause investigation to cease. It would rather only begin in greater earnest. Is mathematics at a standstill, because, generally speaking, it proceeds deductively? Has the investigation of electrical or astronomical phenomena suffered under the fact that simple fundamental laws were found for them? Exactly on the contrary, this fact only sufficed to stimulate investigation. This is easily explained. For even if the law itself is a simple unity, the number of special cases and consequences which it includes are still infinite, and afford the human mind an endless field of activity in seeking ever new combinations and comparing them with experience. Who knows how many, and still unknown phenomena we may be able to explain by means of the electron theory? It is probably only a portion of the whole of reality, and hence itself only a special case, but it already includes an infinite number of single phenomena.

This would be still more true of à system of explanation including the whole of reality. The result would be investigation and discovery, combination and recognition in experience, of which we may form a feeble picture by recollecting the triumphs of the optics of crystals and of modern spectroscopy, the coal-tar colour industry, and wireless telegraphy. Even the converse of this, the 'chance discovery' of new phenomena, is not in the least impossible. For we experience this every day, even in departments already provided with a fully worked out theory. They then can be arranged within the existing system of explanation, but a very considerable amount of mental labour is often necessary for this. The reason for the possibility of the enjoyment of such discoveries and of such problems requiring explanation, is to be found simply in the finite and limited nature of the human mind, which is quite

Rs

unable to foresee and find all the infinite number of combinations and single cases. It is therefore far from our purpose to support an unimaginative self-sufficiency which would be satisfied with the extent of our knowledge and not see the problems which still lie beyond.

But on the other hand, it is true, my intention in these first chapters has been to show as plainly as possible, that, while over-estimation of our present state of knowledge is unjustified, its systematic under-estimation, and the innumerable unfounded criticisms of it, are still less justified. These criticisms are wide-spread to-day, and, unfortunately, are often able to claim the support of great names, which are not altogether without part in them. As we have sufficiently explained, these criticisms them-selves require a great deal of criticism; this fact remains true whatever our epistemological point of view.

If we now finally inquire what is the result of our whole investi-gation as it affects epistemology itself and its two fundamental problems, those of reality and truth, I believe that we may sum the matter up as follows.

In the first place we must recognise that physics, in any case, is never concerned with the actual problem of reality in epistem-ology, and that it may safely leave philosophy to answer the question of the fundamental nature of reality. In this point I am to a great extent in agreement with Carnap, who, in the two essays already mentioned 'On the Problems of Physics,' puts the matter as follows: 'As opposed to a view very widely taken, it is of no importance for physics whether we take the phenomenalist-realist view, and call the first region (he means that of sensation), for example the colour blue as we perceive it, mere appearance, and the second, for example the corresponding electro-magnetic oscillations, "reality"; or, in the converse positivistic sense, we describe the first as the "really given," the second, as "only concep-tual complexes of the sensational content." Thus we do not say: "where this blue appears, there is in reality a certain electron process," and also not: "in place of this blue we use the fiction of such an electron process in order to make calculation easy." Physics expresses itself neutrally by means of a purely formal relation of ascription (Carnap calls it a psycho-physical dictionary), and leaves such interpretations to non-physical investigation.'

This clear statement, which cannot be bettered, cannot be too often made to confront the positivist and conventionalist physicists who, to-day, are fond of putting the matter as if physics and Mach's epistemology were fundamentally identical. On the contrary, molecules and light waves, fields and their tensors, have just the same kind of reality as stones and trees, plant cells or fixed stars. In this sense I spoke in earlier editions of this book of a 'relative realism' with which physics can in any case be satisfied, and I do not retract this judgment in any way. But I feel able to answer for daring to go a step farther in this new edition, and touch here shortly on the philosophical problem itself.

The reasons in favour of 'critical realism' as against mere 'phenomenalism' of the Kantian kind on the one hand, and Mach conventionalism on the other, will not be given here, since this would require an investigation by itself, and furthermore they have been fully stated in the works of Ed. v. Hartmann, Kulpe, Erich Becher, Messer, N. Hartmann.[170] I may refer at this point to my own short statement in *Chief Questions of Scientific Philosophy*.[170] Here, however, we will assume the justification for the standpoint of critical realism, say in Becher's sense, and discuss shortly the question of how, in view of the point at which we have now arrived, it is related to the 'physical realism' earlier referred to, from which it has been developed by further purification, to use Erich Becher's happy expression.[170] The answer appears to me to be, that physics is now on the way to obliterating completely the boundary between the two.

The fundamental thesis of critical realism is the statement that a 'real external world,' that is a reality which transcends consciousness, not only exists (Kant also teaches this), but that within certain limits, we are able to make valid statements concerning this 'thing in itself' (the Noumenon), which Kant denied. Now it is evident, and is also agreed by all critical realists, that this world of things in themselves cannot be the 'rigid lumps of reality' of the older mechanists with their spatial and temporal relationships, and mutual force effects, and hence physical realism in this older form is certainly not identical with what critical consideration now finds to be fundamentals. Critical realists have, nevertheless, striven to save as much as possible of these statements of physical realism; thus Ed. v. Hartmann,[171] the real founder of modern critical realism, and also recently Erich Becher,[172] have been at pains to prove, that we must ascribe a transcendental, and not only an immanent validity at least to statements regarding time, if not also to those regarding space. Both authors have likewise maintained the transcendental validity of the category of causality, although Becher wishes to regard this only as a plausible hypothesis.

We are now confronted with the interesting fact, that physics itself has already progressed to more radical conclusions in this respect than the epistemological realists were willing to concede to it. According to present-day physics, as we have seen above, neither the time-space determinations nor the old categorical ones (substance and causality) are anything more than bases for statements which cannot be further reduced. The physics of to-day, in the first place, does not any longer know separate determinations of time and space, but, as is proved by the relativity theory (also by the quantum theory with its emphasis on action), only the Minkowski union of both. This already leads to the conclusion that it is at least a doubtful thing for epistemology to undertake on its part to measure space and time with different measures, and I was not convinced that Erich Becher, with whom I had a detailed discussion some time ago,[173] could refute my objections in this respect. In the second place, present-day physics, as we have seen, no longer recognises the two categories substance and causality as fundamental, so that all attempts to save the transcendental validity of causality are to-day open to the objection that epistemology is more papist than the Pope, more realistic than realistic science itself, if it attempts to maintain unconditionally this point of view.

What physics still needs – though an exact definition cannot be strictly given even to-day – is rather a something which is at once matter and energy, substance and action, and that is arranged in the world of the relativity theory in a four-dimensional manifold, (this word being taken in the purely abstract mathematical and not in the geometrically concrete sense). Physics thus actually arrives at the statement recently formulated by Driesch as starting point of his own critical-realistic epistemology:[174] 'I have something consciously ordered.' Or Eddington has stated it in other words as follows: '"Give me matter and motion," said Descartes, "and I will construct the universe!" The mind reverses this (in present-day physics): "Give me a world – a world in

which there are relations, and I will construct matter and motion." '
(*Space, Time, and Gravitation*, page 198.)   Only we must not here
overlook the fact that this result of physics is not the result of
epistemological meditation, but the final formulation (with an
element of anticipation, however) of the content of knowledge.
We have already, in discussing the contingency problem, hinted
that this coming together of the final fundamentals, since it is
not arrived at through an *a priori* setting (as the apriorists
thought) but by convergence appearing *a posteriori*, perhaps – or
may we say probably ? – tends towards a rehabilitation of the old
belief in the rational foundation of the objective world (Plato's
doctrine).   In this sense, another thorough critic of the relativity
theory, Winternitz, says at the close of his excellent book[175] on
this subject: 'Light may perhaps be thrown upon this darkness
(the problem of knowledge) when we consider that not only is our
reason a part of nature, but that nature must also in some way be
concerned in reason.   If we no longer see in nature and reason two
realms, worlds apart, which have nothing in common, we are
not therefore obliged to put reason on a level with a mindless
nature, and so reduce it to a mere biological tool, to the "lantern
of the will": we have equally good grounds for believing to be
effective in the world external to us that power for which, and
through which alone, there is a world; namely reason.'

However this may be, this short reference is sufficient to show
that we have arrived at a point at which the sharp boundary,
rightly drawn by the critical realism of the past between itself and
physical realism, begins to vanish.   The only category that the
first needs to postulate, a category which is transcendentally valid
(that is, with reference to things-in-themselves), the category of
order,[176] cannot be very essentially different from the order which
distinguishes the world of the relativity and quantum theories,
for the latter also is not a form of perception, but simply an
abstract mathematical scheme (see above page 219).   It is thus
scarcely possible any longer to find a reason for distinguishing
between the two.   It may be remarked in passing that order, in
this sense, includes distinguishability and hence numerability, for
if anything is to be ordered, it is naturally first necessary to distinguish one thing from another, and what can be distinguished
can also be counted.   But in the transcendental validity of order,
we have then further to imagine as included everything that was

said above concerning the reality of universals. If order is given, relation is also given, and the functional equations of mathematical physics are nothing but the expressions of such principles of order. These – we state this in strict opposition to all mere nominalism – are not put by us into things merely in order to enable us to oversee them and approach them practically, but are found out by us from the thing, although only by way of continual approximation to the truth.

This brings us to the second fundamental problem of epistemology, the old question: What is truth? This question has already been answered in the whole of our previous discussion implicitly, and explicitly in the last chapter. Truth, at least in physics and altogether in science, is not generated by the mind, but only grasped by it, though only, it is true, by way of an infinite process of approximation. In our opinion, Lichtenberg is right in saying that truth is the asymptote of research. Our knowledge of reality is not a conceptual structure, for ever suspended over things, and based solely on conventions and definitions created for the purpose of thought economy; it offers us a picture of reality, that is of an objective body of facts already existing before all knowledge; a picture which is continually becoming more and more adequate.

The decisive reason for this view of the matter is the convergence of the part-results, which we find by experience and which is not settled *a priori*. This can by no means be reduced to the unambiguous definition which is here, again and again, upheld by nominalist positivism. Schlick, and with him the whole of the new Viennese school,[177] maintain that all our scientific statements are not actual statements concerning reality but only concerning our designations of real things (they are supposed all to amount to our being obliged to keep this system of designation in agreement with itself). This theory, as we believe to be sufficiently proved by the foregoing, is refuted by the simple fact that our mode of designation had obviously nothing to do with the constants of the radiation law proving identical with those of photo-electricity, or with the Loschmidt number derived from the elementary electric quantum agreeing with the value derived from the kinetic theory of gases.

There is nothing to be done about it. Here the object itself, the thing meant and not our meaning of it, stands like a rock in the middle point of the decision.   These convergencies depend upon it and not upon ourselves; in the opposite case, when a theory fails to work, no nominalism and no apriorism help us out of the difficulty.   The grandest, and when we think about it, the most overwhelming feature of the whole development is that this apriorism or mere formalism (they are both only different forms of genuine modern subjectivist attitudes), though quite false as starting points, finally receive their rights at the end, namely at the point where reason recognises, as we said above, that it finds itself again when it reaches the end of nature.   But this objective spirit behind nature is denied by positivism itself, which ascribes too great a part to subjectivity within science.   It attempts to mathematicise physics, that is, to make it an affair of pure construction on a freely chosen basis at a point when this cannot be done, because we have first to learn from nature how and what we are to make our basis.   At the same time, the real final aim, the complete rationalisation of physics, the final possibility of which is our hope to-day, escapes it, because, being satisfied with formulae, it attempts to get rid of things too early, and thus misses the last formula which alone really dissolves the thing, or in other words, is able to close the circle spirit-nature-spirit.   In our view, on the contrary, the saying 'man has nothing but what heaven has given him,' is true also for physics.

PART II

COSMOS AND EARTH

I N T H E first part we attempted to state what present-day science can tell us concerning the general laws of non-living nature. If we describe this knowledge, in Dingler's words quoted on page 215, as the 'theoretical prime structure' we have by contrast, as we there explained, the 'historical prime structure,' that is, the sum of our knowledge of what is actually realised within the frame of these general laws. The philosophers of the Middle Ages were accustomed to describe the former as the *essentia* of the world, as contrasted with its *existentia*. The natural sciences are correspondingly divided into the two general sciences of physics and chemistry, and the special sciences, such as astronomy, geometry, and so on.

We may here make a few remarks concerning the classification of the sciences. The Rickert school has attempted to divide the whole of the sciences into the two main groups of the 'nomothetic' sciences, that is those containing laws, and the 'idiographic' sciences; the natural sciences being quite generally placed in the first group, history and the so-called moral sciences in the second group; the latter, according to these authors, being fundamentally concerned only with what has happened once and never recurs in the same form. This position has met on other sides with serious objections, which in my opinion are justified. But in reality the special natural sciences, for example geology, are idiographic in exactly the same sense as history, for example. Geology also deals with a process only occurring once, the history of our earth, and in it also the general laws (in this case those of physics and chemistry, in the former case those of psychology, political economy, and so on) are only means to the end of explaining this unique process, the history of the earth. The difference lies only in the complexity of the objects. Human actions in history are so overwhelmingly complicated that general laws are very much more difficult to discover, and hence much less capable of application than are the physico-chemical laws in geology.

249

The difference is therefore due to the material and not to the method, as has been convincingly shown, particularly by Erich Becher in his profound book *Moral Sciences and Natural Sciences.*[178]

The question of the unambiguity of natural laws does not need to be referred to again at this point. The physico-chemical laws hold in macroscopic dimensions, as we have seen, with practically unlimited accuracy, so that we are able to take as a basis for our present purpose, without noticeable error, the classical determinist world-picture, only reserving the right to enter into the question of departure from strict causality if necessary, in view of a particular case. (We shall see that such a case is given by biology.) According to this classical picture, the present state of the world and the present laws of nature give unambiguously its state in the past and future (see page 202), and in practice every such science, for example, astronomy, amounts to such calculation forwards and backwards.

The question is what, in such circumstances, science can find out, and has found out, concerning the *existentia* of the historical prime structure. There are three practical hindrances in the way of such knowledge. In the first place we do not yet know the fundamental law of all physico-chemical processes, but only single laws, which as such are valid to yield special conclusions in special cases, but for that very reason are only to be regarded as of limited validity. Every law, even the law of energy, regarded from this point of view is an approximation, which holds or would hold with given premises, concerning which we cannot always say with what degree of accuracy they are fulfilled. This is only another expression for the truth discussed above on pages 77, 79, that there is, strictly speaking, no single pure mechanical, electrical, or other similar process. Since we therefore cannot know whether whole regions of physical phenomena still remain for the present entirely unknown to us, we are, on the one hand, entirely dependent upon extensions made by pure supposition, and on the other hand, never absolutely certain in practice, that a quite beautiful calculation may not be illusory, because a factor acting in the process in question has been completely ignored by reason of our ignorance of physical law.

In the second place, we always know only a part, more or less small, of the real world. Our senses and instruments soon reach a limit which cannot be exceeded, and this is true not only as regards spatial extent, but also as regards accuracy in detail. While we are able here on earth to observe singly ultramicroscopic bodies of 5 $\mu\mu$ diameter, details kilometres in size escape our observation on the moon, or Mars, or the sun. Hence in many cases our calculations can only be approximate, and we are obliged to neglect more delicate influences (see above, pages 231 ff.). Another reason also frequently enters in. Cases exist where, theoretically, we might acquire knowledge of all necessary details, but in practice are unable to realise this because we cannot make an infinite number of observations, or even only because we cannot pay for them.

Finally, a further difficulty arises in many cases from the fact that even with sufficient, or more than sufficient, observational material, we are not able to carry out the theory of the processes in question quite exactly, because our mathematical powers fail us.

In other words, our knowledge is limited; 1, in respect of the variety of the physical laws; 2, in respect of the variety of existing states; 3, in respect of our theoretical means for mastering the problems. It is easy to find examples of all three cases.

A good example of the first is found in the attempt to calculate the age of the earth on the basis of its cooling from the sun temperature, taken as initial, down to its present surface temperature. These calculations have been carried out with greater or less consideration of various details. But all these became completely illusory at the moment when we were obliged to recognise that a factor probably affecting the result in decisive fashion had been completely ignored, namely radio-activity. All earlier calculations depended upon the assumption that the earth itself contains no further source of energy, which would be partially able to make up for its external loss of heat, apart from the quantities already included in the calculation (chemical processes, etc.). Then came the discovery of radio-activity, and the proof by Elster and Geitel that the earth contains, at least in its rocky crust, an enormous store of radium and other radio-active bodies. These form a new source of energy, from which it can be supplied in such enormous quantities that other terrestrial sources of energy are almost negligible as compared with it (see page 127).

Meteorology forms a good example of the second deficiency in our means of knowledge, though, here also, all the factors are not within our knowledge, (we know for example even to-day very little concerning the influence of electrical processes in the sun on the weather). But even if we assume for the moment that we knew completely the nature of all factors concerned, it would still be impossible to prophesy the weather exactly in all its details, simply because we are unable to set up as many posts for observation as would be necessary.

Regarding the third point, the difficulties connected with theoretical computation, it is sufficient to remind the reader of the 'problem of three bodies,' that is to say the calculation of the motions of three heavenly bodies attracting one another according to the law of gravitation. This problem is insoluble in its strict form, since the differential equation in question cannot be integrated. Astronomy, it is true, is able to manage in spite of this, since methods of approximation exist, which enable the problem to be solved with any desired degree of accuracy for every single case that occurs (theory of disturbances, theory of the moon, etc.). A similar way out is found in numerous other cases of the kind, approximate mathematics taking the place of a mathematics of precision. But there are still numerous problems which have hitherto found no satisfactory solution, the mathematical difficulties being too great. The theory of the motion of liquids, taking friction into account, is one example, and there are many others. It is then only possible to deal with individual cases, certain simplifying assumptions being made, which more or less correspond to some actual practical case.

Are we then to admit that those are right who are fond of talking of the insoluble secrets of nature, of a limit to our knowledge, and of our present day ignorance, which they believe will always remain ignorance? I do not think that this should be done without further question. The whole history of science is rather a single body of testimony to the fact that nothing does more to remove our limitations in the three respects named, than our insight into the insufficiency of our knowledge in respect of one or other special problem. All research grows with and out of the tasks which are set it. When, for example, mathematics was not able to supply astronomers with the means of integrating the three body problem, a search was made until another way was found which practically did what was required.

Speaking broadly, the general sciences (physics, chemistry, and also mathematics itself) have profited as much from the special sciences (astronomy, etc.) as conversely. History shows us everywhere that insight into the general laws of physics and chemistry has arisen in innumerable cases out of, and with reference to, special problems presented by astronomy, geo- and astrophysics, etc. The classical example is the mechanics of Galileo and Newton in their relation to astronomy. Hence we should have no doubt that the future can and will bring us innumerable scientific achievements of this kind in this field, such as Newton's mechanics of the heavens, the application of spectrum analysis to the heavenly bodies, etc.

The problem for unprejudiced consideration thus remains that of finding the true mean between the Scylla of hyper-criticism and the Charybdis of scientific dogmatism. In this respect, special attention should be drawn to one point: the part played by the hypothesis in the existential sciences. While in physics and chemistry, hypothesis is only directed towards general bodies of facts, it is sometimes necessary in the special sciences to make a plausible hypothesis concerning the initial state; for example, when considering the history of the development of our planetary system. As opposed to the physico-chemical hypothesis, such hypotheses are characterised at the start by not being verifiable in the strict sense, since in the first place, no one was there, and in the second place, documents in the sense of historical documents are not available. The conclusion is frequently drawn by the positivists that these hypotheses are therefore only in the nature of 'as if.'

A view of this kind, however, once again misses the main point. It is in reality by no means the sole object of science to understand the present state of affairs; its goal is rather the whole course of events in the past, present, and future. A mere 'as if' as regards the past, is not in the least sufficient. Science wishes to know what happened, not what we may imagine as an explanation of the present day. Hence for geology, hypotheses concerning states of the earth are not by any means constructed with the object of understanding the present state of the strata, etc.: on the contrary, examination of the latter serves as the means of drawing conclusions concerning past conditions. It is clear that these conclusions are condemned to remain hypothetical, but this by no means changes the direction of our aim. Here again, the premature resignation of positivism spoils from the start the whole object of research.

This state of affairs is still more evident in cases where we are not dealing with causality in time, but with contemporary states in separate parts of space (in the sense therefore, of what was said on page 68 concerning co-existences determined by laws, and not successions). Here the hypothesis made by the individual science, such for example as one concerning the state of the interior of the earth, or of the sun, is in the great majority of cases no more capable of direct confirmation than in the first mentioned case, where an assumption was made concerning the past states. Nevertheless, here again our object is to find the truth about a set of facts not directly accessible to us. Now and then it may quite well happen that, contrary to all expectation, direct examination of the facts becomes possible. For example, a hypothesis concerning the nature of the earth below the surface may possibly be directly tested by boring, so long as it is not necessary to go too deep. But it is obviously absurd to use this practical technical limitation to ascribe to a hypothesis concerning the nature of the earth at a depth of 100 kilometres, a quite different epistemological position from that of one concerning 100 or 1,000 metres depth where practical confirmation is in any case possible.

A theory of knowledge which forces us to such unnatural distinctions, thereby proves itself untenable. It is, on the contrary, evident that the sentence: the earth has such and such properties at a depth of 100 kilometres is in itself meant in exactly the same sense as: it has such and such properties at a depth of 100 metres. The possibility of direct proof in the latter case, and impossibility in the first case are quite secondary matters and have nothing to do with the essential nature of our knowledge, but are simply questions of human technical ability. Nor are matters any different when we now pass on to other quite similar hypotheses, as, for example, concerning the physical nature of the heavenly bodies. Here also, the actual goal of knowledge is, and remains, the proof of a real set of facts, which, since we cannot ascertain them directly, we seek to discover by indirect means. Here too it is quite irrelevant as regards the sense of the question whether we consider a later direct examination (for example by a journey through space) as possible or not. But if this is once admitted concerning hypotheses relating to present-day facts, it must also be admitted concerning past facts, the more so because no sharp distinction can be drawn between the two.

We therefore firmly maintain, as against all positivist and fictionalist denials, that the existential sciences are attempting in the full sense of the word to find out the whole actual state of the world in the past and present, and as far as this is possible, in the future also, and from this point of view, we will now inquire to what extent these sciences have attained their ends. We may divide this question into two parts, namely what science is able to tell us of the world process as a whole, and what it knows of single parts of the same. To both we must answer: much more than nothing, but far from enough to really satisfy our thirst for knowledge.

Concerning the world process as a whole, we can obviously only make statements on the basis of laws, which are to be regarded with more or less probability as generally valid. Up to the present we only know three such laws, namely the law of energy, the law of mass, and the law of entropy.[179] It was repeatedly remarked in the last chapter, that the laws of mass and energy are probably only two sides of one and the same truth, which with Haeckel we may suitably describe as the law of substance, although we are not able yet to define its content with complete certainty. It is finally a question, regarding such a general law of conservation to be found in the future, of the exact physical formulation of the fact, already suspected by Descartes and other philosophers, that no part or property of the world arises out of nothing or disappears without compensation.

From this point of view, it has been supposed possible to draw a very much wider conclusion, which purports, it is true, to be only another form of the same truth: the law of conservation. It has been said to follow from the law of conservation that the world cannot have an end-point in time either forwards or backwards, but must exist from, and to, all eternity. It has then been further said, that this law must be regarded as a physical contradiction of the assumption of a creation out of nothing.

No very deep logic is required to see through the error of this conclusion. On the one hand, we can see at once that the supporter of a belief in creation regarded in this sense is in no way prevented, but rather in every way compelled, to say that God once created the world, and now allows it to exist unchanged in its material extent. This is the experience expressed in the laws of energy and mass, or the law of substance. On the other hand, it is also clear, that the laws of energy and mass, when applied to the world process as a whole, only allow a purely negative statement to be made, namely that this law does not allow any beginning or end of the world to be determinable. As far as it is concerned, the world may always go on as it is at present, and may always have done so.

But this by no means proves that statements concerning these matters may not be made on other grounds. That is also true on a small scale. Let us assume for the sake of simplicity the strict validity of the law of conservation within the mechanical field, and imagine a pendulum swinging without friction or resistance. It is clear that the law of energy then states that the pendulum will continue to swing without limit. As far as this law is concerned the pendulum might have swung from the beginning of the world and continue to swing till the day of judgment. But everyone sees that this does not prevent the pendulum having been made at a quite definite point of time, hung up, and set swinging, and that the physicist who did this can take it down and bring it to rest, in spite of all conservation of its mechanical energy.

To return to the world process, anyone therefore who wishes to assume an end and a beginning of the world in spite of the energy laws, only needs to assume that the world cannot be completely exhausted by physico-chemical laws, but contains forces of quite another kind, or even owes its origin to such forces. The matter is similar to the well known example, that an object inside a closed circle cannot get out, as long as it is compelled to remain in the plane, but can do so immediately as soon as a new dimension of space, the third, is available. The only question is whether we have sufficient reasons for assuming a third dimension; in our case therefore a power, which is able to create out of nothing and annihilate, and for the further assumption that this power has actually done so or will do so. Here we are merely concerned to point out that the energy law as such cannot contribute towards deciding this question.

For the rest, as was already indicated on pages 202 ff., and will be further discussed with special reference to the theory of relativity, the whole statement of the question is wrong from the start if we speak of creation in the sense of a beginning in time of the world, since before that point of time, there would thus be time, but no world, in existence.   For time is to be regarded as a property of the world itself, quite apart from the very definite formulation of this fact in the relativity theory.

The total result of the teaching of the law of conservation can thus be no more than that, as far as the world is regarded purely physico-chemically, neither a beginning nor an end of its process can be determined.   But in no case does the law of conservation tell us anything about what happens.   It simply states that, when anything whatever happens, the energy and the mass, or some concept common to or including both, remain constant. But whether anything happens, and what happens, is evidently not yet completely determined.   The energy-mass law is a necessary, but not sufficient condition of all happening.

We have, however, a further general law which appears appropriate to fill this gap, namely the law of entropy.   So much has already been written and talked about its meaning as regards our general world-picture, that scarcely any other law can be compared with it in this respect.   Many people, for example, have regarded it as a welcome aid to proving the finiteness of the world, and hence the necessity for the assumption of a creator.   On the other hand, Haeckel regarded it as a contradiction of the law of substance, and hence entered into a very ill-founded attack upon it, which brought him a hard but well deserved reproof from the physicist (Chwolson).   The layman, therefore, may be recommended to regard with great suspicion in all circumstance such discussions concerning the law of entropy.   We will attempt to arrive here at a result not open to objection from any side, even if this be only negative.   It is after all a very good thing to know what one can prove and what one cannot.

The content of the law of entropy, or second law of thermodynamics, has already been shortly given on pages 95[180].   It states that, in molar dimensions, perpetual motion of the second kind is impossible, that is, that no apparatus can be made which remains in perpetual motion without addition or loss of energy. The reason for this is the degradation of energy by dissipation, which takes place, if not otherwise, then at least by heat conduction and heat radiation.   This dissipation means, as we saw above,

Ss

in the light of the kinetic theory no more than that the most probable condition of distribution of velocity is automatically set up among the unordered movements of molecules.

We now describe that fraction of the total energy which can be transformed at will into other forms of energy as the ' free energy' of the system. The law of entropy can then be so formulated, that the free energy of any system left to itself, always falls to increasingly smaller amounts in the course of a sufficient length of time, since dissipation of heat causes the difference of temperature, to which the transformable fraction is proportional, to become continually less. In short, the free energy continually decreases. This process takes place automatically without our intervention. But it should be expressly noticed that this refers to processes in dimensions large compared with molecular dimensions. The law of entropy is a statistical law, a law of averages (see above).

From the fact that, as the law of entropy states, the free energy continually diminishes in every finite closed system of bodies, it has been believed possible to draw a very far reaching conclusion concerning the world as a whole. It has been supposed to lead to the conclusion that the world is proceeding towards certain 'heat death,' like any other physical system which receives no energy from outside, and since, on the other hand, the world has not yet suffered this fate we must conclude that it cannot have existed from infinite time, since the end would otherwise have already come. In this way many Christian apologists in particular have believed that a kind of physical proof of creation, or of the existence of God, could be given; many others, for example Ed. von Hartmann, have declared this argument to be inescapable. Let us now consider how matters stand with regard to it.[180]

Leaving aside the tacit assumption generally made here that the law of entropy is really valid, we can nevertheless see that to conclude finiteness in time from it is a fallacy. For it at least assumes that the finiteness in space of the world has already been settled somehow or other. If we assume a world infinite in space, the levelling up of energy in it will undoubtedly take infinite time, even if this process should always take place with the velocity of light, which is by no means the case. Within a universe infinite in space, one could not conclude even for a single definite system in it the necessity for final heat death, since no guarantee would exist that this system would not repeatedly receive energy in an available form from outside.

But this is not, as von Hartmann thinks, a reason for postulating the finiteness in space of the world, in order that the law of entropy may be really exact. The law of entropy itself contains the express assumption that we are dealing with a finite system closed in itself, and receiving no energy from outside. It is true as long as this assumption holds. Anyone who agrees with von Hartmann in concluding that the world must be finite in order that the law of entropy may hold, can equally well conclude that friction cannot exist in order that Galileo's law of falling bodies may be true. This is a pure sophism. To make still another comparison: if a quadrilateral is a parallelogram, its diagonals cross at their middle points. What would von Hartmann have said to a mathematician who concluded from it that all quadrilaterals must be parallelograms, in order that this rule may be an absolute geometrical truth. It is a matter of an equivocal use of the word 'assumption.'

Other defenders of the consequences of the law of entropy try to get round the inconvenient objection of possible infiniteness of the world by somewhat different means. They start from the fact of experience that the universe is divided into a number of single systems relatively isolated fairly completely. Even a sun such as Sirius, the radiation of energy of which to the earth is measurable, nevertheless delivers to our whole planetary system so minute an amount as compared with its own energy balance, that it can be completely neglected without any perceptible mistake. We can, therefore, regard our planetary system as an isolated one, or at least as one receiving no energy from outside. (It rather loses continually enormous amounts of energy, and in so far is not isolated). Now the same is true in greater probability, in greater dimensions, if we take our whole system of fixed stars as representing a cosmic island. We can assume that, even if other systems exist, the distance of these from our own is related to the diameter of one of these systems, about as the distance of two planetary systems to the diameter of our own system of planets (see below page 275). Just as little as one planetary system (one sun) influences another in view of the enormous relative distance, so little will other cosmic islands, if such exist, influence our own system, that of the Milky Way. This may again be regarded as relatively completely isolated, and hence must end with heat death. Furthermore, our cosmic island is generally tacitly identified with the universe as a whole, and the conclusion is then drawn that physics must therefore admit the finite duration of the world process on the basis of observed fact, thereby justifying the doctrine of creation.

But when this crude change of terms is avoided, the whole line of thought proves no more than what we knew already, namely that the exchange of energy takes longer, the larger the system considered. Whether we then consider a planetary system, a cosmic island, a glass globe, or a few cubic centimetres in the laboratory, is a matter of complete indifference. No one denies for the moment that every finite system tends towards heat death. But no one can say what will happen when we proceed in thought to infinity. We can only say, by way of an estimate, that the affair will last longer, the greater the dimensions become, and not only proportionately to the dimensions, but a relatively much longer time. This follows from the fact, that the energy content of a system is proportional, in first approximation, to the volume, that is the third power of the linear dimensions, whereas the levelling of energy is proportional, in the first place, to the size of the surface and hence only to the square of the linear dimensions. Hence the only conclusion, if any conclusion is possible there at all, is that infinite extension in space makes the world process more than ever infinite in time. But it is preferable to realise that such conclusions are better left alone. The entropy law relates, from the very nature of things, to finite isolated systems. Hence it is better not to apply it to the universe, which it is not permissible to rank under the concept.

There is a second objection to be added to this first one. For even if it be admitted that our universe is, say in the sense of the relativity theory, finite in space (see below), it is still by no means a logical necessity to conclude from the entropy law that the world process is finite in time; further assumptions are necessary. The content of the entropy law is exhausted by the statement that the free energy of a finite system continually decreases. But it makes no statement as to the rate at which this decrease takes place. The conclusion that a quantity which continually decreases must finally reach zero after a certain, though very long time, is one of the commonest fallacies into which persons not educated mathematically are liable to fall.   But it is none the less simply wrong.

If we denote the time with t, and the quantity which decreases with time by E, we might for example have $E = a - bt$, or $E = a \cdot e^{-t} = \dfrac{a}{e^t}$ or $E = (a - b)\ e^{-t} + b$, where a, b, e, represent any definite numerical values.   All three expressions have the property, as is easily seen, of diminishing as t increases; for $t = 0$, that is for the initial point of time, we have in all three cases $E = a$, since $e^0 = 1$.   For $t = \infty$ on the other hand, we have in the first case, $E = -\infty$, which is naturally impossible if E stands for the free energy.   Only in this case do we have $E = 0$ when

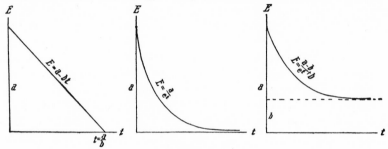

FIG. 41
Decrease in free energy

$a = bt$, or $t = a/b$, that is, at a definite point of time.   For the two other cases on the other hand, we have for $t = \infty$: Case 2. $E = \dfrac{a}{e^{\infty}} = \dfrac{a}{\infty} = 0$; that is, E only attains the value 0 after lapse of infinite time.   Case 3. $E = \dfrac{a-b}{e^{\infty}} + b = \dfrac{a-b}{\infty} + b = 0 + b = b$.   That is, E reaches the value b after lapse of infinite time.   If we represent all three cases graphically, we get the three curves shown in fig. 41.

The first of these shows how E (say the free energy) sinks in a finite time ($t = a/b$) from the initial value a to the final value 0.   The second shows how E sinks from an initial value a, but since its rate of fall is continually slower, it only reaches the value 0 after an infinite length of time.   (E approaches 0 asymptotically).   The third figure shows how E, starting from the initial value a, approaches the final value b, which it never reaches within finite time.

Since similar considerations as regards reckoning backwards (towards the beginning of the world) are possible, two cases being imaginable, we have altogether three times two, or six, different ways in which free energy could decrease with time.   Of these six cases, only one is considered by the usual argument from the entropy law to the finiteness of the universe, and this conclusion is thus shown to be very premature and devoid of sufficient foundation. If it is to hold, it is necessary to add to the statement of the entropy law (that the free energy decreases) a further assumption as to how, that is to say according to what law, it decreases. In other words, we need special assumptions concerning the nature and course of energy exchange, assumptions that can only be founded on facts of experience. If we now inquire of experience, we learn, as the final standstill of all isolated systems practically occurring shows, that the first or the second case apparently represent the type of energy exchange, the greater probability being in favour of the second case, that is the attainment of final zero of free energy after a time theoretically infinite, but practically and measurably, in a sufficiently long finite time. This type corresponds to the chief law concerned, namely the law of cooling. Nevertheless, a generalisation of this special case known to us on earth, to cover the whole of the universe, is a very doubtful affair, quite apart from the unprovable assumption of the finiteness in space of the universe.

To all this we must add a third, and perhaps the most important, objection. We have hitherto tacitly assumed the entropy law to have a quite general validity, and to hold without limitation for all processes and systems in nature. Is this really a fact which has been determined with certainty? As we saw above, it is first necessary to make the limitation, that it concerns only processes in dimensions greater than molecular dimensions, whereas for molecular motion itself (and Brownian motion) the law does not hold in the sense of a continual increase in free energy. If we now clearly perceive that the entropy law is a law of averages, we should not say, if we wish to be really precise, that the free energy continually decreases *because* the temperatures tend to level out. We should rather say that the law of entropy holds when, and so far as, a levelling of temperature takes place which cannot be reversed. Regarded in this way, the question becomes unavoidable whether a reversal of temperature levelling without expenditure of work is really impossible, not merely from the empirical and practical point of view, but also theoretically.

In order to answer this question, we must get at the bottom of the matter, and ask therefore whether the molecular motions which determine the temperature must necessarily lead to this result. The reply will be that this is proved, according to Boltzmann, by the law of probability. But the calculation of probability assumes, as we have seen, elementary disorder; the molecules must move about quite irregularly. Our question thus amounts to whether it is inconceivable, that apparatus or, in general, bodies of any kind may exist, which are able to bring order into the disorderly motion of the molecules or atoms, or their possible constituents (without expenditure of work, be it noted) and which would therefore be able to separate molecules from one another according to their velocities. Maxwell imagined, in order to demonstrate these facts, a 'demon' capable of sorting molecules. But we are also able to form a picture without such an aid. Imagine in a vessel a partition, possessing the property of only allowing molecules to pass through from one side to the other having velocities below a certain limit, and from the other side only those having velocities above a certain limit. The slower molecules would then collect on the one side and the faster molecules on the other, the one side becoming of itself colder, and the other side hotter. In this way useful energy could be obtained, and a perpetual motion of the second kind constructed. Experience teaches us that this cannot be done, but we are here dealing with the question whether it might not be possible, and experience can never prove this, since it cannot prove a negative. If such an apparatus were constructed to-day, it would by no means abolish the law of entropy; the latter would then be found, as has been the case with almost all physical laws, to be limited to definite groups of systems.

We may not say that such an apparatus is opposed to experience, since it is never allowable to describe any imaginable possibility as a contradiction to experience. It can always only be a matter of enlarging or limiting the laws formulating our experience to date. What grounds have we for the conviction that a limitation of the entropy law is for ever impossible? The only grounds we can give are the proved wide field of application of this law. As we have remarked, its applicability is not less, but perhaps even greater, than that of the energy law. Nevertheless, and in spite of all this, a majority of physicists would hardly be quite in agreement with the statement that the laws of entropy and energy are in that respect on a level. The reason for this is easy to find. It lies in the fairly direct relation that exists between the law of energy

and the principle of causality in the widest sense. We can hardly
imagine the world without the law of energy, or one more general
but including it, since the law appears to us simply a physical
formulation of the fundamental rule *ex nihilo nihil fit*. The
thought that energy or mass might arise out of nothing without
compensation or disappear into nothing, appears to us, after having developed physics so far, unthinkable, although this was not
the case with earlier generations.

A similar statement cannot be made concerning the entropy
law. The attempt has certainly been made to relate this law as
well to a general law of thought. It has been said that the law
of energy leaves the direction of events undetermined, and that a
law is therefore necessary to specify this direction. This law is
that of entropy, with its statement that the world proceeds of
itself from a less probable to a more probable state. The entropy
law is thus said to be, as it were, the physically exact formulation
of the general truth that something definite always happens, that
is to say, that physical processes always take place with certainty
in a definite direction, or that the world process is not reversible.
Quite apart from the fact that Brownian motion is certainly not
taken account of in this view, which claims to be of general
validity, the whole argument is fallacious. For the course of
events is certainly not determined by the entropy law alone, as the
sufficient condition. No energetist has yet shown us how all
physical laws could be deduced from those of energy and entropy.
It is also very much to be doubted that this will ever succeed.
For in order to determine a physical process unambiguously, we
require either special laws or the hypothetical fundamental law,
to which those of energy and entropy will, in the most favourable
case, be related in the manner of a general integral to special
differential equations – as, say, Kepler's second law to all sorts of
possible laws of motion about centres. Such a general conclusion
from the special law cannot replace this for all cases, but at the
most, for single, quite definite cases, as we see with respect to
Brownian motion.

For this reason alone, therefore, it is not permissible to regard
the law of entropy as a sufficient physical expression of the fact
that something definite always happens. This expression must
be given by the single physical laws, or by the fundamental laws.
Conversely, it is likewise easy to see that the definite direction of

events would also exist, if the opposite of the law of entropy were true for a system. In the apparatus which we imagined just now, which sorts molecules according to their various speeds, events would also take place in a quite definite direction; in one, however, opposite to that within our experience, namely with increase in free energy. But is that not a definite direction?

In short, however we regard the matter, the entropy law, although resting quite certainly upon just as broad a basis in experience as the energy law, is still wanting in that imponderable distinctive conviction, which would lead us to regard a system of bodies which contradicted the law of conservation of energy as a direct miracle. An apparatus which increased free energy, and thus changed molecular energy into molar energy, would not be regarded by us as a miracle, but only as something very admirable. It would be certainly difficult to imagine a world, even with the imagination of Jules Verne or Wells, in which the opposite of the entropy law held good. It would certainly be remarkable enough; hardly anything could happen in it in any way analogous to our own world. But it would not be such an absurdity as a world in which mass and energy came into being out of nothing.

This being the case, we can understand that many attempts have been made to find physical possibilities for the reversal of the degradation of energy, and hence for the increase of free energy, Auerbach opposes to Clausius's entropy an "ektropy," but he does not succeed in making quite clear in what process an increase of the latter is to be regarded as taking place. It has frequently been believed that living beings could be regarded as systems exhibiting reduction of entropy, but no proof whatever of this has been given.[181] Quite recently, Nernst again attempted to attack the problem from a new side, by placing alongside the laws of energy and entropy a third general fact which has only become clear in quite recent times. If we assume, namely, that all atoms gradually disintegrate as do all radio-active atoms, only most of them so slowly that we have been hitherto unable to observe it (see page 133) we get, as a general principle, an all round disaggregation of mass, and hence of energy, if we regard mass and energy as the same, following the relativity theory. All matter would therefore dissolve, in course of time, into uniformly distributed field energy. This process of dissolution is imagined hypothetically by Nernst as having for its counterpart a converse reformation of atoms, which he regards as taking place in cosmic nebulae in their first stages. The doubtful part is, however, that this new reformation should logically begin with the highest elements (uranium, etc.), which are at the beginning of the radioactive distintegration series. This appears highly improbable in view of the progress from simpler to more complicated, otherwise visible throughout nature.

It appears to me that a more hopeful line of attack would be to use the quantum theory as a proof that the principle of elementary disorder (the equal statistical value of all possibilities) does not hold for atomic dimensions. It cannot be overlooked that in the quantum principle, which obviously governs not only the transference of field energy to matter, but also the internal atomic conditions (this fundamental idea of Bohr's is certainly correct), we have a principle of order exactly opposite to that of elementary disorder, according to which not all possible cases, but only single quite definite ones, are able to occur. As an example, we have Bohr's model of the atom with its special electron orbits. I could imagine that for this reason the very opposite of the law of entropy may hold in the region of processes which lead to the formation of atoms out of some fundamental substance, and that here may be hidden the long sought-for complement of the law of degradation of energy which holds for molar dimensions.

By way of warning to those numerous people who are always trying to solve the riddle of the universe, it should be pointed out that such general ideas as this have no scientific value as long as they are not presented, thoroughly worked out, in a mathematically exact form. If Bohr had done nothing but make the assertion that the quantum principle also holds inside the atom, he would not have become 'the Newton of Spectroscopy.' Someone else would have filled this position, someone who calculated the behaviour of the model of the atom now called after Bohr, on the basis of this idea. Ideas concerning the riddle of matter are today as cheap as blackberries, and not a week passes without a new pamphlet appearing on the subject. But such ideas are not theories, and, at the best, they may stimulate another to produce a scientifically useful theory.

For our philosophical question, however, even the shadow of a possibility that the law of entropy might be balanced by some other fact of nature, already means a good deal. We see from it that it is quite impossible to take the law of entropy as a safe foundation for a proof of the finiteness in time of the world. Even if the objections last brought forward concerning the universal validity of the law are not admitted, the fact that the spatial extent of the universe cannot be determined remains decisive. We are therefore simply obliged to declare that we know nothing positive on the question.   Just as little as the law of conservation guarantees the eternity of the universe, if we admit the possibility of other than purely physico-chemical forces, so little does the law of entropy guarantee the finiteness of the history of the universe. Everything that is written and talked of on this point in popular expositions, often dictated by prejudice in favour of certain views of the world, is entirely unfounded dogmatism.

We must not omit to mention that it has several times been proposed to regard the increase of entropy as the physical reality at the basis of our concept of time, and even simply to identify entropy function and time.   The point of comparison between the two is the irreversibility of both.   It appears to me very daring to base upon this fact the identity of two such totally different notions.   The real meaning of the concept of entropy itself, is, for the present, a much debated question, which would require a whole book for its full discussion.   Clausius introduced this quantity, in the first place, as a simple mathematical adjunct. But what it really corresponds to in physical reality is by no means easy to describe.[180]

We still have to discuss a number of general ideas concerning the structure of the universe, which are connected with the relativity theory.   This theory has opened up quite new perspectives in this direction.   The idea of a world space of definite curvature closed in upon itself (compare pages 143 ff.) was not, however, first put forward by it, but arrived with non-euclidian geometry ; but the relativity theory now opens up the possibility of experimental decision of the question.   It does not follow, as we are sometimes told, from the fundamental equations of the general relativity theory, that space must necessarily be finite; it may also be infinite according to Einstein, and on the average uncurved (as, in two dimensions, a plane with numbers of small humps).   But it might also, according to Einstein, be finite and have a finite average curvature, and the relativity theory allows us to calculate this curvature from the average density of the distribution of matter.

If we assume for this the value derived from the system of fixed stars in our immediate neighbourhood (see page 272), though this assumption is very doubtful, we arrive according to de Sitter at a circumference for the universe of about 100 million light years, and a cubic content of about 10^{77} cubic centimetres. But these are naturally, at present, vague suppositions.[182]

The question as to what kind of experimental proof could be imagined may be answered by saying that it should then be theoretically possible to see the light of a star in two different directions. On account of the enormous cosmic distances, the difference in time of the two ways would however be so considerable, that the certain identification of the two images would no doubt only be possible in the case of stars which are fairly exactly opposite to us. To picture this notion, we may again consider a two-dimensional world on the surface of a sphere. If for example, an observer on the earth's surface could register the direction from which earthquake waves arrive, he would always receive two such records of an earthquake, one after the other from opposite directions, since the waves go round the earth in both directions. Only in the case that the centre of disturbance were exactly at his antipodes, would both arrive at the same time.

But leaving this question, which is far from being capable of proper discussion, we may further ask: How would the Minkowski world-lines run in such space? The doctrine of reincarnation immediately suggests itself. To put it in Weyl's words:[183] 'It is therefore in principle possible for me to experience now events which are in part the effect of my future decisions and actions. Also, it is not impossible that a world line . . . in particular the world-line of my body, may return in the neighbourhood of a world point which it has already once passed by. The result would be a much more perfect double of myself than was ever imagined by Hoffmann. In actual fact such considerable variations of the g_{ik}* as would be necessary for this, do not occur in the region of the universe in which we live; but there is a certain interest in thinking out these possibilities in relation to the philosophical problem of cosmic and phenomenal time.'

I should not be willing to subscribe unconditionally to these remarks. There are two main points to consider. In the first

---

* These are the components of the fundamental tensor characterising the metric field, upon which gravitational effect also depends.

place it is not in itself necessary, even if the region is finite, that every world line should return upon itself. It is easy to inscribe curves upon the surface of a sphere which never return upon themselves no matter how much they are prolonged; they may either repeatedly cut across themselves, or never return at all to the same position. In the second place, even if we limit ourselves to such curves as return upon themselves, it is not necessary that they should all do so after one journey round. Now 'my body,' as well as any event in the world, is made up of a vast number of such world lines. What here appears to us as an event relatively closed in itself (a human life, a thunderstorm, etc.) would therefore be, translated into the language of the Minkowski world, a system of an infinite number of such lines which form a certain four-dimensional figure. It is infinitely improbable for one to be made up only out of lines which all return upon themselves, and further-more, after the same number of circuits. But only this would give Weyl's double. The doctrine of reincarnation is therefore no more probable according to the relativity theory than according to the old view.

A further question to which we are led here is the following: the fact that we are able to imagine various possible kinds of bent surfaces, as well as the plane, depends upon our having at our disposal three-dimensional, plane, euclidian space, in which any conceivable two-dimensional metric field of this kind can be 'embedded,' whereas conversely, it is by no means possible to imagine all kinds of surfaces, for example under certain conditions, no planes, in a curved space. A being therefore, capable of picturing any kind whatever of non-euclidian three-dimensional space, would necessarily itself live in a four-dimensional plane space. Now the Minkowski world of the general relativity theory is a non-plane, four-dimensional manifold of variable curvature. We can therefore imagine this, together with other spaces like it, embedded in a five-dimensional space; indeed one might almost say that if further investigation should prove our world to be actually closed in upon itself, the human mind, which will not tolerate any chance limit to knowledge, will almost of necessity arrive at the idea that this really is the case. The question of possible other worlds then arises, of which we should naturally not experience anything in the empirical and physical sense, since these might like our own, be imagined as completely closed in themselves.

Or may we perhaps after all look at the matter otherwise. Might not our world perhaps be nothing more than the four-dimensional (x, y, z, t) projection of a more general multi-dimensional existence, and every Minkowski world line be only the inter-section of this existence with a single (our) four-dimensional space-time-matter-continuum? Such possibilities make us dizzy, but we may also be thankful to the relativity theory for opening up such prospects. Plato's famous cave image, hitherto only thought out from the subjective and epistemological side, appears in a new light of objective physical possibilities.[184]

But we will not attempt to climb too high, but remain for the present by our own world. If, to use again Dingler's expression (see pages 215, 249), the 'theoretical prime structure' should finally culminate in a general fundamental law, or even general statistics, the 'historical prime structure' culminates, after what we have said above, in what we are able to state in summary form concerning the actual distribution of world lines or quanta of action, as the case may be. Here we again stand before the other side of the contingency problem which we discussed earlier (pages 214), the question of the reason for this particular and actual distribution of world existences. It is, as we saw, completely insoluble, but the point of view of relativity gives us a glimpse in another direction. In it, the whole of existence in space and time appears as 'shape' in the true geometrical sense, and once again Plato's profound ideas appear before us from antiquity the world as a work of art, as idea given form. Only a short step is needed from the work of art to the artist.

But at least we have justification for the question: Are not the units of the world, the electrons, the atoms, the ordinary massive body, the globe, the planetary system, etc., regarded from the point of view of the timeless existence of the world, so many individuals, and at the same time, each lower thing part of the higher; as in the work of art, for example the symphony, every motive, every movement is a well characterised unit, and yet an integral part of the whole?[184a] But above all, the conclusion is certain, which must be unmistakably stated here, namely that alternatives such as whether the world was created by God or has existed from all eternity, are completely meaningless. The assumed endless time in the last part of the question itself belongs, as we have seen above, to the object, to the world. If, therefore, a God exists, that is if the world, as the believers in theism think, has not come into being of itself, but is the deed and effect of an infinitely lofty, self-conscious spirit, it is completely a matter of indifference whether the world thus created is limited or unlimited in time, or whether it returns upon itself.

Every such statement of question obviously brings God into time, which must have been created by Him together with space and matter, which is still greater nonsense than bringing the life and thought of the physicist, assumed on page 200, into the time t, which comes in his formulae. This immediately disposes of such questions as to what God did before the creation of the world and the like, but at the same time the common conception of eternity, as time continuing without end, also vanishes. We shall return to these and other questions connected with them at the close of the book, but we will now break off this discussion and again regard things from the sure ground of experience, though we must not forget, it is true, that experience only extends to sections, which can only be regarded as isolated systems in the sense we have stated.

The greatest object obtainable for our senses, is the starry heavens. The completest possible investigation of its structure and the motions of all single parts, is the goal of astronomy to-day, and one towards which it has made very considerable advances during the last decades.

We will attempt to present, within the limits of a short sketch, the chief advances that have been made and that depend mainly upon the very latest science. In the first place, we know to-day concerning a large number of fixed stars, their distances from our solar system with more or less accuracy, namely for about 1,500 stars, which are not more distant, than about 100 light years. The nearest fixed star, α Centauri, is 4.3 light years, that is 37 billion kilometres away from us. If the solar system is imagined as reduced in size so as to go into an ordinary room, so that the sun is in the middle and of the size of an ordinary millet seed, and Neptune touching the walls, α Centauri, as the nearest sun of the kind, would still be 47 kilometres away, Sirius twice as far, Aldebaran (in the Bull) about seven times, Vega (in the Lyre) about nine times, and the pole-star about ten times as far, (see table on page 275). Naturally all such figures are the more uncertain the greater the distance. These few direct measurements of distance, of course, amount to practically nothing compared to the vast number of other fixed stars.   As regards these we have to depend purely upon indirect, and particularly statistical methods, which space does not allow us to describe here.

The fundamental idea is as follows: If the whole universe were uniformly filled with stars up to any depth, the number of stars in each magnitude would increase, according to an easily stated law, as the brightness diminished.   As a matter of fact, this number decreases considerably in the lowest magnitudes, which proves that the stars are much less densely distributed at greater distances from us than in our immediate neighbourhood.   According to the investigations of Seliger, Kapteyn and others,[185] all the stars visible to us singly form a system of approximately lens shape. Its longest diameter is about 60,000 and shortest about 10,000 to 20,000, light-years.   However, the stars in this system are very unevenly distributed.   There are 'star clouds,' and between them regions relatively poor in stars.   Since we are near to the middle of the system, the edge of it appears very thickly sown with stars, this being the explanation of the bright strip in the heavens called the Milky Way.   The total number of stars in this system is estimated at about five hundred million.

Quite recently, these first and fairly rough results of Kapteyn have been considerably enlarged, mainly by the results of the American astronomer Shapley, which however, are not yet quite certain.   We shall go into them shortly here, not only because they are in themselves remarkable, but mainly because the way in which Shapley arrived at his conclusions affords us good insight into the methods of modern astronomy.   The whole argument starts from the observation made upon a certain kind of variable star, for which δ Cephei may serve as a type.   In the case of these, the duration of the light periods shows close connexion with the apparent brightness, so far as we are dealing with stars of the same larger system, which are practically equidistant from us.   But for such stars, the apparent brightness is proportional to the absolute brightness, and the connexion with the length of period holds also for this.   If this fact is now applied to stars of this type in the neighbourhood of the sun, the distance of which we can determine with considerable exactness, a relation is found between length of the period and the apparent brightness on the one hand, and the absolute distance on the other, the latter being calculable from the former. This is now applied by Shapley to variable stars of the δ Cephei type occurring in the so-called spherical star clusters. This name is applied to star clusters which are so dense inside that they cannot be resolved into single stars, it having long been supposed that they are collections of very large numbers of stars very far away. These suppositions have been entirely confirmed by Shapley's calculations; he found for seven such clusters, by observing the variable stars contained in them, and then for 80 further clusters by the use of similar analogies, distances from 20,000 to 200,000 light years.

Other investigators (Kapteyn and others) ascribe somewhat smaller values to the data for the near stars upon which the whole calculation is based, and this would require these distances to be reduced, but Shapley stands by his figure, and has been able to bring further reasons for the approximate correctness of his ideas. If we make this assumption, our sun would perhaps be a member of such a cluster, and the earlier enumerations of Seliger, and others would then only relate to this cluster, which would be ranged alongside other clusters also containing vast numbers of stars. For, if Shapley's ideas are correct, we can really only see the brightest stars contained in these clusters. At this distance, Sirius would only have the brightness of a star of the 17th magnitude, the sun only that of the 21st magnitude, and thus be just visible only for the best instruments (these statements relate to the cluster M3). But according to Shapley's estimate, the cluster in question contains about 40,000 of these stars, which are only just visible at this distance.

Shapley has further attempted to discover how these star clusters are related to the Milky Way. He finds that they are almost completely absent in the plane of the latter, but are distributed in an approximately egg-shaped space; the plane of symmetry is formed by the Milky Way, our local star cluster with the sun lying upon its axis, about at a point dividing the axis in the ratio 1:4, that is, much to one side, and not in the middle of the whole system. Whether the spiral nebulae belong to this system, or themselves form systems co-ordinate with it, is still an open question. If we make the latter assumption we find on the basis of considerations to be discussed immediately, a distance from us for the nearest nebula of the kind, that in Andromeda, of about a

Ts

half to one million light years, its diameter being about 30,000 light years.

The considerations last mentioned, which for the present are quite hypothetical, were originated by the French astronomer, Charlier. It has been believed since Laplace's time that the assumption of a succession of systems of ever higher order to infinity (planet with moons, solar system, star cluster, Milky Way, etc.) necessarily leads to two impossible conclusions. In the first place, so Laplace showed, gravitation at any point in the interior would acquire infinitely great values; and secondly, the whole of the heavens should be as bright as day. But these two conclusions only hold, as Charlier has shown, when certain limiting conditions – inequalities which connect the number of the terms of two successive systems of the nth and $(n\text{-}1)$th order, their diameter and their mutual distances – are not fulfilled. Conversely, these inequalities also contain the conditions for the possibility of assuming continually progressive structure, and one can then find, from data known concerning the system of the nth order, the lower limits for that of the next order. This value has been found in this way for the nebula in Andromeda, and agrees with estimates made by Lundmark and others on the basis of other methods.

In order to make the whole more easily grasped we add a short table, which gives the real distances and sizes in columns 1 and 2, and in the last three columns reduced distances, which express the relationship between the different values in a more intelligible manner. In column 3, distance is measured in terms of that between the sun and earth taken as 1 millimetre. In column 4 this measure is multiplied by 160, which makes the distance of Neptune about half the diameter of a fairly large room (15 feet), and in column 5 again increased 30 times, so that the earth's orbit will just go into this room.

To this statement of our knowledge concerning the structure of the universe, a knowledge which to-day is in a state of flux, we must now add something concerning the motion taking place, a matter, if anything, still more hypothetical. We know with certainty to-day that a large number of 'fixed' stars move, and with some certainty, also, how they move. The expression 'fixed star' is therefore hardly suitable. The first discovery of this kind, made by Halley in 1817, has been followed by innumerable fresh

	Kilometres	Light time	Reduced Distances		
			I	II	III
Diameter of the Earth .	13,000	0.04 sec.	90 $\mu\mu$	0.015 mm.	0.5 mm.
Diameter of the Sun .	1.4 mill.	4 sec.	9 μ	1.5 mm. (millet seed)	4.5 cm. (small apple)
Distance Sun-Earth . .	150 mill.	8 min.	1 mm.	16 cm.	5 m.
Distance Sun-Neptune .	4500 mill.	4 hours	30 mm.	5 m.	150 m.
Nearest fixed star α Centauri .	38 billion	4.3 years	260 m.	47 km.	1300 km.
Sirius . .	77×10^{12}	9 ,,	0.56 km.	93 km.	2800 km.
Pole Star .	380×10^{12}	43 ,,	2.6 km.	470 km.	13000 km.
Hyades .	1140×10^{12}	130 ,,	8 km.	1300 km.	40000 km.
Open star cluster M. 37 .	13000 yrs.	800 km.			
Nearest globular cluster ω Centauri . . .	21000 ,,	1300 km.			
Globular cluster M.3 . .	45000 ,,	2500 km.			
Diameter of the same .	500 ,,	30 km.			
Farthest globular cluster .	218000 ,,	13000 km.			
Andromeda nebula, dia. .	23000 ,,	1400 km.			
,, ,, distance .	$0.5 \text{-} 1 \times 10^6$,,	60000 km.			
No. of spirals acc. to Charlier	$10^6 \text{-} 10^9$,,	—			
Dia. of the systems of all spirals	about 10^9 ,,	—			

discoveries of the same kind. We are dealing here naturally with apparent motions of the stars, which are not by any means the real motions (see Figure 42). For in the first place, we have in

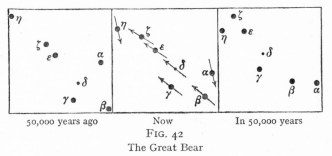

| 50,000 years ago | Now | In 50,000 years |

FIG. 42

The Great Bear

addition to the displacement taking place at right angles to the radius of vision (the direction of the star), a further component of motion in the direction of vision, that is, towards us or away from us. Secondly, all these motions are influenced by the sun's own motion.

The first motion is found on the basis of what is called Doppler's principle, by observing the light sent to us by the stars. If a ship is travelling on the sea in a direction opposite to that of the waves, it obviously meets in a certain time, for example, a minute, more waves than a ship lying at anchor; if, on the other hand, it is travelling in the direction of the waves, it meets fewer of these. The same is obviously true for all wave motion, if the source of the waves and the observer are moving towards or away from one another. Hence the note of the horn of a car moving past us on the road appears considerably higher when it is coming towards us (often by as much as a whole musical tone) than while it is going away from us. The apparent increase in frequency corresponds to a higher note. The frequencies of the light waves correspond optically to the spectral colour (see page 116). Hence if a source of light is moving towards us sufficiently fast, all the spectral lines must appear to be displaced towards the violet end; while if it is moving away from us, they are displaced towards the red end. This displacement can actually be observed and measured in the spectra of numerous stars. From its magnitude it is easy to calculate the velocity with which the star is moving towards us or away from us, and thus the motion in the radius of vision. This, when combined according to the parallelogram law with the apparent motion at right angles to the direction of vision, gives us the true motion relatively to ourselves.

The second point, namely the separation of the motion of the solar system itself from all these relative motions, is considerably more difficult. The following analogy may make clear to us the essential point of the matter. If we are walking in a wood, the laws of perspective cause the tree trunks in front of us to separate further and further, while those lying behind us appear to close up. This would therefore be the case in the star system, if all fixed stars were really fixed and the sun only moved in a certain direction between them. But this is however, by no means the case; if we therefore wish to determine the motion of the sun on the basis of the principle mentioned, we must try to make a plausible supposition concerning the proper motions of the stars as a whole, in order to be able to combine it with the motion of the sun and so deduce the observed motion.

The most obvious assumption is that the proper motions of the stars are distributed irregularly in a similar manner to those of the molecules of a gas. It is however impossible to arrive at the actually existing distribution of motion by combining this assumption with any kind of motion of the sun. The result of the whole extremely difficult investigation is rather that the actual distribution of the proper motions can only be represented, by combining a motion of the sun in the direction of the constellation Lyra (the point towards which the motion takes place is called the apex) with a kind of double 'star drift,' which, generally speaking, follows the direction of the Milky Way. It results from this that the whole system, as seen from outside and reduced sufficiently in size, would appear like a catherine wheel, or a rotating sprinkler with two jets. The well-known great spiral nebulae in the heavens exhibit the same form.

We will turn to some smaller details within the whole system. We cannot go into great detail. We may, however, mention that recent investigations of Hagen[186] at the Vatican observatory, have rendered probable the existence of very extensive dark masses (nebulae) in cosmic space, so that what we see, namely what is self-luminous, is possibly only a small fraction of what is really present. Furthermore, the various kinds of visible objects, double stars, variable stars, open and close star clusters, luminous clouds, and so on, offer material for an indefinitely long period of further investigation. By far the greatest part is played by spectrum analysis, which not only gives us information concerning the chemical composition of the object in question, but also concerning many other matters, such as temperature (see page 172), motion (see above), magnetic fields (Zeemann effect), and so on.

Especial progress has been made quite recently in the matter of luminous masses of gas (fixed stars, sun), by the theoretical investigations of Arrhenius, Emden, Jeans, Eddington, Megh Nad Saha, and others.[187] Eddington deserves the credit of being the first to point out the decisive part played, alongside gravitation, by the pressure of radiation (pressure of light, see page 430) in such masses of gas at a high temperature. Eddington was able to show that, as a result of this force opposing gravitation, there is a certain upper limit to the size of a fixed star for every temperature, so that the enormous giants once assumed could not exist.

Taking it all in all, we may assume that the actual masses of the fixed stars do not vary within very wide limits.

More certain than these still somewhat doubtful suppositions are the present facts known to us about our sun itself. But here also, fallacious conclusions have proved to be easy to make, as is plainly seen by the revolution which has taken place in the last thirty years in our view regarding the nature of the sun. The spectroscopic investigations of Kirchoff and Bunsen led to its being regarded as an almost certain conclusion that the sun consists of an incandescent luminous liquid core with a gaseous envelope (the photosphere). Now it is in itself extremely improbable that at the temperature existing on the sun (about 5,600° C.), any substance could possibly be liquid, since this temperature most likely exceeds the critical temperature of all or almost all gases (for every substance there is a critical temperature above which it can only exist as a gas). This is more than ever true, if, as is probable, the temperature in the interior is considerably greater than on the surface, where alone it can be determined by the radiation curve. Recent theoretical investigations show that in masses of gas of such density and temperature as the sun, the transmission of light waves is so completely different from that under any circumstances on earth, that our former conclusion is by no means allowable. In reality, the sun is an incandescent ball of gas which decreases in density from the outside towards the centre, having by no means a smooth, but a very irregular, surface, which is subject to continual change (prominences, streamers, and spots). It would take us outside the range of this book to go further into these matters. All these investigations are in full swing, and problem after problem appears which will keep astrophysics busy for a long time.

Matters are not very different as regards the investigation of the single members of our planetary system, the only system of the kind which we shall ever be able to observe. For however probable it may be that innumerable other fixed stars are centres or suns of such systems, we are not likely ever to be able to discover anything for certain about them. In spite of the extraordinary improvement in the methods of observation, and in particular the application of photographic methods, no unexceptionable solution has yet been found either of the riddle of the canals of Mars, or of the rings of Saturn, or of many other problems. But we should not forget, in view of all these unsolved questions, the progress that has really been made. It may, for example, be regarded as fairly certain that none of our fellow planets is suitable for organic life in the forms known to us, with the sole exception perhaps of Mars. But even in the case of this one, it is very questionable whether the conditions, at least for higher organic life, are there fulfilled. For though the existence of water on Mars already appears fairly certain to-day, it is still doubtful whether this can be of much use in view of the low temperature, which according to recent investigations (Coblentz) is to be estimated at about – 50° C. to – 60° C.

All in all, Mars gives the impression of exhibiting to us a much later stage of planetary development than that of our earth, whereas conversely, Venus may be about at the point of the earth before the appearance of organic life. The famous canals of Mars can hardly be upheld, in view of all other reasons telling against its inhabitability by higher living beings, as proof of the presence of intelligent creatures. We are still without a quite satisfactory explanation of the phenomenon. Recent investigations[188]appear to show that the canals may be only an optical illusion. But the coarser lines at least are visible even on photographs, and hence must be regarded as real objects.

We cannot leave the question of the habitability of other heavenly bodies without referring to an attempt recently made to show the probability of the earth being the only body in the whole universe of fixed stars which exhibits the necessary condition of life for mankind. I refer to the work of Wallace, the well-known English scientist and friend of Darwin.[189] Starting from the fact that our solar system is nearly in the middle of the system of fixed stars visible to us as the Milky Way (see, however, above page 273), Wallace attempts to show that only in this position can there be a planetary system, the existence of which is guaranteed for a sufficient time. He then shows further for what reasons the earth is the only habitable planet, and how on the earth everything comes together in quite a unique manner to enable the necessary conditions of stability for the process of organic development to be guaranteed for innumerable millions of years. We cannot here go into Wallace's statements. They are not more than suppositions and plausible reasons.

A real proof that any number of other planetary systems

similar to ours cannot exist in our immediate or more distant neighbourhood is, in the nature of things, completely impossible. Hence an impartial observer can only say, in my opinion, that we simply know nothing of a positive nature, though certain recent investigations by Jeans favour the uniqueness of our earth. Hence it remains a matter for individual taste to construct in imagination other inhabited planets as appendages to the innumerable suns which we see, or to think with Wallace, that only a whole universe of fixed stars is good enough to render possible, upon a single planet of a single one of its members, the development of a human race with its history and intellectual culture. For my part, the teleological point of view itself prevents my being attracted by the supposition, since the means used would appear to be altogether disproportionate to the end; even if we are willing to admit that the development of mental life, as we know it in ourselves, certainly cannot be measured with the measures of physics, and that a single Newton must be regarded as more valuable than a whole planetary system consisting merely of heavy bodies.

But what can be the purpose of the infinite abundance of the most varied phenomena in the universe? What is the purpose of double stars and nebulae and star clusters and so on? Perhaps that the Newtons and Herschels here on earth should find enough various objects for their curiosity? The idea appears to me in every way absurd, if the whole world has any sense at all, to find this sense in our little earth's history alone, however seductive may be the analogy which, probably, was Wallace's original intuition; the analogy namely that a nature that wastes thousands of life germs, in order that a single one of them may develop under favourable conditions, might be supposed capable of wasting worlds of fixed stars, in order to produce a single earth, peopled with men. But all this is, as we have said, purely a matter of taste.

The very notable progress made by science in the last few hundred years towards investigating the present state of the heavenly bodies, is out of all proportion, unfortunately, to the insight gained into the development of this whole system, and even of so unimportant a part of it as our own planetary system. In this connexion I may quote here the words of a well-known authority of the present day. The Bonn astronomer, J. Hopmann, writes in his book, *Weltallkunde* (*The Science of the Universe*, which in other respects also may be recommended to the reader in continuation of this chapter[185]): 'In course of time, a large number of cosmogonies of the solar system have been developed. Their originators have unfortunately made little use of the means afforded by mathematics and theoretical physics, and have thus been unable to protect themselves from repeated fallacies. A characteristic example of this is the "doctrine of world-ice" which is in complete contradiction, not only with astronomy, but also with meteorology and geology. F. Nölke[190] deserves the credit for having shown all these theories to be actually untenable from a mathematical point of view, so that we stand to-day, as regards the history of the origin of the solar system, amid a heap of ruins, and must honestly confess that we know nothing positive at all. It will probably be only after the solution of the problem of development of fixed stars has been found [see below], that a useful starting point will be given. But even then it will be necessary to make use of all modern physical and astrophysical theories, the theory of radiation, of the structure of atoms, of radio-active processes, and so on. Previous work in this regard has generally only applied mechanics and, at most, parts of the theory of heat, but not those departments of physics in which, as we know to-day, the greatest energy changes can be observed.'

It is scarcely necessary to add anything to these words of Hopmann, and so there is thus little point in going into the various hypotheses. We may, however, state as probable that the original condition of our solar system was that of a single incandescent ball of gas, and not, therefore, as Kant thought, a cloud of cosmic dust. It is doubtful what were the processes by which the planets, the total mass of which is scarcely $1/_{700}$th of that of the sun, separated from the latter, and we should beware of adopting without further question such hypotheses as the well-known one of Laplace (Plateau's experiment), against which, as Nölke has shown, very important mathematical and physical objections can be brought. A theory which should really explain the way our system has come into being, must above all explain the very unequal masses of the various planets, that is to say, give reasons why the two heaviest and largest planets (Jupiter and Saturn) occupy just about the middle of the system, the smallest lying quite inside, and the three middle-sized ones, Uranus·

Neptune, and Pluto, lying right outside. It would further be required satisfactorily to account for the curious differences in the direction of revolution of many moons, and so on. An advance in this direction is afforded by an hypothesis of Jeans, according to which the passage of another and larger fixed star close by the sun drew out of the latter a tail, at first thin, then becoming thicker, and then thin again, which then divided and drew together, forming single smaller or greater planets. This is at least considerably more plausible than Laplace's theory of the splitting off of rings, but for the present no more than a first tentative supposition. Something quite different might also have happened.

Only one thing may be regarded as certain, namely, the unitary origin of our system. This system cannot possibly have come into being by pure chance, in view of its almost coincident orbital planes, and common direction of revolution (apart from a few exceptions); it must certainly have been produced by some single process. For this reason we may then further suppose, that similar processes will have taken place in a similar manner in the case of other suns (fixed stars), and that therefore other systems like ours probably exist elsewhere in the universe, although we are not ever likely to have definite evidence of this. It is important to make this remark, since the attempt has frequently been made[191] to support lines of thought such as that of Wallace just described, by saying that our solar system represents a special case of the n-body problem, which in itself is immeasurably improbable, but which for this very reason must also be the sole one among an infinite number of others, in which the stability of the system is assured for a very long time.

These considerations are based upon recent investigations of the n-body problem, from which it is concluded that, in general, a system consisting of n bodies attracting one another according to Newton's law, is not stable, single bodies belonging to it again and again receiving so great an acceleration that they fly out of the system, and finally only a two-body system is left. These investigations have been made use of to explain the relatively great number of known double stars. Even if we assume this to be true, we must not forget that no proof is afforded that only one such special case could be realised in the universe among many millions of others, even if the case realised in the solar system (almost coincident planes of revolution, and almost circular orbits) represents, from the purely mathematical point of view, only one among innumerable imaginable cases of possible initial velocities and initial positions of the n bodies. This transference of purely mathematical probability to the physical case would only be allowable if we imagined the solar system as coming into being, in accordance with certain meteoritic hypotheses, by the chance meeting in space of moving bodies. If, however, it was produced by a single process from a ball of gas already rotating uniformly, it would naturally no longer be mere chance that all orbital planes coincide; this would then follow of physical necessity, and would likewise be the case everywhere that a similar process took place. It is, then, pure sophistry to make out of the mathematical special case, a physical one, and so believe oneself to have proved that this case has been realised, in all probability, only once in the universe.

It will easily be understood that we are able to say still less concerning the history of the whole system of fixed stars (the Milky Way), than concerning our own planetary system. According to recent work of Jeans and others, we can imagine the spherical nebula to represent an original state, which then goes over into the form of the elliptical nebula, then the spiral nebula, and the loose 'clouds' (Magellan's clouds). But it may equally well have been quite otherwise. On the other hand, we are somewhat better informed regarding the history of the development of the single fixed star, of which we can form at least an approximate picture on the basis of recent theoretical and experimental investigations in astrophysics.[192] The foundation for these is given by what is called the Russell diagram, which we must therefore shortly describe.

The stars are first classified according to their 'magnitude,' that is, according to the brightness which they have as we see them, and then according to their spectral type, which is essentially a function of the temperature.[193] From the 'relative' magnitude first mentioned, the absolute magnitude may be calculated if we know the distance, this knowledge being possible for very many stars, as we have said above, by direct and indirect methods. We are thus able to make a list in which the stars are arranged according to the brightness with which they would appear to us if they were all equally distant. (This standard distance is usually taken as one parsec, that is the distance from which the earth's orbit would subtend an angle of $1''$).

We now plot upon the x-axis of a co-ordinate system (figure 43) the spectral type. In the figure the designation usual to-day has been chosen, which corresponds to the inverse order of temperature, so that on the left we have the hottest, on the right the coldest stars. On the y-axis we have the magnitudes, which likewise correspond to the inverse order of brightness. so that -5 gives the brightest, and $+10$ the feeblest stars. We thus have,

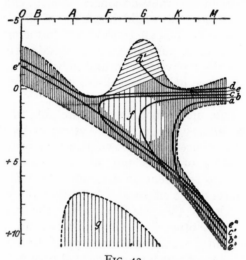

FIG. 43

The Russell diagram. The shading indicates
the density of star distribution

in this diagram, every star given by a point, which corresponds to the numerical values of its type and magnitude. This statistical diagram, first made by the English astronomer Russell in 1918, then shows that the star positions found are strongly concentrated in certain parts of the diagram, so that we get a kind of continuous strip, which on the right hand side at the top (spectral type M, magnitude -5) contains very bright but only red-hot stars, whereas in the middle and towards the left we have much hotter to very hot (white hot) stars of a somewhat small brightness, and at the right and below 'red dwarfs' we have besides a division of the band towards the left top, where a group of very bright and also very hot stars is situated, and finally a spot quite isolated at the bottom in the middle, that is, a group of stars with low brightness and medium to higher temperatures.

This whole picture, which thus gives us a scheme showing side by side the chief different star types, was interpreted by Russell himself as a picture of the succession in time of the single stars. By way of comparison, let us imagine an insect living for one day only and able to think. It would then likewise be able to construct the history of the development of a single human being from seeing children, youths, men and the aged all together. According to Russell, a star begins its development at the top right hand side, that is as a red giant (as, for example, Arcturus), from a previous stage invisible to us (gas nebula). Contraction results in a rise in temperature; that is, the star moves towards the left of our figure, but the absolute brightness decreases, since the star also loses mass by radiation. The temperature rises to a certain maximum (see the curve), which is determined by the total mass of gas present, then falls again, since the energy released by further contraction is not able to cover the loss any longer, and the star finally ends as a red dwarf (at the bottom right hand).

Later investigations, especially those of Eddington,[194] have considerably enlarged and deepened this picture, but as a whole – although Eddington at first obtained other results – have finally confirmed it. The branch upwards to the left, and particularly the stars of the region g, the δ Cephei stars already mentioned, are somewhat difficult to place. From the theoretical investigations of Eddington, Emden, Megh·Nad Saha, and others we have already mentioned, it further appears that the mass of glowing balls of gas must, on account of the pressure of radiation, remain within certain not very wide limits, so that stars above and below certain sizes cannot exist.

Our sun belongs to the smaller stars, and its maximum temperature in the past is estimated at about 9,000° C. It is already on the declining branch of the curve. The largest stars reach 18,000 to 20,000° C. (Sirius, Vega). These numbers naturally only refer to the surface temperature, while in the interior, temperatures up to many millions of degrees exist. The whole process of development is estimated to last about 10^9 to 10^{13} years (according to the magnitude of the star). Nernst[195] calculates for the sun a life of about 10^9 to 10^{12} years, which is approximately compatible with geological estimates.

Regarding the source of energy of the radiation emitted, it has long ago been realised that the source originally envisaged by Helmholtz and others, namely the gravitational energy set free by contraction, in no way suffices to cover the enormous loss by radiation. In the case of our sun, this source would not even suffice for the period of time which geology demands for the history of the earth. Recent physics has pointed the way to understanding these stellar energies, not only through radio-active processes, but also, as explained on pages 191, 194, through the possibility of the transformation of mass into energy. We are probably right in saying that the star, at least at the height of its development, radiates itself away; or in other words, that the energy present in the star in extreme concentration in space, is lost in course of time, in continually increasing diffusion into space, whereby mass itself is dissipated according to the equation $E = m \times c^2$. It is not impossible, that sometimes the 'explosion of matter' occurs, that is, the sudden liberation of the whole energy locked up in the mass, and perhaps in this way we may explain the sudden appearance of new stars (novae), though these have also been explained by the collision of masses hitherto dark. It is, however, hardly to be assumed that this is the typical end of a star, for otherwise the process should be observed more often.

A further question immediately arising in this connexion is, whether the process of development of stars, here described in quite general outline, again gives us, on the largest possible scale, an example of the rule of the entropy law, and whether therefore no automatic reconstruction of energy exists corresponding to this dissipation, the world process being therefore, in this respect also, irreversible, or non-periodic. This question has recently received fairly thorough discussion by authorities such as Nernst, Eddington, Jeans, Nölke and others[196] without its being, in the nature of things, possible to reach a conclusion. There also seems to the author to be little object in attempting to reach a decision at the present time, since we are not quite clear yet concerning the nature of cosmic radiation (see page 191), the final fundamentals of the quantum theory, and many other matters that enter into the question.

So much at least is clear, that here investigation of the greatest and smallest things in the world, meets on common ground in the most surprising manner. The stars must help us to find the way into the secret of atomic physics, and the latter must help us to understand astrophysics. Eddington's fine book on *Stars and Atoms* is plain evidence of this new development. It was a mistake when it was imagined, upon the first appearance of Bohr's atomic theory, that we should straight away find the laws of the macrocosmos (namely, the laws of Kepler) repeated in the interior of the atom. The application of Coulomb's law, corresponding to Newton's law, to the attraction within the atom between nucleus and electron, as was attempted in the original Bohr theory, at first with such great success, has later turned out to be impracticable. It is therefore a matter for further investigation as to how far the analogy really goes.

But we can already see the necessity arising – and this is much more interesting from a philosophical point of view than purely formal analogies – for taking account of cosmic processes in physics, that is, of including the whole of physical happening, if we wish to form a unitary picture. Our laboratory has, so to speak, been enlarged to include cosmic space, as happened once before when Newton founded classical physics. On the other hand, it is to be hoped, that the investigation of the behaviour of matter at extremely high pressures, temperatures, magnetic fields, and the like, will sooner or later shed new light on astrophysical problems. At the present time, energetic efforts are being made in many parts of the world, chiefly in the 'land of unlimited possibilities,' to realise upon the earth, experimentally, such extreme values of physical states. At Monte Generoso, near Lugano, it is hoped to reach voltages of many hundreds of millions of volts, by making use of atmospheric discharges in thunderstorms. In England, Kapitza is working at the present time on an electro-magnet to produce a field of 500,000 Gauss. Some Americans have attempted to produce temperatures up to 18,000° C. by heavy electrical discharges, and already claimed to have observed the transformation of tungsten wires into helium. The immediate future will probably bring all kinds of surprises in the experimental field also. It will then be time to attack again the problem of the development of the fixed stars.

Though we have thus reached a very modest result as regards our knowledge of cosmogony, we are on somewhat surer ground when we turn to the history of our own planet. The beginnings of this are likewise somewhat in the dark, as we have already stated.

We can only suppose for good reasons, that our earth first separated as a small glowing ball of gas from a larger mass of gas, that it then gradually liquefied by cooling, and finally became solid, whereby chemical processes naturally took place on an enormous scale, as we have already remarked (page 251). It has often been attempted in the past to make calculations concerning the age of the earth on the basis of the laws of cooling and radiation, but all these have become worthless, since the discovery of radio-activity has given us a new source of energy in the earth itself.

But the laws of radio-active phenomena have themselves opened up a new and apparently more promising way to the same end. For since, as we have said above (page 127 f.), the radio-active transformation of uranium probably leads finally to lead on the one hand, while on the other helium atoms (α-particles) are thrown off, we are able to calculate from the lead content of such minerals as originally contained uranium, and also from their helium content, the time necessary to produce the quantities of these actually found. Here the helium content gives a lower limit to the time, the lead content an upper limit, since on the one hand helium may have escaped, and on the other lead may have come in from other sources. Calculations of this kind have chiefly been made by Strutt. If we compare them with other calculations based upon the erosion of river valleys, the carriage of suspended matter, the formation of sediment, and so on, we get, according to Franz, the following table:[197]

Beginning of the	Years from to-day.
Alluvial Period	10,000 to 70,000
Diluvial ,,	300,000 ,, 500,000
Cenozoic ,,	3,000,000 ,, 15,000,000
Mesozoic ,,	7,500,000 ,, 37,000,000
Palaeozoic ,,	48,000,000 ,, 240,000,000
Archaic ,,	200,000,000 ,, 600,000,000

while life itself is estimated by the same author to have existed for 48 to 300 million years. These figures are, of course, still very uncertain, but they will approach the order of magnitude in the same way as Loschmidt's first calculations (see pages 22, 23) gave the absolute numbers of molecules.

We are certainly much better informed concerning the merely relative order in time of the layers out of which the greater part of the earth's surface is formed, than regarding the actual duration in time of the various epochs in our earth's history. In this respect, the whole geological work of a century has produced a system so securely put together, that serious changes are hardly to be expected, since it bears the guarantee of its correctness in itself, in the way in which the various parts fit together. It is also noteworthy, that the absolute determinations of age according to the methods given above in the main, entirely accord with the system. The geological system, the division into the well known four great epochs (archaic, palaeozoic, mesozoic, cenozoic) with their subdivisions, is indeed to be found in all school books to-day, which shows in perhaps the plainest manner, how it has already developed from being a result into the assumption for all further geological investigation. And this is the case, in spite of all uncertainty regarding absolute times, since it is certain beyond all doubt, that in any case we are dealing with millions and millions of years, as compared with which all human history appears but a short moment. Whether in any single case, we have ten, or a hundred or five hundred millions of years, is relatively a matter of indifference – at least for those who are less interested in the scientific results themselves, than in their significance for our general world-view.

From this point of view, the question of the causes, the driving forces of geological change, regarding which we are likewise not completely informed to-day, is also comparatively indifferent. To the older and, as far as it went, well-founded doctrine of the power of water, wind, and volcanic action in producing transformation, a whole series of other hypotheses have been added to-day, intended to explain certain phenomena in the superficial structure of the earth, or in changes of flora and fauna; for example, the assumption of the fall of meteorites or even moons, changes in the poles, and the like, and, above all, Wegener's hypothesis of the drift of continents, which appears to be continually gaining ground.

According to this hypothesis, the peculiar shape of the coast, for example, of Europe-Africa, and America, is to be explained by the two continents having originally been together, but having been torn apart by some event, and having then drifted farther and farther away. But one can still clearly recognise by their outlines, how the two continental 'floes' once fitted together (for example, the east corner of Brazil into the Gulf of Guinea, and so on). Space does not allow us to got further into these new ideas.[198]

Us

For our purpose we are less interested in the results as such, than in the general knowledge that, and how, the human mind is busy, here also, in unveiling the secrets of the past. In any case, no one has any further doubt to-day, that we are everywhere dealing with processes fundamentally capable of investigation, though certainly for ever beyond the reach of complete and exhaustive knowledge.

In this connexion special reference should be made to the question of the earth's interior. As we know, it has been said for decades in all scientific and popular books, that science assumes the interior of the earth to be liquid and incandescent. Now against this view we already have the fact known from physics, that the enormous temperatures calculated for the earth's interior from the geothermic gradient (increase of temperature with depth), would lead to all known substances being heated above their critical point, above which they could only exist as gases. However, this also is of no fundamental importance, for we have not the slightest idea how matter behaves at these enormously high temperatures, when at the same time exposed to the extremely high pressures which certainly exist in the interior of the earth. Presumably, gases under such conditions do not behave very differently from solid bodies as we know them. The distinction, therefore, loses its meaning altogether, since it only holds for pressures and temperatures of the order of our experience.

But the more exact study of the course of earthquake waves, teaches us that the composition of the interior of the earth must change as we go down altogether less continuously than was originally thought. There must be in the interior at least one, if not two, fairly sharp boundaries between materials of different nature, since the actual course of earthquake waves cannot otherwise be explained. Consequently, the geophysicist Wiechert, of Göttingen, put forward 30 years ago the hypothesis that the interior of the earth is not at all at an extraordinary high temperature, but that the earth contains a solid core of iron and nickel, over which there is a relatively thin (about 2/9ths of the radius) crust of rock, separated from the core by a 'magma' layer covering either the whole earth or only forming isolated 'hearths.' In the latter we could then find the source of volcanic eruptions.

The fact that the interior of the earth must consist of material considerably heavier than the crust follows without any doubt from the comparison between the specific gravity of the earth as a whole (5.5) with that of rocks, which average between 2 and 3. Wiechert's hypothesis very well explains almost everything at present known concerning the earth's structure, and is therefore generally accepted to-day, at least on its chief lines. But we may once more refer at this point to what was said above on page 253 concerning the meaning of such hypotheses. In the present case a direct confirmation might be imagined, amongst other ways, by sending electric waves into the interior of the earth, where they would be reflected at the boundary layers in question, and would thus show their position and nature. And if a positivist should object here that such confirmation would still be indirect, and that the only confirmation he could call direct would be 'actually seeing' these boundaries or layers, we may permit ourselves to reply by the question: What really is seeing? Is it not also a matter of electric waves, which are reflected, refracted, etc., at certain objects, and then affect our retina, either directly or by means of an interpolated apparatus (lens, telescope, microscope, but also cornea, crystalline body, etc.)? What is the distinction in principle between the two cases? This example shows us very clearly, that critical attacks of this kind are mainly founded upon a very naïve, little thought-out, every-day theory of knowledge (see above page 27). In reality, it is everywhere a question of the enlargement of our senses based upon logical conclusions.[199]

Just as recent investigation has essentially changed our views concerning the nature of the interior of the earth, so have the views generally held in past time concerning the composition of our atmospheric covering, been considerably enlarged and developed. The ascent of balloons carrying only self-registering apparatus, as practised by Assmann and his pupils in Strassbourg before the war, first showed the presence of a definite limit at the height of about 8 kilometres at the pole, to 16 kilometres at the equator. In the layer lying below this limit, the so-called troposphere, there is a continual exchange of all components on account of winds resulting from the warming of the surface, and moving in vertical and horizontal directions. In the stratosphere, above, we have on the other hand comparative calm, and the temperature here remains fairly constant, while in the troposphere it decreases rapidly as we go upwards. What happens further up cannot be determined by direct observation, since hitherto recording balloons have not reached higher than 31 kilometres (Batavia 1913/14). We are therefore driven back upon indirect conclusions.

In the first place, Wegener concluded about 1910 from the kinetic theory of gases, that the composition of the air high up must be considerably different. It is easy to calculate on the basis of the theory, assuming the air to be practically stagnant, that the layer of oxygen and nitrogen, which we know, must be followed by a layer of almost pure nitrogen, then by a thinner layer mainly containing helium, and then by a layer of hydrogen of great depth, but naturally of very low density. But these conclusions have recently again become very doubtful, since observations of the Aurora Borealis have proved that at the very great heights at which this occurs (100 to 1,000 kilometres) oxygen must still be present, for the so-called green aurora line, (557.735 $\mu\mu$) which is characteristic of it, has been shown by recent investigations[200] to belong to oxygen. The nature of the absorption spectrum of the atmosphere further points with fair certainty to the presence of considerable quantities of ozone (active oxygen) at a height of about 40 kilometres.

Further conclusions can be drawn by investigating the reception of wireless waves. If these were conducted round the earth by simple guidance, their range would fall off rapidly, at equal intensity, as the wave length diminished. The extraordinarily good reception found for short waves (about 10 to 40 metres) has proved this not to be the case, but that a layer guiding the waves back to earth must be present in the upper atmosphere, a conclusion already drawn by other investigators in the eighties from research on the earth's magnetism, but first generally admitted when Heaviside used it in 1902, to explain the facts of wireless propagation. What is called the Heaviside layer is about 350 kilometres high on winter nights; during the day, and in summer, it descends to about 90 to 100 kilometres. It probably consists of ionised air, at which the long waves are reflected, and the short waves refracted. Disturbances in it, caused partly by cosmic influences, partly by terrestrial magnetism and atmospheric electricity, are, perhaps, partial causes of what is known as 'fading'; that is, the periodic, or even quite irregular, increase or decrease in loudness of broadcast reception.

Still another means for investigating the higher layers of the atmosphere is afforded by the investigation of the propagation of sound, which shows the peculiar phenomenon known as the zone of silence, and further by the investigation of falling stars, luminous night clouds, the phenomena of twilight, and other matters. Altogether, we shall sooner or later get a clearer picture of the state of affairs in those layers of our atmosphere into which direct observation has not hitherto reached. To-day our picture is considerably more complete, and certainly more correct, than twenty years ago.[201]

This will undoubtedly be greatly to the advantage of another science, that of meteorology, which unfortunately leaves much to be desired regarding the accuracy of its weather prophecies,

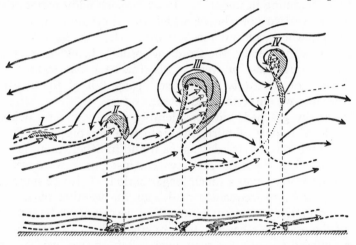

FIG. 44

A succession of cyclones (air-eddies), according to the polar front theory. The shaded area indicates regions of rain

and thus betrays the fact that it has not yet found the final causes of the phenomena of the weather. But in this department as well, we are far from having to give up in despair. Here also, the last two decades have brought quite considerable advances, which are chiefly designated by the general term 'polar front theory' (Exner, Bjerknes).[202] Earlier meteorology saw the chief cause of weather phenomena in the unequal distribution of pressure caused by unequal heating of the earth's surface. The flow of the air out of regions of high pressure and into those of low pressure, whereby the earth's rotation caused the formation of an eddy (anticyclones and cyclones, the Buys-Ballot rule) was a consequence of the distribution of the air pressure.

According to the new theory, the formation of these eddy regions depends, on the contrary, upon the existing currents. They are formed in the air, just as eddies in flowing water, only with the difference, that in the latter there is a downward motion, in the former an upward motion. The eddies arise mainly from the warm equatorial and cold polar air masses, which each in themselves circulate in a closed path, streaming past one another at their boundary, the so-called 'polar front.' Eddies are formed consisting of a narrower 'warm sector ' and a larger 'cold sector,' in which the warm air is gradually raised from the ground by the cold air streaming in under it. In every such eddy region, there are therefore two boundaries, which run approximately radially, (figure 44) where cold air meets warm. Here we have the chief rain zones of the eddy, the front one usually bringing us steady rain with south-west or west winds, the back one squalls with north-west or northerly wind, and heavy fall of temperature, since with it we enter the cold sector.

This easily grasped, but considerably more complete, point of view allows to-day of fairly accurate weather prophecy (about 80% correct, one or two days ahead) and in other, more favourably situated, parts of the earth, we can also put forward weather prophecies to-day over a much longer period. It is not surprising that mistakes are frequently made, for the weather phenomena are determined by so many factors, that we can never be sure that one of the essentials has not been overlooked.

As opposed to such forecasts over a day or two, we have not yet been successful in finding laws allowing the course of the weather to be prophesied over longer periods, such as whole months, or even in a general way, over a whole year; such a possibility would be of the greatest imaginable practical importance. All supposed connexion with the phases of the moon, the constellations of the planets, and so on, belong to the realm of pure imagination, as statistical tests prove. In actual fact, no larger number of changes in the weather fall at or around the changes in the moon than is to be expected, given a sufficient number of cases, according to the law of probability. The ineradicable beliefs of country-folk in this respect are to be explained psychologically. If a remarkable improvement in the weather happens at full moon, the full moon itself becomes particularly noticeable. If it does not take place, the fact is forgotten, since the full moon is not particularly remarked upon.

The way in which large numbers of people react to such proofs of fact, gives a good measure of their capacity for thinking scientifically. The majority of believers in the moon entrench themselves behind the fact that science has so often been convicted of dogmatism, and that the common people have very often proved to be right in spite of it, in medicine for example; and they suppose that in the present case the same will be true. Now the fact that dogmatism in science, particularly of a negative kind, has often played a fatal part, cannot be denied. But in the present case, it is not a matter of dogmas, that is, of any sort of doctrines which are to be recognised as true or false, but simply of the question whether the coincidence between changes in the moon and changes in the weather actually exists to an extent exceeding pure chance, or whether it does not exist. This has nothing whatever to do with doctrines of any kind, but is capable of purely objective determination.

We simply count over a sufficiently long period, say a decade, the changes in the weather occurring (naturally having first defined what we are to understand by the term), and we then, equally simply, count up how often such a change has occurred on a change of moon, and how often it has occurred on the fore-going and subsequent days. We finally discover that if the period of time of observation is chosen long enough, the number of changes occurring are on the average always equally distributed over all fourteen or twenty-eight days of a half or full moon-period. The exact records kept for decades past by meteorological stations allow these determinations to be made on a fully sufficient scale, but no result other than the one given has ever come out. Anyone who cannot or will not admit this must not expect to be taken seriously by anyone capable of thinking scientifically. For he expects science to pay attention to supposed theories of the influence of the moon on the weather, though this influence is proved by statistics not in fact to exist. It would first be necessary for such an influence to be found without any doubt, before the question of its cause could be discussed.

The superstitious ideas which are still quite generally connected with the weather are too well known to need further description here. Unfortunately, religious circles and powers are not quite innocent in regard to the difficulty of getting rid of them. The reasons for savages regarding the weather quite naturally as the direct effect of higher, divine powers are sufficiently clear. On the one hand, they depend upon the capriciousness of the weather, which mocks all laws and all rules, and thus undoubtedly gives the impression of arbitrary action; and on the other hand, upon the impressive power of many weather phenomena, and in particular thunderstorms, and the direct benefit and injury to man, and particularly to the cultivator of the soil. It is here that man experiences most directly and strikingly, and on his own body, his own inter-connexion with the general course of nature. It is perfectly natural that, when at a primitive stage of development, he should express this feeling in the form of all kinds of anthropomorphic ideas. But from the point of view of higher religion, one can and ought to require the abandonment of such inappropriate ideas, which are suitable to a negro or Malay, but not to a modern European.

We cannot leave this theme, as we did six years ago in our third edition, without devoting a word to modern attempts at the revival of astrology, which, as is generally known, continually gain ground. The belief in a mysterious inward connexion between human fate and cosmic events is as old as the hills, and apparently practically ineradicable. Its deepest root is human egoism, which imagines itself as the middle point of the world, and regards itself, or at least the whole human race, as sufficiently important for the shining heavenly bodies above to declare his fate to him. Its second root is unclear thinking, which prefers to be determined by forebodings and feelings, rather than by clear insight into facts. Historically speaking, astrology was an inevitable stage of development in man's thought concerning nature and its connections. But from the moment when real knowledge took the place of pure supposition, its death sentence was pronounced, and everything brought forward by its present day prophets is pure imagination, which has not the slightest particle of reality behind it.

Astrological theories are just as bad as, or still worse than, the pseudo-scientific theories of the weather, of which we have spoken; a discussion of them has no interest for a serious, scientifically minded person, however free he may be from all prejudice, for the simple reason that before all theories, we first require a proof that the relations asserted really exist. It is nonsensical to produce a theory of something which does not exist.

Astrology falls to the ground not so much on the absurdity and fancifulness of its theories, as from the fact that hitherto none of its supporters has been able to bring a shadow of a proof that any tangible parallelism exists between constellations or 'horoscopes,' on the one hand, and the fate of human beings on the other. On an average over the whole earth, a human being is born every half minute. What proof have we that the life histories of those people who are born within, say, the same half or whole hour, or further of those who are born at the approximate repetition of the constellation, show a similarity in any way exceeding that of the average? Until this proof has been given, the whole theory is completely in the air.

A single glance at modern astrological writings proves further that the doctrines of these 'wise men' contradict one another completely, so that the interpretation of a horoscope gives entirely different results according to different systems. Which of them are right? The study of astrology is worth while in one respect, but in this it almost exceeds in interest any other subject of cultural history, namely as regards the spiritual and mental connexions in the history of human development. Anyone who has studied Troels-Lund's profound book on human ideas of the heavens and of the meaning of life in past time, or Boll-Bezold's excellent description of the history of superstitions and interpretations relating to the stars,[203] will have profited richly in respect of his judgment regarding problems of human culture.

The scientist may be thankfully aware that present day Christian theology, apart from rare exceptions, fully agrees in general with the judgment of science on this point, as also on our whole relationship to cosmic events. It is unfortunately less true of religiously minded laymen, and also of many practical theologians. These are often ruled by a fundamental leaning towards mystical or mythical points of view, quite without regard to their content, for the simple reason that they *are* mystical, that is, promise to replace the sober and rational worldview of science by suggestions, acting on the feelings and the will, of a mysterious background of things, which can in this way be brought under human power, without its being necessary to go to the trouble involved in rational knowledge. The scientist cannot protest too sharply against such attempts, of whatever kind,[203] and true religion has no interest in them, on the contrary, it must agree with Goethe in ascribing to the Devil the words:

'Do but despise reason and science,
The highest of man's powers,
And thou art mine for sure !'

Real religion does not need such attempts, for it does not strive to be a substitute for science, but is quite another manner of looking at the same things (namely man and the world), that science looks at in its own way. In this respect is must be all or nothing. Whoever wishes to find in nature or behind nature a divinity, whether theistic or pantheistic, must be consistent and apply his view to the whole of nature, and not to single chosen parts. For this point of view it is a matter of indifference in itself whether we are dealing with a stone rolling down a mountain or the motions of the planetary systems, with the boiling of water in a kettle, or with clouds, snow, and hail; whether a long threatened fall of rock carries away a couple of shepherd's huts rashly built beneath it, or a volcanic eruption destroys a whole town, as happened with St. Pierre and Martinique. If a God, as demanded by theism, exists, this is all his work, the atom as much as the universe of fixed stars, the fall of the raindrop as well as the collision of two suns.

To leave the world in small matters to itself, but to require the intervention of a divine producer for the more important spectacles, such as the formation of a-planetary system or the great changes in the earth, is an insupportable confusion of theistic with deistic, or semi-atheistic, notions. We cannot regard the one as natural and the other as supernatural, and apply religious values to one, which we do not apply to the other, and thus take it out of the connexion with the rest, regarded as indifferent. While primitive man attains to the sense of a divinity by single special things or happenings, the person of insight to-day should rid himself of this tie to the individual event, and on the contrary, rise from the wondering admiration of the whole, in which both theist and atheist are quite agreed, to the heights of religious feeling in so far as he feels the inward need to do so. *The world regarded physico-chemically is a single unit, and must be regarded in all its parts and processes from the same point of view in principle. Our world view can therefore only take in the physico-chemical world as a whole.*

PART III

MATTER AND LIFE

IN ALL our discussions so far, we have used the word 'world' with the invariable addition of the adjective 'physico-chemical,' thus leaving it an open question whether the set of facts dealt with by physics and chemistry, and their branches, astronomy, geology, etc., are the only facts which go down to fundamental reality. First appearances are certainly against this view; we see a mass of beings that creep and fly, that sprout and bloom, that live a secret parasitical life, that reproduce themselves after their kind, that respond to the influence of the external world with movements which are obviously purposive, and that, certainly as regards the higher animals and man, but perhaps always and in every form, are capable of willing, feeling and thinking, in other words are capable of psychical experience, the last being something entirely strange to the physical world. That is life.

Is it really a purely physical phenomenon? Or is the peculiar nature of life only an appearance, and are we really dealing with the same elements of reality, but in new and quite special combinations? This is the fundamental question which inevitably confronts science, and which has formed the subject of centuries of discussion between philosophers. This discussion is now more intense than ever. A belief in the peculiar nature of life and its obedience to special laws of its own, is called *vitalism* and the opposite view is called, somewhat inaccurately, *mechanism*. The last designation is explained by the fact that the mechanical was originally regarded as identical with the physico-chemical, but this view is no longer tenable (see pages 63 f., 203).

In accordance with the principle we are following in this book, of allowing philosophical questions to grow out of scientific results themselves, we shall not deal with this fundamental problem from the point of view of general principles, but shall first turn to the science of life, biology, and see what results it has brought forth. Only then shall we be prepared to turn to more general questions.

I

THE PHYSICO-CHEMICAL FOUNDATIONS OF LIFE

From the narrowly physical point of view, there is for the present not very much to say. The physical phenomena, in the more restricted sense of the words, found in the organism, such as the lever action of the bones, the optical arrangements of the eye, the electrical organs of the gymnotus, etc., have, generally speaking, more the character of things external to the life process and of no great importance with reference to it. The physical laws of importance for the latter are those which deal with objects of very small dimensions, or directly with the action of molecules, for example the laws of diffusion and osmosis, elasticity and cohesion, and further, all the facts which recent physics combined with chemistry has discovered concerning the fine structure of matter, for example, by means of X-ray analysis (page 170). We shall return to this point later, but will deal now with the more immediately interesting question of *biochemistry*. It is self-evident that chemistry is the essential preliminary of all biological knowledge, and more especially, no real understanding of life processes is possible without a knowledge of *organic chemistry*, that is, the chemistry of the carbon compounds. We will therefore start from them.

The first fact to be remarked is, that we only know life as connected with certain quite definite chemical compounds, which consist essentially of the elements carbon, hydrogen, oxygen, and nitrogen; sulphur and phosphorus are also present in small amounts. Other elements, besides these six, found in the organism are iron, magnesium, potassium, sodium, calcium, chlorine, fluorine and silicon; while iodine, bromine, aluminium and others occur occasionally.

Among the innumerable organic compounds out of which the living organism is built, we distinguished from the earliest times certain groups, such as the fats, carbo-hydrates, albumens (proteins), and latterly also the vitamins and other substances. How the actual living substance, the protoplasm (from Greek *prōtos*, first, *plasma*, material), is related chemically to these is a question not yet cleared up. In any case, proteins or compounds of protein molecules with other molecules form the main constituent of protoplasm,[204] but this undoubtedly contains numerous other substances, and is probably a chemical system in a continual state of change, there being a perpetual formation and decomposition of molecules.

This makes it readily comprehensible that so complicated a system is, generally speaking, highly sensitive to external influences (chemical stimuli, temperature), so that life, particularly as regards temperature, is restricted to a fairly narrow interval. However, the lower limit of temperature is, in the case of many low organisms, very far down the scale. Bacteria and the spores of fungi have been kept frozen at the temperature of liquid air ($-190°$) for days at a time without death taking place. Quite recently G. Rahm has shown that animals living in moss (tardigrades, rotifers, etc.) may live in a dried condition for several days at the temperature of liquid helium ($-269°$).

In the upward direction, the limit for active life is, generally speaking, far below the boiling point of water. Algae are, however known to live in hot springs at temperatures up to $93°$, and certain latent forms of life, for example, seeds, spores, etc. are able to stand heating to $120°$ and more.[205] Such organisms as the hay bacillus, therefore, occasionally escape destruction by the usual process of sterilisation by boiling, which otherwise always attains its object completely.

The reason for this fact is probably to be found in the inability of most protoplasm substances to exist at these temperatures in aqueous or colloidal solution.

We must now say a few words concerning chemical processes in the living organism. By the term *assimilation* we understand to-day (in a broad sense) the building-up of complicated substances composing the body, out of the simpler kinds of molecules available for this purpose. The opposite process, the degradation of more complicated molecules to simpler ones, is called *dissimilation*. These two processes are found occurring together in every living organism, but in the plant assimilation has the upper hand, in the animal dissimilation. The latter process is, as a rule, also an oxidation, which, in the case of the higher animals is effected by the oxygen bound to the red blood corpuscles.

Conversely, the chemical processes leading to assimilation are mainly, from the chemical point of view, reductions, that is to say the withdrawal of oxygen (but not always). In particular, this is true of the most important and fundamental of all assimilation processes, the process by which plants assimilate carbon. According to the view at present almost universal and supported by important experiments, this process[206] takes place according to the following scheme:

$$CO_2 + H_2O \rightarrow H_2CO_3 - O \rightarrow H_2CO_2 - O \rightarrow CH_2O \rightarrow C_6H_{12}O_6$$

carbonic acid   formic acid   formaldehyde   sugar

$$(CO_2 + H_2O = H_2CO_3)$$

that is to say, the hypothetical hydrate of carbonic acid is first transformed by the removal of oxygen in steps into formaldehyde, and the molecules of this then group themselves to form the simplest of the *carbohydrates* (sugars, having the formula $C_6H_{12}O_6$). From this, as a rule, starch is formed, which is itself a carbohydrate, that is to say, in form a compound of carbon and water. The oxygen set free by the reduction returns to the atmosphere and forms the foundation of the possibility of animal life.

The remaining elements, particularly nitrogen, are taken in by plants from the earth in the form of soluble compounds (nutrient salts). While, at one time, it was generally assumed that these compounds were assimilated by their being attached to carbohydrates already formed and the derivatives of these, it has recently been regarded as possible that, at least for nitrogen and hence for the formation of proteins, a direct assimilation may also take place, which proceeds from the beginning alongside the formation of carbohydrates. This would mean that carbon dioxide, water, and ammonia derived from the earth, would unite directly to form simple organic nitrogen compounds, such for example as *formamide*, $HCONH_2$ etc. But all these investigations are still in full progress.[207]

The substances formed by assimilation are themselves further transformed in the plant, and carried from one position to another, where they serve for building up the organism or some other purpose. Thus for example, the starch manufactured during the day in the leaves of the plant disappears during the night, being transformed into soluble sugar, which then travels to the nodules of the potato or the seeds of the bean or rye, there to be retrans-formed into starch, and stored up. When germination takes place this starch is again transformed into sugar. In a large plant, therefore, innumerable chemical transformations are con-stantly taking place at all points.

The processes in animals are much more complicated still. The animal, as is well known, can only make use of material which has already been assimilated, and hence must nourish itself by plants or other animals. In the case of all higher animals, the food taken in is transmitted to all parts of the organism by means of a special circulatory system, the blood circulation, and from this nutrient liquid, what is required at each point is manu-factured there, or stores of food are built up, as for example starch in the liver, or fat in bone-marrow.

On the other hand, the activity of the animal organs leads to the using up of the substances, all containing nitrogen, of which they consist. This happens, as already indicated, by virtue of a process of oxidation which leads on the one hand to carbonic acid and water, and on the other hand to urea and other nitrogen compounds. These are removed from the blood by special organs, the nephridia in the case of the lower animals, the kidneys in the case of the higher. Only in the case of unicellular organ-isms do we have direct excretion from the cell itself, by means of so-called vacuoles.

The oxidation taking place in the animal body is also the source of animal heat, since it is a process which generates energy, 'exothermic' process. On the other hand it also supplies the energy required for the independent motion of the animal organism. Plant assimilation, conversely, is a process which requires an external supply of energy, an 'endothermic' process. The necessary energy is supplied to the parts of the plant con-cerned in assimilation by sunlight or diffused daylight, hence plants do not assimilate in the dark.

The chemical compounds formed as a final result of these life processes in the organism are mostly of a very complicated structure. The structural formulae for indigo given on page 18 is a comparatively simple example. The molecules of the higher carbohydrates and fats contain more than one or two hundred atoms and those of the proteins (Gr. *prōtos*, first) are probably still more complicated.[208] It is further certain that every

Ws

variety of organism produces its own compounds, at least its own special variety of protein. If protein of a kind not belonging to the species is introduced into the blood stream, injury usually results. We shall return to this fact again. The question now is, how does the living body manage so to guide all these innumerable reactions, that just those substances required at a certain point for a certain purpose are formed or decomposed? Organic chemical compounds have almost all the property of being capable of transformation in very many different ways, since they all consist of the same few elements (C, H, O, N,). How does Nature manage to select from all these possible reactions just those which are suitable?

A partial answer to this question (but not the whole answer) is given us by the investigation of a phenomenon which has only recently been understood as regards its full importance, *catalysis*. This term covers the fact that certain substances cause others to react chemically, without themselves taking part in the reaction.

If for example, we add to ordinary hydrogen peroxide solution (H_2O_2) a little manganese dioxide (MnO_2), decomposition of the hydrogen peroxide takes place with vigorous effervescence of free oxygen ($H_2O_2 = H_2O + O$), while not the slightest trace of change is to be found in the manganese dioxide. It can be used as often as we please to perform the same experiment. In this case the manganese dioxide acts as a catalyst; in other cases other substances act in the same way. For example, in the contact process of manufacturing sulphuric acid, finely divided platinum brings about the oxidation of the sulphur dioxide (SO_2) to sulphuric anhydride (SO_3). In the synthesis of ammonia by the Haber process, uranium compounds act as catalysts for the combination of nitrogen and hydrogen to form ammonia. The careful investigation of these processes has done a very great deal to advance modern technical chemistry.

The knowledge that catalytic processes play an extraordinary part in the organism should have come to us long ago on the basis of the existing material. It was indeed known to be the fact in many cases, but it remained for a long time neglected since other matters were more pressing and aroused greater interest. The case longest known to us is that of the catalytic action of a substance which is contained in germinating grain, and has the power of changing starch into sugar. This substance is called diastase (from *diastasis*, the Greek word for transformation); it can be extracted by water from malt, and is a yellow powder which possesses the power in question, as can easily be shown by experiments with starch paste. Organic catalysts are generally called *enzymes** or *ferments*, since fermentation is produced by them, as we shall proceed to describe. The better known among them are the digestive ferments, that is to say the substances which cause food to dissolve in the human or animal digestive tract, by decomposing the more complicated molecules of the foods into simpler ones, for example, the ptyalin of the saliva, which, like diastase, transforms starch into sugar, and the pepsin of the gastric juice, which turns albumens into soluble peptones. Many of these enzymes only act in combination with certain other simpler substances, often organic acids, bases, or salts. These are their so-called *complements*; thus hydrochloric acid is the complement of pepsin.

The majority of these catalysts hitherto investigated[209] are group catalysts, that is, they set certain kinds of reactions going in a whole group of more or less related substances. Often, however, the catalyst is strictly specific; it brings about a single definite reaction in the case of a single definite complex substance, which it fits as a key fits a lock. Probably most 'immunity' reactions belong in this class, that is to say, those produced in the organism when threatened by the bacteria of disease. We are naturally confronted with the important question of how it comes about, that the organism always produces exactly the right catalyst at the right moment, for it is fairly certain that in most cases it is formed by the living cell only when required. But we will leave this question aside for the present.

A further important extension of our biochemical knowledge has resulted from modern *colloid research* (see page 19 f.). Most living cells are colloidal systems, that is to say they do not contain true solutions, nor, or at any rate not only, coarse suspensions, but that peculiar intermediate stage of the sub-division of matter, which we have characterised more closely in an earlier chapter. In actual fact, it is only under these physico-chemical conditions that so complicated a play of physico-chemical forces is possible as we find in living matter.

* From *zyte*, the Greek word for cell, not from *zyman*, to ferment.

A colloidal solution contains the dissolved substance in so fine a state of division that it has an active surface millions of times greater than in its ordinary or macroscopic condition; on the other hand, the subdivision is not so great as to abolish all those specific properties possessed by the material in question when in larger pieces, which would be the case if the subdivision were carried to the point of the molecular stage (see page 20). These large surfaces carry very great electric charges, and hence colloids are very sensitive to changes in electrical conditions, but also to changes in temperature, in the concentration of added electrolytes, and other factors.

A recent investigator of colloids, Bechhold, therefore does not go too far when he says that the colloids may be regarded as the intermediate stage between dead and living matter (see also below, page 318). Certainly the spontaneous generation of life can only take place, if it ever does do so or has done so, in this region, and colloid-chemical investigation promises to give us in the future, as it has done in the past, a very great deal of information concerning life processes.[210]

Now that we have noted the most important fundamental chemical facts of biology, we may return to the philosophical questions mentioned at the start, the problem of *mechanism versus vitalism*. It is well known that, at the beginning of the last century, it was regarded as impossible for chemistry ever to manufacture (synthesise) the highly complicated organic substances: it was supposed that 'life force' was required for this purpose. This view, called to-day the older vitalism, was decisively disproved by Wöhler's famous synthesis of urea in the year 1828, and this was quickly followed by the syntheses of other well-known organic substances, such as acetic acid, alcohol, etc.

To-day we know that chemistry, given patience and good fortune, can build up synthetically every kind of organic substance, although many such groups of substances, particularly the proteins, have yielded up their secrets only with the greatest difficulty, or not at all. Recently for example, the synthesis of haemin has been accomplished, this being the main constituent of the red colouring matter of the blood haemoglobin, which is compound of haemin with various proteins.[211] It is well known to everyone that the chemical industry of to-day manufactures artificially a number of important organic substances, which earlier could only be derived from plants or in some cases from animals.

This foundation for vitalism thus proved illusory, but it very soon appeared that behind this first conquered line another and much more difficult one lay. For now the question arose as to what means were used by living organisms for the synthesis of the substance in question. The operations we perform in retorts and distillation flasks are only in the rarest cases copies of the methods used by the plant or the animal. The chemist finds an organic synthesis as complicated as the plant appears to find it simple. The plant produces a highly complicated molecule such as starch from the carbonic acid of the air by a single magic stroke, while the chemist needs hours and days to make a few grams of sugar, and is obliged to observe the greatest possible care as to his conditions. The contrast is at first sight so great, that not a few leaders of vitalism, only a few decades ago, found in it a further argument for their point of view.

For instance, even in the year 1909, J. Reinke of Kiel wrote : 'It would be a wasted effort on the part of a chemist to attempt to transform carbonic acid into sugar in the laboratory in the same way as the reaction goes by itself in the plant.'[212]

It is now certain that this second bio-chemical argument for vitalism is untenable. As a matter of history, the decision in this question has depended mainly upon the investigation of fermentation, which we must now deal with in a few words. Fermentation as we know, is the decomposition of grape sugar by yeast into alcohol and carbon dioxide. The equation

$$C_6H_{12}O_6 \quad = \quad 2\,C_2H_6O \quad + \quad 2\,CO_2$$

1 mol. grape sugar   2 mols. alcohol   + 2 mols. carbon dioxide.

summarises this reaction, but it is only a very superficial representation of the actual process, in which a whole number of bye-products, such as glycerine, succinic acid, aldehyde, and others, are formed. Furthermore, the equation gives us no insight into the way in which the sugar molecule is broken up into four new molecules. Recent investigations, particularly by Neuberg and his pupils, have cleared up the problem, if not completely, at any rate to the point of giving us the most important intermediate steps of the process, and have enabled us to increase the output of bye-products to any extent by changing the conditions, and also to completely suppress the formation of alcohol.[213] In this way, among other applications, the glycerine used in the war for the manufacture of explosives was produced industrially.

The discussion turned at first mainly about the rôle of yeast in this process. Yeast was shown by Schwann in the forties of last century to be a mass of micro-organisms, and Pasteur proved, in face of the opposition of Liebig, that when these organisms were killed no fermentation took place (Pasteur, of course, was the father of modern methods of sterilisation). This seemed to prove that a living organism, yeast that is, was really necessary to the production of fermentation, and hence it is not surprising that we find Pasteur in the camp of convinced vitalists. He would indeed have had to be more than a genius to have guessed at that time the correct solution of the problem.

This lies in the fact that fermentation is a catalytic process, for which as catalyst a substance is required which is itself dead, but which, at present, we cannot obtain from any other source than yeast. This was proved experimentally in the year 1896 by Buchner, who succeeded in extracting from crushed yeast cells, which were therefore undoubtedly dead, a substance which he called zymase and which, when brought into sugar solution produced all the characteristic effects of fermentation. If we could produce this substance artificially, we should be entirely independent of the living plant. The fact that we cannot yet do so, proves nothing in the face of the innumerable organic syntheses which have been successfully accomplished.

This disposes of the second bio-chemical argument for vitalism, and it is useless to assert, after this has happened in the case of fermentation, that matters are otherwise in respect of assimilation and other processes (as Reinke attempts to do in the remark cited above). It is fairly certain that assimilation is a catalytic process of this kind, to which the chlorophyll of the plant belongs as complement, but which obviously requires other special enzymes which we do not yet know.[214] In any case, Molisch has shown that life is not absolutely necessary to the process, since under certain conditions, dead cell material is able to assimilate.

On the other hand, several direct syntheses of sugar from carbon acid have recently been effected by means of ultra-violet light, which, of course, as is proved by the elimination of chlorophyll, cannot be regarded as directly parallel with the natural synthesis of sugar. This problem is still in a fluid state, but no sensible person has any doubt that it will be finally solved.

For the sake of completeness we may add, that the recent investigations of the Neuberg school have shown fermentation to be a considerably more complicated matter than appeared at first sight after Buchner's discovery. Zymase is not a single substance, but contains various ferments, and furthermore, only a small part of ordinary fermentation is caused by it, the greater part being accomplished by other ferments, as Rubner proved some years ago. All this has nothing to do with the fact, that, that, fundamentally, the decision was given by Buchner's discovery; fermentation is a catalytic process similar to others of the kind. It proceeds independently of life when once the catalyst or catalysts are present which cause it to take place.[213]

We cannot leave this subject without shortly referring to another attempt to give vitalism a bio-chemical foundation; an attempt which we may describe as 'stereochemical' vitalism. Organic chemistry tells us of a large number of so-called *optically active* substances, that is to say, substances which, whether as crystals or in solution, rotate the plane of polarised light. According to the direction in which the light is rotated we distinguish between right-handed and left-handed substances (dextro- and laevo-rotatory). The crystals of these substances are always found on close examination to have asymmetrical forms, that is to say, forms which differ from their mirror images as a left-hand glove differs from a right-hand glove.

The explanation of the existence of these asymmetrical forms was given by Le Bel and van 't Hoff in 1874. The substances in question contain one or more *asymmetrical carbon atoms*, that is, carbon atoms whose four valencies are linked to four different atoms or groups of atoms, as we see for instance in the formula for lactic acid, H_3C—CH (OH)—C OOH, the middle C of which is combined with four different groups CH_3, H, OH, and COOH. If we imagine a carbon atom in the centre of a regular tetrahedron (figure 45), and the four groups or atoms at the four corners, we have the possibility of two arrangements which are not congruent, but resemble one another as an object resembles its image in a mirror. These two forms are the two active forms of lactic acid. In an artificial synthesis of lactic acid, as also in a fermentation process, these two kinds of molecules appear in equal proportions, and the resultant product is, therefore, apparently optically inactive, since the two parts cancel one another in their action on polarised light.

The simplest means of separating these two components in this as in other cases, consists in the addition of certain kinds of bacteria which, curiously enough, only act upon one kind of molecule and not upon the other; or in other cases, such a mixed acid may be neutralised to form a salt by an organic base which is itself optically active and hence (sometimes) only forms a salt with one half and not with the other; which salt then crystallises as a single form. This latter method however, assumes, as we

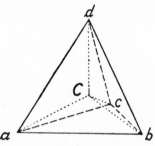

FIG. 45

Asymmetrical carbon atom (C), combined with four dissimilar atoms or groups, a, b, c, d, which form the corners of a tetrahedron

see, that we have already obtained from living nature other optically active living substances, or, on the other hand, that we allow living bodies to act directly on the mixture. It thus appeared at first as if it would not be possible to attain the separation of the two components by purely artificial or synthetic methods.

This was sufficient immediately to bring the vitalists into the discussion. They argued that only the living organism possesses the power of realising one alone of two possible asymmetrical arrangements of atoms, or of destroying one of them alone. This power is certainly very curious, for it is not easy to understand *a priori* why the grape, for example, should always produce only right-handed tartaric acid and the camphor tree only right-handed camphor, and why the latter alone, and not its opposite form which we are able to make artificially, should produce the physiological results.

But since organic chemistry has actually succeeded with syntheses of asymmetric forms,[215] the verdict of the vitalists that this could only be done by vital force, is shown to be wrong. Vitalism has thus again suffered a defeat, and there can hardly be any doubt that it will always have the same experience, if it tries to uphold its position by referring to single and particularly chemical activities of the organism.   We must therefore come to the conclusion that it is hopeless to define life in this way.   There are no chemical substances or processes in the organism which chemistry cannot imitate fundamentally.   If the vitalistic view is to be maintained it must look around for other supports.

We will, therefore, now take our leave of the purely physico-chemical question and turn to specifically biological phenomena. The tools of the biologist are not retorts and test tubes, but microscope and microtome, and his proper field of operation is not the chemical laboratory, but living nature, the incubator and the culture.   Let us now see what he has discovered in his own peculiar field.

II

THE LIVING CELL

Only since the microscope came to the aid of biological research has the latter really progressed beyond the stage of merely classifying existing species. In the year 1660, Leeuwenhoek discovered the world of living organisms in a drop of water by means of the microscope, and about the same time, Malpighi and Grew discovered the cellular structure of plants by the same means. But the fact that the cell is actually the elementary component of the living organism remained for a long time unknown.

It was the discovery by Schwann in 1838 of the animal cell, that opened our eyes to the fact that cellular structure is a common characteristic of all higher organisms, which as multicellular (metazoa) are contrasted with the unicellular organisms (protozoa). If, in most recent times, we have had to set certain limits to this statement, as we shall presently see, the fact still remains that it is correct in by far the great majority of cases. The animal cell was more difficult to recognise as a unit because, in most cases, it does not possess a wall like that of the plant cell, and because the intercellular substance (that is the substance accumulated between the cells as a result of their activity) plays a considerably more important part in the animal organism than in the plant.

Soon after the fundamental discovery of the scientist above mentioned, other investigators, mainly German, showed that all higher multicellular organisms are derived from a single cell or from two germ cells which unite to form one. The multicellular organism is formed by the continual division of the single cell. This fact, which is known to every layman to-day, is a very perfect example of those truths of which Schopenhauer has said, that they are at first just as strange and astonishing as they are afterwards self-evident. Who thinks to-day of the great experimental skill and fruitful imagination that was necessary before K. E. von Baer, Boveri, O. Hertwig and others succeeded in making these matters as clear to us as they are to-day, when they are to be found in every school book?

We must not pass on without pointing out that the boundary between unicellular and multicellular organisms is badly defined, in many cases.   There are colonies of cells, such for example as the globular animal shown in Figure 46 (*Volvox*), which may

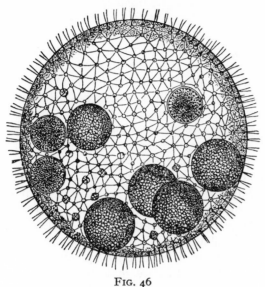

FIG. 46
Volvox colony

equally well be regarded as a single individual, or as a collection of a number of more or less independent cells.   Even the sponges, which are usually counted among the multicellular animals, show so loose a connexion, that they might also be described as mere colonies of cells.

Only when we come to the next group, the corals, do we finally get the impression of individuality for the whole, and here we find a further curious fact that many such individuals again form a structure of a higher order, which obviously functions quite unitedly as a single being.   In particular, the so-called siphonophores are such collections of polyps, which consist of single individuals with various functions to perform, and hence with various characteristics (nutritive polyps, defensive polyps). While in this case the formation of a whole goes beyond the stage of what is commonly called an individual, we find on the other hand, in the case of the multicellular organism, a differentiation of the various organs and organic regions, which likewise distinguishes characteristic wholes, namely cell colonies with quite definite functions (kidney cells, nerve cells, etc.) from the whole individual, which then exhibits itself as the compound result of its various parts, namely the organs.

This hasty survey already allows us to see what is the leading characteristic of living matter; the *combination of parts to form* *wholes*, which again become the parts of wholes of a higher order. In this series of wholes, it is easy to show that the cell is by no means the lowest stage. In this matter, modern science has had to revise its views. The cell is not, as was believed in Haeckel's time, the ultimate elementary part of the organism; on the contrary, it possesses in all cases known to us, as also in that of the simplest living organism such as a bacillus or an amoeba, an extremely complicated structure.[216]

Many unicellular organisms, such for example as the well known infusorium *Paramecium*, even have highly elaborate organic arrangements such as a special mouth, cilia as organs of motion, vacuoles for breathing and digesting, etc. Others have retractive stalks, crowns of cilia for creating currents, and all sorts of other peculiarities; and as well, an internal structure that is anything but homogeneous. To this we may add the most recent discovery, that the intercellular substance[217] of animals is, in numerous cases, by no means to be regarded as dead, that is, as a mere excretory product of the living cell, but undoubtedly as a part in the life of the organism.

Again, the normal structure of the cell with its nucleus and protoplasm is by no means always present in so simple a form, for there are cells containing hundreds and even thousands of nuclear bodies. We thus see that we are not to regard the cell as the essential and indispensable fundamental element of life, but that life itself is only associated with those substances and complexes of substances of which cells are made. This does not however, abolish the rule that the cell is the normal form of this material of life called protoplasm.

The highly complicated structure of which a single cell is capable is shown, apart from the above facts derived from direct observation, by equally impressive indirect evidence derived from *heredity*. It is proved that the germ cells can transmit as many qualities as there are differences between the species and even between individuals of a single species, for otherwise there would be no explanation of the fact that, in the same external circumstances, a hen's egg always produces a fowl, a frog's egg a frog, and a snake's egg a snake. These differences must be contained, in some way or other, in the eggs themselves, although they appear to us, when of the same species, exactly alike, even under the microscope. How this is possible or thinkable, will be discussed later, but the fact is undeniable. But it proves that the cell must have the possibility of fine structure, and must possess one so fine as to be far beyond the reach of our best microscopes.[218]

How is such an incredibly fine structure possible?

Or may it be that we are not dealing with a material structure at all, but with super-material powers, which cannot be captured by the microscope, even if the latter could take us all the way down to the atoms? The size of the cell varies within fairly wide limits. Some attain a length of several centimetres in one direction at least, while others are hardly visible under the best microscope. The majority however have dimensions just about at the limit of visibility with the naked eye. The smallest form that has been measured with certainty is probably the *Spirillum parvum* of von Esmarch, which has a thickness of 0.1 — 0.3 μ; the so-called influenza bacillus (its right to the name is doubtful) is also very small, but 0.4 μ thick and 1.2 μ long.

But it is not impossible, and for certain reasons even probable, that smaller organisms exist which are only visible in the ultra-microscope. It is possible that the agents of certain diseases, such as the lung diseases of cattle, foot and mouth disease, the mosaic disease of tobacco, and others belong to this category, as has been shown by filtration experiments.* However, these will probably not be much smaller than those visible in the micros-cope.[219] We may therefore regard a diameter of 0.1 μ to 0.3 μ as close to the lowest limit of the size of living beings.

If we now calculate for a cell of 0.1 μ diameter the number of protein molecules which it contains in accordance with our present knowledge of the constitution of these substances, we find according to Errera[220] about ten thousand. This is, after all, a number of units which would suffice for the most complicated structure imaginable, when we further consider, that, from the

* This was proved a few months ago by Bechhold.

purely chemical point of view, millions and billions of varieties of protein may exist. But between the molecular structure investigated by chemistry and the coarser constituents of cell structure which are visible under the microscope lies exactly that 'world of neglected dimensions' (see page 20), in other words the world of colloids, which is now in process of discovery. In this region the limits between structure in the ordinary mechanical sense of the word, and chemical structure, that is the building-up of the molecules from atoms, become uncertain. But of recent years, numerous successful excursions into this hitherto unknown territory have been made.

The most important discovery in this field was made by a Frenchman, d'Hérelle,[221] in the year 1917. He placed, day after day, a few drops of the intestinal fluid of a person suffering from dysentery in nutrient broth, filtered this through a filter fine enough to keep back bacteria (Berkenfeld or Chamberland), and observed that the liquid thus obtained acquired the power, after some days in the progress of the disease, of dissolving dysentery bacilli growing in another culture liquid. Now if we add some of the liquid containing such a dissolved culture to another culture, the same effect is obtained, and it increases in strength with every repetition. D'Hérelle concluded from these facts, that he was dealing with an ultramicroscopic living creature capable of dissolving bacteria, and this he called a 'bacteriophage.'

Other investigators, however, believe that the substance which dissolves the bacteria is formed in the bodies of the bacteria themselves as they are dissolved. All that is certain is, that it only increases in amount together with the bacteria, and further, that it is strictly specific, that is, only acts on the particular bacteria by the help of which it was originally obtained (in d'Hérelle's case only on dysentery bacteria). Bechhold, who himself has thoroughly investigated this d'Hérelle phenomenon, proved that the phagic principle can be separated from ultramicroscopic bodies which are visible, and probably fragments of bacteria, by means of ultra-filtration (see page 21 above). Since such ultra-filters only allow albumen to pass in very great dilution, and even the red colouring matter of the blood, haemoglobin, only in traces, it follows that the bacteriophage cannot be much larger than the molecules of these substances.

This leads to the question whether this 'something' may not

perhaps occupy a position between a true living organism and a dead protein molecule, so that one can hardly say whether we are dealing with a living creature or dead matter. This 'something' has perhaps only one of the two fundamental properties of living matter, assimilation and reproduction. It has the property of assimilation, while reproduction is left to the bacteria. However this may be – the case concerning this phenomenon is far from being closed – we certainly see enough to say that we are in a border land where it is possible that a kind of transition between living and dead takes place.

A recent attempt has been made to approach this region from another direction, by means of X-ray analysis (page 170). The investigations in question were initiated by the industries dealing with fibres, which obviously have an interest in gaining knowledge of the fine structure of plant and animal fibres (cotton, silk, etc.) which far exceeds anything attainable by human methods. By means of the Laue diagram mentioned above, we have succeeded in getting on the track of this and many other structural secrets of living cells.

The result has been to confirm the hypothesis put forward sixty years ago by the botanist Nägeli, at any rate to a considerable extent. Nägeli assumed that the living cell consists of smaller units which he called micelles, and which, in his opinion, were separated by other intermediate substances. According to the investigations we are discussing, it seems to be a fact that a cellulose fibre has such a micellar structure.[222] The cellulose fibre consists, as X-ray investigation combined with chemical has shown, fundamentally of grape sugar molecules, which are in the first place united in pairs to form a double sugar, cellobiose. These double molecules then form a regular (crystalline) arrangement, which is rendered evident by optical polarisation phenomena and X-ray analysis, and we have thus, primarily, a linear elementary structure. If we now imagine a larger number (some hundreds or thousands) of such linear structures with their axes parallel to one another, but otherwise arranged at random along side one another (like a large number of lead pencils of unequal length in a long narrow box, considerably longer than a pencil) we have an approximate picture of a micelle of cellulose fibre (figure 47).

In other similar cases, similar, half regular, half irregular arrangements of plate-like structures have been found (Haber,

Naturwiss., 1926, p. 854). We see how in this case, the chemical fine structure, that is, the arrangement of the molecules made up of atoms, passes over without a sharp transition into mechanical fine structure, that is to say, the arrangement of the molecules to form greater aggregations. We need hardly point out that the forces acting from one molecule to the next are not fundamentally different from those which act within the molecule between the atoms. It has already been shown[223] that a similar crystalline or semi-crystalline arrangement exists in the substances contained in the cell nucleus and its sub-divisions (for example, the so-called chromosomes, see page 336), but here matters appear to be much more complicated. It has been quite generally proved, that in colloidal solutions, such semi-crystalline

FIG. 47
Micelle structure of cellulose in fibres

structures are found very frequently, and probably the distance of the single units from one another is regulated by quantum or wave mechanical considerations, just as in the internal structure of the molecule, but we have not yet any certain information on this point. Hence these few remarks must suffice.

So far we have only discussed the structure of living cells, and we must now say something about their functions. These are three in chief: *Assimilation and growth, reproduction, and reaction to stimulus*. All these are very closely connected. An object very often used to demonstrate these various functions is the amoeba. We see this minute being under the microscope, appearing like a naked fragment of jelly, stretching out its pseudopodia, and moving with their help towards another tiny living creature or other desirable piece of food; we see how the pseudopodia close around the prey, how it is gradually drawn into the body of the amoeba, and how the pseudopodia, their duty having been performed, are drawn in again; we see the prey wholly or partially dissolved in a juice quickly produced for the purpose, the indigestible residue thrust out of the body, and the whole process soon repeated with another piece of food.

We are further able to observe that such minute beings answer to certain stimuli, for example light or shade, by definite behaviour, hence they must obviously be sensitive to the stimulus in some way or another. Indeed, the feeding process itself is sufficient to demonstrate this susceptibility to stimulus, for we can easily notice that this fragment of jelly distinguishes exactly between the substances which it meets; that the amoeba, for example, takes no notice of indigestible grains of hard mineral.

In short, we are confronted with a being, that, in unmistakable analogy with our own actions and those of the higher animals, is apparently gifted with the power of voluntary motion, of receiving and making use of sense impressions, and of selecting and digesting suitable food. Apart from these readily observable activities of the cell as a whole, more exact investigations show that continual changes and eddies are going on inside the cell itself. It is probable that, in the case of multicellular plants, an exchange of plasm is continually taking place from cell to cell, the cell walls being provided for this purpose with minute canals. And all these movements take place, not by any means according to blind physical laws, but obviously in the service of the whole, and thus form a part of the great problem of life, which is that of explaining how the living organism comes to function as a single whole.

We shall therefore now devote our attention to this problem, limiting ourselves for the present to what can be said from the standpoint of the investigation of unicellular organisms and their life. We will deal later on with the question of multicellular organisms and its bearing on the problem. It is necessary to proceed in this way, otherwise we shall not be able to see the whole clearly. It is further necessary to realise at this point, that the problem of life has, in another respect as well, a two-fold character. We can first of all ask to what extent physics and chemistry help us to understand the processes which take place in the given living cell or multicellular organism. We may then further inquire how far they are able to go on to explain how such existences have come into being.

It is easy to see that the latter question goes a great deal further, when we consider that, for example, the structure and functioning of an electricity generating station can be explained from the physical forces acting on the machines in question, but that no one would assert that such a station could have come

Xs

into being as the result of the blind play of such physical forces. In other words, we must be careful to distinguish the problem of the *phenomena* of life from that of the *generation* of life, if we wish to gain a true understanding of the position.

Let us therefore first confine ourselves to the life phenomena of the unicellular organisms. At the time of the greatest popularity of mechanism (1880–1900) it was often supposed that physics and chemistry were already on the way to explaining the functions of the living cell, particularly assimilation, by the application of known physical and chemical laws. The naïve opinion that the simplest elementary unit of life had been discovered in the cell, encouraged this view; and certain experiments, at first sight very surprising, appeared to give a direct proof of it. Since these 'proofs' of mechanism, although they have long been recognised as erroneous have not yet entirely disappeared from popular literature, it is necessary for us to devote a few words to this point. Actually, analogies to almost all life phenomena of the single cell are known in physical chemistry, and they often simulate life with an extraordinary accuracy.

If a drop of potassium ferrocyanide solution (yellow prussiate of potash) is carefully introduced into a solution of copper sulphate (blue vitriol), there is immediately formed around the drop a thin skin of insoluble ferrocyanide of copper according to the equation:

$$K_4Fe(CN)_6 + 2\,Cu\,SO_4 = Cu_2Fe(CN)_6 + 2K_2SO_4$$

a skin of this kind forms a 'semi-permeable' membrane (like parchment, animal bladder etc: see above), and allows water and sometimes salts, to pass through. Under suitable conditions of concentration in this case, only water passes into the interior of the drop or 'artificial cell.' The result is an excess of pressure inside the drop, and the membrane finally tears. At the point of tearing, the liquid within pours out, but the same thin skin is immediately formed around the extruded drop, and the same process now takes place with this 'daughter cell,' as with the original 'parent cell.' A process thus takes place before our eyes, which strongly reminds us of the growth of an organism by cell division. Exactly similar experiments can be made with other pairs of suitable salt solutions, for example, water glass and zinc vitriol (Traube, Reinke, and others, see Figs. 48 and 49, opp. page 323).

Another very pretty series of experiments is due to Rhumbler. He introduced drops of chloroform (for example) into a liquid of the same specific gravity, in which they remained suspended. He was then able to reproduce a number of phenomena which remind us of the intake of food

PLATE V

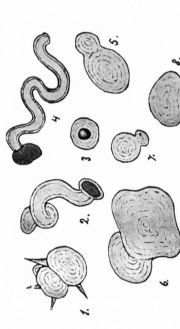

FIG. 48. Artificial 'cells' formed from Copper Sulphate and Potassium Ferrocyanide solutions

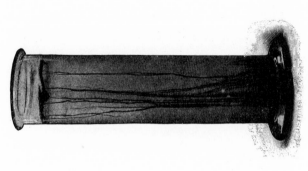

FIG. 50   Leduc's 'Artificial Algae'

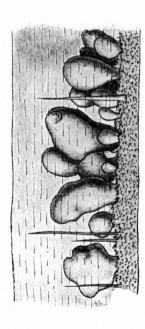

FIG. 49.   Artificial 'cells' from Water Glass and Zinc Sulphate solutions

by unicellular organisms such as the amoeba. Thus the drops of chloroform enveloped small grains of sealing wax in exactly the same way as an amoeba grasps a fragment of food. These phenomena are easily explained by means of the laws of surface tension, and we do not need to discuss them further here. They are thus fairly simple phenomena.

Lehmann observed, in connexion with the 'liquid crystals' which he discovered, a whole series of phenomena which have an astounding likeness to the processes of cell-conjugation (union of two cells) and cell division.

Quite recently Leduc has obtained really wonderful results of this kind by similar means to those first mentioned.[224] He succeeded in producing structures which were externally closely similar to a colony of algae: indeed he could imitate the turf-like coating, which we see on water surfaces covered with algae. The annexed illustration, Fig. 50, gives us an idea of these 'artificial plants.' Here again, of course, we are merely dealing with a peculiar and cleverly arranged play of known physical and chemical forces, the explanation of which offers no difficulty.

These two examples will suffice to make clear the essentials of the analogies in question. We should add, it is true, from the point of view of many mechanists, a whole series of more general analogies between inorganic and organic processes. W. Bölsche, for example, has stressed the analogy of assimilation, the intake of the surrounding inorganic material into the cycle of plant life, with the chemical processes of flame, and has even found, in his poetical manner, a connexion with the ancient idea of a relationship between fire and soul (Heraclites). Similarly, Preyer, in his book on scientific facts and problems, compares the sea with a mighty organism, the 'quiet breathing of the sea' (the ebb and flow of the tide) is made to do duty as the 'heart-beat of the earth,' etc.

Sober consideration will, of course, lead everyone to agree that all this is nothing more than very loose comparison and superficial analogy, which does not advance us a single step in the investigation of actual living processes. For no one disputes that the latter, even when they are explicable on pure physicochemical lines, are certainly not to be referred to those forces which produce the apparently analogous phenomena in the inorganic systems in question. For example, when one of those artificial cells first mentioned bursts on account of endosomose, and thus apparently 'buds off' in a new cell, this simple physicochemical process has certainly nothing, or scarcely anything, to do with the extremely complicated processes which have to be assumed, also on the mechanistic view, for real cell division, and the same is true in all the other cases.

The general public cannot possibly know enough about the subject to estimate the importance of these experiments, and anyone describing them in popular language without adding that they are nothing but superficial comparisons, is consciously or unconsciously throwing dust in the eyes of the public, and the vitalists are quite right in protesting. Anyone who supposes that these experiments prove that the living cell is nothing but a physico-chemical machine, might equally well assume that the existence of automatic electric pianos proves that Liszt and Chopin were also musical automata. To this extent, vitalism has nothing to fear from these analogies.

On the other hand, we must point out with equal definiteness, that it is no less wrong for vitalism, on its side, to use the contrast between real vital phenomena and the physico-chemical processes mentioned to draw the conclusion that biological and physico-chemical processes are fundamentally unlike. It is true that the 'growth' of a crystal in a solution is something entirely different from the growth of a living cell, for the former takes place by the addition of layers upon the outside, the latter by so-called intus-susception. Likewise the flowing of a chloroform drop around a fragment of sealing wax, on account of surface tension, is certainly something different from the flowing of the pseudopodia of the amoeba around the algae. But this by no means proves that biological processes are something entirely different from physico-chemical processes. There are innumerable types of the latter; the processes of organic chemistry are also, in almost all points different from those which we study in inorganic chemistry, and yet no one doubts that they are fundamentally alike. The intussusception of the amoeba may perfectly well be a physico-chemical process although one entirely different from the growth of a crystal. Speaking quite generally, when a vitalist has said: 'analogies may never be misused in order to prove the similarity of different phenomena,' this sentence must be extended by adding: 'that contrasts also must never be misused in order to prove the fundamental and absolute dissimilarity of different processes.'

Among the false arguments of this kind coming from the vitalistic side we may note particularly the emphasis laid upon the irreversibility of life processes and upon the mortality of organisms. It is of course true, that simple chemical processes such as the reaction $Mg + O = MgO$, can be reversed without difficulty, whereas life processes cannot be reversed. But a thunderstorm or the outburst of a geyser also cannot be reversed, nor can any other physico-chemical process of any higher degree of complication; indeed this is already expressed in the law of entropy. What guarantee have we, therefore, that the irreversible phenomena in question, the life processes, are not also highly complicated physico-chemical processes of this kind?

Furthermore, when we consider death (we shall go into this matter at greater length later on) it is of course true that a machine is theoretically capable of operation for an unlimited length of time, if its parts are so constructed as to renew themselves continually from surrounding material, whereas experience tells us that life always ends in death. Also, a machine can (theoretically) stand still for an indefinite time, whereas experience teaches us that life, if it stands still for too long, invariably becomes extinct. But even these contrasted statements do not in reality carry weight, for: 1, every machine also rusts in practice when left alone, and then does not work any longer; and in practice also, every machine becomes too old in spite of all repairs; 2, but in the case of life processes, it is by no means proved that their cessation or death follows from absolutely inward necessity: on the contrary, we are almost certain to-day that this is not the case (see below), but that in reality it is only external injuries, just as in the case of a machine, which bring about the end.

As regards the durable forms of life (spores, seeds, etc.), we can assert with complete certainty that an absolute isolation of them from the external world by no means exists, and hence we can never say for certain whether chemical changes have not taken place within them, to however small an extent, or whether the process of life has really come to a standstill. It is true that the latter view is supported by the astounding durability of such forms of life. It has been proved without question, that seeds can remain capable of germination for sixty years and bacteria for as long as ninety-two years. Although the famous Egyptian wheat has been shown to be a deception practised by the natives, we still have sufficient evidence to demonstrate a duration of quiescence, which could only be compatible with an extraordinarily small transformation of energy within the cell. But who is to prove that no transformation whatever takes place? And even if that were the case, we should still have to reckon with external influences, which are also finally effective in the case of a machine which is out of action for a long period, and can be no more isolated from the external world than the seeds or spores.

These arguments for vitalism are therefore just as weak as those, referred to above, derived from biochemistry. *The final conclusion is, that it is obviously hopeless to base vitalism on the demonstration of certain processes or groups of processes which are supposed not to occur in inorganic systems.* It, is not too much to say that every assertion of this kind can be countered by the mechanist with a striking inorganic analogy such as we have seen above in connexion with artificial cells, and that this has again and again compelled the vitalists to limit their assertions. Anyone who has a general and unprejudiced view of this discussion cannot escape the impression that, taking it all in all, vitalism has not played a very fortunate part.

Nevertheless, we are aware of an inward voice which tells us that it cannot be a complete illusion, when living organisms again and again appear to us as something totally different and higher as compared with inorganic nature. Even when we intentionally refrain for the moment from considering psychical phenomena, we still have a general impression that even the most primitive living being is something quite other than the most complicated inorganic system. What is the reason for this, if it is not to be found in the individual processes which are similar in the two cases?

The answer is obviously to be found in the fact of the 'wholeness' character of life processes, which always take place in such a way that they act in the service of the whole (as far as this is possible, that is to say). Everything that a living being does, is always done (apart from artificially produced exceptions) in such a way as to maintain and reproduce it and its species. All the enormously complicated transformations in the living cell, indeed, the separation from the living plasm of the enzymes which effect these transformations, take place in the service of the whole. This is just as certain as any fact in physics and chemistry, and it is the fundamental fact of life that everything about it is related to the whole. However, we may look at it, we are here confronted with something which does not happen in physics and chemistry, even if all single processes are comprised in these latter.

Vitalism has performed the unquestionable service of repeatedly drawing attention to this fact, which mechanism, in the heat of the battle, has too often lost sight of. The only real refutation of it would be to show that this relation to the whole, or putting it honestly and plainly, this purposiveness (teleology) of life processes, can be explained by purely physico-chemical principles. We are at present very far indeed from such a possibility, and we cannot even imagine how this state of affairs is to be altered.[225]

We can also express this point of view by saying that, even if we could explain by physico-chemical means all the single actions of a finished organism, we should still be unable to explain how such an extraordinary complicated physico-chemical machine, capable of all these functions, could come into existence. As we see, we are now confronted with the problem not of the phenomena of life, but of its origin, and we must therefore devote a few preliminary words to this question. Our final attitude can, as in the former case, only be defined, when we have considered the whole range of life phenomena.

We may anticipate the final conclusion as follows. The successes of the mechanistic methods of investigating life processes have themselves removed the grounds for regarding the origin of life from a similar point of view. If the cell, as is the mechanist view, is really nothing but an enormously complicated piece of physico-chemical machinery, the probability of its production by physico-chemical forces becomes smaller and smaller, the further we penetrate into the fabulous complication of its structure. While twenty or thirty years ago the doctrine of spontaneous generation was regarded as agreed upon, its position has been entirely changed since we have been able to gain closer insight into the finer differentiations inside the simplest cells by the use of increasingly delicate methods. Most scientists are unwilling to discuss this matter at all, while a few still support – not for experimental, but for general reasons – spontaneous generation; a large number incline to the so-called doctrine of panspermy or the idea of the eternal existence of life (see below). Let us attempt to make clear the chief objects and counter arguments.

The complexity in the structure of a living cell affects both the *chemical material,* and also the *structure of the cell-body* made from it.

(*a*) The chemical substances comprised under the main constituents of protoplasm, or, more specifically, nuclein, globulin, albumen, are, as already mentioned on pages 302, 308, all of so complicated structure, that we have hardly made a beginning of unravelling it, in spite of the work of E. Fischer, Bergmann and others. But let us suppose that these investigations have been brought to a successful conclusion, and that we have an exact picture of the more important constituents, and that we have even succeeded in making them synthetically. It will not be denied that all this lies within the realm of possibility. But now let us imagine what an enormous amount of trouble and knowledge would be necessary to prepare, say a definite enzyme, what a vast number of exactly maintained conditions, both simultaneous and successive, would have to be fulfilled in order to effect the result. For the greater the number of atoms in a molecule – and in the case of the proteins this amounts to hundreds and thousands – the greater is, generally speaking the tendency of this complicated molecule to decompose.

In the laboratory the guiding mind is of course the chemist, who mixes together the right liquids in the right concentration, at the right moment and at the right temperature, and so on. But who is to take his place in non-living nature ? In the living cell, if we assume for a moment the mechanistic standpoint of the machine theory, this part is played by the machine conditions, that is, such things as the enzymes present. But here we are dealing precisely with the problem of how these themselves came into being.

It has been answered that there was a gradual progress in the direction of higher and more complicated groupings, but there is a serious objection to this idea, since *we know of life only as connected with the most complicated substances, and have good reasons for supposing that the life functions only become possible in the case of these complicated substances.* But where have these come from when, after all, far simpler substances than they exhibit only a tendency to decompose, rather than to build up ? Does not this reduce to zero the possibility that even the chemical material alone which forms the essential foundation of life might have come into existence spontaneously ?

This conclusion has often been drawn by vitalists, and it cannot be denied a certain power to carry conviction. There can be no doubt at all that the nut is a very hard one for the mechanists to crack. But justice demands that we should not neglect the exceptions to the general rule, and should not ignore those cases, however rare they may be, in which fairly complicated products are formed under comparatively simple conditions, which might have existed, and still might exist somewhere and sometime in inorganic nature. In view of the far reaching importance which every small detail may assume in this matter, it is necessary that we should devote a few remarks to this point.

There are two main groups of facts which must be mentioned. In the first place, many groups of organic substances have long been known, in which the above mentioned destructive tendency is by no means observable, but which, on the contrary, have a very decided tendency towards the formation of more complicated molecules from simpler ones. The chief of these are the aldehydes (and ketones) to which belongs the very substance, which is probably the first product of the assimilation of carbonic acid in the green parts of plants, formaldehyde CH_2O: this substance can easily be formed anywhere in non-living nature (for example from carbon monoxide and water, $CO + H_2O$). It has been shown that the simple action of lime, for example, and generally of basic substances upon formaldehyde results in the formation of complicated mixtures, among which at least one true sugar, acrose $(C_6H_{12}O_6)$, is found (Butlerow, Fischer, Löw.), but this is not the only reaction of the kind, for all aldehydes show more or less tendency to 'polymerise' in this way, that is to say, their simple molecules unite to form multiples.

The same is true of the compounds of cyanogen, the monovalent atomic group CN (the hydrogen compound of which, HCN, is prussic acid). Cyanogen compounds are formed with comparative ease from inorganic materials, and could certainly pass over without much difficulty into organic substances such as formic acid, although Pflüger's hypothesis, that protoplasm was formed in this way, is more than daring. We may nevertheless maintain, that syntheses of the two kinds mentioned, or also perhaps of other kinds, might take place even under free but suitable natural conditions, and that they have therefore probably done so at some time or another on this earth. In particular, we should once more refer to the synthesis of carbohydrates by ultra-violet light, already mentioned on page 310. By this means, for example, whole tons of sugar might have been formed 'of themselves.' If further investigation should produce more results of this kind, above all, as regards the inclusion of nitrogen in the upward series of syntheses, the improbability of the spontaneous formation of protoplasm would become considerably less. But we must not overlook the fact that the affair would become more and more difficult the higher the stage reached.

At present nothing can be said for certain on this point, but we may learn again from it, how extremely careful one must be in drawing conclusions about the momentary position of scientific knowledge, when it is a matter of proving something to be impossible. The older vitalism of 1828 had also a certain right to conclude from the results of science that the artificial preparation of organic substances was impossible or improbable; nevertheless, Wöhler's synthesis of urea abolished the conclusion with a single blow.

We can easily see the reason for this. It was wrong to put forward as a positive result of science, something that in fact was no result at all. The two sentences: 'It has been shown, that organic compounds cannot be made artificially' and 'it has not been shown, that organic compounds can be made artificially,' were confused, the one with the other. Only the latter was true up till 1828, and at any moment further progress might have corrected it. The first sentence is a logically unjustifiable transformation of the practical absence of a result into the dogmatic result of a permanent impossibility; the old but perennial mistake of confusing *ignoramus* with *ignorabimus*.

If we are to avoid this mistake we must not, so far as I can see, declare dogmatically the impossibility, or even merely the infinite improbability, of the synthesis of natural proteins, but must only put on record that, so far, very little result in this direction has been obtained. Whether the future will bring more, is a matter concerning which we can make no assertions; we can only wait and see.

(b) The second step necessary for mechanistic spontaneous generation, appears still less probable than that of the spontaneous formation of the constituents of protoplasm. This step is the production of an organised structure in these substances, which is required to make them into true protoplasm or structural material.

That a mere assemblage of such substances is a long way from being a living cell is best proved by considering the dead cell.

On the contrary, the production of the materials would only be the first and smallest step towards the manufacture of living beings. The materials only would be available, and the main task would now become that of distributing this material in the necessary extremely fine and correct arrangement, and of bringing into existence that peculiar state of dynamic equilibrium, which constitutes life. A mere heap of wood, brass, iron, etc., does not turn into a machine, and we should have no better reason to expect a few proteins to turn into a cell. Even if we assume that the existence and functioning of the cell is a mere play of physical forces, the formation of the cell is thereby no more shown to be also a play of such forces than, as we said above, the construction of an electricity station can be explained by the forces which take part in its operation.

Now, all that we know at present, concerning the living cell, is that it must have a much more complicated structure than the closest microscopical investigation has hitherto revealed. The phenomenon of *regeneration* alone is sufficient to show us how superficial the comparison with a machine certainly is. Imagine a machine which, when cut in half, regenerates the whole from surrounding material – we are unable to begin to form a picture of it. We can merely record the bare facts. Progress of our knowledge in this respect, as in every other, can only be attained by systematic biological experiment. What has been done in this respect is merely a beginning, although a very fine beginning, and it is entirely insufficient to allow any general conclusions to be drawn.

All that we can assert with confidence is, that even the most primitive cell known to us is already an organism in the fullest sense of the word, a whole, in which every single part serves a definite end; a whole, which on the one hand possesses an astonishing extensibility, adaptiveness, and power of resistance, but on the other hand so complicated a structure, that this latter cannot be abolished without destroying the most essential feature, that is, making life impossible. How an unimaginably fine internal structure of this kind could come into being by itself as the result of physico-chemical forces, is for the present completely outside the range of our imagination, in spite of all assertions of certain dogmatic mechanists. It appears to us at first sight just as improbable as that a poem should be formed by shaking letters together, a picture by pouring colours together, or a watch by melting brass and iron together.

In spite of all this, a cautious critic will not agree with the vitalists in asserting that the mechanist supposition of spontaneous generation is simply impossible. We should rather consider, that a much more advanced study of the substances forming protoplasm might perhaps one day lead to an insight into the necessity of certain arrangements and structures of molecules. If the forces acting from molecule to molecule, which are still entirely unknown to us, lead in the case of the simpler chemical compounds to the simpler crystal forms, if these molecular forces produce, in the case of many more complicated organic compounds the peculiar phenomena of Lehmann's liquid crystals, it is after all not entirely unthinkable that, in the case of the enormously more complicated plasm substances, the cell itself may result as the, so to speak, natural form as soon as the correct conditions occur.

This is the direction taken by the investigations of the d'Hérelle phenomenon and the micellar structure of cells, discussed above on page 318. They lead to the conclusion, that between the living cell and the inorganic crystal it is necessary to insert a hitherto neglected intermediate step, the investigation of which is necessary before the abyss between matter and life can possibly be bridged. This intermediate stage would consist in complexes of organic substances which, simply united to form larger or smaller aggregates without cellular form, would exhibit forms somewhat like those of the micelles in question. In such complexes of material, at first devoid of any characteristic of life, the power of assimilation must then have appeared at some time and somehow, that is to say, they must have contained enzyme-like components, which enabled the complexes of substance to form materials of the same chemical properties as themselves out of surrounding material.

Although such systems are not yet known outside living nature, they are nevertheless imaginable.[226] It has often been supposed that such a discovery has been made, but these reports have always turned out to be erroneous. Nevertheless, without some such supposition, a transition can hardly be imagined. Anyone not prepared to make it must decide to call entelechy or the like to aid at the right moment as *deus ex machina,* a point of view which even arouses some doubts in many vitalists to-day.

There is not much more to be said at the present time about the problem of spontaneous generation than what we have said already. We ought perhaps to mention some considerations put forward by the mechanists in order to refute the vitalists' argument that spontaneous generation, if it has ever occurred, ought to be observable somewhere or other to-day; that one should be able at least to observe the spontaneous formation of organic substances, somewhen and somewhere.

The reply to this is, that such a free synthesis of higher carbon compounds could hardly be observed to-day, since life is already present everywhere, and infinite numbers of living organisms see to it that the smallest amount of organic substance which is anywhere obtainable is immediately drawn into the life cycle. If for example, a considerable quantity of sugar were to be formed by chance anywhere (say by the irradiation with ultra-violet light of water containing carbonic acid), this sugar would be immediately devoured by yeasts and other organisms, and hence we should never be able to observe its formation. Life is present everywhere to-day. Apart from the very coldest regions at the poles, the depths of the earth and perhaps also the lowest depths of the sea, there is no spot on the face of the earth where living beings, mostly of very many kinds, are not to be found. This argument thus removes one difficulty out of the way. But it follows from what we have said that the hypothesis of spontaneous generation is, nevertheless, confronted at the present time with the greatest difficulties.

We shall return in a different connexion to what is called the hypothesis of *panspermy* later on (see page 430). For the present we will try to draw a general conclusion from what has been said, as far as this is possible. The conclusion is, that the discussion between mechanism and vitalism is so far undecided, unless further arguments or counter arguments are produced. The discussion is very much in the air, so long as we are not able to say exactly what physics and chemistry are able to achieve in the matter of explaining the phenomena and origin of life. Biochemistry would have to have finished its task before we could decide whether it alone was capable of explaining life. Since this is out of the question to-day – since, indeed, we are obviously at

the very beginning – there is nothing left for us but an honest
non liquet. The most we can say is that, as regards the phenomena of life, mechanism has hitherto made several victorious
advances against vitalism, but that, on the other hand, the
prospects of mechanism as regards the origin of life become less
and less, the deeper becomes our insight into cell structure.

Those who are not satisfied with this very indefinite result, had
better turn to those who are already sure about the matter to-day.
There are enough of these in both camps.   One party, as Frenssen
neatly remarks, 'were there when the first cell was married' while
the other party were watching when 'God knelt down and, smiling
sadly, created man's soul.'   We for our part are on the side of
Mach when he says: 'A satisfactory philosophy of life cannot be
presented to us, we have to win it . . . the highest philosophy of the
scientist is to endure incomplete knowledge, and prefer it to the
apparently complete, but inadequate' – though we should, it is
true, wish to replace the more general term 'philosophy of life'
by the limited one 'philosophy of nature.'   We shall show later
on that the philosophy of life as a whole may very well allow itself
certain conclusions, even although many single points have to be
left open.

Having thus reviewed the most essential problems that already
arise from a consideration of the simplest forms of life, we will
now turn to those which originate in the phenomena of the higher,
multicellular organisms.   Here also, we will first glance over the
facts themselves.

III

THE DETERMINATION PROBLEM

The development of all higher organisms starts, as everyone knows, from a single, or two united cells, the germ cells, which in the latter case are almost always different from one another; one, the female, having accumulated a reserve of protoplasm, and hence being in a resting condition, while the other, the male, usually consists only of the nucleus, but is mobile, and hence has the task of seeking the female cell. This fundamental fact contains the whole problem of sexual difference. It is one of the most incomprehensible mysteries of nature that the polarity of the sexes, which apparently rests upon so simple a foundation, later attains such general importance.[227] We shall later turn to this point in another connexion. For the present we will follow the process of development of the fertilised egg cell, confining ourselves for the sake of simplicity to animals. In plants there are many points of difference, but the process is similar on the whole. The most suitable objects for the study of these processes are furnished by the eggs of certain lower animals, such as namatodes and sea urchins[228] For the study of the further processes of development of the embryo, use is often made of lower vertebrates, particularly amphibians (newts, toads).

We distinguish, as already stated above, the nucleus and the cytoplasm (cell protoplasm) in the structure of the cell. The nucleus itself, however, is again plainly differentiated internally into various parts which can be exhibited by suitable methods of staining. One of the easily stained portions is called chromatin (from *chroma*, colour; but the substance itself is colourless). The chromatin fills the whole space of the nucleus like a network with fine meshes (see figure 51a). Alongside the nucleus filled with chromatin, we are able to distinguish in the animal cell a small body which, on account of its function in cell division about to be described, is called the centrosome or central body.

When a cell starts to divide, the first thing we observe is, that the centrosome divides into two parts (figures 51a, b, opp.

this page), which are first of all close together and then gradually move to two opposite positions close to the cell wall. While this happens, the boundary between nucleus and cytoplasm becomes less and less clear, and at the same time there is formed in the latter a set of lines radiating out from the two centrosomes, while the chromatin draws together into a few loop-like threads, the number of which is different, but always constant for a given species. The whole conformation at this moment has a surprising likeness to certain diagrams of magnetic lines of force.

The process of division now takes place. The chromatin threads, which are now called chromosomes, all arrange themselves in the equatorial plane (see figure 51e), and then divide longitudinally, the parts separating, drawn as it were by the centrosomes. The whole body of the cell then contracts between these two halves, the boundary between nucleus and cytoplasm is soon restored in each half, the chromosomes again disperse into nets (now two in number) of chromatin, and we now have, after the last connexion is broken, two cells, each of exactly the same character as the original cell. Each possesses exactly half the chromatin, in the form of one longitudinal half of each chromosome.

In the case of sexual reproduction, where the egg cell or ovum has first been fertilised, the process is essentially the same; but a very important observation has been made in this case, namely that each of the daughter cells receives one half of the chromatin mass of both egg and sperm nucleus, so that at every further division, *every new cell always shares with its twin cell the chromosome substance from both father and mother.* The series of diagrams in figure 52 (opp. this page), will thus be intelligible without further explanation. The union of two reproductive cells does not lead to the number of chromosomes being double that proper to the particular species, because the germ cells always have half as many chromosomes as the body cells. The germ cells are formed from cells originally contained in the sex glands, by the two so called maturation divisions, which, on account of a peculiar mode of division different from the normal, result in the ova or spermatozoa produced having exactly half the normal number of chromosomes (figure 60, page 364). This fact most probably accounts for the Mendelian rules of hybridism (see below).

The development of all multicellular organisms from a single

PLATE VI

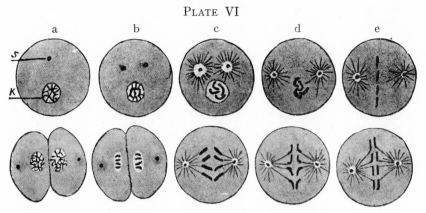

FIG. 51. Division of the Nucleus (Mitosis)

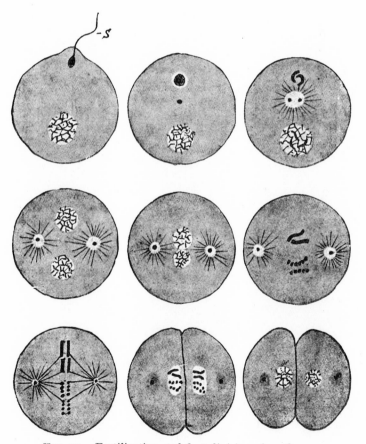

FIG. 52. Fertilisation and first division of an Ovum

cell, or from two combined cells, is effected by this process of cell division. From one cell first two, then four, then eight, then sixteen, and so on, are produced; finally we have a spherical mass of cells, which is called a *morula* (mulberry). In animals this becomes a hollow sphere by secretion of liquid, and is called a

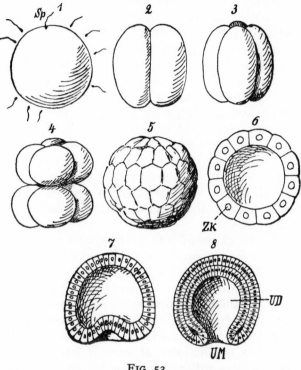

FIG. 53

Scheme of the first stages of development of a multicellular organism. (From Hesse, *Abstammungslehre und Darwinismus*, Teubner, Leipzig-Berlin)

blastula (bladder), this then collapsing to form the so-called *gastrula*. The latter process, called gastrulation, results in the formation of two different layers of cells, the *hypoblast* and the *epiblast* (the internal and external embryonic layers); while in all higher animals a middle layer, the *mesoblast*, is also present. The resulting cavity is called the *archenteron* (figure 53). The whole structure has a surprising likeness at this stage to the shape of the simplest coelenterates, which likewise consist only of two
    Ys

layers of cells, the *endoderm* and *ectoderm*, forming a cavity (with appendages). It is not necessary at this point to follow the further development in the case of higher animals. In the case of plants, the development goes on different lines from the morula stage onwards.

It is a general truth that the differences in embryonic development are observed the earlier, the further the organisms in question are separated in their systematic relation. The part played by the male germ cell in starting the development of the egg appears at the first sight so mysterious – particularly before the modern science of heredity had taught us the equality of the two germ cells – that there was a tendency to see in this process a most impressive exhibition of life forces, dominants, or entelechies, and to lay great emphasis upon it. This somewhat incautious vitalistic argument (vitalism is, as we have seen, always incautious when it lays hold of single selected processes) received a severe blow when it was shown that the inception of development is, in all probability, a relatively simple chemical process. For it has been shown that *artificial parthenogenesis* is possible in a great many cases.

The first to be successful was J. Loeb, who was able to cause unfertilised sea urchin eggs to develop by adding certain salts (particularly $MgCl_2$), to the water in which they were living. Various investigators (R. Hertwig and G. Hertwig) then showed that vigorous brushing was capable of causing a part of such eggs to develop; and finally a very skilful French experimenter, Bataillon, was able to cause even frogs' eggs to develop by piercing them with a fine needle. One of these at least even developed into a complete larva with tail and hind legs (most of them unfortunately died during the experiment on account of the injury caused through the experiments).

These experiments strongly suggest the hypothesis that, in the case of ordinary development, the sperm cell entering the ovum perhaps brings with it certain special substances (enzymes) which have the same effect as the external stimuli in the case of artificial parthenogenesis. This hypothesis received the best possible support from an observation of O. and G. Hertwig, that the development of the ovum can also be set going by sperm nuclei which have been injured, shortly before entering the egg, by radium or X-rays, to such an extent, that although they still enter the ovum, they die immediately thereafter and do not unite with the nucleus of the ovum. The creatures thus produced are found to have only the qualities of the mother, which also goes to prove that the nucleus is at least the most important carrier of inherited characteristics (see below page 371). But the injured nucleus still obviously contains those substances which bring about the division of the ovum.

We may add that the long sought opposite of parthenogenesis, androgenesis, the development into an organism of a single male cell, has now been realised by this means. For if, conversely, the uninjured sperm cell is allowed to enter an ovum injured by X-rays, there results a being with only the qualities of the father. In this case the nucleus of the ovum dies. But it supplies the necessary protoplasm, for the sperm cell cannot develop on its own account owing to the absence of this material.

From these experiments we may draw a very strange conclusion. If an ovum is able to produce a complete creature on its own account, and if the same is also true in principle for the sperm cell, it follows that, in the case of normal fertilisation, two lines of development which would otherwise be divided, are united to form a single one, or in other words, that two (presumptive) individuals become one. This view is now further confirmed in the most curious manner by some recent experiments in the field of *developmental mechanics*, a branch of experimental biology founded by W. Roux, which has set itself the task of investigating the causes of the processes which take place during the formation of a higher organism.[229] One of the first and most famous experiments we may mention, is that by Driesch.[229] He divided the germ developing from a sea-urchin egg at the twin cell stage, that is, after the commencement of the first division, into two single cells (blastomeres) by careful tying, and then allowed them to develop further without interference. The result was two complete larvae, but of half the normal size. The experiment was also successful with the four cell stage, and even at the eight cell stage the development of a complete larva could be observed in its beginnings.

The extension of these experiments to other varieties of animal resulted very quickly in the discovery of a bewildering number of new facts. In many cases the process was the same as in Driesch's experiment, but in other cases, for example medusae, the two divided cells gave only half-individuals. In other cases again, to which, as after appeared, the sea-urchin belonged, the beginning of the development took place as if the part-cells were going to remain part, but a process of completion, then occurred which was obviously analogous to the process of regeneration well known in other cases; the final result was a complete individual. But it must be expressly stated that this result by no means always occurred when complete individuals arise by 'merogony' (development from part-cells), but that in certain cases, for example medusa and lancet fish, the development of a part-cell is exactly like that of a normal cell from the very beginning.[230] The opposite of these experiments in division is seen in an experiment by Zur Strassen, who succeeded in fusing two eggs together so completely that they remained together and produced an individual of twice the normal size.

What do these experiments teach us? First of all, obviously, they show quite generally that the developing embryo has, in most cases, the power of taking several different lines of development. The decision as to whether one of the two part cells in Driesch's experiment is to give a whole or a half embryo is brought about by a complex of causes, which, as regards this one cell, are external to it. If it remains alone, it becomes a whole embryo, if it remains together with the other cell, it becomes half an embryo.

All the possibilities of development lying in a cell or cell complex are called their 'developmental potencies,' but this is not to be understood as the introduction of an occult quality, but simply as an expression of the undoubted fact just stated, that a cell is in general capable of becoming more than it usually does become. Secondly, all these experiments illustrate the power of the organism to re-establish its form under all circumstances, its 'integrative power'; a power possessed to some extent by every part of a living organism. It is as though every single cell carried within itself a plan of the whole to which it normally belongs. This at least appears to be so in cases where the parts develop straight into a complete whole. Later on, the phenomena of regeneration will afford us convincing proofs of the presence of this tendency to wholeness, but we will first glance shortly over the history of the matter, as it is essential to a proper understanding of the present position.

In past centuries, as a rule, no objection was raised to the assumption that living beings had arisen from lifeless material; e.g. gnats and frogs in swamps, and maggots in cheese. Even so deep a thinker as Aristotle unhesitatingly took this view, which appears grotesque to us to-day. It was the experiments of Pasteur, described above, which finally disposed of this primitive form of the doctrine of spontaneous generation (*generatio aequivoca*). Hence it was a considerable step in advance when, in the eighteenth century, Swammerdam, Spallanzi, Leeuwenhoek and others, defended, at least for the higher animals, the doctrine of *omne vivum ex vivo*, although they believed that their view had to be combined with what was called the 'preformation' or 'evolution' theory. They assumed namely, and believed that they had proved by direct observation, that the future organism was already present in miniature form in the germ cells (ova or spermatozoa), and hence only needed to be 'unfolded.' The defenders of this doctrine developed among themselves, after the discovery of the male spermatozoa by Leeuwenhoek,[231] a bitter controversy as to whether the future being existed in the egg or in the 'seminal animal' (ovists and animalculists). Pictures of that period show a complete little human being, 'rolled up' in the head of a spermatozoon. Serious investigators even asserted that they had observed this.

Some even went further to the absurdity of the so-called *emboîtement* theory, (Malebranche, Bonnet) which asserted that in every miniature human being, the countless spermatozoa of the future man must be contained, in these again those of the next generation, and so on, 'right back to Adam' (Eve, in the case of the ovists). The whole life work of a distinguished investigator was necessary to destroy this nonsensical idea. Kaspar Friedrich Wolf carried the doctrine of *epigenesis* (new formation) to victory; the germ (the fertilised egg) is not yet organised. Organisation first results through the 'vessels and bubbles' which are formed, one after another, from the egg, and this organisation simply consists in the simultaneous presence of these parts, which are arranged according to a certain plan. The parts are as little organised in themselves as are lumps of inorganic material.

There is no need to show that this doctrine, too, is entirely erroneous in this form, for although it is certain that the fertilised ovum does not contain the future organism in miniature, it is equally certain that it is by no means unorganised, but rather that, as we have already seen above on page 316, it must be just as finely differentiated as the future organism itself. The egg of a frog is just as different from that of a hen as a frog is different from a fowl (Nägeli), since, as already remarked, not only the specific character, but also the individual character, is inherited. It is quite certain that all the differences that living beings, even of the same species, show among themselves, let alone the differences between different species, must also be defined in some way or another in the germ cells.

But this does not by any means make it necessary to assume, in the simple manner of the old evolutionists, that these differences already exist in a finished form in the germ cell. There need only be as many differences, but these have to do with quite other matters in the egg than in the adult individual. The matter is obviously analogous to the curve traced by the needle on the gramophone record. This curve must, if the record is a good one, contain every shade of the spoken sounds, but of course not in the forms of sound, but in a geometrical form.

There is no need to lay further stress on the fact that the formal notions of the old evolutionists and epigeneticists have been quite given up in modern biology, but we may still speak of a certain divergence of view, inasmuch as some scientists are inclined to underline those arguments which are in favour of absolute preformation in the germ cell, in particular the facts of heredity, while others, basing their view on the above described experiments in developmental mechanics, emphasise the variability of organic material, which is capable of a large number of possibilities, when the corresponding conditions are given. This dispute is often linked to-day with that between mechanism and vitalism. Mechanism will always be allied with preformation and vitalism with epigenesis, for the former wishes to stress strict causal connexion, and the latter the relative freedom of the organic world from physical determinism. We will now see what new experimental facts promise to help us in penetrating further into this secret.

We may first consider the phenomena of regeneration, which are among the oldest experimental results of biological investigation. This is easy to explain, as regeneration effects are very striking (for example the well-known case of the lizard's tail), and have been used from the earliest times in cultivating plants (grafting, etc.). The first result of the numerous investigations carried out by recent biological experimenters in this field,[232] has been to confirm completely the result of general experience; namely that the power of regeneration decreases the higher we rise in the organic scale.

Hence young individuals regenerate more easily than older ones, and, generally speaking, lower animals more easily than higher, although in the case of plants a corresponding rule is not so easily recognisable. In this case we find, even with the higher plants, a quite surprising capacity for regeneration. Thus, for

FIG. 54

Regeneration in the case of the mono-
cellular organism *Stentor*

example begonia leaves may be cut into pieces of almost unlimited smallness, and yet all will regenerate a complete plant, the only condition being that the fragment carries cells from both sides of the leaf. In the animal kingdom, a fabulous power of regeneration is possessed by, among others, planarians, which Morgan showed to be capable of reproducing the complete animal from all of seventy-two parts into which a single individual was cut. And we should take especial note of the fact that a similarly remarkable power of regeneration is even possessed by many unicellular animals (see figure 54).

It is of particular importance to note that regeneration by no means occurs only by a certain tissue forming new cells of the same kind, but that there are many very well investigated cases in which the regenerated tissue is furnished by cells which belong to entirely different organs, and have even been separated very early in the course of development from the cells of the organs regenerated.

The most famous case of this kind is the regeneration of the eye lens of the newt. The eye grows during embryonic development from two quite different parts, one of which, the optic cup, originates in the brain and provides the future retina, choroid, sclerotic, and iris; while the other, which sinks from the skin into the optic cup, provides the cornea, the lens and the vitreous humour (figure 55). When the lens of the eye of a salamander is

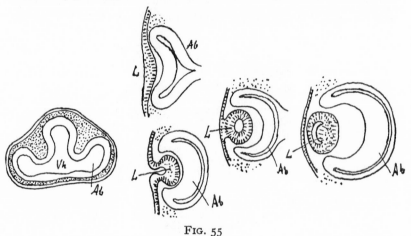

FIG. 55

Formation of the eye.

Vh = Fore-brain L = lens Ab = Optic track

removed, it is not regenerated from the cells of the epidermis, to which it belongs as regards its origin, but from the iris, that is from a tissue which was already completely differentiated at a very early stage of embryonic development.

This behaviour is found to an even greater degree in certain regeneration processes in earth worms, in which parts, originally formed from two different embryonic tissues, endoderm and ectoderm, are regenerated from cell layers which only belong to one of the two. In such results, the whole shows a mastery over the single parts which seems to point unmistakably to the existence of 'dominants' or 'entelechies' which regulate everything. And nevertheless – how hazardous it would appear to make such a supposition, which is nothing more than a refuge of ignorance. For if the lens is not entirely removed, but merely pushed into the crystalline fluid, the edge of the iris nevertheless forms a new lens, and if this is injured at several points, two or more new lenses are formed (figure 56); that is, an entirely useless result. Indeed, it has been shown quite recently that in certain circumstances, a lens may be 'regenerated' at quite another position in the eye, although the original lens is working quite correctly in its original position.[233]

In other cases too, useless regeneration has often been observed. For example, by a slight alteration in the treatment of wounds, animals and plants may be caused to produce structures as useless as a second head (figure 57), or the like. Even in the case of the

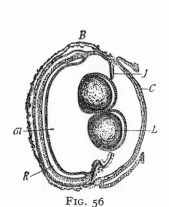

FIG. 56

Duplicate regeneration of a newt's eye-lens. *Gl* = crystalline body. *R* = Retina, *I* = Iris, *C* = Cornea. According to Fischel (from Weismann, *Vorträge über Deszendenztheorie,* Vol. II. G. Fischer, Jena)

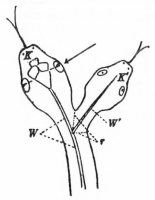

FIG. 57

Two-headed snake (regeneration), from Thesing, *Exper. Biologie* (Teubner, Leipzig and Berlin)

famous earth-worm, the decapitated head grows a second head in place of a tail, when the cut is suitably performed. These results again give the impression of a blind causality, which has nothing to do with the pursuit of a definite aim, but merely operates as it must.

As against these cases, which appear to make every kind of teleology ridiculous, there stand not only the above mentioned cases of regeneration from totally different tissue, but even a large number of cases in which the new formation can only come about by a complete 'fusion' of the surrounding tissue, and often of large parts of the whole body, before regeneration can commence.

Driesch, who cites a large number of experiments in his *Philosophy of the Organism*, has made them the especial foundation of his proof of the autonomy of life, that is, of vitalism. In actual fact, a process of this kind so obviously gives the impression of purposive behaviour, that one can scarcely escape from it. It is obviously analogous to the processes which take place in the pupa stage of higher insects, where, likewise, a great part of the cell material of the caterpillar is, so to speak, melted down, afterwards to be used in the formation of the imago or complete insect. In this case also, one almost seems to be able to watch entelechy at work.

But before we come to a decision let us consider some further facts. A second group of experiments deals with the *influence on the process of development of all kinds of external factors*, such as temperature, chemicals, etc. The experiments of Klebs[234] on the barren privet (*Sempervivum tectorum*) are particularly famous; this plant sends, from a thick rosette of leaves, a fairly long stem carrying an umbelliferous flower at the top. By changing external conditions, Klebs was able to produce at will remarkably varied forms, for example, plants having several rosettes of leaves, one above the other, others with several stems, and he was able to show that certain substances passing up and down with the sap, caused the production of the new formations at the points in question.

Numerous experiments have also shown that the local varieties of many plants are also due to similar sets of causes, and the same is true of what is called 'seasonal polymorphism,' to which we shall return in another connexion later. In Klebs' experiment as well as in other experiments of the kind, it was further found that the action of external factors upon the developing cell material was not always possible, but that there are certain 'sensitive periods' in which an influence of this kind takes place preferentially, or even solely. In the case of butterflies, the pupa stage is particularly sensitive, whereas the fully grown caterpillar can hardly be influenced. The total result of these experiments is a new and impressive confirmation of the law of the 'pluripotence' of by far the great majority of cells and groups of cells; they are capable of becoming a great deal more than they become under normal circumstances, and the selection among all the possibilities is made by external circumstances.

This leads us immediately to the third and most important set of experiments in developmental mechanics, namely, *transplantation* and *implantation* experiments, which have carried us farthest into the secret of the process of development.[235] The rule that external circumstances make the decision as between the various developmental possibilities, is not only true in respect of such external factors as temperature, irradiation and the like, but also, as we have already said, as regards the rôle played by each part of a developing organism towards the other parts. For the one part-cell of Driesch's experiment, the other forms the decisive, 'external' circumstance. Its presence causes the first to become half a sea-urchin, whereas it would become a whole one if the other cell were not present. The problem was presented of showing how, by what means, and according to what laws, this 'determination' of individual developing parts by other parts takes place, and in this direction, the latest research has brought quite surprising results, which go so deep that we may already say to-day, that we are on the way to a complete solution of the problem, though it is far from being actually solved.

We may mention a few of the most interesting and important experiments, due chiefly to Spemann and his school, and to Ubisch, Braus and others. For example, if the eye sockets of a grass frog (*Rana fusca*) are destroyed before the formation of the lens which sinks into it, the formation of the lens does not take place, and in this case we therefore have a dependent differentiation, which is regulated in some way or another by the cells of the eye socket. If, on the other hand, the same experiment is performed with the closely related edible frog, (*Rana esculenta*) we have eye formation in spite of the missing eye socket, this being self-differentiation. In other words, the cells in question are already so far fixed as regards their developmental power, that a change in their normal surroundings no longer makes any difference.

A quite analogous result was obtained from numerous transplantation experiments which were carried out mostly on newts. If we take from the body of a young animal that group of cells, which later on becomes the tail, and replant them at the point where the leg usually grows, they actually become a leg and not a tail. But if the same experiment is carried out later, when the group of cells in question has already been attached to the tail end of the body for a longer time, they form a tail and not a leg when transplanted in the same way. The longer period in the normal position has, therefore obviously already determined them to the point of being able to carry on their development only in the normal direction.[236] If further, a leg stump was transferred to the other side 'round by the back' that is, in such a way that front and back remained front and back, but above and below were interchanged, there appeared in the new position a correctly placed leg (not that of the other side), and the surroundings were therefore able to reverse the change from above to below. If however the transplantation was made 'round by the mouth' at the same age, that is to say that above and below were unchanged, but back and front interchanged, the new leg grew the wrong way round (that is to say forwards), and hence this change could not be reversed by the new surroundings.

These experiments are already extremely instructive, since they strikingly illustrate the complete obedience of these phenomena to fixed laws, but they are excelled by the very latest series of experiments, which also come from this Spemann school. It was first shown that the fixation of the single parts of an embryo occurs the earlier, the nearer these parts are to the blastopore, that is, to the point at which the gastrula invaginates. At a stage at which those parts nearer to it were no longer capable of being influenced when transplanted by their new surroundings, those more distant were still capable of change, and as time goes on, the limit of this pluripotence still existing moves away from the blastopore. Spemann drew the conclusion that the centre of organisation must lie in the latter.

This bold hypothesis receives a further and most important support by an experiment performed by H. Mangold. She transplanted a piece of gastrula into another quite indifferent tissue, namely into that which later becomes the wall of the stomach, and obtained at this point what appears at first sight to be an astounding formation, namely an entirely new embryo, which however was not made only out of the transplanted cells, but to by far the greatest extent *out of the cells of the animal host*, which cells were now completely redetermined.

The reader may ask how one is to know the difference between the two kinds of cells. The answer is that transplantation experiments of this kind are carried out by taking the cell group in question from another species, naturally one closely related, but one which can be clearly distinguished, by colour for example. We are then able to tell at once to which of the two animals the cells in question belong. Formations of this kind are called by the general name of 'chimaeras,' and many other very curious experiments have been carried out on the formation of chimaeras which we will pass over here, as they would not lead us very much further. [237]

Let us remain by the transplantations. The Mangold experiment which we have described above proves that the cells of the blastopore actually possess the power of compelling other groups of cells surrounding them to produce the form in question (that of the guest animal). Particular note should be made of the fact that a corresponding experiment made later by another investigator, Bautzmann, even succeeded in the transplantation of a piece of the blastopore of a toad into newt tissues, in the case of two animals therefore, belonging to the two entirely separate groups of newts, urodeles and anurans. Further experiments of a similar kind have shown that in certain cases, even cells of one of the germ layers (the ectoderm) are able to serve for structures which are normally furnished by the endoderm.

We are thus directly confronted with the fundamental question of biology; we are almost able, so to speak, to lay our hands upon it. It is an assured fact that an organising effect on neighbouring groups of cells is exercised by the blastopore, and later again by the individual parts, such as the eye sockets, and it is also certain that this effect spreads like a current, as it were, outwards from the centre in question. It is clear that these facts strongly suggest the hypothesis that we are dealing in the case of these organisers with quite definite substances which spread out from the centre in question.

However, a whole series of objections immediately arise to this mechanistic view. How are we to imagine that the cells of the eye socket produce substances which cause other cells to form lenses, since these cells themselves do not produce lenses, but sockets? Why does not this lens-determining agency act first of all upon itself? And how, further, are we to explain the fact that such an organiser is able to produce a highly complicated structure, such for example as a nervous system ? The mechanist of to-day has no answer to these questions, and hence the fundamental question still remains undecided. It is certainly true that these results, taken together, represent a strong argument in favour of mechanism. For if the 'entelechies' behave in so many respects *as if* they were certain substances spreading out from a point, we are naturally inclined to expect that they really *are* such substances, or something very much like them.

We add to these experiments a few more which have also made an essential contribution to the determination problem. Carrel and others have succeeded in keeping alive *tissue cultures outside the living organism* for an unlimited time;[238] that is to say, in growing such tissues as nerve or kidney or connective tissue in nutrient broth, so that they divide normally, and form a whole colony of cells. It was then found, as a rule, that these emancipated cell cultures acquired distinctly different properties from those which they exhibited in the living organism, where they are associated with other groups of cells. They very frequently develop into so-called amoeboid cells, that is, they become independent and swim about freely like an amoeba in the culture medium, in an analogous manner to the white blood corpuscles, which swim about freely in the blood, and, according to Mechnikov, perform the function of capturing intruding bacteria and rendering them harmless.

It is probable that malignant tumours (cancer and sarcoma), depend upon a similar degeneration of otherwise harmless body cells, which suddenly refuse obedience to the control of the body as a whole. The exact reasons for this are not yet completely known, but it has been found, amongst other things, that in cancer tissue, normal breathing (oxidation) is replaced by fermentation-like processes, with exclusion of oxygen. According to this, the enzyme composition of the cancer cells must certainly be distinctly different from that of normal cells. These investigations have, however, not yet led to a final result, which is to be regretted in the interest of suffering humanity.[239]

The investigations last mentioned have already taken us away from the field of the actual problem of the determination of form, of the causes effective in embryonic development. They nevertheless throw some light upon this problem, for they have again demonstrated impressively the mutual interaction of all cells of the body and the conditioning of structure one by another. This correlation of all parts of the organism with one another continues of course in the adult individual, and we must now consider shortly the experimental results, which led to further light on this point.

In the case of the lower multicellular organisms, and of plants in general, there appears to exist a connexion between all the individual cells by means of direct connective threads between the plasm of neighbouring cells, as has already been stated (page 321). It is certain that, at least in numerous cases, contact and other stimuli are transmitted along these connecting paths. The well known *Mimosa pudica* is an example of the fact that something can occur even in the higher plants, which bears a strong external resemblance to the conduction of stimulus in the nerves of the higher animals, although it is produced on an entirely different anatomical basis. The nerves themselves are, fundamentally, nothing more than such connecting threads from cell to cell, only in this case we have cells which have specialised in this conduction of stimulus and have given up other functions. The nervous system of the higher animals renders possible extremely rapid communication between the single parts, the speed of communication in the human nerves being about 100 feet per second. The actual nature of the process is still not yet quite understood,[240] but so much is certain, that here also the micellar structure of the fibres, and the electro-chemical processes taking place between the individual micelles, play a decisive part.

Alongside this direct protoplasmic connexion, there certainly exists in the case of all higher animals, and very probably in plants, another kind of communication between the different parts, namely, *transmission by chemical influences.*

At one point substances are formed which, when carried by the sap or blood, as the case may be, to another point, exert specific effects there. These substances are called *hormones* (from Greek *hormao*, I stimulate) or internal secretions. Their discovery in plants is of very recent date and is still doubted on some sides.[241] In the case of animals, on the other hand, numerous relations of this kind have been proved with complete certainty.[241] We need only refer to the results of the removal or hypertrophy of the thyroid or thymus glands, to the importance of the pineal gland (in the brain), to the regulation of blood pressure by adrenalin (the effective substance in the secretion of the suprarenal cells); one or other of these examples will be known to everyone.

The most striking proofs are given by the physiology of the organs of reproduction, and we may give a few of these in order to illustrate the far reaching importance of these investigations. If a substance obtainable from dog-embryo shortly before birth is injected into a bitch, which is not and never has been pregnant, its lacteal glands swell and finally yield milk. Steinach removed the sex glands (ovaries or testicles) from rats and guinea-pigs, and planted in the animals thus neutralised, the sex glands of the opposite sex. The result was a complete transformation of the animals so treated, a feminisation of the former male, and a masculinisation of the female. There was not only a change in all external, so called secondary sexual characters (colour, form, size, growth of hair, etc.); the animal underwent a complete psychical change, the masculinised animals being savage, aggressive, jealous of one another and attracted by the normal female. In the case of animals operated upon when young enough, there was even a partial transformation of the external organs.

In later experiments by Steinach, Lipschütz, Sand, and others, which were also in part extended to human beings, numerous equally remarkable results of this kind were obtained as regards the formation of hybrids, homo-sexuality, etc., and some of these have already been made use of in medicine. There can hardly be any doubt that these effects throughout the body are caused by substances produced by the sex glands, for it was possible to show in many cases, that not only the transplanted organs themselves, but an extract from them alone, was sufficient to produce the effects.

It has frequently been observed that internal secretions of this kind act in pairs as antagonists, that is for example, the secretion of one gland will raise the blood pressure, while that of another will reduce it. In the same way, according to Steinach, male and female sex glands are so antagonistic that when planted together in the same organism they hinder one another by their secretion. While Steinach and his first disciples believed they had found in the sex glands the fundamental centre of these actions, very remarkable recent investigations (Zondeck and others) have shown that the sex glands themselves are again regulated by hormones which originate in the *hypophysis*, a small gland lying at the lower side of the brain. This discovery is already being used in medicine for diagnosis (pregnancy), and therapy.

It is not unnatural to suppose that the hypophysis is not the last centre, but that it again is regulated by other secretions; and this leads us almost necessarily to the idea that a definite centre does not really exist, but rather that the whole is an extremely complicated chemical arrangement in very accurate adjustment in which every part depends upon every other part.[242] The few results hitherto obtained only give us a feeble conception of the enormous complexity which exists in the interplay of internal secretions in higher animals.

New and unlimited possibilities are opened up for medicine; every new advance in genuine insight promises to supply us with valuable new methods of treating disease, or to place things long known and practised empirically upon a firm foundation of genuine scientific knowledge. The ancient doctrines of so-called humoral pathology are certainly not pure fancy, but have as a real basis the fact that the whole internal chemical cycle may be very different in different individuals, and that we may be able to distinguish certain types which, in some way or another, hang together with the bodily and mental types (see also below). But there is still nothing known with any certainty about this matter, and it will probably be a long time before any clear results are reached in this field of research into personal constitution.[243]

If we now combine the recently acquired knowledge concerning the decisive rôle played by the hormones in the body with what we learned above concerning the problem of determination, we shall obviously incline to the hypothesis, that, the development of the embryo and the determination of one cell by others, must also be due to substances like hormones, which transmit these effects. The experiments which we have already discussed above on artificial parthenogenesis, strengthen this supposition, for they also point to its being due to certain 'developmental hormones,' which are brought by the sperm nucleus into the ovum, but are capable of being replaced by other stimulants, such as those produced in the Bataillon experiment by the prick of the needle. Numerous other experiments render it very probable that such 'wound hormones' exist, even in the case of plants

Zs

(Haberlandt), and they also appear to play an important part in such cases as the regeneration of the lens (see above). It would however be obviously mistaken to declare adherence to a *panhormonism*, that is to say, to ascribe anything and everything to the effect of such internal secretions. Other means of correlation between the different cells and groups of cells certainly exist, means perhaps quite different from the three hitherto discussed.

The last sentence is not a vague supposition, but assumes a quite definite content from quite new and very astonishing data afforded by the experiments of a Russian, Gurvich.[244] Gurvich found that cells when dividing exert a quite specific action at a distance on neighbouring cells, division in these (irradiated) cells being excited or accelerated. For example, he counted the number of cell divisions which took place on a root while growing, one side of it being exposed to this mitogenetic (mitosis=cell division) influence, the other not, and he thus found a decided increase in the divisions on the irradiated side.

These statements were at first received quite generally with great scepticism, but they have later been fully confirmed, and other investigators found that the rays in question could be detected by other objects such, for example as yeast cultures, and even inorganic structures, namely systems which form what are called 'Liesegang rings,' the latter means of course promising more certain results. If the cells undergoing division are brought on to an object-glass above a metal plate provided with a slit, under which there is an arrangement for producing Liesegang rings, the latter are interrupted in a characteristic manner at the slit (Stempell).

Hence there can no longer be any doubt of the existence of such an influence. We may even assert with great probability that it is due to ultra-violet radiation of a definite wave length, $\lambda < 260 \ \mu\mu$; some Russian experimenters found the most powerful action between $\lambda = 206$ and 220 $\mu\mu$.[244] The direct action of living objects upon the photographic plate[245] which has often been asserted but always disputed, is probably to be explained in this way. Indeed, it is imaginable that radiation of this kind, hitherto unnoticed, accompanies all, or at any rate a good many, chemical processes, and not only biochemical processes. The phenomena of 'chemiluminescence' was hitherto known only in the case of certain chemical processes, to which also belong the processes taking place in luminiferous organisms (luminous bacteria, 'phosphorescence' in the sea, glow worms, etc.) These discoveries open up a new field for investigation.

We have reached the conclusion of our short survey of recent investigations into the problem of determination and matters connected with it. The amount of work that has been done is enormous. If the reader will take the trouble to look through any volume of, for example, the *Journal of Experimental Zoology* or *Biological Abstracts*, he, like the author of this book, will be filled with admiration of the industry and skill, the ingenuity and fruitful imagination of the investigators who have shown a way in this field. And nevertheless we are obliged finally to admit that what we really wish to know is not told us even by these series of experiments. In spite of the genius with which they have been thought and carried out, they do not yet afford any definite means of deciding the fundamental question of biology, that of mechanism versus vitalism.

It is doubtless true that physico-chemical explanations have been found, or seem likely to be found, for a very great deal that appeared to be inexplicable even 30 years ago. Mechanism is everywhere making victorious progress in the field of life phenomena. But in this field the same is happening as in the case of the problem of the single cell. The more progress we make in solving mechanistically the riddle of the organism's functions as they now exist, the greater becomes the riddle of the original production of so extremely complicated a mechanism. Furthermore, we find crass instances in which any mechanistic explanation seems hopeless. When an organism, injured in some way, proceeds, instead of simply repairing the damage in a way which might be explained mechanistically by means of hormones and the like, to first 'melt down' the whole adjacent parts, and then begin development again from the beginning, so to speak – for this is actually the case in numerous instances – every mechanical explanation seems finally ridiculous. We shall return to this fundamental question later.

We may, however, raise one or two other philosophical questions which naturally occur at this point. The first concerns the idea of the 'individual.' Consider the earth-worm which has been divided for a regeneration experiment. At the moment after the division (which, by the way, only results in a complete regeneration of both parts in quite closely defined circumstances), we have unquestionably two half earth-worms, for one part has no forepart, the other no hind part. When both parts have regenerated what is missing, we have again two 'whole' earth-worms. At what moment did half a worm become a whole one ?

To state this question is to see that it has no sense in it. But what is the reason for our arriving at so senseless a question ? The answer is this: that we have tied ourselves to a concept – the concept of the individual, the indivisible whole – which is not applicable in such cases. It is also clear that in the case of cell conjugation, or the fusion of ova (see above), which form the opposite case to such divisions, the concept of the individual requires a similar extension.

But when we have once grasped this point, the further conclusion is not far off, that we must not in any case regard the organic individual as an absolute unit divided from all others, but as a wave, more or less well defined, on the great stream of organic happening, for which also, it is not possible to give an exactly determined beginning and end-point. This latter consequence in particular will appear strange to many who are accustomed, in accordance with our usual language and with our ruling ideas, to regard life and death as absolute contraries. But no analysis of concepts will help us to avoid recognising that such absolute limits and opposites do not really exist, but that in this point also the rule: *natura non facit saltum*, holds good.

The fact that, in the first place, an initial point cannot be strictly defined, but that it is a matter of agreement whether we reckon from the moment of birth' (which, as we know, is sometimes a day in length), or the 'moment' of cell conjugation (fertilisation), or another point of time, becomes quite clear when we take into consideration the numerous cases of parthenogenesis, in which everything takes place simply in a continuous process. Take, for example, Bataillon's artificial parthenogenetic frogs. The vitalist will be the last one to maintain that the experimenter breathes life into the èggs when he pricks them with a needle; he will, of course, be convinced that the dormant life was simply awakened. Very well, but then, in the case of normal fertilisation of the frog's egg, the fertilisation cannot have been responsible for the beginning of a new life. What then are we to say about it ? The egg comes from the mother's ovary, it was originally there as a living cell, which itself resulted from other living cells by division. The same is also true of the sperm cell. In short, whichever way we look at the matter, it is easily seen that every attempt to define unambiguously the beginning of individual life is met with hundreds of contrary facts.

These considerations lead us to very noteworthy results when we once more turn our attention to the divided sea-urchin's egg. The experimenter here obviously made one individual into two by his interference, and it is worthy of note that nature herself performs this trick (if we may call it so) in many cases on her own account. On the one hand, namely, there are certain species of animals, e.g. armadillos among the vertebrates, in which a single fertilised egg regularly provides several embryos by subsequent division, and on the other hand it happens even in human beings (and, indeed in the case of almost all mammals) that in special and exceptional cases, two embryos are formed from a single fertilised ovum. These cases are known as 'identical twins,' for, as everyone knows, such twins are almost completely alike in every possible respect ; cases of this kind are of great importance for the science of heredity. The similarity goes so far that such twins have been known to die almost on the same day of the same disease, in spite of having lived entirely different lives. In any case, they are as a rule almost completely identical in all bodily and also mental characteristics.

Alongside of these we also have twins which are no more alike than any other children of the same parents; they may also be of different sex. These obviously arise not from one, but from two separate fertilised ova, which have separated from the ovary at the same moment, in an abnormal manner. They have no further interest for us at this point. The identical twins, on the other hand, are a highly instructive example of the insolubility of such academic questions as that of the true moment of commencement of a human life, as will be evident without further discussion.

The second question, as to when life ceases, is equally insoluble – assuming that we do not confine our attention to partial phenomena, but take a general view. There is no doubt that an individual life, both in the case of a unicellular and of a multicellular organism, can be destroyed in so short a time, for example by crushing or burning, that we are able to speak of a 'moment' of death. But it is equally unquestionable that in very many cases it is simply impossible to define this moment. In the case of unicellular organisms when they divide to form two new individuals, we have an equal right to say that the mother individual is dead or that it lives on undiminished in its children, and in the case of multi-cellular organisms there are enough forms of death in which the process is completely continuous, and no one can find a definite moment of death unless he is absolutely determined to do so.

This is even true of human beings, although the higher the degree of organisation, the more inevitably, doubtless, does death assume the character of a sudden and general catastrophe. When, for example, a human being is drowned, and the heart-beat and breathing have ceased, it is possible in certain cases, as is well known, to ' call them back to life ' by artificial respiration and other means. Are we then to say, in order to insist upon a limit which no observation is capable of determining, that the person in question cannot therefore have been dead? But supposing he had been left alone, and the arts of restoration had not been applied, what new event would have *suddenly* taken place? To say that this sudden event must have happened unbeknown, is too easy a way of getting out of the difficulty, it is simply a hypothesis *ad hoc.*

We know perfectly well what happens when nothing is done to the person drowned. That which has already begun simply proceeds further, more and more changes take place, and when these become too great in extent, no later efforts are sufficient to make the machine work.

Matters are very similar at the termination of many diseases, or in such cases as when an injection of camphor is used in a dangerous crisis to stimulate the heart which is ceasing to beat; the complete breakdown of the organism, otherwise inevitable, being thus prevented. This view received the best possible support from the experiments, such as those of Winterstein, who succeeded in reawakening to life animals killed by freezing three hours after all signs of life had ceased, by irrigating the heart with salt solution containing some adrenalin.[246]

We may therefore say quite generally, that, the stage in which successful 'restoration to life' is possible, is a purely relative question. It is merely a question of the state of medical science at the moment, and our posterity will be able to do many things which we regard as impossible to-day, when we say with resignation: nothing can be done, he is already dead.

Recent biological investigations have thrown a great deal of new light upon the problem of death in another respect also.[247] The much contested doctrine of the (potential) immortality of the cell has been rendered very probable by the experiments of Woodruff and others on *Paramecium*. These experiments succeeded in keeping the protozoa in question alive for thousands of generations, by continual renewal of the culture medium, with complete avoidance of cell conjugation, so that the assertion frequently made, that even a unicellular organism (or more correctly, a strain of unicellular organisms) will finally die of senile decay without the rejuvenating effect of conjugation, can no longer be upheld. Indeed, if we accept the interpretation, which is not quite certain, of recent experiments by Hartmann, Goetsch, and others, this statement may even be extended to include certain multicellular organisms (planarians, fresh water polyps), for these authors succeeded in keeping such animals permanently alive and active by cutting away parts of them from time to time. The regeneration which resulted had the same rejuvenating effect as normal reproduction.

That multicellular organisms in general cannot escape death, is naturally not altered by these experiments. Although we have not hitherto been able clearly to understand why progress in the differentiation of cells into specialised functions must be paid for by final wearing out and degeneration of the system as a whole, so that only the germ cells remain to carry on life, we may still regard it as probable that a causal connexion exists between differentiation on the one hand and the probability of inward changes taking place which are not reversible and result by accumulation in death – quite apart from the fact that sensitivity to external conditions naturally increases with differentiation. The fact can also be understood from the point of view of the doctrine of selection; mortal multicellular organisms would obviously have the advantage (regarded phylogenetically) over immortal, since each new generation would give them the chance to improve progressively their structure. The succession of generations has, as it were, a rejuvenating effect, and allows of continual new adaptations.

But apart from this, the biological necessity of death is perfectly obvious, and no new and mystical laws are needed to explain it. The totality of all inward and outward life activities presupposes an enormously complicated play of interlinked conditions, such that a perfect periodicity of phenomena is completely impossible. Irreversible processes are bound to reach a point, after a time, at which further functioning becomes impossible. This view has been notably supported by new experiments, which have shown that by suppressing flowering, for example, and other means, the life of plants can be considerably increased. Annual plants, which normally die at the latest in autumn, could thus be kept alive for three and a half years. Hence here also we are not dealing with a mysterious law of the duration of life, but simply with the combined effect of numerous causes, which can each be varied singly. It is therefore possible to change the whole sequence within certain limits, which themselves are determined by other partial conditions.

The numerous determinations of the normal length of life of certain animals and plants, and the more definite relations which exist between the length of life and the period of development, are therefore not to be regarded as an expression of direct and simple causal relationships, but simply as, so to speak, statistical data concerning the average period in which death occurs under normal conditions. It is perfectly imaginable, and is sufficiently confirmed by the well known experiments of Steinach,[248] often unjustly ridiculed, that sooner or later we shall be able to increase the average length of life, in the case of both animals and man, by suitable interference with the normal course of events.

In practice such an attempt soon reaches its limit in face of the complete impossibility of ever grasping, even approximately, the enormous complication of the conditions. But here as in other cases, it is quite wrong to make the theoretically unimaginable out of the practically impossible. The fact that every living thing dies sooner or later is not the consequence of a special and mysterious law founded in metaphysics, but simply a fact of experience, the explanation of which is to be sought in the combined action of numerous other single facts of experience which are either known or remain to be discovered.

When we consider all this, we might be inclined to regard the 'organic individual' merely as a subsidiary concept serving the ends of thought-economy, and to say that, in reality, what we call by this name is only a wave on the stream of events, without a definite beginning or end, forming a unit for our abstract thinking, in truth an incredibly complicated aggregate of thousands and thousands of things and processes. We shall see later to what extent such radical relativism overshoots the mark, but for the moment we will turn to a problem which we have hitherto only touched upon, that of heredity.

IV

THE PROBLEM OF HEREDITY

This problem, on account of its great importance, also in connexion with later considerations, deserves special discussion. The study of inheritance has developed in recent times into a special branch of science,[249] and we will first glance over the most important groups of facts, and then discuss shortly the theories of inheritance.

According to Weismann's doctrine, now fairly generally accepted, those cells which are later to become the parent cells of the germ cell, separate at the beginning of embryonic development. They form what Weismann calls the 'germ path,' leading directly from generation to generation, while the body cells or *soma* are only, as it were, budded off from them every time, just as new plants arise from time to time from a root lying in the earth. The usual view, illustrated by the diagram of figure 58, according to which the germ cells are products of the body, the latter being the product of the germ cells of the previous generation, is therefore false as regards the first half of it. *The germ cells of the living generation are not the product of the body in which they are embedded, but the direct product of the germ cells of the previous generation,* the latter germ cells having also produced the *soma* cells of the living generation. The diagram of figure 59 represents this more correct view.

In spite of the attacks and denials to which Weismann's doctrine of the continuity of the germ plasm has been subjected it has remained victorious, especially since the isolated path of development of the germ cells has been directly followed by what is called 'cell lineage' research. It is of fundamental importance for our whole view of heredity and also of numerous sociological and even political questions, that we should be clear about this new and revolutionary discovery which we owe to Weismann. But we may

362

remark, that in the case of the higher plants, the separation between germ path and *soma* is less sharply defined, since plants have
cell groups of embryonic character, which therefore must be
regarded essentially as germ cells, not only in their actual organs
of reproduction, but all over them, in what are called the vegeta-
tion points. But these embryonic cells in plants are also quite
distinctly separate from the other (and already differentiated)
soma cells.

We must now describe the formation of the germ cells, and shall
deal with animals, since in these relationship is clearest.[250] The
organs of generation of the higher animals do not contain the germ

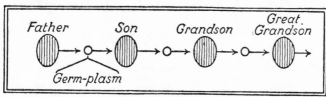

FIG. 58

Older (erroneous) idea of the linkage between generations

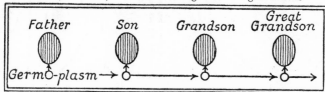

FIG. 59

Recent and more correct representation of Weismann

cells ready made, these are formed within the organs as required
(hence, generally speaking, only at certain periods) from the
'primordial germ cells,' which form a part of the tissue of the ovaries or testicles.[251] The formation of the actual germ cells from
the parent cells is a special process, which takes place always more
or less in the same manner, but with many detailed variations.[252]

The essential point in the process is that, in two successive cell
divisions of the parent cell (which are called the first and second
maturation divisions), the *number of chromosomes in the original
cell is reduced to half* (see page 336). The chromosomes, the special
function of which in cell division has been shortly discussed above,
first collect in peculiar groups of four, and then separate twice in
succession from one another fairly rapidly, in which process,

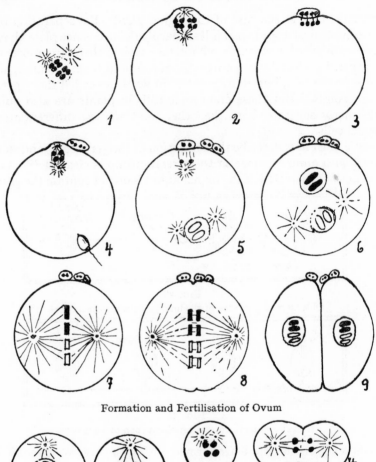

Formation and Fertilisation of Ovum

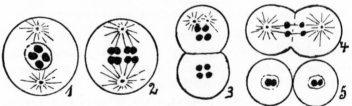

Formation of sperm cells

FIG. 60

however, they only once divide longitudinally in the usual manner while on the other occasion, one half of the chromosomes simply enter one of the daughter cells, and the other half the other. This is called the 'reduction division,' and is usually the second, but on occasion the order is reversed.

The lower part of figure 60 gives a diagram of these events in connection with a sperm cell. The formation of the ova (upper part, No. 1-3) takes place fundamentally in the same way, but in both divisions the whole of the cytoplasm always remains in only one of the daughter cells, so that of the four cells produced three are without cytoplasm. These three degenerate, and are expelled. It is particularly characteristic of the causal character even of biological processes, that the second division takes place in the case of those cells produced by the first division containing no cytoplasm, and hence condemned to death, and this even happens after the main cell has been fertilised (figure 60, 5-9). This division has not the slightest use, but takes place, as it were, from habit; but that is by the way.

The later union of a sperm cell with an ovum (fertilisation: figure 60, 4-9) results in a cell having again the full number of chromosomes. The human body cells have, for example, 48 chromosomes, the germ cells only half this number (see however below) and the same is true in all other cases that have been investigated. The only exception occurs in the case of natural parthenogenesis, where the reduction division does not occur, so that the daughter cell receives the full number of chromosomes. What happens in the case of artificial parthenogenesis in still not properly understood.

From the point of view of evolution, the behaviour of the higher organisms can be traced back to the simpler relationships shown by the lowest living organisms such, as for instance, green algae and mycetozoa. Here a male and a female cell adhere, with the production of a zygote (united cell) possessing double the number of chromosomes. In this cell the chromosomes then arrange themselves together two by two, and in two quickly succeeding steps, there is division into four daughter cells, which then again exhibit the haploid stage (with the normal number of chromosomes). From this low stage we can imagine the higher stages being developed, the 'diploid' stage being no longer confined to a single cell, but extended to several cells, which then form the multicellular organisms.

To-day it is quite certain that these chromosome phenomena are all closely related to the processes of heredity. In order to understand this, we must first shortly review a second complex of facts, those of recent investigations of hybridism, which form the core of the whole science of heredity. This was founded by the Augustine abbot Gregor Mendel of Brünn, in 1865. His writings, which now have long been reckoned among the classics of science, received little attention in his own day. The time was not ripe for them, and the work appeared in a small and little known periodical, the communications of a local scientific society. Only when Mendel's results were rediscovered about 1905 by the biologists, De Vries, Tschermak, and Correns, did Mendel attain well deserved fame, unfortunately too late.

A hybrid is a product of the sexual union of two individual plants or animals belonging to different races or varieties, as a rule; but sometimes to different species, genera, and in a few rare cases, even to different families. Generally speaking, hybrids of widely separated nature as regards the parents are sterile. In a few cases it was supposed that a constant inheritance of the hybrid qualities had been proved, but we shall return to this point later. For the present we shall deal with the simpler cases of the mixture of species or races differing from one another in only a few characteristics.

Mendel said quite correctly that it is here that investigation must commence, and he also saw clearly that the first requirement of all exact research in this field must be the purity of the initial stocks. Experiment is therefore only possible with any hope of success in the case of such species of plants and animals as pass on their characteristics, or at least the characteristic to be investigated, such as the colour of the flower or the like, with perfect constancy, each on its own account. He chose plants (peas) for his experiment, because these are easier to obtain in large quantities and to manage, than animals. It has however been shown that the laws which he found and which have been named after him are also true for animals. The best introduction to an understanding of these laws is obtained by studying, not Mendel's peas, but a case investigated later, namely, the Japanese *Mirabilis jalapa*.

This flower blooms in two varieties, a red and a white, which are otherwise exactly alike (figure 61). If the two are crossed, that is to say, if the pistil of one flower is fertilised with the pollen of the other we obtain a 'first filial generation,' 'F_1',* which has a pink colour; it forms a so called intermediate hybrid.

If this generation F_1 is fertilised within itself, we get a second filial generation F_2, in which one quarter of the plants are red, one quarter white, and one half are again pink. The first two remain, when bred pure among themselves, constant, and hence have simply thrown back to the parent forms Pr Pw. The other two

* The parents are designated by P, and the subsequent generations by F_1, F_2, etc.

quarters divide up in the following generation (F_3), in the same manner as F_2, $^1/_4 + ^1/_2 + ^1/_4$ and hence have exactly the same intermediate character as generation F_1. A similar formation of intermediate hybrids is known in the case of many other characteristics such as size and form of leaf, etc.

A second case, however, is more frequent: the formation of hybrids with 'dominant' characters (one-sided hybrids). If we use for the above experiment a red and a white pea, all the individuals of the first generation F_1 are red, the white is completely suppressed ('recessive' character), and the red is dominant. But this

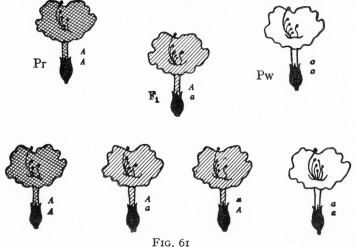

FIG. 61

Simple Mendelian division in *Mirabilis jalapa*

suppression is only apparent. Further in-breeding leads to the appearance, exactly as before, of one quarter of white plants, the other three quarters being all red, though not in exactly the same manner. Only one quarter namely, is pure red and remains so on further in-breeding, the two others are, like F_1, only red by dominance, and hence, as in the case of F_1, again give one-quarter of white plants in the next generation.

In other words, the position is the same as in the first example excepting for the fact that individuals which are actually hybrids are exactly, or almost exactly like one of the parents (closer investigation often discloses small differences). Apart from these two cases, a third is known, the hybrid being neither intermediate nor one-sided, but possessing characteristics different from those of either parent (mainly throw-backs to genetic forms). We will leave these aside. We can sum up the facts described in three rules.

1. When two different pure races are crossed all hybrids have like qualities, (law of the *uniformity* of hybrids).

2. As a rule one of the two differing characteristics dominates over the other (law of *dominance*).

3. On in-breeding the hybrid generation, the second generation breaks up in the portion: 1 : 2 : 1 (rule of *division*).

We have now to add to these rules an important fourth rule, namely, the so-called rule of *independence*. It relates to hybrids which differ in more than one character (so-called di- tri-. . . . polyhybrids), and is as follows:

4. In the case of polyhybrid crosses, the various characteristics 'mendelise' independently of one another (independence rule). The meaning of this rule must first be made clear by a simple example. We will assume that we have a variety of pea which gives yellow and smooth seeds, and cross it with one that gives green and wrinkled seed. Experiment then shows that the third generation produces also peas with yellow and wrinkled seeds, and others with green and smooth seeds. The numerical results (see below) prove that the division has taken place in the $1 + 2 + 1$ quarters for both characteristics independently of one another.

We will proceed to explain this curious rule, which Mendel himself already saw quite correctly, and we will consider first the simplest case of the monohybrid of the *Mirabilis jalapa*. The pink hybrid here results from the union of a 'red' with a 'white' germ cell (these are of course not themselves coloured in this way, but contain the quality in question). When we now get in the third generation pure white and pure red flowers, the only explanation can be, that the *germ cells of the hybrid plants do not contain characteristics mixed as do the body cells, but only either one or the other characteristic*. These characteristics must therefore be separated in the formation of the germ cells.

If one half of all the germ cells of all the hybrids contain red, and the other half white, there are four possible combinations for the third generation namely, RR, RW, WR, WW, if we indicate the character derived from the male by the first letter and that derived from the female by the second letter. Since all four combinations are equally probable, we get $^1/_4 + {}^1/_2 + {}^1/_4$ as experiment shows. We will now write this result in the usual formulae of the science of heredity, or as it is called, genetics. The characteristics which correspond to one another (homologous characteristics), are designated by large and small letters, for example, the red colour of the flower by A, the white by a. The hybrid F_1 then has the formula Aa, but its germ cells are only either A or a and we have:

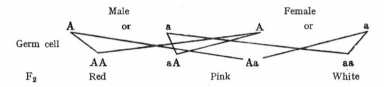

	Male				Female	
	A	or	a	A	or	a
Germ cell						
	AA	aA		Aa		aa
F_2	Red			Pink		White

In this way of expressing the matter, we can also easily indicate that one characteristic dominates over another (pea type). We then write $A > a$.

A di-hybrid cross, in which for example, A represents the green, a the yellow colour of the seed, and B the wrinkled, b the smooth skin of the latter, and $A > a, B > b$, may be expressed according to the following scheme.

Parents: green (A) wrinkled (B)—yellow (a) smooth (b)
Hybrid generation: AaBb (*somatic* cells)
Germ cells of the latter: AB, Ab, aB, ab.
Possible combinations:

	AB	Ab	aB	ab
AB	AABB	AABb	AaBB	AaBb
Ab	AABb	AAbb	AaBb	Aabb
aB	AaBB	AaBb	aaBB	aaBb
ab	AaBb	Aabb	aaBb	aabb

This gives, as we see, nine different combinations altogether, but on account of dominance, these only lead to four different types. For all of those are green and wrinkled which contain the factor A and B once at least, that is to say AABB, AABb, AaBB. AaBb, or nine of the sixteen members of the scheme; while all those will be green and smooth, which contain A at least once, but never B, that is AAbb, Aabb, or three members of the table; those will be yellow and wrinkled which contain B at least once but no A,

Aas

that is aaBB, aaBb, hence again three members; and finally, those will be yellow and smooth which contain aabb. We thus get a division into $^9/_{16} + {}^3/_{16} + {}^3/_{16} + {}^1/_{16}$: and this in fact is what is regularly observed in such a case. The still more complicated relationships in the case of a trihybrid (where we have division into 64ths) we leave to the reader to work out. Since all consequences have hitherto been confirmed by experiment and since some apparent exceptions later proved to be the best possible confirmations of the rule, we may regard Mendel's law as one of the best attested results of science. Mendel's 'hypothesis of the purity of the gametes' (germ cells), is quite generally accepted to-day.

We may also remark that specimens containing mixed inheritance in respect of one of the characteristics are called 'heterozygote' with regard to this factor, while those which have received the same factor from both parents are called 'homozygote.' Of the nine possible combinations in our scheme above, the four diagonal members from the left below to right above (AaBb,) are heterozygote in regard to both factors, while the four diagonal members from top left to bottom right (AABB, AAbb, aaBB, aabb) are homozygote in relation to both factors, the remainder are homozygote with regard to one factor and heterozygote in regard to the other. The parents themselves are naturally also to be regarded as compounded according to this furmula, since they are exactly like the homozygotes AABB, aabb; and this means that *we must assume every individual not a hybrid* to contain every inheritance factor twice over (homozygote), but only once in the germ cell. In the row of our scheme, therefore, we ought not really to designate the parents with AB and ab respectively, but with the formulae AABB and aabb. We now see that the Mendelian combination has given us two new pure races in the third generation, AAbb, aaBB, which we did not possess before. These will continue to transmit their characteristics unaltered so long as they are not again crossed.

An immediate test of the correctness of these theories can be made by backward crossing. If the hybrid Aa is crossed with the parent AA we get the following combination:

Germ cells of the hybrid	of the parent
A or a	Only A
possibilities : Aa or AA.	

that is 50% of the pure parent form or the hybrid form. Experiment proves the correctness of this conclusion. Since in the whole of the animal and vegetable kingdom the rule holds that the two sexes are produced in almost equal numbers, it was early supposed that *sex itself might be such a Mendelian character*, in regard to which one sex would then be heterozygote, and the other homozygote. The objection was however made to this supposition, that the proportion of the sexes often varies greatly from equality, and even when normally equal, may be greatly disturbed by external influences. We shall see immediately that the hypothesis is, nevertheless, true in all probability, while the apparent exceptions are susceptible of another explanation, and become a remarkable confirmation of the hypothesis.

We now come to the *explanation of the Mendelian rules by means of the chromosome theory*. The 'hypothesis' of the purity of the gametes may itself be regarded as one of the most certain in all science. Without it, the whole body of Mendelian research, which has become a large branch of science on its own account, would be as incomprehensible as chemistry without the atomic theory. This is not quite the case with the explanation we are about to give of the distribution of the inheritance factors in the germ cells, derived from the theory of cells (cytology).

The supposition was expressed quite early on that this division was in some way connected with the reduction to half the number of chromosomes in the process of genetic division. If we imagine, in an homozygote individual every inheritance factor, which is present in duplicate (AA), localised in a pair of related chromosomes, then the germ cell, since it receives only one quarter of the pair, would obviously contain the factor in question only once. But if the individual is heterozygote, that is contains in its body cells the combination Aa, and if the factor A is present in one chromosome and the factor a in its partner, the reduction division will supply one half of the germ cells with chromosomes containing A, and the other with chromosomes containing a.

Almost all further facts of genetics since discovered tell in favour of this hypothesis. First of all, very careful anatomical investigations with the highest magnifications have shown that the chromosomes of a normal diploid individual do really belong together in pairs. To-day we are actually able in many cases to individualise the single pair. Figure 62 shows the chromosome set of a human being as figured by the Japanese workers Oguma and Kihara. The hypothesis of Mendelian inheritance of sex has been particularly confirmed by the fact that, as a rule, *the two sexes differ in respect of a pair of chromosomes.* Either (this is the rule) one sex – generally the female – contains the so-called X-chromosome double, the other sex only single; or the other sex has, instead of the second X-chromosome, a differently shaped Y-chromosome.

In the case of man, to which example we shall keep for the present, the first is the case, that is to say, all cells of the female body have two X-chromosomes each, and all ova one only. On the other hand, all male body cells have only one X-chromosome, that is 47 instead of 48 chromosomes, and the male germ cells

FIG. 62
Set of chromosomes of a man according to Oguma and Kihara.
(From Fetscher, *Erbbiologie und Eugenik*, O. Salle, Berlin)

(spermatozoa) have either an X-chromosome, that is 24 chromosomes, or are without an X, having thus only 23 chromosomes. If one of the latter kind fertilises an ovum, the result is a male 24 + 23 = 47); if a sperm cell with 24 chromosomes fertilises an ovum the result is a female child. The small excess of male births over female is obviously to be referred to secondary effects. We have to suppose that the 'male determining' spermatozoon (that is, the one without an X-chromosome) has some kind of small advantage as regards the others in its movements towards the ovum.

Experiments on plants and animals have shown that this subsidiary hypothesis is not a mere evasion, but is justifiable and even necessary. It was shown that by arbitrary changes in the secondary conditions, the ratio of the sexes could be changed within wide limits. Some well-known experiments were performed by Correns on the Campion (*Melyandrum*), a so-called 'dioecious' plant, that is one the individuals of which are unisexual. There are many more male than female examples of this plant, as one can see on any meadow in spring. As Correns was able to show, this depends upon the fact that the male-determining pollen cells move some-what more quickly through the stylus than the female-determin-ing. For if he cut the style off early, the buds only contained seeds which produced male plants. The longer he waited with the operation, the greater was the number of female seeds. The

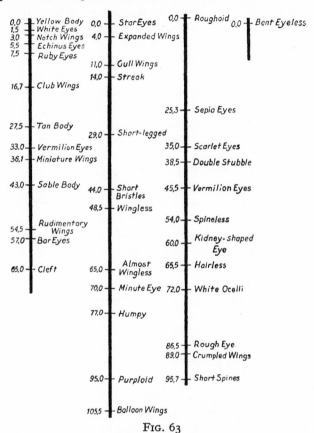

FIG. 63

Arrangement of genes in chromosomes of the fruit-fly
Drosophila (Morgan)

investigation of the chromosome number gave the expected result.[253]

A further proof of the truth of this theory of inheritance of sex is to be found in the facts of what is called 'sex-linked' inheritance. In order to understand this, we must first consider what we call 'coupling of factors.'

The fourth Mendelian rule given above, the rule of independence, is not unlimited in application. On the contrary, it appears that certain inheritance factors mendelise more frequently together than would be expected from the rule of independence. This phenomenon is called the *coupling of factors*; a measure for the degree of this coupling is given by the departure from the normal numerical relations. The hypothesis at once suggests itself that such factors are to be regarded as localised in a single chromosome, but as, on the other hand, there are very many more independent groups of Mendelian characters than chromosomes, this law is not reversible. That is to say, all factors lying in the same chromosome are not coupled, and we are obliged to make the further assumption that the chromosomes themselves consist of certain sub-units, each of which (called a 'chromomere') contain a coupled character group, which can however be exchanged independently of one another (hypothesis of crossing over).

Further investigations, however, which would take us too far to describe, led Morgan to conclude that the individual inheritance factors or 'genes,' as they are generally called, are arranged on a single line in these chromomeres, and this bold hypothesis has also been confirmed by recent investigations in a most surprising manner.[254] We reproduce without further explanation the diagram, which Morgan gives of the chromosomes of the banana fly (*Drosophila*) which is the classical object for inheritance experiments (figure 63).

In this connection, the phenomena of sex-linked inheritance are to be explained as follows. If a disease, such for example as bleeding or red-green blindness, is only transmitted through the female sex, and always only to sons, we must suppose the inheritance factor in question (shown by black colour in figure 64) to be contained in the X-chromosome, and further, that it is recessive as compared with the corresponding health factor. If a man has the disease in question, it means that he has a diseased X-chromosome. One half of his germ cells then contain this diseased chromosome, the other half containing no X-chromosome and hence no disposition to disease. If he marries a healthy homozygote woman, his sons inherit their single X-chromosomes only from the mother and hence are all of them healthy. His daughters, on the other hand, which have one of their X-chromosomes from the father and the other from the mother possess a diseased and a healthy chromosome alongside one another. Since the latter is dominant to the former they are all apparently healthy, but transmit their disposition to disease through one half of their germ cells, which, when they meet with a male determining germ cell (without an X-chromosome), therefore give diseased sons, while the daughters again are recessively healthy.

The annexed genealogical tree (figure 64) shows the state of affairs in four successive generations of such a family (we also show the case of the marriage of a woman carrying the disease,

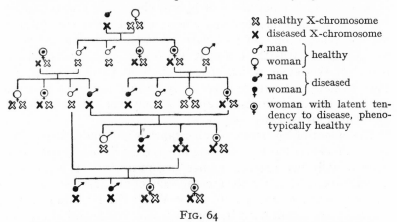

healthy X-chromosome
diseased X-chromosome
man
woman } healthy
man
woman } diseased
woman with latent tendency to disease, phenotypically healthy

FIG. 64

Pedigree of a family with sex-linked inheritance of a recessive tendency to disease (haemophilia). (From Fetscher, *Erbbiologie und Eugenik*)

with a diseased man, which the reader may follow out for himself).

This tree shows an exact correspondence with actual observed cases.

We have attempted to show in this way how the new science of genetics is serving our knowledge of human, and naturally also of animal and plant, diseases. It is not too much to say that it has rendered comprehensible for the first time numerous hitherto entirely mysterious facts of pathology, which have become completely clear at a single blow, whereas in former times they offered a complete riddle.

We must however refrain from going further into these very interesting matters,[255] and proceed to discuss the question as to how far we are dealing in modern genetics with assured knowledge, with more or less well founded hypotheses, or with purely

temporary working hypotheses,[256] and what consequences follow
from these results as regards our fundamental biological question.
Their much wider consequences for human physiology, ethics
etc., will be discussed in another connection.

We may first say a word or two concerning the range of validity
of the Mendelian laws. The controversy as to whether there are
other forms of inheritance than the Mendelian has been long and
bitter. On the one hand, cases were brought forward which appeared to contradict the law of division, and thus showed a more
or less constant inheritance of hybrid characters ; on the other
hand the attempt has repeatedly been made to prove inheritance
via the cell protoplasm (*cytoplasm*) as well as via the chromosomes. As regards the latter, it now appears from later experiments of Wettstein, which space forbids us to discuss here,[257]
that in many cases there is certainly an inheritance of parental
qualities via the cytoplasm, and not only via the cell nucleus
(Wettstein investigated leaf-moss).

The extent of this mode of inheritance is still under discussion.
We can however hardly go wrong in ascribing little importance to
it, especially as regards higher organisms.[258] It is already
certain to-day that far the greater number of characters investigated in animals and plants show Mendelian inheritance, and
therefore are conditioned by the nucleus. We should also point
out especially that the same is true *a fortiori* for human beings,
and, as far as investigations have been made, this has been confirmed. Naturally the difficulties of genetics in the production of
assured results as regards human beings are greater, since experiments cannot be made. But the statistical method gives us a
whole series of very useful results,[259] and we may safely regard it as
proved, that far the great majority of characters of mankind that
can be defined, are inherited in exactly the same Mendelian manner
as in the case of animals and plants.

This is also true for mental qualities, though these are, naturally,
less easy to define exactly. We have, however, very trustworthy
statistics as regards musical and other gifts, which are more
easily and plainly distinguished from the mental constitution as
a whole, and these statistics show without question that the
Mendelian laws also hold for such mental qualities. The fact
that many other mental and spiritual powers, such as intelligence or will powers, are much less easy to define quantitatively,

is no reason for the purely dogmatic assertion that these qualities do not obey the laws of inheritance. Although we have no direct proof of the affirmative, we have still less proof of the negative, and all analogy is in favour of it.

We now come to the second of the two points mentioned above: the question as to whether there are not hybrids which show constant inheritance of the hybrid qualities, such for example as have been believed to be shown in mulatto populations. In order to understand the solution of this riddle, we may look once again at the scheme of di-hybrid crossing on page 369. We easily see from this that even in the third generation the cases of pure homozygote combinations are relatively very rare, they each only occur once (AABB, AAbb, aaBB, aabb), while the partially homozygote, partially heterozygote cases occur each twice, and the completely heterozygote case (AaBb) even occurs four times. As it is easy to prove by figures, this is the truer, the more pairs of characters differ. In the case of a pair of parents having, say, twenty different pairs of characters, the number of complete homozygotes in the third generation is only $(\frac{1}{4})^{20}$, an almost unimaginably small fraction, and far the greatest numbers are found for those combinations in which most qualities are heterozygote.

What is therefore the result of such populations as arise, say, from marriages between negros and whites, which are certainly different from one another by not merely twenty, but hundreds of inheritance factors? In the third generation (assuming that the mulattos resulting in the second generation marry one another) there will be only a few individuals almost completely homozygote in all qualities, with very rarely a specimen having almost all the qualities of the white or almost all the qualities of the negro. Far and away the greatest number are partly homozygote, partly and mostly heterozygote, and will therefore show on the average a middle type with the qualities of the original races in a mixture, which on close inspection will always show varying composition, but on the whole will give the impression of a mixture of negro and white. This gives the impression of constant inheritance of the 'hybrid race' (the expression itself is a contradiction, for a hybrid is never a 'race').

Careful statistical investigations recently made on various sides regarding such mixed populations, particularly the famous investigations of Eugen Fischer on the Rehoboth hybrids have fully confirmed these explanations of modern genetics. Certain cases from animals, such as the well known hare-rabbit cross, and the Meisenheimer butterfly hybrids,[260] are not quite so clear. Difficulties are increased by a circumstance which we have hitherto neglected, but which plays a part in all Mendelian experiments. *The externally visible qualities of an individual are not inherited as such. What is inherited is the aptitude for producing the quality in question under certain external conditions.* Now an external quality of the kind may very well be the product of the joint effect of several inheritance units or genes. If in such cases a Mendelian experiment is made, we get relations which appear at first sight to depart completely from the rule; we do not have simple Mendelian numbers; the third generation may in some cases produce quite new individuals, with qualities departing from those of both parties and the parents.

For example, by crossing white mice with ordinary grey mice, you get in the third generation, besides white and grey mice, completely black individuals. The numbers which appear here prove, however, quite definitely that we are dealing in the case of the parents with a higher hybrid than was apparent; (in the example given we have a dihybrid, instead of a monohybrid for exactly one sixteenth of the animals are completely black. The grey colouring of the mouse depends upon two factors and not upon one). We can therefore say quite generally that an external character does not tell us at once whether it depends upon one inheritance factor or several; for *homomery,* that is the production of a single character by several inheritance factors, is established without doubt in many cases.

This easily leads to another possibility of apparently constant hybrid inheritance. If, for instance, the difference between the colour of a negro and a white is homomeric – and this appears to be the case – then for this reason alone the pure homozygote individuals will be very rare in the progeny, since we are dealing with a polyhybrid cross; the great majority will consist again of partly homozygote, partly heterozygote individuals, which will consequently take up a middle position, this being, in the case in question, a middle skin colour (in so far as dominance does not shift the average in one direction or the other). In spite of this, pure white or pure black individuals will also appear here and there, and this is exactly what we find where such a population has been examined.

We now see how extremely difficult it is to exclude this explanation where apparently constant inheritance of hybrid characters takes place. Very extensive hybridisations with thousands of individuals are necessary in order to arrive at certain results. In the whole of Mendelian research, the simple numerical relationships only appear in reality according to the laws of statistical probability; in other words, here as everywhere the law of large numbers holds, and the theoretical values given by probability calculations are verified the more closely, the greater the number of cases investigated. In order to obtain a simple monohybrid scheme fairly exactly, it is necessary to count at least a few hundred cases. If in the nature of things (for example too small a number of progeny or too high a price of the individuals in question) these figures cannot be reached, it is necessary to manage as best one can with more troublesome and less certain methods, which we cannot deal with here.[259]

A whole branch of science has arisen on this matter, which is indispensable, particularly as regards the application of genetics to human beings, where experiment is not possible. Matters are the same here as in all other sciences, even the inorganic: we must be glad if we succeed in establishing the principle of the matter in simpler and more obvious cases, and be satisfied, as regards the more complicated cases where our mathematical powers or practical experimental means fail us (see above), if we are able to show fairly plausibly that these more complicated cases can be referred fundamentally to the simpler ones. It is no more likely that we shall ever completely clear up the complicated inheritance of human beings, than that we shall ever settle the exact structure of the higher elements from protons or electrons, or the Milky Way from stars. We may, however, be able to see that the more complicated problem differs from the simpler one only as regards its complication.

The total result of this discussion is therefore that *non-Mendelian inheritance* (apart from the cases of inheritance via the cytoplasm, which we are not considering) *has never been proved in any case beyond doubt.* But should this ever be done, as in the case of cytoplasmic inheritance, there is not the slightest reason for speaking of a total failure of all previous genetic research or the like, which has already happened several times as regards such investigation. *There is no permissible doubt that Mendelism determines the matter* in the widest possible sense, and the whole of our animal and plant breeding to-day, with its previously quite unexpected successes, is a further proof of the fact that there is truth in these notions. In the case of animal or plant species, the genetic units of which are more or less known, such as, for example, rabbits or banana flies, a new race can be 'ordered like a pair of shoes.'

That we are still at the beginning of things, and have still an infinite amount to do, is evident in so young a science: genetics is not more than 25 years old. It is also self-evident that such far reaching hypotheses as that of Morgan's of the *linear arrangements of the genes in the chromosomes* – Morgan even calculates their distances from one another – cannot be regarded for the present as any more than useful working hypotheses. A final decision can only be arrived at after long and laborious research.

This need not prevent us from thinking out such hypotheses and it would be entirely wrong to dismiss them from the start as being too mechanical. There is no doubt that the Morgan hypothesis is mechanistic, but so is the whole Mendelian doctrine, although its originator was an Augustinian abbot. There is something depressing, particularly as regards humanity and human family relationship, in realising clearly that the purely chance play of Mendelian genetic units in the meeting of sperm cell and ovum, or previously during ovarian cell division, finally decides which of the qualities of the father and the mother are received by the child as its inheritance. But has not the experience of generations taught us this very fact that we cannot make our children as we would have them ?

The religious person may perhaps answer that we have hitherto believed, and were able to believe, that it was the will of a wise Providence which gave us a particular child. The answer is that there is so much else in life which is pure chance in exactly the same sense, and hence equally destructive of religion, unless the latter is able to deal with the whole matter in quite a different way. The fact that the production of children depends on chances beyond our control is really not decisive, for the same applies to our nationality, position in life, and hundreds of other things. For a religious person, nothing of this is chance, and not a sparrow shall fall from Heaven without God's will. All is divine dispensation, the great as well as the small, both what concerns us closely and what does not matter to us.

The mistake is once again that of the religious person who so often wishes to take certain definite things and events and give them a providential character of quite a special kind; who wishes to reserve for God, in the whole of creation, a few private corners, into which no rational knowledge can peer. Attempts of this kind have always led to failure and to fresh attacks upon religion itself; we make no progress by following this road. The insight which we owe to the modern science of heredity is not more or less theistical or atheistical than all other scientific knowledge. The whole of science shows us humanity caught in the wheels of laws, pitiless if we will, which are none the less implacable because they are statistical laws, that is laws of chance. According to what we have said earlier, perhaps their statistical nature is quite fundamental.

How religious belief can, and ought to, deal with the recognition of this fact, is another story. In any case, not by allowing 95% of determinism in the world, and keeping 5% for the miracle of a special Providence, but only by recognising in the whole contrivance a higher Will, to which it can and will submit itself with faith and trust. Whether we are dealing with the laws of falling bodies, the electrodynamics of the thunderstorm, the Mendelian rules, or anything else, we must no longer attempt to except this or that from the common method of interpretation because it touches us particularly closely.

Let us take a further glance at the controversy between mechanism and vitalism from this point of view. It is undeniable that mechanism obtains a very powerful ally in the whole of Mendelism, particularly if it should be found possible to show that the genes are, as many genetic theorists already assume, nothing more than certain substances analogous to enzymes. It then becomes very tempting to connect them with the development hormones of Spemann (see above), all the more as it has been found possible quite recently to show the probability of a quantitative arrangement of Mendelian genes. Goldschmidt's famous investigations concerning the *intersexuality* of certain species of butterflies, which we unfortunately cannot describe for want of space,[261] render this quantitative view of the genes very plausible; at any rate, this view gives the simplest interpretation of the experimental results.

We are far from putting these doctrines forward as proved. Here again, for the present we are only dealing with a useful working hypothesis. We must however, clearly recognise that mechanism is gaining ground steadily in this way also in genetics, while the vitalists, here as so often before, have to play the thankless part of repeatedly pointing out the gaps and contradictions that still remain. Although this is not a very stimulating spectacle, we ought, in order to be fair, not to leave the subject without pointing out a few of these unsolved problems.

According to the doctrine of chromosomes, the genes are supposed to enter into all body cells (and the fact that they do this appears to be proved by the phenomena of regeneration described above, according to which all body cells contain all potentialities of development, the latter being the genes). But why then does the red colour of the blossom, for example, contained in one species of *Mirabilis*, develop only in the flower and not in the remaining cells of the plant? How are the cells differentiated when they all contain like inherited characteristics?

Again, Mendelism can be tested by its very nature only on crosses between races, that is to say, when there are not very great differences between the two parents. But there are a few cases of hybrids between different species, even genera and families. These are unfortunately almost always sterile, so that Mendelian experiments cannot be done with them. Even if this were possible it would still be difficult to decide whether we were dealing with true constancy of the hybrid, or with an apparent constancy resulting from the homomery of the characters. Hence it is quite an open question whether specific and genetic characters are subject to the Mendelian scheme, or whether these may not be carried by the cytoplasm according to quite different laws, or in quite a different way.[262]

It has been pointed out on several sides[262] that all this leads to the necessity for a whole new branch of science alongside genetics, which would have the task of finding out in what way, by what means, and according to what laws, the Mendelian genes are transformed into the actual characters as they appear. We have already said that it is not the character as such, but the disposition to it that is inherited. But how this change from a disposition to a character takes place is a problem, and without its solution the whole idea of Mendelism remains a torso.

A further objection, which is often raised in popular discussions against the genetic science of to-day, may be mentioned here if only to show how untenable it is. This is the supposed experience of breeders, that female animals when once crossed by a male of impure race, will never again produce pure progeny. This belief in what is called *telegony* (action of a conception at a distance in time), is actually so ingrained, that dog, horse, and other breeders separate animals to which this misfortune has occurred from their stock as no longer first rate, and the buyers of pedigree animals usually demand an assurance that it has not happened.

A critical and historical investigation, which is reported by Goldschmidt in his book on genetics, resulted in showing that the belief is derived from a supposed experience of the kind which an English horse breeder had with a true Arabian mare. The mare had been once covered by a zebra stallion, but produced later on, from a genuine Arab stallion, foals which were partly striped. Investigation has shown that the progenitors of the mare in question, or the second stallion, had been partly of striped races, and the stripes which appeared in the offspring were much more simply explained as the appearance of hidden Mendelian characters, than by the absurd hypothesis of telegony. All careful attempts to prove the result – and innumerable ones have been made – have never given a shadow of a reason for believing the hypothesis to be correct. The common practice of the breeders is therefore unjustified in its strictness, and the objection in question also.

In all probability, the innumerable stories current among the people regarding the effect of certain things upon pregnant women and the child they are carrying, belong to the same category. If the child has any striking peculiarity such as red hair, a hare-lip, or the like, popular superstition searches for some event which may have occurred to the mother when pregnant, and the sight of which may have caused her to confer the quality upon the child. For example, the sight of a fire or a fox (coupled with a shock), is supposed to have produced red hair or a red birth mark, the sight of a hare a hare-lip, etc.

Certain vitalistic theories, to which we shall return later on, have taken up these popular ideas and coined a learned Greek word for them – ideoplastic, that is to say what has been imagined or thought is supposed to have introduced a bodily impression on the growing child in a way as incomprehensible as that of the whole problem of body and soul.[263] In reality, all such characters are decidedly inheritable, and their Mendelism has been tested by numerous statistical investigations. There is thus not the slightest ground for making use of another explanation in this case, which is the common occurrence of the appearance of a Mendelian character hitherto recessive. We shall return to this later in connection with race hygiene, and hence shall not go further into the matter here.

Strictly speaking, we should now take, as directly associated with the science of inheritance, that of the variability of organisms in general, and furthermore a discussion of their history, that is to say, the doctrine of evolution. We will however, interrupt at this point our purely scientific description, and turn to the philosophical problems which we have perceived to be in the background all through the preceding chapters, but have not yet discussed, the most important of which is the problem of body and soul. Our discussion so far has principally turned about mechanism and vitalism. We may also describe this as the discussion concerning the idea of purpose or teleology in biology and will now view from this position the whole standpoint of mechanism and vitalism.

V

CAUSALITY AND PURPOSE

In the following discussion we shall regard physical causality, as it is generally regarded to-day, and in the sense of what we said on page 205, as holding strictly, and hence shall take no account of the present debate in physics concerning the results of the Heisenberg and other investigations. Our justification for this is that, at any rate in macroscopic dimensions, the unambiguous course of events is guaranteed with such enormous probability, that this probability may be regarded as equivalent to certainty. We naturally reserve to ourselves the right to come back in case of necessity to the question as to what consequences for biology would result from a doubt as to strict determinism in the field of physics. We shall have something to say about this later.[264]

This problem is without bearing on our discussion for the present for this reason, that even the vitalists have never yet denied the strict validity of physical determinism, and have not even denied that, if their doctrine were right, biological happenings would still obey strict (though super-physical) laws. For if for example, two apparently equal eggs placed under the same conditions of warmth, pressure etc. yet produce two somewhat different animals, no one will ever hesitate for a moment to conclude that either the internal circumstances in the egg, or the external conditions, were after all not so exactly alike as they appeared to be.

But the question as to what the circumstances are which make the difference is quite a different one. The principle of causality says nothing about this, and can say nothing about it, since it contains a purely formal condition. Every series of events in the real world which we observe as connected together of necessity, that is of law, we formulate into a law of nature. The principle only tells us that if once the circumstances A, B, C . . . always carry with them the results X, Y, Z . . . that this is always the case under the same circumstances. But it can neither say how many and what categories of such circumstances and phenomena exist altogether, nor which of them are connected together individually. Otherwise the whole of physics and chemistry should be deducible from the simple principle of causality.

It is thus clear that the principle of causality alone cannot help us to any *a priori* decision as to the self-determination of life, that is to say, cannot tell us whether real things exist which are incapable of being described by physico-chemical methods alone. A comparison will best explain this. The whole system of *plane* geometry being developed, we might try in vain to understand by its means the relationship of the elements of three-dimensional bodies, since in these a new dimension is added to the two existing in a plane. Nevertheless, the stereometric rules do not in any way abolish the planimetric; on the contrary, they comprehend them as special cases.

Might not a similar thing be possible in the case of physics-chemistry and biology? Might not the laws of the latter include those of the former, and thus add a new dimension to the purely physical? It is certain that no principle of causality in itself can give any decision. It is just a matter for direct experience. Investigation itself must finally yield it as a result. *The principle of causality cannot be given as a decisive reason for mechanism against vitalism.*

I fully agree with E. Becher, who says that it is a mistake to believe that the introduction of vital factors makes biology inexact. The fact that biology is less exact than physics and chemistry depends, as he quite rightly states, not upon the introduction of particular vital factors, but upon the enormous complication of organic happening, and is not improved by mechanism nor made worse by vitalism, any more than physics is made worse by introducing new factors such as Faraday's fields of force, or electrons. The argument often put forward on the mechanistic side, that vitalism is working with unscientific and mystical conceptions, because it refuses to reduce everything to physico-chemical forces and laws, is therefore beside the point in this connection. An 'entelechy' need not be in itself any more mysterious than a physical 'force.'

It is merely a question of what science can actually do with these concepts, that is to say whether it makes real progress in knowledge by using them, or whether they are in reality merely a refuge for ignorance – a bare name behind which the actual problem remains unsolved. We will now attempt to come to a final decision on this point.

For this purpose we must try and present two points to ourselves with the greatest clearness. The first is, whether there are really facts of first rate importance, the explanation of which only appears possible by vitalism and in no other way. The second is this, as to how the relationship between the entelechies or dominants required by vitalistic doctrine and the generally recognised physico-chemical concepts and laws is to be imagined.

Let us take the first question for the present. We have reviewed above a large number of facts which have played a more or less important historical part in the controversy. We have seen that a great deal of what vitalism at one time used as a decisive argument for its position, later appeared as a triumph for mechanism, and we have to ask whether any facts may have been overlooked which even if the arguments failed might yet decide the quarrel finally in the favour of vitalism ? The question must be put – thirty years ago we should have turned away from it with a smile – because vitalism actually claims to-day to be in possession of such decisive facts, which I must therefore not overlook or I shall myself be open to the reproach of having repressed some of the chief arguments for vitalism.

Driesch, who is the recognised spokesman of neo-vitalism to-day, has several times collected these facts,[265] and describes them directly as proofs of the autonomy of life. The first and most important proof appears to exist for Driesch in the facts above described concerning organic 'regulators.' It is not necessary to deal with all that once more, what is decisive for Driesch and all his followers is the fact that the regulation in question (for example the regeneration of whole animals from portions of quite arbitrary shape and size, with sometimes the melting down of the remainder) always lead to the same results, the typical form, although the initial circumstances are variable to an unlimited extent. A machine, says Driesch, out of which one can cut pieces of any kind from any part, and which will then restore the damage by utilising either materials lying around or the material remaining, is unthinkable. Such machines do not exist and can never exist, they contradict the meaning of the word machine itself, which is only a machine because it forms an indivisible whole.

The following replies to this argument may be made: we do not at present know the internal nature of all construction and the effects which, as were shown by Spemann and others, are exerted by the cells on one another. We know only so much from these experiments, that organising effects are exerted by certain groups of cells, particularly in the neighbourhood of the blastophore, upon other cells, and that this appears to take place according to quite strict laws, so that the result may be calculated ahead if the conditions are correctly controlled. This whole gives the impression by no means of purposiveness, but quite on the contrary, of purely mechanical causality. It is certainly true that the organism always restores its typical form even if the initial conditions are varied within wide limits, and it is also correct to say that it would be difficult for us to imagine machines of this kind.

But it is by no means certain that a rejection of vitalism necessarily leads to a machine theory. Perhaps, or even in all probability, the machine is a conception which cannot be used under any circumstances, and also not for the biological mechanism. We are able to draw upon quite other physico-chemical systems for comparison, which meet the essentials of the case much better than the comparison with the machine. Consider the kinetic equilibria met with in physical chemistry, for example a suspension (see page 20). In such a case, perpetual movement of the individual particles exists while the total state, the distribution of the suspended substance, remains constant on the average. This state may be disturbed in many different ways. We are also able to remove parts of any size from the system which will nevertheless always reconstitute itself in the same way on a smaller scale, simply because it is in the most probable state (see pages 97, 258).

Might not the 'harmony' of the organism, together with the 'equipotentiality' of all cells composing it, be nothing more than the necessary state of equilibrium associated with this system of potentialities, which therefore always restores itself, however much the whole is disturbed or mangled? We do not know whether this is the case, since we are at present entirely ignorant of the actual relationships inside the cell, but for this reason we should be equally careful in making the negative assertion, that it cannot be so. Only deliberate experimental investigation, on the lines so happily laid down by Spemann, can really lead to further progress.

A second proof of the autonomy of life, which is to be regarded as independent of the first, is found by Driesch in the facts of heredity. He regards it as positively absurd that the machine should have the power of dividing, and that each part should then possess the power of reconstituting the whole as before (proof from the genesis of equipotential systems). Here we must say that Driesch is again unjustifiably tying mechanism to the idea of the machine, in order to show that this idea is insufficient. Physico-chemical systems, which may be divided again and again and still retain their form, that is, the relative functions of their parts, may be imagined to any extent, though not in the form of machines. The suspension mentioned above, for example, fulfils this condition. It does not, it is true, divide on its own account. But that is not what we are discussing here. Lehmann's liquid crystals showed, conversely, the power of fusing together with production of the original form; the opposite, that is, of cell division, which was realised biologically in Zur Strassen's experiment (see page 340).

It is quite evident that here, and in the problem of form, we are really dealing with only a single fundamental problem, for, as we have already remarked, the genes of inheritance, and the 'organisers' of Spemann, are in all probability very closely related. How they bring about their determination of form is known to us just as little as is the manner in which they divide up quantitatively in chromosome division, but remain in existence qualitatively, and then again renew themselves quantitatively.

We may, by way of illustration and not to anticipate anything, construct the following hypothesis. Suppose that each gene is a quite definite enzyme, that is, a certain catalytically active substance, and that each of these substances possesses the power discussed on page 319, of increasing itself from surrounding material related sufficiently closely to it chemically, for example, sugar and lower nitrogen compounds. Let there further exist between a considerable number of these bodies a state of 'cellular equilibrium,' that is, let us suppose that the formation of further quantities of each of these substances is limited when a certain maximum volume is reached, on account of the change in the electrical and other forces occurring as the relation between surface and volume changes. Within this limit we assume that the competition of the substances in question decides the part played by each in a kinetic equilibrium. We may now assume that when this limit is reached, the process which we observe as cell or nuclear division takes place on account of certain physical quantities exceeding a given limit.

Such a system would already have a whole series of properties which we are accustomed to regard as specifically biological. It grows, divides, and restores its condition automatically on account of the 'cellular' equilibrium, provided a certain limit to its possibilities is not overstepped (a condition which Driesch also requires for his entelechies). I do not think of asserting that it is so, or even of demanding that biologists should try this hypothesis. The biologists themselves know best what hypotheses they need to help them along, and philosophers should not attempt to interfere with them, or with any other of the special sciences.[266] I only state it here, in order to make clear that one may be a 'mechanist' without adhering to a machine theory, and thus cause the main objections of the vitalists to collapse.

They will naturally find other objections, the hypotheses just suggested, for example, that only colonies of exactly similar cells could be produced in this way, but not differentiated structures. This would bring us into the middle of the problem of the development of the higher organisms from the lower, and we naturally admit that mechanistic explanation here meets new and perhaps even greater difficulties. We will deal with this point later on. For the present, what we have said is sufficient to show that Driesch's second 'proof' is not so convincing as he and his disciples believe.

A third proof of the autonomy of life is found by Driesch in animal action. Since the psychological question plays a part here, we will leave it to be dealt with in the next chapter, which shall be devoted to this side of organic processes, a side which we have hitherto intentionally ignored.

We may now further inquire how the vitalists of to-day picture to themselves the action of their entelechies upon the physico-chemical systems undoubtedly present in the organism. According to what has been said on pages 205 ff., such a system is to be regarded as unambiguously determined by the initial conditions and physical laws. How is this process, which is already determined, to be changed as regards its operation by an entelechy or the like, in the sense of organic purpose ?

Vitalism gives the following reply to this question.[267] Physical processes are caused by forces which, as far as we know, obey the law of energy. That this law is also true for living processes is not exactly demonstrable, since the energy balance of an organism can only be followed with great difficulty, but it may be assumed to be true, for experiments (by Rubner, Atwater, etc.) have actually shown it to hold within the limits of possible accuracy. It has been shown that the energy contained in the food intake is actually equal to that given up again in the form of heat, external work, and excretory products. If therefore the entelechies are to interfere, or rather regulate, the play of the physical forces, they can only do it, just as man does in running his machines, under the condition that the law of energy holds good.

Now this appears to be possible in several ways. First of all, entelechy might be regarded as *workless force*. If namely, a force acts on a physical system at a direction at right angles to the momentary direction of motion, the work done by it is zero (the mathematical expression for the work is, generally, $F . s . \cos \alpha$ where F is the force, s the distance moved, and α the angle between the two). The sun draws the planets (assuming a circular path for the sake of simplicity) continuously in a direction away from that of the tangent, and towards itself, without doing work in the process. An electron moving as a cathode ray is continuously diverted by a magnetic field without its energy of motion being increased or diminished, since the force exerted is, according to the Biot-Savart law always in a direction at right angles to the momentary direction of motion, and hence work is neither gained or lost. If we imagine such workless forces exerted as required by the entelechies on physical systems, all particles subjected to them will be diverted in other directions without there being any change in the energy balance.

A second possibility of influence without change of energy would consist in the entelechies introducing resistances for certain lengths of time, against which the energy of motion would be transformed into potential energy, which would then be released again at a suitable moment.   Just as a miller is able to store up energy in the stream for a certain length of time by inserting a weir or dam, and then to make use of this energy later, the entelechy also might do likewise, and would obviously be able to alter radically the course of events, without altering the energy balance.

A third mode of interference would consist in the entelechy actually injecting into the organic cycle a certain small quantity of energy, which it would afterwards take to itself again.   In this case also the balance would remain unaltered.   Driesch himself does not regard the latter hypothesis as particularly happy. We shall not therefore take any further account of it, for it is obvious that one would only have recourse to it if there were no other way of understanding the life processes.   What are we to say to the other two hypotheses ?

They, and particularly the first, sound so very plausible that it is difficult to retain one's power of sober criticism.   If it were only a matter of maintaining the law of energy, the question would be settled, and vitalism's right to these arguments would have to be admitted.   The error lies in the fact that the arguments only take account of the law of energy, and suggest that it is the same thing as physical law altogether.   But we already know that the energy law alone does not completely determine the course of events (compare page 85).

It is therefore not a matter of our being able to imagine interference which does not contradict this law, but interference in which physical causality as a whole is not contradicted.   A physical system takes its course according to certain differential equations, which connect the initial state with the changes in the magnitudes in question (page 76).   The law of energy is only one integral of these differential equations, but contains much less than they do, and it in no way allows the whole course of events to be followed forwards and backwards.   If, therefore, vitalism is to make the power of the entelechies comprehensible, it has not only to tell us how change in the energy balance is avoided, but it must also show how the entelechies manage to alter a course of events unambiguously determined from the initial condition and the law.

The comparisons which the vitalists make at this point all avoid the actual difficulty. They all appeal fundamentally to the fact that mankind, by means of its machines, also guides physico-chemical processes for its own purposes without thereby infringing physical laws. That is true, but we do it by introducing at suitable points small quantities of energy, 'release energies,' into the course of events, or by so constructing the physico-chemical systems in question that they are obliged to take a certain course. The first would, when transferred to our problem mean that the entelechies would be obliged to generate energy, although in very small quantities, from nothing. The second would have no other meaning than that the production of the organism is only to be explained by the entelechies, but that its functioning is purely mechanical, which view is rejected by vitalism.

Furthermore, the analogy of human direction contains a vicious circle. For every time a human being switches on the release energy at the right moment, he makes use of his muscles, and these are brought into action by the nerves, which are made to function by the brain. But it is obvious that it is in the brain that the direction takes place which we are trying to explain (the problem of body and soul, which is behind this fact, will be dealt with later). Our 'explanation' has thus only taken us round in a circle.

The vitalist further makes some such statement as the following: I can change a course of events by switching in certain arrangements, without infringing natural law. I can for example divert a ball moving in a straight line, without diminishing its energy, by placing at a certain point a wall on which it can roll and which it meets tangentially; it will run along this for a certain distance only and then leave it in free motion, having retained all its energy.

This is undoubtedly true. But the vitalist forgets that this wall itself belongs to the physical system which follows a certain unambiguously defined course. It is not in the least an immaterial entelechy, but a very real physical body. But in the organism, the entelechies are supposed to set up walls of this kind which cannot be perceived, for the physical system is expressly imagined as being completely given. The ball without the curved wall, and the ball with it are two different physical systems.

The vitalist on the other hand maintains that the organism

with entelechy and that without entelechy are the same from the physical point of view. He expressly points out the phenomena of decomposition which take place at death as happening according to purely physical laws when the entelechy withdraws, and hence actually regards the living organism as the combination of a 'dead' physical system of this kind with one or more entelechies of a non-physical nature. But this can only mean that he postulates immaterial curved walls, and hence every comparison of the kind ceases to hold.

To sum up, we may say that vitalism has not yet succeeded in giving a plausible account of the way in which the entelechies can really regulate physical processes. The comparisons which it makes use of either contain a vicious circle, or they contradict the fundamental assumption of the immaterial nature of the entelechies. *The fact that compatability with the law of energy is possible is not sufficient to remove the real difficulty, namely the unambiguous course of physical determination.* Is there any other way which is open to vitalism ?

It would appear as if modern physics leaves open a way of escape, physical law itself not being, according to our most recent view, absolutely unambiguous ; instead, as we have seen above, it consists in a 'continuous succession of probabilities' (*Reichenbach*). If this is the case—and we know that we approach this state the more nearly, the smaller the objects which we consider—the possibility remains that the entelechies make use of the choice which physics as such leaves free. If certain deviations are actually undetermined on the subatomic level, the entelechies might choose from among these those which would produce the desired result in the course of the further chains of action, which would then be statistically calculable.

I imagine that it will not be long before vitalism seizes upon this new argument, which is not yet so ready to its hand, since physics itself has only just been led to these new ideas. This idea might be followed up further and result in a more definite formulation of Auerbach's suggestion that in organisms we are dealing with 'ectropic' systems, that is to say such in which the free energy does not always diminish of itself, but increases instead (see above page 258). This line of thought certainly is open to doubt, since exact measurements have shown that the efficiency of the living engine is not sensibly higher than that of other thermodynamic machines, that is to say, that the fraction of the energy of the chemical food material transformed into useful muscular work is not sensibly greater than if this food had been burned in a heat engine. We may however say that even if this is true for muscular work, free energy is nevertheless gained in other ways in the organism, or speaking more generally, elementary disorder may pass over into a greater degree of order, and hence a more probable state into a less probable state.

However this may be, the reader will probably have the same feeling as the author, that this way out for vitalism, although it cannot be refuted for the moment, is still open to serious doubt. This line of thought could only acquire real importance if it were successfully developed into a more general philosophical hypothesis, capable at the same time of throwing new light on the problem of body and soul. The future may perhaps bring us a construction of this kind, and then we shall be able to discuss the matter further. We shall return shortly to this problem once more, but we will first ask ourselves what is the position of the problem of organic purposiveness, if we are not willing to take this last way out for the vitalism of Driesch. We shall see that the loss is, as a matter of fact, less than is commonly supposed.

The nature of organic purposiveness or adaptation is so obvious to everyone from daily experience, that it is unnecessary to say very much about it. We already find in the simplest beings, and to a much higher degree in higher organisms, arrangements and activities which are quite obviously there for the purpose of attaining certain ends, which always ultimately relate to the preservation of the individual or the species. We may leave aside here the question as to whether there are, alongside useful arrangements and activities, other which are useless or even the opposite, such, for example, as the absurdly large and numerous branches in the deer's antlers or the excessive prolongations of many shells and the like; or whether these are of hidden utility. In any case, the useful, that is to say that serving the ends in question, is so much more in evidence, that we are unable from the start to regard all arrangements and activities of human beings from any other point of view.

Everyone who observes organisms in any way always asks involuntarily: what is the use of this or that? Why does the animal or the plant do this or that. If, for example, we see potato sprouts or the branches of a plant bend towards the light, our first thought is that the plant does it in order to obtain the greatest possible amount of light necessary for assimilation. If we consider the matter more closely and find that we cannot well ascribe intention and reason to the plant, the idea of purpose takes the form of considering the plant as constructed in such a manner that the reaction that is useful to it takes place. On the other hand, in the case of an animal, we shall in general even tend to regard purposive processes of this kind as quite 'intentional,' in a manner fully analogous to our own action.

On the other side, we have the fact that investigation has shown without doubt, in many such cases in which this 'teleological' view was quite common, that a causal explanation can be found; the causes from which these activities must of necessity result have been discovered, quite apart from the question as to whether they bring an advantage for the plant or the animal. A particularly instructive example of this is found in the construction of bones. In the spongy bone substance which forms the ends of the large hollow bones, for example the thigh bone, the lines follow almost exactly the same course as a structure of this kind built according to modern technical methods, with the object of obtaining the maximum of rigidity with the minimum amount of material (figures 65 and 66). Here we have a case of teleology, and it has frequently had to serve as a proof of a governing intelligence behind nature.

The causal explanation of this problem was given in two steps. It was first shown that the girder structure of the bone matter in question results from the fact that the bone cells (like most cells which secrete solid material) always deposit lime in the direction of maximum tension or pressure. This disposes at a blow of the necessity for ascribing the geometrical construction of the structural lines to a special guiding intelligence, for this construction is in any case only a necessary consequence of a quite general state of affairs which already holds for single cells. But the defenders of a special entelechy could naturally still ask who had taught these single cells to behave in this obviously very useful way. But this question also has been answered by causal research. In all probability it is a matter of the natural consequences of laws which hold quite generally for all colloidal solutions, as P. Weiss has shown quite recently.

Vitalism is thus again driven from this second position, in so far as it maintains that such teleological cases are not susceptible of a causal explanation, but only of a teleological one ; in other words in so far as it places purposive causes in the place of real causes. The position is much the same as regards the heliotropism of plants mentioned above. The so-called Blauuw theory,

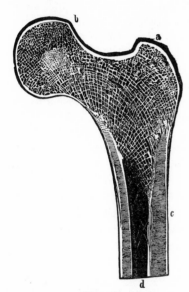

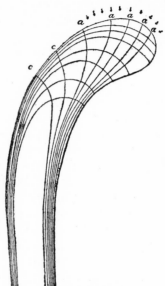

FIG. 65
Structure of bones
(hip bone)

FIG. 66
Upper end of a crane, loaded in a manner similar to the hip bone, with lines of pressure and tension indicated. (From Erdmann)

according to which the light itself automatically effects the bending of the stem by hindering the illuminated cells in their development, has, it is true, recently been once more challenged, but there can hardly be any doubt that if not this, then some other causal explanation will be found for the process, and something similar is true in numerous other analogous cases.

The question upon which everything depends is whether such causal explanations, already existing or probable, have really abolished teleology. This is the point at which the *sins of mechanism* begin, whereas we have hitherto spoken almost exclusively of the sins of vitalism. Does not purposiveness hold good to the fullest extent in spite of all causal explanations? And *is there such*

a thing as biology in which this purposiveness is not mentioned ?[268]
Anyone who will not admit that heliotropism is of advantage to a
plant and the structure of its bones an advantage to a vertebrate,
must be an idiot, and likewise anyone who will not admit that
eyes are made to see with, and that the organs of generation of a
plant are adapted to insects. A single glance into any text-book
of biology whatsoever (even if written by a fanatical mechanist)
is sufficient to show that the whole of biology is saturated by this
idea of purposiveness in organisms, and no causal analysis, however extensive, will ever alter this fact.

We can therefore come to no other conclusion than that *the
causal and teleological views have an equal right to exist.* In some
form or another, this statement is admitted on almost all sides.
Many mechanistically minded biologists, however, interpret it as
meaning that they only ascribe to the teleological point of view
a provisional value, as stating summarily the problem for causal
analysis. They therefore deny that the idea of purposiveness
has any more than methodological importance, and that it corresponds to anything real; they maintain rather that in 'reality'
only those complicated causal relationships exist, which we summarise under the concept of purpose for the sake of economy of
thought. It is at this point that I believe that a vitalism valid in
all cases may be founded.

Vitalism has a right to exist as soon as we ascribe to organic
purposiveness an objective meaning going beyond that merely
imagined. But this by no means implies the entelechy as a
hidden causal factor, such as vitalists always believe it necessary
to defend. We can very well secure an objective meaning for
purpose without it.

In order to perceive this fact, it is only necessary to see clearly
that the concept of purpose in this respect is fundamentally in the
same category as the concept of causality (and the 'law of nature'
etc.) which do not themselves 'run about free in nature,' to use
E. Becher's expression. Both of them, like other relations such
as comparison of magnitudes, arrangement, etc., are primarily
concepts or categories (forms of thought, formulae) by means of
which we, the subjects of knowledge, make reality into an ordered
system. According to the positivist view (Mach), as well as that
of idealism (neo-kantian), they are nothing more, and there is
nothing in reality which corresponds to them.

As opposed to this, we have already taken up earlier the view of modern critical realism, which regards these concepts of relation (*universalia*) as having a real content (see pages 232 f.), and hence must regard the concept of purpose and also that of causality in the same way. The purposiveness of the heliotropism of plants or the construction of the eye of a vertebrate is just as 'real' as the laws of refraction of light or conservation of energy, or the causal connection between bacteria and disease. No biology will ever exist able to avoid taking a teleological view of the facts as well as a causal.

There can therefore be no question of a purely provisional view. *The mistake of the mechanists lies in the unjustifiable narrowing of the concept of science down to the causal-mechanical method of physics.* The successes of the latter so dazzled the eyes of the philosophers of our classical period (Kant himself could not quite escape from it) that they believed themselves able, and indeed compelled, to regard physical science as identical with the science of reality.

This was false. *Science is every logical connection of facts in thought, whether it be etiological or teleological.* Anyone who describes in detail how, for instance, the accommodation mechanism of the eye serves its purpose, or how incredibly ingenious the mutual relations of the life habits of the Yucca moth and the flower of the palm lily are (see also below), is doing work quite as genuinely scientific as that of one who discovers the causes of a thunderstorm, or makes experiments in developmental mechanics. *It is pure prejudice to suppose that only causal connection can form the object of science.* If there are purposive connections in nature, they are just as much a part of nature as causal connections. Any one wishing to understand nature must pay attention to both, otherwise he will become one-sided, and place obstacles in his own path towards true understanding of nature.

In this connection a word must be said about an entirely new branch of biological research which is generally referred to to-day by the somewhat unsuitable name of 'ecology.' This term means, or originally meant, the relations of living beings to their environment (*oikos*, the Greek word for a house): to the factors of the external world, therefore, such as light, air, temperature. To-day it has been widened to include also the relations of living beings to one another. This latter question is one on which we must now make a few remarks.[269]

We already mentioned, at the beginning of our biological discussion, the fact that the formation of organic 'wholes' undoubtedly goes beyond what are otherwise usually termed individuals, at some points of the animal kingdom. This is the case with certain groups of coelenterates, namely, the siphonophores or colonial polyps. Here numerous and differently formed, hence clearly differentiated, individual polyps, form an organic totality, which behaves like a single animal. The movement polyps propel it by their common labour, the nutritive polyps feed it, but this does not prevent individual polyps separating off from time to time and living an independent life as 'medusae,' and also reproducing themselves.

It is also clear that the animal state so common in the insect world (ants, termites, bees) forms a functional whole which is not quite so closely united, but in general similar to the polyps. Direct bodily connection does not exist, but it is replaced to a greater or less degree by the intercommunication of sense impressions. We may add two other examples to these.

In the lower range of plant life we find in the lichens so close a life association of algae and fungi, that a long time was needed for scientists to discover that they were not dealing with a single living being, but with two quite different species. When on the other hand we consider the symbiosis of the hermit crab with the sea-anemone, we have a case of looser community, which nevertheless can be plainly recognised as such. Between these two cases we have other intermediates of all possible kinds, and there are naturally also instances of looser community than that of the crustacean mentioned.

What does all this teach us ? We see that those forces and powers, which 'integrate' the cell to an organism, do not stop at the individual, but are also able to unite many individuals of the same or even different species to form closer or looser units. And if we look a little further afield, we easily see that this principle holds for still more comprehensive aggregations, namely for whole associations of plants and animals in certain typical forms, such as a moor, a wood, a steppe, a lake and the like. In such associations we have 'biological equilibrium'; that is to say, the relative numbers of the individuals of the various species concerned vary around certain mean values, which remain in general constant. The oscillations depend upon the effect of external factors, particularly temperature, rainfall, and the like, but compensate one another on the average, in so far as climatic conditions do not change fundamentally.

It cannot be denied that the discovery and exact investigation of such forms of life – this word being taken in a much wider sense than as relating only to a single individual – belongs to the legitimate objects of biology, but in this case there is obviously very little to be done by causal-analytical methods. In this case the question of the value of the single components for the whole is of much greater importance, in other words the teleological point of view.

In the biology of to-day we can find everywhere unmistakable signs of a re-appreciation of this second side of biological investigation, best described as *synthetic biology*.[269] The Americans have coined the expression 'emergents' for every new form of life appearing on earth in the course of development.[270] Something new appears, as it were, above the surface of the stream of development, and maintains itself at least for a time, in its typical form. We may also speak of *Gestalt*,[271] following the German psychologists, Wertheimer, Koffka, and Köhler,[271] or of 'wholes' following Driesch – the matter is always the same. We are dealing with the fact that out of existing units on a lower plane, units on a higher plane are formed, from atoms molecules, from molecules micelles, from these the chromomeres and other constituents of the cells, from these the cells themselves, from them the multicellular organism, from these again symbioses, associations etc., and in the case of human beings, finally, families, states and alliances, which are all (to use Driesch's expression) 'more than the sum of their parts.'[272]

Although this principle may be employed very generously, we must not go so far as to declare any collection of organic beings brought together by chance anywhere and anyhow to be such a 'whole.' To mark off the meaning of this concept from chance complexes, two conditions are necessary: first that equilibrium must exist in the complex in question (although it may be capable of variation) and that it always restores its own typical form, when the external conditions remain the same. And second, that it really forms a *Gestalt* in Köhler's sense, that is, a system in which each part contributes towards the existence of the whole,

---

* This word has been adopted in English ; it means ' form.'

Ccs

just as in a modern bridge, each part takes its share of the load and supports the others. Hence it must not be possible to remove any part (in the case of a plant complex, for example, one of the plants) without the whole equilibrium being destroyed, just as in the case of a single organism we cannot in general take away any part of importance to life without endangering the whole (the limitation, that certain less important accessory parts, such as the appendix or tonsils in the human body, may be dispensed with, may also be allowed in respect of the higher order of units in question).

It must also be admitted, that ideas of this kind readily degenerate into useless playing with words and vague aesthetic considerations which cannot naturally form the object of serious science; but an insight into this danger must not blind us to the fact that a real understanding of the living world is impossible without an open eye for such formations of higher order units. We must, as we said in the first part of this chapter, free ourselves from the prejudice that the cell and the individual are the only scientifically permissible units in biology. They are only the most important and striking of a whole hierarchy of such concepts, which continues in both directions, up and down. If we seriously consider this point of view, we readily see where the point of application of a vitalism tenable in all cases lies, and what meaning we are justified in giving to the much discussed concept of entelechy.

Vitalism is nothing but the method of investigation suited to this synthetic biology, and the entelechy is – exactly as it was defined by its original inventor Aristotle – simply the idea of the whole which exists *in re*, or in the Platonic view *ante rem*. If it does not 'run about free in the world,' it is all the less a mere chimera, all the more a very real something, just as real as the causal relationships which analytical biology strives to discover. There 'really exists' such a thing as a living community, an animal state, a people etc., and these are not, as positivism and nominalism would like to persuade us in this case as well, mere *flatus vocis* for human ways of looking at things, or categories, or schemes, or whatever else we might call what man uses to make nature intelligible.

We say here also, as in the analogous case in physics (see page 234), that it is not we ourselves who have read these things into nature, but nature herself which has forced them upon us. We have read them out of her, and never ceased to learn from her in this respect. But then, naturally, there 'really exists' the entelechy of the single individual in this sense, and we are entirely justified in pointing out the wonderful phenomena of regeneration as examples of its operations in nature. Only we must not believe that this frees us from the necessity for investigating the question by causal analysis, and that we may set the entelechy in the place of a cause. This is the 'lazy teleology' which Kant quite rightly objected to, but he very clearly saw that true teleology has every reason to exist alongside causal analysis.[273]

When we take up this standpoint, the question arises as to how it happens that the improper use of the concept of purpose in place of that of causality is always slipping back again. The answer is, naturally, that man knows the mentally pictured purpose in himself as a driving force, a motive of his action, and hence knows it as a causal factor; and he is thus inclined to judge other natural phenomena according to it; in other words, to view nature as analogous to himself. Since without doubt this view has a high degree of probability for the actions of the higher animals, we are led to consider at what point down the scale the limit is to be set. Where and how does purpose in general become conscious (intention) ? This brings us to a new problem, the most important of all problems in the philosophy of natural science, that of the relationship between mind, soul, or spirit, and body.

VI

THE FUNDAMENTAL PSYCHOPHYSICAL PROBLEM

I have purposely avoided hitherto all reference to the spiritual, also the later form of vitalism called psycho-vitalism. We must now make good this deficiency.

First of all we most note that spiritual processes must probably be assumed in all living beings. Even in cases where the organism is so completely different from our own that we are unable to imagine what the 'inmost soul' of such a being must be like, the principle of continuity leads us to assume some form of spiritual activity. For it is quite impossible to assume that a dividing line can be drawn anywhere in the organic series between 'with a soul' and 'without a soul' ; the whole system of zoology is against it.

At one time it was attempted to draw this line between unicellular and multicellular organisms. But as we have already shown above, the difference between the two is, in many respects, far from definite. To ascribe to a loose collection of cells on the lowest level, for example certain sponges, the honourable title of individual and a common soul, while denying the latter to a much more unified simple cell such, for example, as the slipper animalcule, appears impossible, since in the latter case the single cell exhibits a much stricter organisation than that of the whole in the former case, the parts of which are single cells.

An attempt has also been made to draw the line between animals with, and those without, a central nervous system, but here the analogous difficulty again arises. The difference between the highest animal without a central nervous system and the lowest possessor of this structure is infinitely smaller than the difference between the latter and the highest animals (insects – mammalia, birds). In short, whichever way we look at the matter, the fact remains that every attempt of the kind has the appearance of an assertion made *ad hoc*, of the intrusion of artificial boundaries into the almost continuous series of the animal world.

If we are thus compelled for the reasons given to ascribe psychical experience even to the lowest animals, such as amoeba, the radiolaria, paramecium, the foraminifera, although this experience be dim and quite incomprehensible to ourselves, the same principle of continuity leads to the further consequence that we must again pass up the scale from the common root of the animal and plant kingdom, the unicellular organisms, to the higher plants, although we may feel that here at least it is impossible to form any conception of their spiritual life.   In recognising this we do not advocate the application of human and animal psychological concepts such as feeling, pain, pleasure, etc. to plant life.   A poet may be allowed to do so.   But the sober investigator must say to himself that according to all analogy the 'soul' of a higher plant must be something fundamentally different from that of a higher animal, just as different as the body of the plant is from the body of the animal, and that the further we go in either direction from the common basis of the unicellular organism, the less the possibility of a comparison between the spiritual and the bodily organisation, and that we must leave it open for the moment whether we shall ever succeed in reaching any kind of understanding in this respect of organisms so far removed from ourselves.

In spite of this fact, we may be allowed to reckon with spiritual processes also in the case of plants, although we must pursue plant psychology with the greatest circumspection, and for this reason it is not permissible to pass over without comment material recently contributed by R. Francé, Haberlandt, and others, however critically we may be inclined to regard Francé's experiments in plant psychology.   These experimenters have at least succeeded in finding many phenomena hitherto quite neglected, and these show us that plants possess, to a much higher degree than we have hitherto believed, the power of response to stimulus and utilisation of stimulus, and that they possess special organs for this purpose.   A few of these phenomena have of course been known for a very long time, for example, the sensitivity of the leaves of nolimetangere and of insect-eating plants (sundew, *Drosera*).

We may also notice a fact which is often mentioned in connection with this question, but still more often forgotten—the relativity of psychological time.   We are accustomed to regard actions as conscious or willed, when their speed is approximately of the same order as human action, but if we watch a cinema film showing the growth of a plant, and arranged so that the real time is shortened from a thousand to ten thousand times, we are really amazed to see the complete analogy of the movements observed with those of animals. When, for example, the sticks of asparagus shoot up from an asparagus bed we seem to see the battle for light and air between neighbours almost as a hand-to-hand affair.

Only a little critical consideration is required to prevent our believing that a spiritual life can only be one which takes place at the rate proper for mankind. From the standpoint of the theory of relativity, all that happens is, of course, fundamentally timeless existence (compare page 202) only certain parts of which, namely the processes taking place in our brains (more correctly the 'world-lines' forming its history) being associated with sensations etc. This is obviously equally possible in the plant processes we have mentioned. From this point of view, also, we are led to the result that a plant psychology is quite imaginable, however great our difficulty might be in imagining it.

At this point the question naturally arises whether reasons of continuity do not enable or even compel us to pursue the concept of mind further back, beyond unicellular organisms, and to ascribe a mental or spiritual counterpart also to the inorganic processes of nature. The doctrine that one and the same will is manifested in the fall of a stone to the earth, in the play of atoms, in the growth of plants towards the sun, and in the movements of animals, the doctrine of 'hylozoism,' was not first invented by Schopenhauer, but is as old as philosophical speculation itself. It has undoubtedly had something seductive and also poetical about it. The thought that, everywhere in the world, we are not dealing with rigid and dead 'lumps of reality' but with life and feeling, will and sensation, must appeal to everyone who has seriously tried to enter sympathetically into the great life and movement of nature, and not merely to dissect it with the scalpel of intellect. But this attitude of sympathy must not be allowed to interfere with sober and serious judgment, which tells us that, at any rate for the present, we have no grounds for applying the concepts and categories of living matter to inorganic.

Life is, in the sense of the American philosophers mentioned above (page 401), an emergent of the earth's development, that is to say something new, even if it is linked with the inorganic by continuous transition (see page 332). As long as we do not speak of assimilation and growth, of reproduction and response to stimulus in connection with inorganic systems – and so far at least no one has felt the need to do so – we have no real grounds for speaking of spiritual experience in the case of such systems. This is in no way an attempt to deny that, seen from a higher and more comprehensive point of view, contrast may not perhaps, in this case also, be resolved into higher unity (see also below page 433). Let us therefore leave this question for the present and bear in mind, that, at any rate in the field of living matter, the spiritual may and must be assumed all through.

We are thus confronted as regards the whole realm of the organic, with the difficult problem of the nature of the relationship between mind and body. What have the bodily processes which we can observe in living protoplasm, to do with the sensations, impulses, etc. which we are obliged to suppose as their accompaniment, if the world is not to remain incomprehensible to us ? It is perfectly obvious that interactional relationships exist, and this has been demonstrated, if demonstration were necessary, by thousands of experiments, and particularly by psychiatrical observations. It is surely superfluous to give examples. Palpitation as the result of fear, irritation caused by bad digestion, the numerous brain injuries in the war, with their sometimes quite extraordinary mental consequences, prove these interactions between body and mind to the most obtuse. But in spite of some centuries of philosophical work, the great majority of the laity has not progressed beyond the most primitive view of these relationships, a view which goes back directly to the animism of savages. The 'soul' is imagined as a kind of dweller in the body, a being that escapes at death etc., and even scientifically educated people are not always perfectly clear, in our unphilosophical times, as to what is the real problem for philosophy in this case.

We may therefore consider a few examples to make the matter clearer. Take a tuning fork in vibration. Physics and psychology teach us that waves in the air proceed from it, and that these set into similar mechanical vibration first our ear drums, then the auditory ossicles, and then the endolymph. These vibrations next excite, in the auditory nerve which traverses the labyrinth, some kind of processes which are at present very obscure to us, but probably belong in the field of electro-chemical phenomena.

These are transmitted as such through the auditory nerve, as a sort of wave impulse, to the brain, where likewise some kind of electrical changes take place in certain parts of the grey matter of the cortex. And then – we hear a sound.

If we now take instead of the tuning fork radiating a musical note, something else, such as a hot object, which excites in us a sensation of heat, or a sharp-edged object which stimulates our nerves of touch, or a lamp which radiates light to us, matters are always the same as soon as we follow the stimulus in question as far as the nerves are concerned. From these we always have a conduction of certain processes, at present little known to us, as far as the brain, and there also a doubtless very complicated group of equally physiological changes. All that is probable is that what goes on in the optical nerve when it is stimulated by light is not essentially different from what occurs in the auditory nerve when it is excited by the sound waves etc., and that the brain processes also will be qualitatively the same in all cases. For all nerves are quite similarly constructed (or at least are similar in principle) and the same is true of all the ganglia in the cortex.

How can these processes in the nerves and brain, which only appear to permit of differences in quantity, intensity, and arrangement, lead to the completely incommensurable, and qualitatively entirely different sense impressions (not to mention will impulses)? Ganglia are ganglia, in one case as in another, but a note and a colour are something absolutely incomparable, and the sensation of hot or cold is again something quite different from either.

When once we have really grasped this apparently hopeless question, we have only a short step to the realisation, that sensation itself (apart from the differences in its various qualities) is something quite different from the physiological processes taking place in the brain, whether these are electro-chemical or anything else. If we had an 'astronomical' knowledge of the processes on the brain, we should still only see a mass of moving electrons or a system of some other physical magnitudes, in which certain changes are taking place; but what in all the world has what we should see there to do with the sensation experienced by the possessor of the system in question?

This, therefore, is the psycho-physical problem – the complete incomparability of mental processes (sensations, will impulses, etc.) with the physiological processes which we have to imagine as their accompanying corporeal phenomena in the brain; and it is easy to see that this problem does not, for the present, become any clearer, when we descend from the highest organism to the most primitive. We may assume, after all that has been said above, with some certainty that even in an amoeba, when it gets hold of a piece of food, some kind of dim feeling of pleasure occurs – we know nothing about it, but we must draw this conclusion on grounds of continuity. And so the problem always remains the same, namely, how this 'feeling' hangs together with the bodily states and processes, with which we imagine it associated.

We might even say that the matter becomes more obscure, the further we go away from ourselves. For while in the case of mankind, we know mental phenomena from self-observation, they are, in the case of other beings, eternally removed from our direct experience, and must be imagined on the basis of analogies from the bodily phenomena observed, which are all that we can observe. Hence – we might conclude – it will certainly be best to start with man. But this conclusion is perhaps not quite correct, for even though it may be difficult to form a picture of the mental life of the lowest animals on the basis of mere analogical judgments, this does not alter the fact that it is in the case of these that the simplest relationships certainly exist, and that here, if at all, a beginning must be made with the solution of the problem.

We must, however, keep in mind from the start the very great difficulty in this case, which lies in the fact already mentioned, that the mental life of animals is not directly accessible to us by communication, as in the case of other human beings, but must be deduced solely from the directly observable bodily processes. Animal psychology[274] is therefore in its initial stages, and we may even say that it is only in the last few decades that the problem has been properly attacked, for it was previously the practice to draw uncritical inferences with reference to man, and hence many wrong conclusions resulted. A well known example of these difficulties is given by the many years of controversy concerning the colour sense of bees, which has been finally settled by von Frisch's researches in favour of the existence of such a sense, although one partially colour-blind, in all probability, as compared with that of man.[275] In the days when Sprengel and Darwin discovered the relations between insects and flowers, it was at once assumed that the flowers were coloured in order that the insects, especially the bees, might recognise them. Doubts were then raised, and it was even asserted that insects are completely colour-blind, and hence many of the beautiful biological relations of flowers discovered by the older workers were purely imaginary.

We have at last to-day a really secure basis for the investigation of these relationships, which fact does not prevent its being necessary to make new special investigations in connexion with every further step, for example as to how the bees manage to communicate to one another the presence of food supplies.[276] The worst distortion of the truth produced by humanisation has been in the matter of the psychology of the highest mammals, especially the domesticated animals, dog, horse, elephant, etc. Nine-tenths of all the touching or amusing stories of the older biological descriptions depend upon such uncritical ascription of human mental processes to animals, and hence are to be taken with the greatest reserve.

Recognising these difficulties, the animal psychologists who have come to the subject from science, have made the fundamental demand that the science should limit itself strictly to observing the external actions of the animals, and refuse absolutely to consider the internal (mental) states and processes, since these cannot be observed. In America, this school of thought has recently, under the name of 'Behaviourism,'[277] even claimed its right to be applied to human psychology as well. But even decidedly vitalistic workers such as Driesch[278] have, curiously enough, expressed similar opinions. Empiricism is obviously so deeply ingrained in scientists that they will not even abandon it when their whole attitude would otherwise point beyond it.

The fundamental reply to this attitude is, that since nature actually does contain living beings with mental processes (which is quite without a doubt) these processes must belong to science just as well as the so-called bodily processes. There is not the slightest reason why the scientist should be forbidden a priori to answer the question why a dog runs away from its master when he threatens it with a stick, by saying that it is afraid. This fear is just as real a fact as the running away or the movements of the molecules which constitute the heat content of a body. I am also unable to see the latter directly. We have shown above that they nevertheless belong to the entirely legitimate objects of science.

Behaviourism is thus like every positivism, a negative dogmatism; it erases quite arbitrarily certain sides of nature from the book of science, because these cannot be immediately fitted into its all too narrow conceptions of science, and thus blocks its own road to the decisively important problem. Just as positivism once wished to decree the abolition of atoms from physics, so now it would like to remove mind from psychology – with the only result that nature takes no notice of its dogmas. *Atoms exist in spite of Mach, and mind in spite of the Behaviourists*, and the real problems begin at the very point where, in their opinion, we should stop asking questions.

The fundamental problem of animal psychology is and remains the connexion between the bodily and the spiritual. And this is in its simplest form in the case of the lowest animal, and then, in ever increasing complication, up to the high and highest, and finally to man, in which case we are clearly conscious of it, since we are, as Bohr puts it, both audience and actors in the great drama of life.[279]   We will not allow ourselves to be diverted from this by the pseudo philosophy of positivism (see also below, pages 426 f.).

If therefore we first ask ourselves what phenomena, in the case of the simplest of all living beings, lead to the assumption of a spiritual counterpart, we are led to what has already been mentioned above under the functions of the living cell, to the sensibility and contractility of protoplasm. We observe that even simple unicellular organisms respond to certain stimuli, for example light, by performing certain actions, for example moving, turning, unrolling, etc. Both these two elements, sensibility and contractility of the protoplasm, are found everywhere, also in the case of higher beings, to be active whenever, as far as we know, mental phenomena are taking place. In the very motion of our muscles, for example, which is caused by our will, such stimulation of single cells and consequent contraction of protoplasm takes place.

We may therefore assume with good reason that these are really the elementary processes, at which we should begin in investigating the psycho-physical problem. If it is to be solved, it is certain that we must first comprehend this matter fully, where the condition of affairs has not become unmanageably complicated on account of the enormous tangle of the effects of so many single cells. How far, therefore, has our understanding of this simplest stimulus action of unicellular organisms progressed ?

Unfortunately, we have to admit that there can at present be no question of an understanding, but rather that all we have up to the present is nothing more than certain 'brute facts' for which we do not know any explanation. In other words, we have nothing but a large number of single observations. The movements that the living cell carries out in response to certain stimuli have been divided into two groups. As regards movements carried out by freely moving (swimming) cells, in moving say towards the force of stimulus, we speak of 'taxis' (chemotaxis, phototaxis, etc.); as regards changes in the shape of stationary cells, such as those of the higher plant which turns towards the light, we speak of 'tropisms'—heliotropism, geotropism etc.) But this is not much more than a classification with the help of learned Greek names, however interesting and surprising the facts that have been discovered may be. We may give a few examples. In all higher plants, the stem always grows away from the earth and the root towards the earth (negative and positive geotropism); we have already mentioned that the cells of the part above ground contract in such a way that the plant bends towards the sun.*

Infusoria mostly show a striking sensitivity for light; they are even able to distinguish the colours of the spectrum, and hence collect, when illuminated by the different colours, in a certain region of the spectrum (phototaxis). If an electric current is sent through the water of an aquarium containing jelly fish, these always turn with their heads towards the anode; on the other hand, root threads always grow towards the cathode. Certain protozoa set themselves across the current while others move towards one pole (galvanotaxis, galvanotropism). Hundreds of such facts have been investigated, but the profit hardly corresponds to the labour so far.

One of the main problems which has remained a subject for controversy up to to-day[280] is whether the simplest actions of the lowest forms of life are causally conditioned in an invariable manner, that is strictly according to an invariable law, or whether they already show something of the higher human faculties of

---

* The higher plants, however, probably have special sense cells, which then send the stimulus out to the other cells (Haberlandt).

choice and learning. It at first appears as if the former were the case. If in an experiment with the jelly fish just mentioned the current were suddenly reversed, all the animals immediately turn round, 'they make a right-about turn at the word of command.' The root grows towards the earth with the same absolute certainty,[280a] the stem towards the light, etc. It therefore appears as if we may ascribe to these simplest actions, along with which we have a good reason for assuming spiritual experience, a causality entirely equivalent to that of physics. And it further appears that this would then have to be postulated also for those processes which at first give us the impression of arbitrariness, for example the intake of food, which does not take place at all times, but only at certain times, alternating with those of rest. There is scarcely anything to be said against the assumption that the alternate inception and cessation of these processes depends upon the fact that the interior of the cell must first be in a certain state, before the process in question begins, and that this state is each time gradually brought about when the previous portion of food has been digested.

But as against these results apparently favouring strict causality, there are others again which show that even the lowest organisms possess a quite surprising capacity for learning from previous stimulus, and hence for undergoing experience and making use of it. If, for example, we allow grains of sand to fall from above upon the crown of cilia of *Stentor polymorphus*, it contracts each time, and rolls up spirally with its long stem. If this is done several times, the animal finally has enough of it, lets go, and swims away. All this might of course be explained by a machine theory. But if the same experiment is done several times with the same animal, we obtain the quite astonishing result that the tiny creature takes flight on the very first occasion. It has therefore actually learnt from the previous events that rolling up is no use, and hence does not even try it. This would at least be the way in which we could interpret the matter at first sight by analogy with our own actions in such a case.

There are a whole number of similar experiments. Their total result is that the actions even of the lowest animals take place on an 'historical reaction-basis,' that is to say, they are determined not only by the momentary stimulus, but also by the whole previous history of the animal, and always take place on this basis, in such a way that – within the limits of possibility – there is a better adaptation to the stimulus at each successive action, in the sense of the action being carried out in a manner advantageous to the animal. This is, to be sure, difficult enough to imagine as the characteristic of a machine, and one would have to think out incredibly complicated mechanism in order to produce effects of this kind, but it has actually been found possible, at any rate approximately, to build models of this kind.[281]

But the matter becomes apparently hopeless when we rise up in the scale of animal life. Here there is added to the power of learning, and of trying and remembering, something also without doubt new, namely the consideration of the purpose. We shall go into this matter in further detail in connexion with the relationship between man and animals, but we may anticipate at this point. It is certainly established that the higher mammals, for example dog, cat, elephant etc., and above all the anthropoid apes, are able to carry out useful actions which they have never carried out before, and which they therefore cannot have learnt, at least in this form, by 'trial and error.' Every dog owner can give dozens of such examples, and although not a few of them appear on closer inspection as less intelligent than the dog owner generally believes, there remain enough unquestionable cases of animal intelligence, which are distinguished from the actions of lower animals by the fact that the higher animal knows how to find the right means even in the case of combinations of stimuli which it has never previously experienced (naturally this only happens within the limits of the animal's powers, which are very narrow as compared with man).

How a machine is to perform all this is really quite unimaginable, but since on the other hand we are again unable to draw any sharp line in passing down the scale, but rather find that the 'intelligent' action obviously passes gradually into the pure 'reaction,' we may naturally ask with justice, whether there is any point in adhering to a machine theory for the latter at all costs, when it is not sufficient in the case of the higher levels.

But let us remain for the present by the lowest animals and plants, and consider those of their actions which give the impression of being unmistakably conditioned by law, or of being automatic. Even when we admit this strict causality, the psychophysical problem is by no means solved in that way, but only taken at its lowest stage. For we are still left with the question as to what happens mentally or spiritually, when we observe this or that bodily process (reaction stimulus). How is the one connected with the other?

If the application of the principle of continuity, as more fully described above, is regarded as admissible, we must surely suppose that an action of the kind is accompanied, even in the case of the simplest protozoon, by some kind at least of feeling of dim pleasure, unpleasure, or the like. It would be certainly permissible to risk the further hypothesis that, at this lowest level, a differentiation of such mental events into the great categories of sensation and thought, of will and feeling has perhaps not yet taken place. We may rather suppose that just as the whole process of response to stimulus exhibits an entirely unit character, so also will it be accompanied by a correspondingly simpler and more elementary mental process. But all this, the greatest possible simplification of the problem that is, in no way removes the fundamental difficulty of the problem, which consists in the complete incompatibility of the observable bodily processes with the mental experience which we suppose to accompany them.

It has frequently been believed and stated that this mystery can be evaded by adoption of hylozoism, that is, the ascription of a soul to inorganic nature (see above page 406). But it is easy to see that the question is not rendered any clearer by this, but merely pushed back one stage further. For if it is difficult to explain what a cell contraction has to do with a supposed simplest form of will impulse, it is no more comprehensible when we get down to the molecules and atoms and electrons. The dualism of matter and soul yawns before us unbridged as ever before, just as it did for Spinoza, who simply ascribed to his 'substance' the two 'attributes' thought and extension.

Or is, after all, the whole problem which lies here before our eyes, a single huge deception, a catch, depending upon a totally false statement of question? There are two directions in philosophy which maintain this.

Materialism maintains that soul has no reality, that it is a mere appearance, a function of the bodily which alone has real existence. Unfortunately the materialist does not tell us how the matter which we call brain, has become possessed of the totally wrong idea, that there is in itself something more than purely material, and indeed how it comes to imagine such a thing. It is like Münchhausen, who pulled himself out of a bog by his own pigtail, for, as Du Bois Reymond has said, it is a complete riddle why a mass of moving atoms (electrons, etc.) should not be for ever indifferent to what happens in it and to it. Materialism simply denies the psycho-physical problem – *car tel est son plaisir.* But the same thing will happen to it as regards spiritual phenomena as happened to the learned man in the Walpurgisnacht in the matter of the Spirits :

> *Still there? Such impudence has never yet been known!*
> *Begone! We're all enlightened nowadays.*

Materialism has been refuted too often for it to be worth while going into the matter here.

The second philosophical view, on the other hand, deserves to be taken seriously. This view regards the problem as merely apparent, and depending on a question which should not be asked. This is the positivism of Mach and Petzold, to which we have often referred, and to which must be added the 'empiriocriticism' of Avenarius and the 'psychomonism' of Verworn. We have hitherto considered this philosophy, which to-day is also frequently called 'conscientalism' (because it recognises as real only the content of consciousness) more from the epistemological point of view. But apart from this, it is easy to see that it also offers a solution of the psycho-physical problem.

Matter, the 'thing' is as we have seen, nothing more according to Mach than a thought symbol for a complex of sensations of relative stability. What is truly real are the psychical elements, (colours, tones, pressures, space, time) and what is constant in the world are the relations between these elements which are expressed scientifically in functional equations, $F (A,B,C \ldots)=0$. Among the elements A, B, C, ... we easily distinguish a certain group a, β, γ ... which hang together more exactly and which we call our ego, to which also belong our memories. Opposed to these in a certain sense is another group of elements K, L, M ... which we call the physical world, and we attempt to find out their relations, independently of the elements a, β, γ ... when we pursue the science of physics. The investigation of the relation of a, β, γ ... between themselves is the business of psychology, that of α, β, γ, to K, L, M, by means of a certain group of middle elements K', L', M', which we call our body, is the task of physiology.[282]

The object of science is to represent all these relations as completely as possible, and in the simplest manner (thought-economy). The idea of 'things' is a fiction, a self-deception, caused by the relative constancy in time with which certain relations of elements appear. A piece of ice for example, is nothing but the combination, constant as regards time, of the sensations (elements) hard, cold, transparent, etc. The constantly repeated experience that these elements appear in this particular combination, is shortly expressed by us with a view to economy of thought, by the single word ice. The dualism of body and soul must, according to this view, necessarily be an apparent problem only; according to Mach 'it has been introduced artificially and unnecessarily.' According to him 'the whole strangeness of the situation as described above (see page 409) is an illusion.'[283]

The following is a short statement of what is to be said against this philosophy (more complete arguments are to be found in the works of Hönigswald, Külpe, Messer and others.)[284] One of the main objections to the doctrine in question is the impossible double part played by the ego, on the one hand as 'a complex of elements of relative stability,' on the other hand as a subject of scientific knowledge, and also *of the positivist and other philosophies.* How a complex of elements can be subject, not only to the monstrous deceptions of the psycho-physical problem, but to any deception at all; and how the same complex of elements can then put itself right, and in Mach's philosophy – push itself back again into the liquid mass of the elements themselves, appears to me to beat the record of the materialistic Münchhausen twice over. Closely connected with this contrast between truth and deception, which runs through the whole of Mach's work, but should be impossible according to the principles of his own system, is the contrast, already discussed earlier (page 32), between the 'actual' and the 'hypothetical.' Both are quite impossible on the basis of positivism. 'The correctness of the theory presumes,' as Husserl says in quite an analogous case,[285] 'the unreasonableness of its premises, the correctness of the premises presumes the unreasonableness of the theory.'

We do not need to lay stress here on the want of clarity in the

DDS

statements of the positivists, as to whether it is the elements or the relationships which are real.[286]

On the other hand, a very doubtful point appears to me to be the twofold rôle played by time in this philosophical system. The 'thing,' also the ego, are thought symbols for sensation complexes of relative stability. But stability means the same as constancy with regard to time

$$\frac{d}{dt} \, (K, L, M \ldots) = 0.$$

Now space and time are, according to Mach, themselves elements. These equations $\frac{d}{dt} = 0$ differ in no way from other functional equations between the elements, how is it that just the equations $\frac{d}{dt} = 0$ give the opportunity for the 'hypostasisation of things-in-themselves'?

Finally, conscientalism does not explain to us what is here the main point, namely the wonderful fact that we do not merely regard, without any justification, our own complex of elements $(\alpha, \beta, \gamma \ldots)$ as a fixed pole 'ego' in the flux of phenomena, but above all that we imagine as added to certain groups A, B, C, of the so called external world, namely the bodies of our fellow human beings, and of the higher animals, quite definite elements $\alpha', \beta', \gamma', \ldots$ regarding which we are furthermore convinced that *we shall never be able to perceive them directly*, but which we nevertheless assume just as certainly to be 'real,' as our own $\alpha, \beta, \gamma \ldots$ Now there is nothing in itself unusual in our adding in thought to observed elements others which do not represent a real, but only a possible observation. This happens everywhere in physics in order to enable us to comprehend the course of events.

But let us now consider what this means in our case, where the elements added are not, as in physics, possible objects of experience, but (for me), are simply inexperienceable. Here is a complex of elements L_x, M_x, $N_x \ldots$ of my experience, which I call the brain processes of the person X (assuming that such experience is possible with complete exactness). I now imagine as necessarily connected with it a certain element ρ_x, for example 'red,' as the sensation felt by the person X, that is to say as part of the complex $\alpha_x, \beta_x, \gamma_x \ldots$ Nothing could be easier than that, says the psychomonist. Alongside the many other relations between elements, there exists also that between L_x, M_x, N_x . . . and ρ_x.

Very good. But I think we knew that already. The very question that we are asking is why and in what way does this relationship exist, the problem as to how the brain processes hang together with the corresponding sensations does not in other words become any clearer for us, if I analyse the brain processes themselves into a sum of sensations of a third party. We followed above the way taken by a sound from the ear drum to the brain. The sound wave is, according to the psychomonist, nothing but an expression based on economy of thought for certain complexes of sensation of the physicist. Let this be so. The same is true of the concepts: ear drum, cochlea, nerve, brain molecules etc., all these are only symbolical expressions for elementary complexes, if you like. But what in all the world, then, compels us not to remain, in this case, as we otherwise always do in physics, by these elementary complexes, but to think into them a certain ρ_x, which is for ever inaccessible to our experience? The psycho-physical has not been got rid of, but merely translated into another language.

We may leave the matter at this point. Psychomonism or conscientalism is not a possible road to the psycho-physical solution. Realising the impossibility of evading this fact, many people to-day are contenting themselves with the simple admission of *parallelism* between body and soul, which is then generally expanded into hylozoism. The central point of the doctrine of parallelism is the denial of so called *influxus physicus*, that is to say, direct action of the body upon the soul and vice versa. The same body acts only on the bodily and the spiritual on the spiritual, but both move together inseparably and the appearance of interaction results from the fact that we sometimes observe one side, and sometimes the other of such a causal series. The parallelist is therefore compelled to demand as an addition to the bodily (physical) processes of stimulus, mental or spiritual counterparts, and conversely, to every mental process in the brain, a bodily counterpart, in order that both series may remain unbroken.

As opposed to this, we have the *theory of mutual interaction*, which says that interaction between the two series does actually

take place.   A third theory, the identity theory, strives to regard
the physical and spiritual as two different sides of one and the
same thing, which is then generally regarded, in the sense of the
'spiritualism' opposed to materialism, as of the nature of mind.
What position are we to take up with regard to all these ?[287]

So much is clear for the present, that all dogmatism on this
most debated of all philosophical questions is out of place.   Let
us say first of all, honestly and plainly, that we do not know.   In
each of these theories we are feeling our way only; no one has
explained the relationship we are discussing by means of any one
of them.   And the arguments which are put forward for one or
other views are always more or less convincing, but never so
demonstrative as their supporters usually believe.   On account
of the importance of the matter, it is necessary to discuss a few
of these at least.

First as regards the denial of the *influxus physicus*, this depends
fundamentally on the same arguments, which we brought forward above against the action of the entelechies upon the physico-
chemical organisation of the structure.   We cannot quite see
how a transphysical factor, whether it is called entelechy or soul,
can act here in any way at all.   It is, by the way, a very short
step to assume, that the entelechy required by vitalism actually
is identical with the spiritual in some form or another.   Accordingly, Driesch first introduces alongside the entelechy the 'psychoid' as a special factor, but only to conclude by expressing
the opinion, that both after all are ultimately identical.[288]
Altogether, the whole of the vitalism of to-day is psycho-vitalism
in this sense.   But naturally, all the arguments given above
against vitalism also affect the theory of interaction, which will
not need any further exposition.

On the other hand, it is not to be denied that a direct inward
voice, which is difficult to silence, speaks against parallelism.
Let us consider a little more exactly, what according to it, is supposed to take place in an everyday psycho-physical process.

Suppose, for example, a letter or telegram arrives with news
of some event.   Thereupon the recipient performs a whole number
of thinking processes and conclusions.   He gets out the railway
guide, decides to travel by a certain train, and then actually
travels by it.   Parallelism compels us to suppose that the light
waves meeting the eye, and the process set going in the optic

nerve and the brain on the basis of a historical basis already present, that is to say a bodily disposition of the brain atoms etc., set going according to pure physical laws just those physical processes alongside which we have to range, as their spiritual complements, the wishes, conclusions, choices etc. in question. How this is to be imagined as possible in view of the infinite variety of mental combinations, which often lead, as a result of the smallest differences in stimulus, to absolutely opposite results, is quite incomprehensible. The telegram 'William wed' differs from another 'William dead' only very slightly, but very fundamentally from a mental point of view. On the other hand, it may reach us in a physically totally different form *Guillaume décédé*, and the result is the same as in the English, assuming that we understand French.

Even if we could find in case of necessity a parallelistic explanation of all this, what are we to imagine according to this theory, when anyone, as the result of a certain piece of news, tells a lie to someone else? Are there machines, or other physico-chemical systems, which can lie? Furthermore, all our direct inward experience speaks in favour of the physical letters in the case in question, having first set going a spiritual process, and that this again sets a bodily process going (movement etc.). Parallelism remains a hypothesis which has the words *ad hoc* written all over it. It is made in order not to admit the *influxus physicus*, but what do we really know as regards the possibility or otherwise of the latter?

Closely related to parallelism is the *theory of identity* which, as we have discussed above, maintains that one and the same 'unknown third' appears to us at one time as the material, at another as the spiritual. The question naturally arises at once why this double mode of apparition exists. The answer of the identity theorists is, because we look at the matter in question at one time from inside, at another from outside. One and the same circular arc appears different to us (concave or convex) according to the side from which we look at it, and nevertheless it is always the same line. This sounds very plausible and apparently clears away the difficulties connected with interaction in a very happy manner. But unfortunately the whole theory still remains unclear as to how we are to imagine the 'unknown third' and how it comes about, that we are able to look at it in two different ways.

It therefore gives us, like its antithesis the dualistic interaction theory, only a programme, but no real explanation.

Further, the identity theory, as we can easily see, amounts in the end to a form of spiritual monism. For since the mental or spiritual is the only thing that we can really understand directly and from the inside, this will always claim first place (also for epistemological reasons [289]). And hence the 'unknown third' will after all be finally completely identified with it. This would then lead to the bodily being finally only a mode of appearance of the spiritual for the spiritual, and we may perhaps say that development is strongly tending in this direction to-day. If, namely, matter can finally be analysed into purely formal conditions concerning a Something in itself quite indifferent, as we have discussed in our physical section (pages 198, etc.), there is apparently nothing in the way of regarding the spiritual itself as this final substance, as the English physicist Clifford remarked forty years ago, and as his more famous compatriot Eddington has pointed out quite recently. [290]

In Clifford's case, it is true, the statement had a conscientalist sense as with Mach, but the idea may better be turned in a metaphysical direction; the postulated spiritual would then be the universal world substance (not the spiritual in the sense of individual consciousness as in Mach's case), and the complex presence in the single individual would only be a minute fraction thereof. This would then bring us to a metaphysical idealism (to be carefully distinguished from an epistemological idealism) lying directly on the line of Platonism, and it is noteworthy that the epistemological realism, which we set forth in the first part, leads to this consequence.

It is true that there remains unexplained, for the present, the relationship between the part complexes in question (the individual souls) and the whole, and for the present also, the reference-back of the material to purely formal conditions of an indifferent substance, is naturally far from being carried through. We shall return to the first question later in another connexion, but biology also offers some suggestions, which we will now shortly discuss at this point.

We saw above, that we have to recognise in the higher 'wholes' which go beyond the individual, for example animal aggregations and animal states, but also in the biocoenoses, *Gestalt* in a wider sense, the reality of which, from the point of view of critical realism, must be just as fully admitted as that of the single individuals composing them.   This is naturally true, first of all, for the characteristic example of such a whole of higher order, the cell-state forming the multicellular organism.   But we did not then discuss the question as to how this works out in the spiritual domain, and we must now make good this intentional omission.

We have already seen at the beginning of this chapter, that the principle of continuity compels us to ascribe spiritual experience also to the single cell.   This must then be true of the single cell of the multicellular organism, and this brings us to the question of how this 'cell soul' is related to the soul of the whole.   When we state the question in this way, we recognise that an exactly analogous question exists as regards the higher units above mentioned.   All analogy requires us to ascribe not only a unity in life, but also a unit spiritual experience, to the polyp colony.   How is this related again to the spiritual unit of the single polyps which compose it ?   Other questions again appear in such cases as that of the earth-worm cut in half, or the planaria divided into seventy-two pieces.   What happens there on the spiritual level ? And what happens in the case of regeneration ?

We do not know, but the questions themselves are sufficient to teach us that this is a domain in which we have to be very careful in applying concepts and rules derived from our own spiritual life.   Animal psychology, as we have already set forth, is in its initial stages, and we cannot yet speak of the real foundation of a plant psychology.   Hence we can say no more than that, in view of the biological facts mentioned, it is entirely unsound to assume, as is commonly done, that the spiritual is always attached to a brain, or at least to a central nervous system.   Since we know absolutely nothing at all about how and why it must be attached in this way, the only valid argument remaining for such an assertion is, that we only know of its existence in such a connexion. But this is just what is rendered doubtful by the biological facts we have mentioned.   Stating the question quite generally, it runs: what biological systems give us the impression that spiritual experience (sensations and will impulses) is operative in them ? Honesty compels us to reply that this is far more true of one of the animal colonies in question, than of a sponge or a volvox colony, and at least as true as for one of the polyps belonging to the same class of animals.   Indeed, the same is even true for ant and bee states.   It may be replied that this is admitted, and that the animal colony may be therefore given a unitary comprehensive spiritual identity, although we know nothing at all about it.   But how is this to come about in the case of an animal state, the individuals of which are completely separated from one another, while in the case of the polyp colony they are at least closely connected together ?

This objection must certainly be taken seriously, but it also is not fundamental, since complete isolation between two material objects does not exist.   In the physical world everything is connected with everything else; that is a fact beyond all doubt, which follows with certainty from the whole system of modern physics. Who can guarantee that the spiritual does not also reach out beyond the limits of the individual ?   We know nothing in detail about its connexion with the bodily, we only know that in Minkowski's language, it is peculiar to certain 'world-line' complexes, that these serve as 'guides' or whatever we like to call them for the spiritual (this is certain as regards the physical processes taking place in our brains).   Who is to tell us where this possibility ceases ?   It is, obviously, again only a question whether the observed facts in a certain case compel us to suppose that this reaching out of the spiritual beyond the individual limits is actually present or not – no judgment can be made on this point in any other way.

But if we regard the matter from this point of view, biology at once shows us a number of striking facts, which tell very much in favour of the enlargement of the limits.   The riddle of the migration of birds, the fact that a train of caterpillars, say the processionary moth or the like, when interrupted at any point, is immediately upset at all points with lightning speed, similar phenomena in the case of ant heaps, and then certain other matters that will be discussed below, are much more easily understood if we assume that a spiritual experience reaches out beyond the individual.[291]   A mechanistically thinking biologist may perhaps object at this point, that all this is none other than the psychovitalism so sharply criticised above.

The answer to this is that the same must be true when a biologist such as von Frisch teaches us to understand animal actions by making use of spiritual factors such as sensation etc.   As a matter of fact, no one doubts that the actions of the higher animals at least may be quite legitimately explained with the aid of psychological notions such as fear, hunger, sexuality, and the like.   We have already pointed out that these are as entirely real as bodily processes themselves, and as I do not wish to introduce, by the assumption of a spiritual connexion reaching beyond the individual, anything that is not already quite generally accepted, on the other hand, I also do not wish to prejudge the fundamental problem by this supposition.

Extended spiritual connexion proves just as much and as little for vitalism as any everyday explanation of an action by psychological motives.   We cannot wait for the physiological explanation which the mechanists promise us, any more in one case than in the other. What is animal psychology to do in the meantime ? There remains for it, at least for the present, nothing else than to see how far it can get with specifically psychological motives such as instinct, sensation, association, etc.   What we propose, therefore, means only that this method, which is generally approved and applied in the case of individuals, must necessarily be applied to the extended units.   But since we are obliged to return to this question twice again in other connexions, we may leave it for the present and we will now devote a short time to considering what finally results from all that we have been discussing.

VII

METAPHYSICAL CONSEQUENCES

Before entering upon the subject of this chapter, we will make a few preliminary remarks concerning the concept of metaphysics in general, which has unfortunately an evil sound for most scientists even to-day, although development has long gone beyond the purely negative attitude of the late nineteenth century. Eduard von Hartman[292] is known to the majority of people only as the author of *The Philosophy of the Unconscious*: that is to say, as the originator of a metaphysical system very vulnerable at many points. A much greater achievement of this philosopher was the foundation of 'Critical Realism,' in which he discovered an entirely new position, a 'Metaphysics *a posteriori*,' and thus finally disposed of the Kantian view of metaphysics, which started from an untenable conception of science (compare in this connexion pages 243 f.). Of recent years, Erich Becher has continued his work along the same lines, and we owe to him the highly illuminating definition of metaphysics as the 'science of total reality.'

It is in fact easy to show, that the existence alongside one another of the various sciences leads to certain new concrete problems (and not as certain philosophers have maintained, only to new epistemological problems) of which no single science can find a solution, since they lie outside the field of work of any one science. The physicist is not concerned with mind, nor the psychologist, at least for the present, with physics, but a problem arises as soon as the results of both are confronted with one another: namely, the problem under discussion of body and soul, which therefore forms one of the chief objects of metaphysics. It is a self-evident consequence that such metaphysics, since it deals fundamentally only with the ultimate results of individual sciences, and hence often with results more intuitional than actually proven, will be considerably less definite and more uncertain even than these results themselves.   This is, indeed true to a less degree even for the theories of each field singly; the more general and comprehensive the question that is put, the more difficult it will be to answer it, but anyone who has followed the brilliant development of modern physical theory, for example, will not condemn the striving after theories of the greatest generality (see also above pages 192 f.).

The same right may therefore be claimed for the most general theoretical problems, that is, those of the science of 'total reality,' so long as we remain conscious of the difficulties of the task, and are particularly careful to avoid hasty and dogmatic conclusions.   In other words, we are seeking in metaphysics nothing more than a general view of the world and the relationship of its various types of phenomena (body and soul, nature and civilisation etc.), to one another, which on the one hand allows of direct connexion with the results of the separate sciences, but on the other hand also brings this theoretical knowledge of ours into a rational connexion with the other factors of our cultural life. For after all, science is only one of the many civilised activities of mankind, and must suffer if it is permanently isolated from all the others.   Here also synthesis must finally follow analysis, which of course is unavoidable.

We have already actually touched upon a whole number of metaphysical problems, and discussed some of them, the chief of which were the problems of contingency, teleology, and psychology; the last and highest of all metaphysical problems, the problem of God, has also been shortly mentioned at various points. It therefore only remains for us to see what can be said on these matters from the biological point of view by way of conclusion.

In this connexion we may point out for the present the following facts.   In popular literature, concerning the controversy as to the meaning of life, it is still fairly usual to identify the contrast between the mechanist and the vitalist biology with the contrast between *atheism* and *theism*, without further question. The falsity of this is shown at once by the fact that the third main category of belief, *pantheism*, is ignored.   It is therefore useful to gain first of all a clear view of the essentials of these three views of life.   Under the word atheism we understand a view which regards the world as something unpersonal (generally also purely physical), a neutral thing or a machine or something even less:

a mere conglomeration of single machines isolated practically completely from one another. Pantheism on the contrary, regards the world as an organic whole, which is usually, in the great majority of cases, imagined as having a spiritual essence. According as this essence is looked upon as a more or less unconscious driving force, or as more conscious, or even as something far exceeding the human consciousness, pantheism approaches nearer either to atheism or theism, which latter of course regards the world as an organism also, but not as an organism on its own account, but by virtue of a highest Will, which is completely conscious of itself, and is free in the highest sense; thus possessing the main attributes of a 'personality,' self-consciousness and freedom of will.

It is obvious from what we have said that the boundaries are far from sharp. We shall not discuss further the various varieties of these views, but shall only remark that a certain variety of theism, namely that called deism, again approaches closely to atheism (extremes frequently meet). For according to deism (particularly popular in the 18th century), the whole world is supposed to be an enormous machine, which plays its tune mechanically (see page 220), the deist merely adding God, at the beginning of all things, as the builder and winder-up of the great world-clock. In this respect, deism and atheism are diametrically opposed to pantheism and, properly understood, theism, for these are agreed that God cannot be separated from the world in this drastic way, but that, as St. Paul says, 'In Him we live and move and have our being.' While pantheism identifies God entirely with the world, theism on the contrary regards the world as contained in Him, but by no means regards Him as entirely represented by this world symphony.

After these preliminary remarks, we may inquire as to what the relations are between the mechanist-vitalist controversy, and these various philosophies of life. It is first of all clear, that mechanism in biology, including the doctrine of spontaneous generation, contains no direct reference to any of the three categories of world-view. For nothing compels even the theist, let alone the pantheist, to adhere under any conditions to a dualism of matter and life in the world. If the theist regards the world as the act of God from the electron and atom up to mankind, and furthermore as timeless and eternal (see page 271), it is a matter of indifference whether God performs this act in one or in two stages. Theism does not consist in God having created life, while the world of dead matter goes on by itself, without him – anyone who is of this opinion is half a deist, and not logically a theist, at least not a Christian.

Theism simply consists in regarding the fundamental origin both of the inorganic and of the organic, whether these are two kinds or only one kind of things, as lying in a personal, that is, self-consciously acting Being. Any theist who imagines that his theism would be endangered if he no longer assumed that a particular act of creation by God of the first living being took place at a certain point in time of the earth's history, may equally well regard every physico-chemical process as an instance which tells against his belief in God. No one whatever doubts that the latter process takes place by itself on account of forces immanent in matter. But mechanists do not hold any other view than this concerning spontaneous generation. Hence if the latter contradicts theism, any physical or chemical process does the same (Dennert). If this is true of theism, it is still more true of pantheism; *biological mechanism is thus consistent with any world-view whatsoever.*

It might however appear as if vitalism, on the other hand, does not possess such neutrality, which naturally must always be regarded as a recommendation. It rather appears to involve theism, or at least pantheism, even if vitalism is limited to the problem of the origin of life, in other words, if it is limited to the rejection of spontaneous generation. This was what Haeckel meant when he said: 'to deny spontaneous generation is to announce a miracle.' We must however be careful here.

Let us therefore first assume that the origin at least of the first living cells cannot be explained by physico-chemical forces, and the question then arises: when did the first living cell come into being? The theist answers perhaps rather hastily: God created such cells once at a certain moment, which he believed to be the right one. But there is another answer, namely this, that life did not arise at a definite point of time upon this earth at all, but was there long before the earth was capable of sheltering living beings. It was always there, life is just as eternal as matter, and the question of its origin is entirely one with the question of the origin of the world itself, of material reality.

The doctrine that life exists everywhere in space, and that our earth was colonised by living germs from cosmic space, is called the doctrine of *panspermy* or the *cosmozoic hypothesis*, its chief defender in most recent times being Svante Arrhenius.[293] But while the earlier defenders of this doctrine had great trouble in rendering the transport of germs through cosmic space plausible, and were obliged to call in the aid of volcanic eruptions and all sorts of other things, Arrhenius has at least succeeded in pointing to a force which might perhaps be capable of transporting them. This is the pressure of light, calculated by Maxwell from the electro-magnetic theory of light and later proved experimentally by Lebedev. It is the pressure exerted by electro-magnetic waves on the object upon which they fall. Since this pressure of light changes proportionally to the surface, while gravitation varies as the volume, hence the former according to the second, the latter according to the third power of the linear dimensions, it is obvious that below a certain definite limit of size, the repulsive power of light pressure will exceed the attractive force of gravitation. Calculation shows that, for a spherical body of unit density, the two forces, assuming the sun as the source of attraction and radiation, will be exactly in balance when the body is about 0.00015 mm. in diameter.

So far, all this seems very plausible. But there are a whole number of serious objections to the hypothesis of Arrhenius. First of all, as far as we know, very few living beings exist, the smallness of which approaches this limit. We spoke above of Esmarch's *Spirillum parvum* having a diameter of 0.0001 to 0.0003 mm., and the existence of ultra-microbes considerably smaller than this, though often put forward as an hypothesis, is very doubtful, since there must have been many opportunities for discovering them, if they were fairly plentiful.[294] Nevertheless, we have a certain amount of right to assume that such beings capable of transport by radiation pressure exist. But there are other objections which we will state briefly:

1. The cold of cosmic space would kill bacteria on their long journey. According to Arrhenius, they would need about 9,000 years for their journey to the next solar system (α Centauri). It is true that, on the other hand, we know the lowest forms of life to possess a quite astonishing power of resistance to cold (see page 303). Bölsche regards this fact directly as an adaptation

to the extreme cold of cosmic space, and Arrhenius explains it by the fact that cold slows down all chemical reactions, approximately in a geometrical series. When the temperature falls by 10 degrees a reaction is slowed down by half, and a fall of 100 degrees slows it down $2^{10} = 1024$ times. This point is nevertheless doubtful.

2. Bacteria spores would die in absolutely dry cosmic space from loss of water, even if they were capable of enduring a journey lasting 9,000 years.

3. Intense ultra-violet light, with no atmosphere to act as a shield, would kill all bacteria. Arrhenius's counter-argument that this action only occurs in the presence of oxygen, is of doubtful validity.

4. The formation of such spores assumes highly organised beings, whereas they are supposed to be the most primitive germs of life.

5. Special hypotheses are necessary in order to explain the first escape of these germs from a planet peopled with life.

6. This assumption would lead to the atmosphere still containing germs coming from outside. But all experiments hitherto have shown that it is sterile above about 6,500 feet, while the few germs occasionally found at a higher level prove to be varieties known on earth.

From all that we have said, it does not appear surprising that numerous scientists, including those who have no prejudices as regards a view of the general question of life, reject the whole doctrine of panspermy. There is a further point to consider. It will be said, namely, that the whole doctrine is not even a solution of the problem which it intends to solve. For transferring the origin of life from the earth into cosmic space does not make it any more explicable. If we cannot understand how the constituents of protoplasm could unite of themselves on this earth to form a living cell, this union does not become any more probable in cosmic space. It is also no answer to say that we do not know what are the conditions elsewhere – on the contrary, they are certainly as favourable as possible with us. For it is not a matter of other conditions, but of the highly improbable coincidence of a vast number of necessary conditions (see pages 328 f.). The production of a poem by shaking up letters in a bag, is no more probable in space than on the earth.

All these objections are perfectly valid. The doctrine of

panspermy merely pushes back the unsolved problem of the
origin of life, it does not solve it. But it certainly gives it a
different complexion, inasmuch as it puts the problem of life
directly alongside that of matter in the whole problem of contingency (see page 214). If vitalism is right, the world is
contingent again in another relation, and this, and not the
temporal beginning, is the chief thing. If we regard the matter
in this way, it is more easily seen that vitalism by no means leads
directly to theism. To conclude the existence of God from the
proved duality of the world would be to adopt the old 'cosmological' proof of the existence of God, the essential of which is:
if the world is as it is and not otherwise, it must have had a Creator
who made it as it is and not otherwise. As is well known, this
conclusion can easily be carried to an absurdity by inquiring as
to the creator of this Creator, and so on *ad infinitum*. The whole
proof therefore only shows in the plainest possible way, that
thought finally stops at some datum or another (see, however, the
limitations to this pointed out on pages 215 f.), and it is then a
matter of taste whether this final datum is assumed to be the
world itself, or a blind original urge, or a highest consciousness,
unless something further can be made of it from an entirely
different point of view (namely from the point of view of judgments of value, see pages 594 f.). We may here remind the reader
of Faust's four different translations of the first verse of St. John's
gospel.

The doctrine of panspermy deserves the credit of having made
this state of affairs particularly clear, even if it may have
to be given up. In the usual version of theistic doctrine of
creation, the only permissible standpoint for regarding these
matters is more easily lost sight of than is compatible with a
rational view of the world. Believers tend involuntarily to
regard the creation of the world as occurring at an unimaginable
distance of time previous to the creation of life, and hence to lose
sight of it altogether. Life appears as not only a new and higher
stage, but also directly as one more divine and worthy of God,
as compared with dead matter. The latter, the cold, empty,
rigid *hyle*, assumes an ungodly, even an impious, appearance.
Motives from older, even the oldest, dualistic philosophy* enter

---

* Persian religion, Neoplatonism, Gnosticism, mark the chief points on this
stream, part of which finally runs into Christianity.

in, and the contrast is complete between the divine 'life,' and the godless 'dead matter.' There is really no need to point out that this is entirely wrong from the point of view of theism properly understood; nevertheless, it is repeatedly forgotten. It is therefore good that the doctrine of panspermy makes the whole situation clear at a single blow. The fundamental question, as regards our view of the world, is not when did life arrive, but whence did life and matter, whether these be one or two stages in the world, come. *The usual form of the question, namely, whether life is eternal, or whether it was created by God, is just as false as the analogous question mentioned above* (page 270) *regarding the physical world.*

Having thus arrived at the result that science is justified in entirely omitting from its purview the metaphysical matters in question, and having thus freed ourselves from all sidelong glances after such consequences, we may on the other hand now take the liberty of following a few thoughts on the matter a little further, without thereby attempting anything of the nature of hasty dogmatisation. What is the total impression which the world makes upon us, after what we have heard about its material and its living side ? I think we may surely say this, that it may also be regarded as a whole, as a 'psycho-physical' system which, both from the physical and from the psychical point of view, shows a structure in stages, which lead from lower to higher units. In this sense it is a cosmos and not a chaos. *From the atom to the world of fixed stars, from amoeba to humanity, there is an almost uninterrupted series of steps in the formation of ever higher and more comprehensive wholes.*

How this is actually to be understood in the psychical field, we do not know, at least for the present; the problem of the relation, for example, of the cell soul to the soul of a whole individual is, as we have said, inaccessible to us for the present. We can only suppose that the psychical side of the lower stage in some way or other contributes to that of the higher stage, but certainly not by a simple law of addition like that of forces in physics, but otherwise; perhaps in some such way as, to use a metaphor, single notes together constitute a melody which, without doubt, is also something new and whole as compared with the notes singly, although it is made up of them.[295] It is this new thing which was designated above by the expression 'Emergent'; but when we

EES

regard it from the point of view of timeless eternity (see page 271) it may equally well be described as 'entelechy.' In this great world symphony, man finds himself immersed as a minutely small fraction of it; but on the other hand he is nevertheless able – and this is almost the greatest mystery of creation – to recognise his own part in it. Alongside himself he finds the other living beings, 'each after his kind,' that is, each as a peculiar and characteristic idea, if we regard matters for the moment from the point of view of theistic faith. *The whole world thus becomes a hierarchy of objectivised ideas.*

The question whether, and to what degree this system might possibly be reproduced *a priori* by human logic, has already been discussed at length above. No further proof is necessary that this is quite hopeless at present in view of the multitude of living forms. For the present, we still stand like children before a table full of Christmas presents, and wonder ever and again at the infinite variety of forms which life has brought into being on the earth. But when we also remember that the same will to existence is also present in all this infinitude of living beings, the statement made above, that the world is a system of objectivised ideas, cannot but be felt as much too abstract and lifeless. The world is not *Logos* alone, it is at the same time, and perhaps in its deepest nature, *Eros, it is reason and will in one* – this truth has become clear to most metaphysicians who have thought about the cosmic problem.

It appears to the author, who in this matter follows entirely in the footsteps of von Hartmann, to be wrong to lay especial stress on one rather than the other. Hegel and Plato were too exclusively logicians; Fichte, Schopenhauer, and Bergson were too voluntarist. The truth is that there is no reason without will behind it, and no will without at least some glimmering of reason. Both are abstractions; reality, wherever we are able to take hold of it, is always both at the same time.

At a later point, we shall once more return to these ultimate metaphysical questions, and to the problems of value which they raise. For the present, as this short discussion will no doubt have shown the reader, there remains only one question to discuss, namely what we know concerning the logic of organic forms. In other words, can we attain to any kind of knowledge, as to why these and no other forms of life exist upon our earth ? This last question may be attacked from a purely speculative point of view, that is to say, as a search for possible purely logical reasons that, if living beings exist at all, these must have precisely the forms which exist, or at any rate their general types. However, the vain speculations of earlier centuries have shown that no real knowledge is to be gained in this way[295a]. Such knowledge first arose in the moment when the question was no longer attacked systematically, but historically; by asking whether we may possibly find an intelligible view of the origin of species, genera, families, etc., as a process taking place by natural law. To this question we must devote a few sections of our book, on account of its importance.

VIII

THE PROBLEM OF SPECIES
AND THE DOCTRINE OF DESCENT

In the discussion of this problem, we now assume the existence of life, and we only have to deal with the question whether we can form a picture of the development of this life into the overwhelming multitude of individual forms. A few years ago, this problem formed the centre of all general philosophical discussion concerning biological matters, and the first editions of this book were written under this influence. This may perhaps have influenced too strongly the manner of statement, but we will return to this point later on. Anyhow, an almost sudden reaction has taken place in biology, and the number of biologists who are busy with these questions is at the present time very small as compared with thirty years ago, when all the world was constructing genealogical tables, and giving 'evolutionary explanations' of the biological phenomena which we observe to-day. That has happened which always happens, when a matter is approached with too great expectations; the subsequent disappointment is greater, the higher the original hopes, and the result is that the child is all too easily emptied out with the bath. No good whatever is seen in the thing that before was praised up to heaven. This is true of political constitutions, as well as of man and of scientific theories.

In the above we have already mentioned a small part of the history of our problem, its very latest phase. We need to say a few words concerning its previous history, if we are to understand to-day's position.[296] The first who tried with success to bring order and system into the apparently overwhelming multitude of living forms was the well known Swedish biologist Linnaeus. His exact observations appeared at first to confirm everyday experience that children are always like their parents apart from small and unessential differences, and so Linnaeus put in the forefront of his systems the law: *Tot sunt genera et species quot ab initio creatae sunt* (there are as many genera and species, as were created from the beginning). Although he later noticed a very definite mingling of the boundaries in many families of plants (for example, in the case of blackberry bushes), he held in principle to the law of the constancy of species, and his successors, for whom he very soon became an authority, followed him in this respect.

A new and decisive impulse was given to the problem only when palaeontology developed in the course of the 18th century.[297]   While hitherto the fossil remains of former plants and animals had been regarded merely as the play of nature, it was now gradually recognised that they were really the remains of living beings.   But for the moment people were satisfied with the explanation that they had been drowned in the Deluge and that in this way, for example, shells came to be found at the summits of the Alps.   Cuvier (1769–1832) was the first to alter entirely the view of this question, and he must be regarded as the real founder of palaeozoology. Cuvier, who had a gift unheard of in those days, for identifying the bones of animals, recognised that fossils were mostly beings totally different from those at present living, and that the earth therefore must have been peopled by a succession of quite different plants and animals.

Unfortunately, he associated this entirely new discovery with an untenable theory of the history of this lost world, the 'catastrophe theory,' according to which there had been successive periods of creation, at each of which an entirely new flora and fauna had been produced, to be radically destroyed each time by great catastrophes – Cuvier was thinking of volcanic events – the earth being re-peopled after settling down, by a new creation.

An immediate and energetic opposition to this doctrine arose in geology from the 'neptunists' (Lyell), who ascribed the transformation of the earth's surface in the main, not to volcanic action, but to the slow but continuously effective activity of water (in which, as we know to-day, they were right); and in biology, Cuvier was now opposed by the founders of the theory of descent, Geoffroy St. Hilaire, Erasmus Darwin, Oken, Treviranus, and above all Lamarck.   It is well known with what interest Goethe followed the development of this battle in the last years of his life. We shall have to return to the most important work of this epoch, Lamarck's *Philosophie zoologique* (1809), later on.   For the moment the new doctrine proved to have too small a foundation in experience to make its way into science itself; while in philosophy, the speculative school of Fichte, Schelling and Hegel reigned alone, and was the last to be willing to listen to such a realistic interpretation of the connexion between living forms (which they regarded purely ideally); an interpretation which seemed to them crudely materialistic.[298]

A change of view was first brought about by the work of Charles Darwin, *On the Origin of Species by Means of Natural Selection* (1859).   The further development is generally known and does not need to be discussed here. In these circumstances, the victory of the doctrine of descent meant the same thing as the victory of Darwinism, that is to say, the theory of selection.   It was only much later that it was seen that one does not follow directly from the other.   And while the last decades have brought a fairly sharp criticism of Darwinism itself, that is, the doctrine of natural

selection, to which we shall refer at length later on, the fundamental idea
of descent has, in general, maintained its position.[299]

With this we may leave the historical question and turn to the
matter itself.   The facts to be dealt with are as follows: The
species of organism known to-day show an undoubted systematic
connexion with one another, which is described by the words
'nearer or more distant relationship.'   In other words, there is
a 'natural system' both of plants and animals, although it is not
always so easy to find straight away.   This systematic relationship requires a reason to be found for it, just as every other single
fact of nature.   We cannot be satisfied to say that it just is so,
that we are naturally compelled to include the beast of prey, for
example, in the families of dogs, cats, bears etc.; we are obliged
to ask why these various species within a certain genus or family
are so like one another that we are obliged to collect them
into groups of this sort (here again, mere nominalist 'thought
economy' does not get us any further).

On the other hand, palaeontology teaches us with equal certainty, that at earlier periods in the earth's history, other quite
different varieties of plants and animals peopled the earth,
though the general style of them began to approach that of our
present-day flora and fauna in the tertiary period.   Not only
families, but whole classes and orders have died out, and new
ones have arisen, and taking it all in all, in the earliest periods,
the systematically lowest groups were as much in the majority as
are to-day the higher and highest.   These facts of palaeontology,
which are beyond all doubt, require an explanation.   We are
promised an explanation by

The Doctrine of Descent

This explains the correspondences of systematic relationship
by a descent from common ancestors, by common inheritance
therefore, and the differences, which also exist, by assuming
that along with the conservative effect of inheritance, other factors
were at work, which produced changes.   Although at first only
small in degree, over long periods in the earth's history these
finally result in very great changes, as is shown by the facts of
palaeontology.   This fundamental idea of every doctrine of
descent must be firmly laid hold of.   Without it the whole

doctrine is purposeless. For what it has to explain, is on the one hand the facts of systematic relationship, and on the other hand the transformation of species in the course of the earth's history.

We cannot make it too clear from the start, that these fundamental assumptions of the doctrine of descent involve no special presumptions concerning the way in which the transformation of species has taken place in individual cases, and concerning the causes of the changes. These two questions, the problems of *genealogy* and of *factors*, are questions by themselves, and although every theory of descent is naturally compelled to raise them, the doctrine of descent itself does not stand or fall by the answers that are given to them.[300] This remark is so important, because as already mentioned, the doctrine of descent was at first regarded by general opinion as, so to speak, identical with Darwinism, which in reality was a special theory of the factor problem, namely the theory of selection.

Hence even to-day, both friend and enemy – at least outside the ranks of biologists – regard all criticism of Darwinism as a criticism of descent itself. Obviously this is totally false, since it by no means follows *a priori*, that the transformation of species assumed by the doctrine of descent must have taken place by means of the factors assumed by Darwin (selection etc.). We shall therefore separate these three problems here strictly, that of descent as such, its various modes, and its cause: for the present we will consider only the first.

What reasons can the doctrine of descent produce in its favour, and what objections have been made to them. We will begin with an objection which has been raised frequently enough, and may even be raised to-day by rabid opponents of the whole doctrine of descent. It is said that this doctrine starts from two mutually incompatible assumptions. On the one hand, it calls upon heredity to explain the correspondences between related species. On the other hand, it denies the validity of the laws of heredity by allowing species to be variable.

This objection is obviously pure scholasticism, useless playing with concepts. The doctrine of descent in no way maintains, any more than does the whole science of heredity, that there are absolutely no factors which change inheritance, and that no processes of change in inheritance ever take place, but rather presumes the presence of such factors (what they are is a second question).

By at the same time maintaining that the law of inheritance usually holds, nothing more is meant than that the causes of change act at any rate slowly and perhaps also only occasionally, so that over short geological periods of time, no apparent change will take place. Matters may be much the same as with human manners and customs, which are frequently repeated for centuries from one generation to another, but nevertheless slowly change. Here also, conservative forces and the forces of change act side by side without there being any question of contradiction. In any case, therefore, the doctrine of descent is not to be refuted in this simple manner. What positive reasons has it to show in its favour? We may fairly begin with the material of palaeontology, since

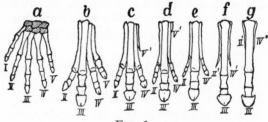

FIG. 67

Development of the hand in the predecessors of the horse. (From Hesse, *Abstammungslehre und Darwinismus*, Teubner, Leipzig)

this, in the nature of things, is the most direct means of proof. Palaeontology proves, as we have already shortly pointed out, a general upward development of species on the earth. The fact that single groups of a higher order appeared very early, and that conversely a few low groups that appeared very early are still living to-day, does not alter this general picture. But beyond this, palaeontology can show, at least in individual cases, direct series of steps in the transformation of certain organs, the best known being that of the horse, where we have a series of limbs of the ancestors of this animal (Fig. 67), and further, the famous Steinheim snails.

It is true that great care is necessary, for in numerous cases of this kind, in which a well ascertained series of forms was believed to have been found, the subsequent investigation showed that matters were considerably more complicated. We now distinguish, following Abel,[301] ancestral series, and step series. The former refer to the whole animal or plant, the latter only to a single organ. We can only expect to find an ancestral series when the whole development takes place in the same locality. But this will only rarely be the case, especially with animals which wander about, or are passively carried to other places. In such cases – which form the rule, the case may be, for example, that several branches proceed from one ancestral form, which we may express by the following scheme

Supposing now that a certain organ, for example the foot, changes in the first of the branches slowest, and in the last of them quickest, the change taking place approximately in the same direction, (see below page 475). c is then the most backward, g the most advanced in development, and we should have then, in comparing this organ only a step series, a—c—e—g. But if at the same time, c is most advanced in a certain direction in the specialisation of teeth, we should have an entirely different order a—g—e—c: any other organ may again give us another series of steps.

$$a \left< \begin{matrix} b-c \\ d-e \\ f-g \end{matrix} \right.$$

Such cross-overs in specialisation makes it very difficult to reconstruct the original course of affairs. To this must be added that, according to Mendel's laws above discussed, the formation of new Mendelian combinations in any case lies within the bounds of possibility, and that there is then no further possibility of an actual genealogical table; we have rather a network of Mendelian heredity factors all mixed up together.

These reasons are quite sufficient to explain why the great majority of palaeontological documents which were originally taken for ancestral series, have turned out on closer investigation to be merely step series. As far as I am aware, no uncontested example of a true ancestral series is known, even the one longest believed to be genuine, that of the Steinheim snails, has been challenged on several sides.[302] We are here plunged involuntarily into the problem of the genealogical tree. The reader will recognise why this is; in a scientific investigation it is impossible not to go into the detailed facts, since the total result is really determined by an exact and careful valuation of them. Hence all that we are able to say for the present is that palaeontology certainly gives us proof of the slow transformation of single organs in numerous step series. But what changes have taken place in the whole animal or plant, at the same time, must be specially determined to each special case.

A further objection to the doctrine of descent follows immediately upon this. If, as the older (Darwinian) form of the doctrine of descent hastily assumed, there is an almost continuous process of change, in which species made small steps from one millenium to the next, the sum of which small steps only became evident after periods of time – hundreds of thousands of years at least – palaeontology, it was objected, should present an entirely different picture from the one which it actually does present. For in palaeontology we cannot really find any traces of continuous transformations. We find again and again dozens or even hundreds of fossils of the same sort, but never all possible connecting links between different species.

When for example, the famous fossil *Archaeopteryx* was taken to be the link between reptiles and birds, one might really expect, if enough fossiis were found, to find all the possible intermediate stages of this transformation. But one might wager a hundred to one, that, if a dozen new examples were found, these would all belong to the type already known; or perhaps, in part, to some other similar type, but then again to the *same* other type; for the whole of palaeontological facts are of this nature.

This is undoubtedly true, and hence the conclusion is also true, that the process of the formation of species cannot have taken place in the simple way assumed in the primitive form of the doctrine of descent. It is practically certain, assuming descent to be true, that the transitions took place in comparatively short time, if the change from one type to another was continuous at all; or that the transition was not continuous, but definitely discontinuous in nature; in other words, that the alterations must have taken place in fairly large jumps. The latter is the more probable, as we shall see later. But it is obvious that we can only speak with certainty when we have a fair grasp of the problem of factors, and we likewise must recognise, that while in principle the doctrine of descent as such is independent both of the latter problem and of the problem of genealogy, it cannot in practice be argued in detail without reference to thèse problems.

Taking it all round, it must be admitted that palaeóntology leaves the doctrine of descent, as regards the points last mentioned, very much in the lurch, although one ought not to assert that what has been found contradicts descent. But descent receives an argument in its favour from a further group of facts, to which we must now turn, namely the investigation of the present form of organic life. We shall see that our palaeontological knowledge is still very faulty, so that the absence of connecting links, as is very often the case over very long stretches, is easily understood.

This second argument consists in the citation of a large number of anatomical and morphological facts which cannot be understood in any way than by the assumption of a previous extensive reconstruction of animal or plant organs. These facts may be divided into three sub-groups which are, however, connected together by links. The first group is formed by the facts of comparative anatomy, the second by the so-called rudimentary organs (which may also naturally be reckoned as a part of comparative anatomy), and the third by the facts of embryology. A few short remarks must suffice here, since these facts are in all biological school books. If we compare, for example, the fore-limbs of the various groups of vertebrates, particularly the mammalian orders and families, it strikes us at once that they are all built on the same lines, no matter whether they form ordinary legs for walking, or wings (birds), flying membranes (bats), flippers (penguins), excavating organs (mole) etc. We find always the same bones in them, although often misformed and degenerate, the toes only varying, according to necessity, between one and five. These relationships cannot be understood in any other way than by assuming that the specialised organs have developed from a more or less fundamental type.

These transformations or organs may, in certain circumstances, also take place in a negative direction, the organs becoming almost or quite incapable of functioning, if they are not used. We then speak of rudimentary organs, such for example as the eyes of the mole (in the case of south European species they have quite disappeared, in our mole they lie under the skin), the shoulder bones of the blind-worm, the pelvis of snakes and whales, the legs of numerous varieties of lizards (chalcis, chirotes, scheltopucik), etc., etc. It would be simply incomprehensible from the point of view of a strict constancy of species, how these animals could have come by these poor remnants of otherwise useful organs, if they were not the remains of the former possessions of their ancestors.[303]

As soon however as we admit this, we have to admit, if we are consistent, that organisms are capable of exhibiting much greater variability than we should be inclined to suppose from our short experience. If a four-footed mammal can develop into so comparatively perfect a marine animal as a whale, no chance, however great, can be regarded straightaway as impossible. Between what the forefathers of the whale must have been, and the whales of to-day, there is a considerably greater difference than between the forefathers that the doctrine of descent has to assume for known species, and these latter themselves. Up to a short time ago, we actually had no palaeontological evidence of such a transformation of the whale. Only a few years ago, skeletons were found in Egypt of animals which must be regarded as the ancestors of the present whale, and show a fairly complete series of steps in the degeneration of the limbs. And yet, previously to this discovery, an examination of living animals positively forced us to conclude that transformation had taken place. Are we not justified in drawing by analogy from this case a conclusion concerning the incompleteness of our palaeontological knowledge?

The phenomenon of rudimentary organs is certainly the most convincing argument in favour of descent, since transformation in this case—and in some cases a very thorough transformation—is entirely obvious. The most convincing case of this argument is where the reduction of the organs takes place, as it were, under our eyes during embryonic development. Thus for example, calves have as embryos the normal number of teeth in the upper jaw, which however disappear before birth (the upper jaw of full grown cattle is only covered with a plate of horn, which has quite a different origin). In the same way, the whalebone whales have as embryos at least sockets for teeth, whereas the grown animals have whalebone, that is to say, horny projections from the gums, which serve as sieves and have nothing to do with teeth.

We have a further particularly striking example in the rudimentary hind legs of the embryo viper, which in this case likewise entirely disappear, whereas adult giant snakes still retain rudiments of them. Furthermore, the development of the circulatory system (to be found in any text-book of embryology or general biology) in the higher vertebrates (mammals, birds) shows in a very characteristic manner the first beginnings of the gill arteries of a fish with two or more branches on each side of the aorta, of which subsequently only one remains on one side only.

These facts of embryology are more generally formulated in Haeckel's fundamental law of biogenesis. This states that *ontogeny is a repetition of phylogeny*; more correctly that during the development of the individual we frequently find striking echoes of the ancestral stages postulated by the theory of descent.   The most famous example of this kind is that of the transformation

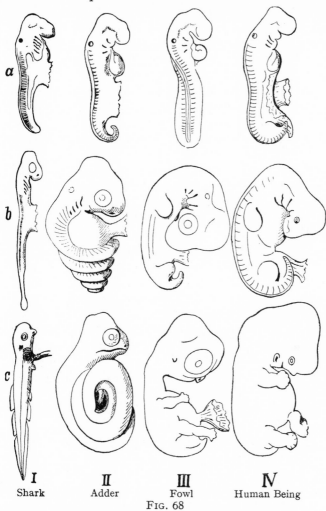

a

b

c

I	II	III	IV
Shark	Adder	Fowl	Human Being

Fig. 68

Embryos at three different ages (from Hesse)

of the gill clefts connected with the above mentioned reconstruction of the circulatory system, which is also plainly seen in the case of the human embryo (figure 68).

Connecting all these facts – a true impression of their full implication can only be got from closer study, and not from the few references which we have space to give here – we receive, taking it all in all, an overwhelming impression of the indispensability of descent as the only possible explanation of the facts. This is not changed by the discovery that the 'biogenetic' law above mentioned is not (as Haeckel thought) a strict law, but really only a rule, to which there are many exceptions. Cases analogous to those mentioned are much too numerous to depend upon chance, and we are led by this road to far-reaching conclusions concerning the possible ancestral stages of the beings in question, though we have to bear in mind that such conclusions, since they are for the present hypothetical, need further confirmation by other means. In this respect, the doctrine of descent forms no exception to the general rule that scientific knowledge can only be reached by the convergence of results to some degree of certainty.

What we have said hitherto practically exhausts the direct grounds for believing in descent. To these must be added a whole number of more indirect and general grounds. Among the chief of the indirect (not direct, as many suppose) grounds is the so-called 'biological reaction' or 'blood reaction'. We have already spoken several times of the fact that every species possesses its own kind of protoplasm. It is not possible to give experimental proof of this by the usual chemical methods, since we have too little knowledge of the protein constituents of protoplasm; but we owe to Uhlenhuth, Friedenthal, Nuttal, and others, methods which allow individual forms of protoplasm to be recognised no less sharply than by a chemical reaction. These methods are, by the way, also used in forensic medicine, to distinguish, for example, between human and animal blood. The process is, shortly, somewhat as follows.

It had been observed, that, when an animal is injected with blood from another species, more or less serious toxic phenomena appear, similar to those produced by the invasion of microorganisms (pathogenic bacteria). We know to-day (see above, page 307), that, in the latter case, the toxic effects are produced by certain substances (toxins), secreted by the bacilli. In all these cases the body fights against the attack by producing what are called 'antibodies' (antitoxin, precipitin and bacteriolysin, agglutinin etc.), which in some way or other render the poison introduced, or the bacteria producing it, harmless. Since these antibodies are mostly produced in excess, the organism is then armed against further injury of the same kind, it acquires 'immunity.' We are not yet able to isolate these antibodies from the blood of the injected animal, but they can at least be obtained in a clear colourless solution, when the blood is freed from its suspended red corpuscles, say by contrifuging. The resulting colourless liquid is called blood serum (hence also diphtheria serum, for example).

Let us now suppose an experiment carried out as follows. A guinea pig is injected with a dose of human blood, naturally not large enough to cause its death from the poisonous properties of human blood for guinea pigs. It then overcomes the poisonous effect by the production of antibodies in its own blood. This is several times repeated with ever increasing doses. The animal finally endures injections of a quantity of human blood sufficient to kill ten or even more animals not accustomed in this gradual manner to it. Its blood, which is now rich in antibodies protecting against human blood plasm, is freed from its solid and we have these antibodies in a clear serum. If this serum is now mixed with equally clear human blood serum, a turbidity is immediately formed and a precipitate falls down. This reaction does not take place when the guinea pig blood is taken from animals which have not been previously treated with human blood.

The reaction is to be regarded as the effect which the protective antibodies also have in the blood of the living animal upon the proteins from another species (man). They throw it down in an insoluble and hence uninjurious form. The same experiment has been carried out with a large number of different animals, and it has been found that the reaction is always specific, that is to say, that it only indicates the particular foreign blood with which the experimental animal in question has been previously treated (in our example, human blood).

And all the same, the reaction is not strictly specific. If the blood serum of a species closely related systematically to the species used for treatment is taken, the reaction takes place in a less degree. If the guinea pig, for example, has been treated with dog blood, the blood reaction also takes place with wolf blood, and also to a less extent with fox and jackal blood ; indeed, the degree of a turbidity indicates, generally speaking, the degree of the systematic relationship. Thus the reaction takes place almost as strongly with the blood of the anthropoid apes as with human blood. It is true that there are certain exceptions to this general rule.[304]

It is naturally impossible to derive from these blood tests, as the popular fanatics of descent have done, direct proof of the relationships of organisms. For the blood relationship which they prove, that is to say the chemical similarity of blood, is of course something quite different from the blood relationship of the doctrine of descent. To confuse one with another is to be guilty of a *quaternio terminorum*. It is equally inadmissible, also, to regard it, as certain anti-descent fanatics have done, as a direct proof of the fallacy of descent, the argument being that the blood reaction proves the chemical difference even between closely related species as regards their protoplasm. There is of course no continuous transition between two different kinds of molecules, thus grape sugar is grape sugar, and vanillin is vanillin, and nothing else. Hence, it was argued, there is no transition from one species to another.

This conclusion is absurd, for the reason that the whole of chemistry has to do with nothing else but the transformation of one substance into another. And even continuity might be retained in the process, if a molecule A were transformed in increasing proportions into a molecule B, so that between 100% A and 0% B, and 0% A and 100% B there would be a continuous range of steps. We must even quite seriously assume such transformations in the mutations which we shall discuss later.

A further indirect proof of the doctrine of descent – when considered more closely, one of the most striking – is the way in which it illuminates numerous facts of the geographical occurrence of animals and plants, more especially the flora and fauna of isolated islands. Also, the geographical distribution of animals and plants at sharp boundaries such as are formed by high mountain ranges (the Andes), or seas or arms of the sea, offer very cogent examples. A well known case is the correspondence between the American and African mammalian fauna (camel – lama, lion – puma). Much more fertile in results is a comparison between the mussels in the gulf of Mexico with those on the other side in the Pacific, or a comparison of the flora on the two sides of the Andes. The doctrine of descent, combined with geology, easily explains all these matters by the assumption of common ancestors at a time when the cause of separation did not yet exist.

Let us content ourselves with these indirect proofs, although there are many others of the kind, and ask ourselves finally, what general grounds can be brought forward in favour of, or against, descent. Here we are more than ever compelled to say, that every general and unprejudiced consideration speaks on behalf of the only doctrine which could lead to a scientific comprehension of the way in which the world of living organisms has come into existence. Since the whole of biological research has proved the absolute validity of the law *omne vivum ex ovo*, and since spontaneous generation could only be held as, at the very most, a hypothesis concerning the first production of the simplest cells, we are left with no choice regarding the origin of species but that between descent or continued acts of creation. There can be no doubt which of these possibilities most appeals to science.

But we may also consider without prejudice, what advantage religion might be supposed to gain by sticking to the second supposition. Is it really so particularly religious to imagine God as continually interfering afresh with creation ? Is theism supposed to consist in God entering as often as possible as cause, where our knowledge fails us, and have therefore the defenders of theism nothing better to do than to track down as many such ' inexplicable' secrets as possible, and proclaim them to the world ? This is not in reality theism, but multiplied deism. We have already said above, all that need be said about this matter, and will spare ourselves the trouble of repeating it here. But I cannot suppress the observation that it appears to me an idea little worthy of a Christian conception of God, when it is supposed that he created the aforementioned whales, snakes, calves etc., with all their rudimentary organs and their embryological degeneration, exactly as they are. What is supposed to be His purpose ? Perhaps to lead the scientists of our day into certain and unavoidable error ?

If there are no other reasons in the narrower sense of Christian dogma which oppose the adoption of descent, particularly for mankind, it is not easy to see why so much opposition has come from this side even up to to-day. There is not the slightest reason for theism in general to dictate to God in what way He is to produce individual species: the decision on this point, like any

F Fs

question regarding special things in the world, may be left to science.

Our whole previous discussion has been concerned solely with descent as such; that is, with 'whether or not', but not with 'how,' and 'by what factors.' Let us consider these questions somewhat more closely. As regards the 'how' of genealogy, a few words are enough. It is relatively a matter of indifference, from the general philosophical point of view, how we are to arrange individual genealogies – it may be left to science to produce the results which appear most probable. At present things are in confusion in this respect, as soon as we leave the narrower circles of the families and certainly when we go beyond classes; the results of the embryologists, the systematics and the palaeontologists not infrequently contradict one another directly.[305] For example, there is still no agreement yet on the question as to whether birds descended from reptiles, or whether both descended in parallel from common ancestors, although the first hypothesis is preferred.

It is perhaps false to make the common assumption of a tree (genealogical tree) branching in the usual manner, as already pointed out on page 441, for we might also imagine that the common characters of two different species (say lion or puma) go back to original like inheritance, but that the differences between them by no means come from the same source, modified by circumstances, in one case in one direction, in a second in another. These differences might very easily be inheritances coming from different sources, which have combined in one case in one way, in another case in a different way, with the common inheritance. The genealogies of the two species, when followed backwards, would then converge as regards common qualities, but would diverge in respect of other qualities, so that there could be no question of a simple division.

In addition to this branching and crossing over of the heredity, we should have the fact that both what was common, as well as what was not common to the two species, could be altered by environment in a manner to be discussed below. A multitude of possibilities thus exists, which have not been considered by the ordinary form of the doctrine of descent, and we must therefore admit that the genealogical trees, once so cheerfully constructed, give a picture which has little correspondence with reality. But we may leave this question to take care of itself, as also that as to whether the whole organic world has sprung from one or several roots (mono- or poly-phyletic hypothesis). For the present we cannot say anything at all with any certainty. *Ignoramus sed non ignorabimus.* But the most important question of the theory of descent is the second question, which we will now have to consider more closely.

IX

THE DRIVING FORCES
IN THE FORMATION OF SPECIES

All investigators to-day (with scarcely an exception) are agreed concerning the truth of descent. But when we come to the causes of transformation, opinions differ entirely. Only the total neglect of these, as well as all other deep biological questions in the curricula of most of our higher schools,* explains the depressing fact that the great majority of educated people are not in the least clear concerning the great differences between the question of descent generally and that of its causative factors. For most of them, the doctrine of descent means the same as Darwinism, that is, a quite definite answer to the particular problem which we are now about to discuss. Still worse, naturally, is the definition, still very widespread, of descent 'or' Darwinism as the doctrine that men are descended from monkeys. As if there were only monkeys and men in the world of living things – quite apart from the fact that hardly any theorist of descent to-day would uphold such a view.

We take up again at this point the short historical summary given above. The first attempt at a more general foundation of the theory of descent was made, as already mentioned, by Lamarck in his *Philosophie Zoologique* in the year 1809. He attempted to give an account of the manner and the means used by nature in the process of transformation. Lamarck's explanation starts from the observation that the use of organs strengthens them and their disuse weakens them. We may think of the powerful arm muscles of gymnasts and athletes; on the other hand of the weak muscles of lazy people, etc. These observations lead Lamarck to imagine the production of the giraffe's long neck as follows: at a time when a great drought existed in parts of Africa, the ancestors

* But see, as regards England, *Biology and Mankind*, by S. A. McDowall, Chaplain and Senior Science Master at Winchester College. [*Trans.*]

of the present day giraffes stretched their necks as far as possible in order to reach the higher leaves of the trees. Those which were most successful in doing this had the best chance of survival and of producing offspring, other things being equal. The latter inherited a small part of this power of stretching the neck which had been produced by practice, and the process being repeated in like manner, the length of the neck continued to increase until it finally reached its present abnormal size. Along with such a change, also, as Lamarck likewise pointed out, many other necessary changes (co-adaptations) also take place, in the case of giraffes for example, changes in the build of the body and the forelegs, which then are also bred in the same way.

The reasons why Lamarck's work did not have wide-spread success have been shortly indicated above. Success was first obtained for the doctrine of descent by the work of Darwin, which appeared exactly fifty years later. Darwin's fundamental idea is the principle of the selection of the fittest by the struggle for existence, shortly called the selection principle. Illustrating this in connection with Lamarck's example of the giraffe, Darwin does not start with the direct adaptation of the individual, but with the fact that among young of the same parents there will always be some animals having longer necks than others, since none of the descendants is ever exactly like any other. Those having the longest necks have the greatest chance to survive. And thus in the next generation there is already a slight increase in the average length of neck, and this increase is maintained by inheritance, and still further enlarged by selection.

The observational material upon which Darwin based his work is extraordinarily rich, and this accounts in part for his success. Individual variation is an unquestioned fact, and it is equally unquestioned that, of the enormous excess of offspring produced in almost the whole of the plant and animal world, only a very small fraction can survive and attain to reproduction, for otherwise the offspring of a single species would very quickly fill up the whole earth. There must therefore be wholesale destruction by injurious influences of every kind, and therefore, the assumption that those individuals best suited to the circumstances would on the average have the best of it, is certainly not to be dismissed straight away (see below, page 479). The conception of the struggle for existence is also to be taken as widely as possible.

It may exist also in the case of entirely pacific animals, and of plants, for the simple reason that one may take light, air, space, and nourishment from another.

Apart from these two fundamental facts, Darwin also brought forward a large number of particular observations, above all, the experiences of the breeders of animals and plants, who likewise obtain their results by the process of selection ; and furthermore a large number of individual adaptations and geographical facts concerning plants and animals, etc., into which we cannot go here. Almost at the same time as Darwin, and independently of him, another English worker, Wallace, also thought of the principle of natural selection.

The incredible rapidity with which Darwinism – and in those days this meant the same thing as the doctrine of descent – won the day is only to be understood, when we take into account the whole intellectual position of that time; the collapse of speculative *Naturphilosophie*, and the penetration of pure empiricism into science, with a resulting return to materialistic tendencies; the universal progress of industrialisation among European nations, with the corresponding appearance of the 'social' question. Taking it all together, it was a new philosophy of life which appeared to be then in progress of formation, a philosophy of the pure 'here and now' of pledged enmity to all speculative philosophising, to all belief in transcendental reasons or purposes of the world, – at the time of a determined struggle for existence in economic life as well.

Friends as well as opponents of this view of life recognised the importance to be attached to Darwin's principle, which undertook to explain the production of useful adaptations by mechanical means, by the blind play of chance and automatic necessity. In a short time therefore the battle was being conducted with a fierceness which the intellectual history of mankind can only show in rare cases, and its repercussions still continue up to to-day. It is not surprising that both main branches of the Christian church attacked the new doctrine in the sharpest possible manner, since it had been greeted with jubilation from the very beginning by all opponents of ecclesiastical teaching, although this was certainly far from Darwin's intention.

On the other hand, this resistance of the Churches to what was in any case a scientific doctrine deserving of serious consideration, produced a rapid desertion by educated people of organised religion, while at the same time the working classes accustomed themselves to regard Darwin's doctrine as the scientific foundation for the teachings of their two prophets Marx and Lassalle (economic materialism).   It is well known what an important part was played in this whole development by popular philosophers and scientists such as L. Bückner, Vogt, Moleschott and above all Ernst Haeckel.   The few in the scientific camp who stood out against this general current (Wigand, Nägeli and others) were passed by or dismissed with a shrug of the shoulders.

The doctrine of descent, and Darwinism, melted into one as far as the public was concerned, and Darwin's original scientific hypothesis became overnight a world principle, which was not only to explain the formation of species, but finally, almost the whole world process.[306]   In particular, it was soon regarded as settled, that the whole of human civilisation was also a matter of development with selection of the fittest.   Art, morality, and last but not least science itself, appeared as the products of selection in a struggle for existence.   Pragmatism, an American doctrine, was the final development of these ideas in epistemology.   (According to it, 'taking anything to be true' is a biological reaction; we hold anything to be true, if and because this behaviour is useful to us or to the species.)   We shall return to these general philosophical conclusions later, but for the present will remain on the purely scientific ground.

Criticism dared to attack this apparently firm structure only timidly and slowly.   We must deny ourselves the interest of following it historically; we will instead deal with the questions purely on their natural order, and thus at the same time become acquainted with the other attempts to solve the problem of descent which exist to-day.   The fascinating apparent simplicity of Darwinism has long ago given place to an enormous complexity. Innumerable investigations, above all in experimental biology, but also in pure morphology, embryology, ecology, etc. must be taken into account to-day.   It is an extremely foolish and unjustifiable procedure, when in popular works matters are so stated, as if Darwin's principle of selection really settles all fundamental points. Without considerable knowledge, first of all of modern investigations into heredity, true insight into the problem is not to be thought of. We will first consider the problem of the inheritance of variability.

X

VARIABILITY AND ITS INHERITANCE

Naturally, we must limit ourselves to the most important matters.[307] Darwin's doctrine takes, as material upon which selection acts, mainly the so-called individual variation. Industrious investigation in the last decades has shown that this concept is not by any means unambiguous, but on the contrary, includes a whole number of different phenomena which depend upon completely different causes.

First of all we must note, that a large number of individual differences are to be explained, as discussed above under Mendelism (see page 366), as the breaking up or recombination of Mendelian characters. We then saw that entirely new species might arise by the production of new combinations of the inheritance units or 'genes' as is the usual expression. Differences of this kind are obviously transmissible in the same manner. On the other hand, the laborious investigations of Jordan, Johannsen, and others have brought to light a result of fundamental importance, that what we regard without question as a species of animal or plant, such as we have in nature in a bean field, or a flock of sheep, simply represents a collective notion. In reality, every such population is made up of a large number of elementary species and these again of so-called 'pure lines,' there being also a large number of mixed or hybrid forms. Since on account of sexual reproduction and the resulting continual mixture of inherited factors, the state of affairs in the case of almost all animals and most plants is almost impenetrably complicated, these experiments have been done by preference on such plants as are purely self fertilising, and hence do not produce hybrids.

In order to understand Johannsen's experimental results, which are also of importance for practical breeding purposes, we must discuss the notion of so-called 'fluctuating variations.' If we compare the progeny which result from a single bean after one or several generations (with strict inbreeding), in regard, say, to the length or the weight of the seeds, by dividing these into classes with a difference, say, of 1 mm. between them, we get the following figures, for example.[308] Of a total of 450 beans the following results were obtained:

Length .. 8 9 10 11 *12* 13 14 15 16 mm.
Number of
this length .. 1 2 23 108 *167* 106 33 7 1
(10 mm. means here between 9.5 and 10.5)

We see directly from this, that the middle length (12 mm.) occurs the most frequently, while those greater or smaller, the plus or minus variants as they are called, are the rarer the further the length in question departs from the average. Series of figures of this kind are also obtained by pure chance according to the laws of probability, for example in the apparatus shown in figure 69.

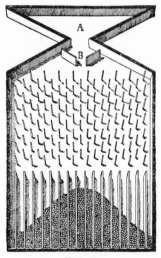

FIG. 69
Galton's chance apparatus.
(From *Hertwig*)

In this apparatus marbles are allowed to roll through the funnel at the top, and each one is able to go to the right or left of every nail that is in its way. The number of marbles which collect in the compartment at the bottom are those corresponding to the law first discovered by Quetelet (1871) in connection with variations of human characteristics. This law is easily stated in mathematical form (binomial law), and can be deduced from the theory of probability.

These fluctuating variations have for their prime cause the unequal amount of food obtained by each individual. But they are in no way inheritable, for the progeny bred from minus variants always gives the same curve of distribution as that derived from the mean or the plus variants, assuming the food conditions to be the same; this is true even of the most extreme variants at either end of the series, and is not altered by selection continued through generation after generation, as Johannsen showed in experiments extending over many years. More or less plentiful food supply naturally shifted the whole distribution curve in the direction of the plus or minus variants, but nothing was changed as regards inheritance.

A series of this kind is called by Johannsen a pure line, and he was able to show that an ordinary bean population consisted of many such pure lines mixed together, so that the individual variations that we observe in them are actually the result of two quite different phenomena mixed together. A certain length of bean may then be equally well a minus variant of one pure line, a plus variant of another, or the mean of a third, and conversely the difference in the length of two beans may either be due to their being fluctuating variants of the same line, or to their belonging to different lines. If now in the case of plants and animals which are not self-fertilising we consider the effect of the innumerable crossings of the pure lines, we see how extraordinarily complicated the position of affairs is, which Darwin regarded simply as the single notion of individual variation.

A further point must be considered. Darwin had noticed, that alongside apparently arbitrary variations which take place in every possible direction and which he assumed to be inherited, although he could not prove it, particularly large and one-sided variations occur. Darwin gives two well ascertained cases of this, which occurred in the years 1791 and 1828, namely the birth of two rams with quite abnormal legs in one case and abnormal hair in the other, their qualities in each case passing unchanged to their progeny, which they produced in great numbers, with the result that quite new races were formed. Such cases have been observed still more frequently in plant breeding; thus for example, the first double petunia is known to have appeared in 1855 in a palace garden, and it became the progenitor of all subsequent examples.

Sudden variations of this kind, which are inherited in a constant form and include many inherited malformations, are called to-day 'mutations,' following de Vries. He observed in a number of evening primrose plants (*Oenothera Lamarckiana*) in the neighbourhood of Amsterdam, the sudden appearance of quite a number of sub-species, the characters of which were apparently transmitted with constancy (figure 70, opp. page 461). He called these mutations, and built upon this observation a new theory of descent, distinguishing sharply inheritable changes of this kind and fluctuating variations which are not inheritable. The struggle for existence was only supposed to take place between the first. Following upon this, de Vries became one of the rediscoverers of Mendel's rules (1905).

As a matter of fact, later investigation of de Vries's original evening primrose has shown that in the case of this plant it was not in any way a production of new inheritance factors that took place, but a Mendelian rearrangement (Renner), and every case of the kind naturally gives rise to the same suspicion. But the possibility exists of deciding the question by careful hybridisation experiments, and in other cases true mutation, erroneously assumed by de Vries originally, has been found as a genuine and sudden change in the inheritance factors of single individuals forming part of otherwise constant stock.[309]

For these reasons a strict distinction is drawn to-day between the three following kinds of variability; first the fluctuating variations or 'paravariations' above discussed, which are not inherited, but simply different forms of the 'phenotype' (see page 460), resulting from different environmental conditions, the 'genotype' remaining the same. Secondly we have 'mixovariations', that is, Mendelian combinations; these are inherited, it is true, but they likewise offer no sufficient basis for a theory of descent, since they obviously never produce anything really new, but always new combinations of what already exists, though these combinations may be almost unlimited in number. Thirdly, we have real mutations or 'idiovariations,' that is real changes in the germ plasm, and these changes are inherited. It is clear that a theory of descent can only operate with inherited variations. The question which first concerns us is, therefore, how such new inheritance units can be produced, and how we are to imagine their production in the course of the earth's history. This question is of such fundamental importance that we must discuss it at greater length than in previous editions. The central point of it is the problem of the *inheritance of acquired characteristics*.[310]

In order to be able to regard this problem from all sides, we must bring forward some further facts, that might equally well have been given above under the heading of inheritance. We may first of all once more point out expressly that the genotype, that is, the totality of the inheritance factors of an individual, does not represent a sum of certain final characters, but a sum of 'potencies' which enable the individual in question to react in a certain way under certain definite conditions of environment.

We may give an especially impressive example of this. The Chinese primrose blooms, as we all know, in our part of the world in a white and a red variety (which can be used like Mendel's peas for hybridisation experiments), but by keeping the red variety in a hot house at a very high temperature, it is possible to obtain a white instead of a red flower. Its *phenotype* is thus changed ; it is now, apart from the small differences between the two varieties, like the phenotype of the other race. Nevertheless, the *genotype* is completely unchanged, and remains so for however long cultivation at high temperature is carried on. As soon as the progeny of the normal plant is brought back to normal conditions, the red flower appears again. We see therefore that the inheritance factor in question (the 'gene' as the Mendelian experimenter calls it) is not the red flower as such, but only the power of forming red at normal temperatures, but white at higher temperatures.

At the same time, the experiment is an impressive demonstration of the fact that, just as in Johannsen's beans and other hundreds of other similar experiments, the changes produced by environmental conditions (paravariations) apparently make no change in the genotype. It remains as it is, for however long a time the phenotype may be altered by abnormal conditions. This would appear to lead directly to the non-inheritance of acquired characteristics of the phenotype. But let us first see what further facts are known.

A vast number of experiments have been made concerning the influence of external factors, such as temperature, moisture, illumination etc., upon development, chiefly of insects, but also

PLATE VII

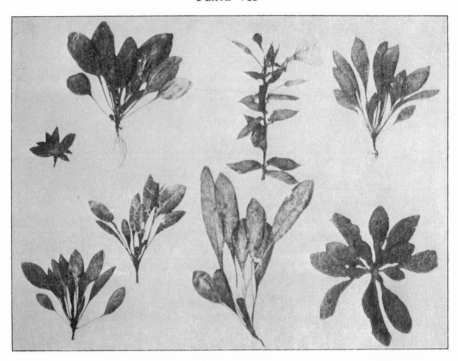

FIG. 70. Mutations of the Evening Primrose *Oenothera Lamarckiana*, according to de Vries. 1. Parent form. 2. *Oe. gigas*. 3. *Nanella*. 4. *Scintillans*. 5. *Oblonga*. 6. *Rubrinervis*. 7. *Lata*

FIG. 71. Temperature aberration of *Arctia caja*. A. Normal form. B. Pupa kept at −8°C. C. Offspring of B bred under ordinary conditions. (From Hertwig, *Werden der Organismen*)

of other classes of animals and plants. Kleb's experiments with barren privet have already been mentioned above. In the case of butterflies, the pupa stage is a particularly sensitive period (see above page 346). By means of abnormal temperature it is easy to produce so great an alteration in this case that the animals, when they emerge, would undoubtedly be considered as belonging to a different species, if their origin were not known (figure 71, opp. this page). Also, many cases of so-called 'seasonal polymorphism' are to be regarded as due to heat or cold, the most usual example of this being the well-known tortoise-shell butterfly. (*Vanessa prorsalevana*) which occurs in two quite different forms in spring and late summer (see figure 72, opp. page 466).

Other cases of the same kind can be referred to different conditions of nutrition and other causes. In the same category we can count many local modifications of plants and parallel phenomena in the animal world. But it is difficult to suppose that everything must here be referred to direct action. While such phenomena give the impression that the environmental factors produce direct chemical changes in the living substances which show themselves in the changed development, we must consider, alongside this, changes which are produced in the complete organism by such influences, although these are not usually very extensive. It is known that the chameleon is able to change its colour more or less in accordance with its surroundings; a similar effect is produced by their surroundings on many other animals, especially reptiles and amphibians.

Kammerer,[311] in particular, has tried to show by experiments with salamanders, which he kept on both black and yellow earth, that these animals adapt themselves in the course of some weeks more closely to the colour of the ground, the animals kept upon a yellow ground increasing greatly the size of their yellow spots, while these decreased conversely in the case of animals kept on black ground. These changes in colour did not occur when the animals were rendered blind, hence they must be produced by the sense of sight, and other observations of this kind have been made.

How these stimuli are able to influence the body cells in question in a suitable manner, we do not know. But the explanation will undoubtedly belong to the same chapter as the well attested fact that a hypnotist is able to produce in his subject by the suggestion of a blow a perfectly characteristic swelling, and by the suggestion of a burn, a genuine blister, which however heals more quickly than usual. The riddle of the psycho-physical problem is particularly evident in cases such as this. Probably also, the so-called functional adaptations belong to the same category. If, for example, a human kidney is destroyed, the other kidney increases very greatly in size automatically, and in a similar way, parts of the brain that have hitherto served for other purposes take over the duties of parts destroyed, say, by a stroke, so that often the patient may afterwards learn again to walk, talk, and so on, quite satisfactorily.

The question upon which the main interest is concentrated at present is therefore the following: *how far may the theory of descent reckon with the passage into the heredity, that is the genotype, of changes produced by environmental conditions?* Is this at all possible or is it possible for certain groups of such changes, and not for others? And if not, how have the changes presumed by the theory of descent in the genotype come about? In view of the enormous importance of this question, which, as we shall see further on, is in practice far and away the most important question of the whole of biology, it is absolutely necessary to define our concepts in the closest possible way, and not to exclude any possibility by which doubt might subsequently enter.

We may first recollect the separation, discussed on page 362 between the 'germ path' and the soma. As we then saw, it is certain that the germ cells remain isolated from the remaining body cells from the beginning of embryonic development, and that therefore they must not be regarded as an actual product of the body in which they are carried, but rather the latter as a kind of distegration product of the germ path, as is shown by figures 58, 59, page 363. Weismann, to whom we owe the first clear formulation of this state of affairs, already concluded that the direct transference of properties acquired from the germ cells to the body cells is unthinkable, that in other words an inheritance of acquired characteristics did not exist. He attempted to prove this conclusion by a large number of experiments. He cut off the tails of mice for many succeeded generations, without finding the slightest change in the inherited length of tail. It has already been mentioned that a similar result has been found in the case of modifications produced by nutrition. The question may, however, be raised as to whether the final results of these experiments settle the problem.   It might be that these experiments have been done in cases in which success was not possible.

The doctrine that, in spite of the negative result of these and so many other experiments, the inheritance of characteristics produced by environmental influence is possible in a certain degree and must be postulated, is called to-day Lamarckism, and Lamarckists are usually divided into mechano-lamarckists and psycho-lamarckists.   The first assume that, in a physico-chemical manner at present unknown and inexplicable, the changes produced in the soma affect the germ cells, in such a way as to cause the individuals produced from the latter to exhibit by inheritance the same changes as are exhibited in the bodies of the parents.

Psycho-lamarckists, on the other hand, consider this is only possible in cases where the changes are connected with psychical processes.   When an organism comes into a new environment of any kind, certain needs arise in it, which it feels, consciously or unconsciously, and these feelings, or whatever we like to call them, produce first of all in the body, and in certain circumstances in the germ cells, means calculated to meet the new conditions.   How this comes about we do not know – so say the psycho-lamarckists – the phenomena of hypnotism prove clearly enough that precisely in the unconscious there is extremely intimate contact between the bodily and the spiritual, and that the latter may act so as to produce direct changes of form in the former (compare the blister produced by suggestion); and conversely, the numerous ascertained cases of the prevision of death, for example, are to be referred to organic sensations already present in the unconscious, which are not yet present to the consciousness, but already indicate pathological changes.   If we think out these psycho-physical connections further we come to the Hering-Semon hypothesis,[312] that inheritance is nothing more than what is called, psychologically, memory. Hering's chief work is called 'Memory as a general function of organic matter.'

To sum up, we may say that the psycho-lamarckists reckon with an inheritance of acquired characteristics in cases where this acquisition depends upon some activity of the organism. It would thus be self evident that Weismann's mice showed no decrease in the length of their tails.   Weismann ought rather to have tried to find out, whether the repairs of wounds taking place in the animals appeared more quickly in subsequent than in earlier generations, in other words whether they had 'learnt' anything in this respect by inheritance. That is all that is present by way of an active achievement of the organism, the cutting off of the tails being a purely passive mutilation which is naturally not inherited, since the organism has not been actively concerned in it.[313]

These doctrines appear already doubtful from the fact that they are excluded of themselves in the case of the above mentioned experiments of Johannsen with beans, where without doubt the paratypical changes which took place depended upon an active reaction of the organism to changed environment; also when the Chinese primrose, grown in a hothouse, gives a white flower, this is undoubtedly not a simple passive change of form, but represents an altered reaction of the whole organism to altered external conditions, which should therefore be inheritable according to the psycho-lamarckists. The psycho-lamarckists may naturally say that we do not know why a result may be obtained in one case and not in another; we know in any case nothing at all about the way in which the mind acts on the body.

If this is so, we are then left with the purely empirical question as to whether any single well ascertained case of such inheritance of acquired characteristics has been found. The same is equally true of mechano-lamarckism. It is certainly conceivable that the body cells might transfer their changes in a way at present unknown to us, say by hormones or nerve-currents, to the germ cells, although we cannot imagine such things at present. But here also, the first thing is to produce some facts which do not admit of doubt. Richard Hertwig is quite right when he says in his excellent description of the theory of descent in *Kultur der Gegenwart* that the question can only be decided finally by experiment. What is the use of theories which only exist on paper, when reality refuses to realise the imagined cases?

Have we any experimental proof of the inheritance of acquired characteristics? At the time of previous editions of this book, in common with numerous other scientists and philosophers, I believed that this question was to be answered in the affirmative, though not unconditionally. I must now specifically retract this statement. As far as I can see to-day, this experimental proof does not exist and never has existed. What was then considered to be and believed to be proof, was in part error, and in part –

it is necessary to say so – conscious deception. I will deal with the latter first. In recent times the experiments of Kammerer, mentioned above, have been regarded as decisive, and his newts have been referred to in all text-books, and even in dictionaries of science.[314] Kammerer claimed to have proved that his animals which had become changed in colour by the influence of environment, actually transmitted this change of colour to their descendants, which then showed from the first the abnormal lighter or darker colour when placed upon normal ground.

Kammerer did further similar experiments with the midwife toad, whose peculiar manner of reproducing itself, is described in any text-book of zoology. This toad performs its reproductive functions on land as a rule. By keeping it in water continuously, Kammerer claimed to have caused the male to form pigmented nuptial pads on its thumbs similar to those which other allied species used in the process of reproduction, and this change was likewise stated to be transmitted to the progeny. The only example which Kammerer produced to show this effect of inheritance was however shown to be a forgery by the American zoologist, Noble: the dark areas on the feet had been produced by Indian ink, and there was no sign of a true callosity with the characters shown by other species. The newt experiments have also turned out subsequently to be more than suspicious; not only did one of Kammerer's co-workers give a very unfavourable account of the precautions taken in these experiments against mistakes and deception, but repetition by Herbst and others showed no signs of the inheritance of acquired changes in colour. In one case even, this change in colour itself could not be found.[315]

Kammerer evaded by suicide, in autumn 1926, the further judgment of his colleagues and the world in general. I for my part must regret having been, with others, the victim of methods which are otherwise not known in science. We shall see later what tendencies were behind them, but for the present we are only considering the question of descent.

The convinced Lamarckist will naturally say: what does it matter if a single experiment turns out to be wrong or even forged? Others might still lead to the same result. Very well, let us see. Among the experiments that figure as proofs for the inheritance of acquired characteristics, a particularly important place is held by Tower's experiments on Colorado beetles (figure 73,

Ggs

opp. Plate). Tower bred temperature aberrations of this well known potato pest, and could prove that these changes were inherited under certain conditions, but not under others. The first was the case when the environmental factor in question (for example, great cold), had acted during the 'sensitive' period (see above).

Is this the inheritance of acquired characteristics? Yes and no. A characteristic has been acquired and this has also been fixed for inheritance, and the process is nevertheless quite different from what the Lamarckists are trying to prove, for here cold has acted on the one hand on the body cells, but on the other hand likewise on the germ cells, producing a change in both, and chance, and perhaps also certain necessities, which we do not know for the present, may have caused the individuals produced from the changed germ cells to afterwards show similar changes in the parent animals which were influenced somatically.

When we think of the Goldschmidt hypothesis discussed above, that the genes are definite complexes of matter, this so-called 'parallel induction' will appear more or less natural to us. The proof that we are really dealing with the latter has been given by later repetitions of the experiments, in which it was shown that in many cases the change in the soma did not occur, while that in the germ cells, and hence in the next generation, did occur. [316]

This parallel induction may, it is true, be also described as the inheritance of acquired characteristics. However, it is obviously by no means really what the Lamarckists actually want. For they require that characteristics taken on by the body should be passed on to the next generation, and hence pass first into the germ cells contained in the body; they therefore require not parallel induction, but so-called somatic induction, and Tower's experiments give no more proof of this than do any others. To this must be added that it is a pure stroke of fortune when on occasion, as in the present case, the influence of the environment of the germ cells is such as to cause them to produce individuals the changes in which lie in the same direction as those in the soma of the parents.

It may equally well happen – and enough cases of the kind have been observed – that an environmental factor may cause the parent animal, say, to produce a stronger growth of hair, but at the same time act on the germ cells in such a way as to cause changes in size or form or colour in the individuals produced from

PLATE VIII

FIG. 72. *Venessa prorsa-lenana*. The Spring form (top left) is connected with the Summer form (bottom right) by a series of artificially produced transition forms (from Goldschmidt, *Vererbungswissenschaft*)

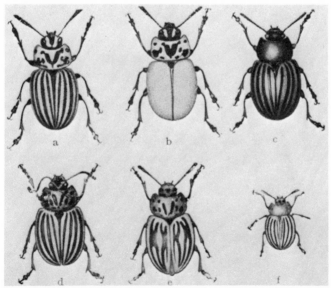

FIG. 73. Mutations of the Colorado Beetle. (*Leptinotarsa*). a. *L. undecinlineata*. b. Mutant of a. c. Mutant of *L. multitaerriata*. d. *L. decemlineata*. e. and f. Mutants of d. (From Tower)

them.   The best known example of this kind in the science of human heredity, is that of the progeny of fathers or mothers given to drink.   Alcohol, the effect of which in the father may be to produce a disease of the liver, enlargement of the arteries etc., likewise affects the germ cell contained in his body (at least this is possible),[317] in such a way that his child may be an idiot or epileptic or otherwise defective in some respect.   This also is parallel induction; with nevertheless totally different effects, and this will in all probability form the rule.   There is also, a priori, no reason to suppose that matters should be otherwise.

Hence these cases are only apparent proofs of the inheritance of acquired characteristics, though they prove a type of behaviour of the organism, that may also bear upon the theory of descent in certain cases.   It is perfectly imaginable that in nature true mutations are occasionally produced in the same way as in Tower's experiments and that they may then, if favourable, maintain themselves in the struggle for existence.   But, as we have already made sufficiently clear, this is by no means Lamarckism.

Still less is proof to be found in its favour from other phenomena which are often adduced, which in reality are nothing but permanent modifications.[318]   If for example, female animals are kept for years in a state of starvation, their bodies thereby being very much weakened, it is self evident that the germ cells also suffer, and the young are therefore weaker in the next and perhaps also in the second generation, than the normal ancestors kept under normal conditions.   In a similar way, we may occasionally have in the case of mankind a disposition to tuberculosis, apparently inherited, whereas it may really be due to insufficient nourishment of the mother during or previous to pregnancy.   Likewise, certain diseases are known which are transmitted directly from mother to child during embryonic development (for example, syphilis).

Uncritical people are accustomed to bring forward such cases as proof of the inheritance of acquired characteristics.   The reader is, we may hope, quite clear by now on the point that they are nothing of the kind.   A change which is lost in the course of a few generations again, and in any case does not show Mendelian characters, is not an inherited character, but a paravariation or permanent modification, and its transmission does not take place by the method of chromosome division characteristic of the genes, but in some other way which is no doubt quite different from case to case. These are therefore of no value as demonstrating Lamarckism.

I omit some other cases once cited in this connection now regarded with equal doubt, and come to the final result. This is as follows: *There is not the slightest evidence of a real experimental proof for true Lamarckism, that is to say for an actual transmission of characters acquired by the body to the germ cells.* The result of all well attested experiments in this direction has hitherto been negative. Although we have expressly protested on several other occasions against turning the absence of a result into the impossibility of the same (compare page 330), we must nevertheless not fail to notice, that in certain cases this, call it if you like 'negative,' dogmatism has also resulted in very positive conclusions. This has been the case, for example, with regard to the laws of energy and entropy, which state nothing more than the impossibility of a perpetual motion of the first or second class, and also in the case of Einstein's theory of relativity, which postulates the impossibility of discovering an absolute system of co-ordinates in space. We are therefore justified at least in assuming as an hypothesis for further investigation, that there are reasons in the nature of things which render Lamarckian inheritance impossible; in other words, that there is no connection between the body and the germ cells which would enable changes in the first to be carried over to the second in the same sense.

As against this negative result, an objection, which must not be taken too lightly, is raised by the theory of descent itself.[319]   It is undoubted that characteristics such as callosities, or cusps (on the molar teeth), which are produced under our eyes by use, are inherited.   Camels, for example, have already as embryos the plain beginnings of their knee callosities, which we must after all assume to have been produced by continually kneeling on the sharp sand of the desert.   But if that is the case, they are certainly acquired characteristics, and nevertheless they are inherited.   A logical opponent of Lamarckism would have to make an accessory hypothesis; and one not very plausible, that in all such cases a process of selection must have taken place.   Such camels must have been selected as containing, among their heredity factors, those which effected a particularly strong thickening of the skin under pressure (the heredity factors are, as we have seen above, potencies, and not characters).

At a first hearing this does not sound particularly plausible, since we are directly aware of the fact that the skin produces this reaction without any regard to hereditary disposition. It happens in our own case as well, although we do not kneel in desert sand, and it happens in the case of almost all mammals. The opponents of Lamarckism can naturally reply that this fact is even to be referred to inherited qualities common to all mammals, since they are useful for all. Nevertheless, this argument is not very convincing. On the other hand, the Lamarckists produce no proof as to the way in which the production of callosities on the knee is to influence the germ cells in the corresponding manner, and hence we are again confronted with the fact that the question must be decided by experiment. But even if we assume for the moment, in the distant future some few pieces of evidence should turn up in favour of that supposition, it must be clear from all that we have discussed that such somatic inductions certainly require a very long period of time to become really fixed in inheritance. We might therefore make use of it as regards the doctrine of descent, but it would be obviously hopeless to attempt to get any results by these means within humanly possible periods of time in the breeding of plants or animals, or in other matters later to be discussed. Here only selection holds out any promise of success.

And as a matter of fact, which may be mentioned by way of conclusion, the whole astonishing success of our present methods of breeding plants and animals, which we have already spoken of in connection with Mendelism, depends in no way upon Lamarckian principles, but always and everywhere on the selection of existing mutations (pure lines, sports, etc.). Breeders have long ago discovered that it is quite useless to attempt to produce, say, a breed of cattle producing a larger yield of milk, by feeding the animals for several generations on special food. That makes a great change in the phenotype, but none whatever in the genotype.

Apart from these practical consequences, we may say that Lamarckism might be taken into account as a possibility with regard to the theory of descent, though with very great caution, but that it is extremely doubtful how much may really be put to its account. So much appears certain, that selection among favourable and unfavourable mutations can effect changes in a few generations that would require by Lamarckian means at least dozens if not hundreds of years. We are thus again brought back to the view that the mutations, and their selection, and not direct action, will have to supply the main support for the doctrine of descent, and it is therefore necessary to devote a few words to the question of mutations themselves.

How do they happen ? Can we produce them artificially and therefore imitate on a small scale the origin of species ? And might we not be able to produce mutations in this way for practical purposes of breeding ? The answer is that we know at the present time nothing whatever as to how the mutations which have undoubtedly been observed in free nature, also including Darwin's single variations, have come about. Undoubtedly mutations, like everything in nature have a cause; we do not however know where to look for it. It is quite certain that the factors of the external world, as well as those of the body cells (see above), are able to influence the germ cells in many ways.[320] In the first place, X-rays have been shown to produce mutations; alcohol has been mentioned above; it is probable that many other poisons also, and also high nervous strain of human beings, and also certain products of disease, may injure the germ cells.

Mutations produced in this way are, it is true, almost all so called 'loss mutations', that is, certain inheritance units are lost which ought normally to be present. In a few individual cases, really new inheritance factors have been produced experimentally. But all that has so far been done is so little, that there can be no question of the application of this method in the breeding of animals or plants.[321] Still more unlikely is the hope that we could produce improvements in human inheritance by this means (this is unfortunately not the case with degeneration. The same is true here as in the whole of life, that it is much more easy to destroy than to build up). Coming generations may have more success in this direction.

We must not however, in seeking the causes of variations, look too exclusively to outward circumstances. It is perfectly possible that within the organ itself, there may occur from time to time states in which the system of inheritance factors situated (as we assume) in the chromosomes, fall over as it were into a new position of equilibrium. Galton, the founder of the modern science of heredity, once said that the fluctuating variations may be compared to the oscillations of a polyhedron standing on a surface, which has received a light blow. After a few oscillations it will again come to rest on the same face.   In the same way the fluctuating variations always oscillate about a certain average, to which they always return.   The mutations, on the other hand are to be compared to the rolling over on to a new surface, which takes place when the blow exceeds a certain limit.   The comparison is very illuminating.   To this we must add that as we know from present day physics, variations are continually taking place in every material system.   The cells, to say nothing of their subdivisions the chromomeres, or the genes contained in these, are already small enough to make deviations from the average moderately possible.   It is thus an obvious step to attempt to introduce into the science of heredity, the hypothesis that, simply on account of these oscillations, which are characteristic of every physico-chemical system, large disturbances of equilibrium must occur now and then in a certain percentage of cases, which then appear as mutations.

This hypothesis, first put forward by Freundlich,[322] has not yet attracted the attention which it deserves.   The reason will no doubt lie in the fact that we are at present unable to say anything in detail concerning the oscillations in question, since the whole constitution of the genes is at present unknown to us.   Furthermore, biology has hitherto been so wrapt up in a blind belief in physical causality, that it has, so to speak, become more papist than the Pope himself, while physics has long ago made use of the statistical method, and quite recently, even regarded it as possibly its own fundamental basis.   Biology, which quite certainly has to do with only statistical laws, is still clinging rigidly to the pure dynamical conception of law.   If we assume the hypothesis mentioned above, we obviously do not need, for part of the mutations, any external cause at all, since they lie in the organism, and might possibly even be referred to final physical uncertainty itself (Heisenberg's relation, see above, Bohr).[323]

In the above, the main facts concerning the variability of organisms have been sufficiently indicated, and we must now turn to the question how Darwin's doctrine of natural selection appears from this point of view.

XI

THE DOCTRINE OF SELECTION

This theory has shared the fate of so many other scientific doctrines, of being at first blindly believed, then sharply criticised, then almost completely abandoned, finally to be rehabilitated within its true limits. The belief in the supreme power of natural selection (Weismann) was followed by the expression of the impotence of natural selection: we know to-day that both attitudes were exaggerated. Selection does not explain everything, and much less the whole world, as its enthusiastic prophets tried to make us believe fifty years ago, but it explains a great deal, and more than its rabid opponents were willing to admit. Hence it is a factor of importance in the explanation of descent.

Before we attempt to criticise the idea of selection as a whole, we must first mention some groups of facts which must be taken into account. First we have sexual dimorphism, the fact that the bearers of the male and female germ cells in the animal kingdom have almost always considerable differences in form, size, colour, etc., differences which are frequently great enough to have sufficed to cause male and female, had their relationship been unknown, to be regarded as quite different species.[324]

In very many cases, above all in the case of the lower animals where such extremes are often observed, it is merely a question of male and female having each specialised in their own way for their duties (impregnation, hatching of eggs), and the general grounds which have rendered possible the production of these numerous adaptations must be regarded as sufficient for the understanding of these phenomena, although we cannot quite see through them for the present; also, there is a further special question as to how the characters of one sex can be inherited by means of the germ cells of the other.[325] But along with sexual characters of this kind, there is further a large number of others which cannot be explained as the adaptations to the new conditions of life, these

PLATE IX

Fig. 74. Left above, *Danais limniacare* (avoided by birds); left below, *Papilio clytia La. var. dissimilis L.* (imitator). Right above, *Euploea core Cr.* (avoided); right below, *Papilio clytia L. var. panope L.* (imitator). From Hauri

being generally characters of the male. The courtship adorn-
ments (decorative colour, unusual bodily processes, such as horns
or pincers, song of birds, the instinct for play and dance etc),
Darwin assumes that these characters were produced by 'sexual
selection.' The choice of the male by the female is supposed to
play in this case the same part as in a general way the struggle for
existence.

A second and equally important phenomenon is that of a special
group of adaptations known under the general term of *mimicry*,
first carefully investigated by Bates. In a further sense, all
protective colouring, such for example as the grey brown colour
of the hare and the lark, the yellow colour of the lion, the white
of the polar bear or the weasel in winter costume, is to be reckoned
under this heading, and also certain instincts such as shamming
dead by many animals. In the narrower sense, mimicry is the
imitation of quite definite natural objects, for example leaves,
twigs, stones, and the like; in the narrowest sense it is the imitation
of other animals by certain species to such a degree of perfection,
that even the eye of a practised observer can be completely
deceived.

Some of the most astonishing cases are those shown in figures 74,
and 75, 76 (opp. page 474) or that of *Papilio merope*, a South African
variety of butterfly[326] related to the swallow-tail. The female
of this species – only the female, not the male – occurs in three
different forms apart from the normal form, which differ from one
another entirely, both as regards colour and shape of wing, but are
each precisely like another species of butterfly belonging to totally
different families, which are not eaten by birds on account of their
evil taste. The unusual forms live, as in this case, almost always
only where the immune species also lives, and always accompany
their prototypes, so that Bates, who first discovered such cases in
South America, only had his attention drawn to them by pure
chance. These phenomena have been put forward by upholders
of the theory of natural selection as particularly strong evidence
of its truth. Latterly, a very strong critical opposition has arisen,
to which we shall return almost immediately.

We may now consider critically the theory of selection.[327]
What can it do and what is it unable to do? It may be taken
as agreed to-day, that fluctuating variations cannot be pushed in a
certain direction by means of selection (compare page 458). The

experiences of breeders, which were used by Darwin as the main support of his theory, are to be explained by the fact that selective breeding is not dealing here with fluctuating variations, but with a mixture of pure lines or elementary species, of which the Linnaean species is in reality made up. Even when, the breadth of fluctuation of one line overlaps that of another (see page 458) selection (say that always of the largest beans) must always increase, in the following generation, the representative of the largest line, so that this line often comes out quite pure. In this way a new and constant species has been 'bred', but in reality, it has only been isolated from the mixed population in which it originally existed.

There can be no doubt that the older Darwinism needs correction in this respect. But it is a great mistake when many opponents of selection believe that they have completely refuted it by this argument. What has really taken place, is so far simply a more exact formulation of Darwin's premises. As regards the essential point of selection, it is a matter of indifference whether we regard this as applied to Darwin's indeterminate notion of individual variations, or the more precise ones of elementary species or pure line. Variations suitable for selection are thus these elementary species which are constant in the ordinary sense, but obviously not invariable. For as far as I know, the most rabid opponent of selection has never yet maintained that the 200 elementary species of the corn marigold (Jordan) or the 19 pure lines of beans (Johannsen) were originally created as such. They can and must be regarded as mutations in the De Vries sense, and selection takes place fundamentally only in connection with such qualities, which are inherited, but capable of changing suddenly from some cause or another. From this point of view, the mutation theory thus amounts to no more than an improved Darwinism.

The matter takes on a different complexion when we investigate the question of the *reason for such mutation*. It is here that the paths of the pure theorists of selection and of the Lamarckians in their newer form, separate. Darwin's chance individual variations could first be substituted by irregular mutations, these would come about for any reason whatever, for example, in the case of Tower's Colorado beetles on account of cold, in other cases perhaps on account of special conditions of nutrition, or new combinations

PLATE X

FIG. 75. Left above, *Danais chrysippus L.* (avoided); left below, *Hypolimnas misippus L. var.* ♀ *diocippus Cr.* (imitator). Right above, *Danais dorippus Klug.* (avoided) ; right below, *Hypolimnas misippus L. var.* ♀ *inaria Cr.* (imitator)

FIG. 76. *Hypolimnas misippus L.* ♂

of Mendelian factors (as was probably the case according to Renner's results with de Vries's mutations), and selection would then choose between such mutations. We thus still have a theory depending on pure chance, which is a point which has been the cause of both praise and blame for Darwinism.

But the new Lamarckism protests against this, maintaining firmly that the change is not irregular, but has a definitely recognisable connection with existing needs. In other words, variations in a definite direction are required. In this connection it is important to note that the direction is to be that of the desired adaptation, not of a definite direction in general. This remark is necessary, since a group of facts and theories exist that go to show that variations are by no means irregular, but that they only move along certain definite lines. The phenomenon is described by an expression coined by Elmer, orthogenesis, but ideas belonging in this category had already been expressed before Eimer, by K. E. von Baer and Nägeli. According to these investigators, the cause of change is not to be found in environment, but in the germ cells alone. And certain phenomena do in fact render this supposition likely.

The most prominent of these is the hypertrophic development, for example the antlers of the stag, or the horns of the stag beetle, the curious prolongations of many shells etc. The attempt to explain these phenomena by selection in general, and sexual selection in particular is especially unsatisfactory. Are we really expected to imagine a female stag beetle having her aesthetic feelings aroused by a male with particular horns? ·Monstrosities of this kind, which are very common in the whole animal world, rather make the impression of some principle of change in the organism having been, as it were, ridden to death. The result also is that many such species are becoming extinct or are so already, since formations of this kind finally become altogether too unpractical. Hence in cases like these, in spite of all natural selection, development does not escape from the blind alley into which it has erred. Indeed, the whole of palaeontology with its extinct species really shows that not only the power of selection, which favours adaptation, but also a persistence of the process of development, which is opposed to adaptation, are both active, and perpetually at war with one another.

Further strong reasons for the activity of such orthogenetic factors are given by recent exact determinations of natural or artificial variations which actually occur. It has been shown that these do not occur by any means irregularly and in all possible directions, but always only in a few quite definite ones (Eimer, Tower, R. Hertwig). We must mention also 'Dollo's Law,' which says that organs which have once become rudimentary or have disappeared entirely, never return, but are replaced by other new organs if the need for them recurs. If natural selection alone is at work we cannot see any reason why development should not go backwards when the conditions are reversed. Hitherto this has never been observed.

A more difficult question is, how we are to imagine the production of these orthogenetic tendencies. A possible solution might be found in a new form of the doctrine of selection, suggested by Weismann,[328] the fundamental idea of which is as follows. The genes contained in the chromosomes, or determinants as Weismann calls them, are themselves subject to a fierce struggle for existence, every change in the organism favours or opposes one or other kind, and this has the result of increasing or diminishing their effect in the next generation. In this way, however, the injuriously affected determinant in question arrives in the new generation with an initial disadvantage as compared with the others, and again retreats, and it is finally defeated, unless personal selection, that is the elimination of individuals affected by this phenomenon of loss, puts a limit to the process. But conversely, every plus variant will favour the corresponding determinant, and thus automatically increase the character in question unless individual selection regulates this effect of germ selection.

Weismann's object in making this supposition, which in his time was purely hypothetical, was to give an explanation of the fact, inexplicable on the original Darwinian lines, of the disappearance of useless organs on the one hand and of hypertrophic development on the other. The double struggle for existence (between the determinants in the organism and between whole individuals in nature) results in every tendency to variation that is once begun striving automatically to continue, unless it is uprooted by individual selection. The organism then no longer resembles an inert mass which must be set in motion from outside, but rather a stretched spring, which may come loose at any moment, or better still, a bicycle in motion, which can only be kept in equilibrium by motion of the steering wheel, and falls over when this motion no longer occurs.

Considerable objections have been urged against Weismann's theory by Darwinians, and many, Plate for example, have even declared that they would rather give up selection altogether than assume it in Weismann's form. Nevertheless, one cannot avoid the impression, especially since similar ideas have recently arisen on a pure basis of genetical theory, (see Goldschmidt's hypothesis pages 381, 383), that there may nevertheless be a kernel of truth in it, even though its detailed formulation by Weismann may be in need of improvement. This kernel of truth would then be the quite general law, that the fundamental heredity of an organism forms a system of such instability, that it continues in a direction of change once entered upon until it either reaches a new state of equilibrium with the environment, or the species dies out. A hypothesis of this kind lies entirely within the region of scientific discussion.

If we are thus able to make orthogenesis theoretically comprehensible in this manner, particular attention must be drawn to the fact that this is by no means orthogenesis in the Lamarckian sense. Just has made a very happy distinction in this connection between causal, and final or purposive, chance.[329] If orthogenetic theories are right, variations are not, as Darwin supposed, purely accidental, but have a definite direction. But this does not tell us what the direction is, nor that it must be directed towards the goal of adaptation. These theories only exclude causal chance; chance from the point of view of purpose remains. More far reaching suppositions are necessary, if the latter also is to be excluded. This must be understood if we are not to draw blindly a conclusion in favour of Lamarckism from the facts discovered by Eimer and others. Lamarckism would only be right if it could be proved that the orthogenetic variations in question stand in direct relationship to the goal of adaptation; but as we have seen, we have yet no proof of this.

Certain other theories of descent carry orthogenesis much further than the Lamarckists, by excluding the external world altogether, and supposing that changes take place simply through an inward progressive drive. The older nature philosophy spoke of a *nisus formativus*, Nägeli spoke of a 'drive towards perfection,' and Bergson of the *élan vital*. All these mean the same thing, an entirely mystical principle active within the organic being itself, which principle is directly parallel to the entelechies, which are supposed to direct individual · development.[330] The general objections to such teleological 'explanations' have been discussed above. We only need to add here that the parallels usually drawn in such discussions of the history of development, between the development of the individual and that of the species are quite mistaken. For the development of the individual actually takes place under the influence of heredity, which is so to speak, its entelechy, while the development of the species has to explain the production of these inheritance factors.

If we are not going to assume here true 'epigenesis' but once more only 'evolution' (see page 341) we arrive at the absurd result that the first and simplest living cells already contained human beings and lions, herrings and ants, and that the whole process of the development of species was nothing but an unfolding of potentialities already present at the first beginning. This abolishes the theory of descent itself, since it becomes completely without any purposes. What it had to explain was on the one hand the correspondence between related species (through common inheritance); on the other hand, the divergence between them on account of individual variations. If everything is already determined beforehand, both human being and bean may be just as well supposed to fall finished and ready from heaven. What would be the use of passing through the stages of the primitive cell, worm, fish, etc., if we were in reality man from the beginning, and hence entirely different from every other being even in that disguise, as Dacqué[331] and others have once more been trying to persuade us.

So much is true in these ideas, that the older writers on descent were inclined to set the points of divergence much too high as a rule, and even to have living species or animals directly descended from one another (for example human beings from apes). A more exact investigation of the problem of genealogy leads of necessity to the view that the division of forms, originally one and the same, into the present variety of species, must, as a rule, have started much earlier. We cannot, for example suppose that an original mammal first descended from reptiles (as Haeckel supposed) and then divided up into all the various forms (carnivora, rodents, etc.). If a genetic relationship really exists here, we must suppose that the various groups of mammal must already be ascribed to definite groups of reptiles as their descendants (for example Steinmann[332] has proposed as original ancestor of the primates, 'metareptile' *Delphinognathus*, already existing in Permian times). No objection need of course be made to such hypotheses, only they must not be driven to the absurdity of not allowing divergence at all, but of tracing each single line back to the very beginning. If polyphyly is driven so far, the whole doctrine of descent becomes superfluous.[333]

It is evident, after what we have said earlier concerning the reality of the teleological side of biology, that we do not intend for a moment to deny a tendency towards perfection in this whole process of the formation of species. And we do not therefore need to give any special arguments in its support at this point. In this sense we might even agree with Dacqué's thesis, that the whole world of animals is merely split off, so to speak, from the development which from the first aimed at man. We only need to oppose in this connection the attempt to put such teleology in place of an investigation into causes, an attempt which will certainly be read into Dacqué's works, even if their author had not this intention. Teleological considerations can always only accompany causal but never takes their place. If this is not forgotten, there is even no objection to ascribing to the process of this formation of species, the character of an action, that is to say, giving it a psychical counterpart. In this sense the *élan vital* of Bergson may likewise be readily admitted, whether we regard it as a driving force working unconsciously or whether as the outflow of a creative wisdom exceeding our power of conception.

Our criticism of the theory of selection would be too incomplete, if we did not mention three objections which are very often brought against it. These are: 1. The want of relationship between numerous lethal factors (disease, weather, enemies, etc.) and the qualities which are supposed to be bred by natural selection. 2. The insufficient value of the small variation. 3. The so-called panmixy (general mixing).

Regarding the first point (compare also page 480), we must say that the opponents of the principle of selection are right in pointing out that numerous lethal factors, in particular all kinds of injuries to brood, are generally in no direct relation at all to any kind of advantageous qualities, but strike the adapted and unadapted indifferently.[334] But this does not prove anything against the possibility that selection may also be active, since this only needs a single one of the numerous lethal factors to stand in a direct relationship to an adaptive character. Almost super-fluously many experiments have been latterly made which prove quite without a doubt that when mixed stocks of plants or animals are exposed to natural conditions, some are very seriously, others very slightly decimated, so that in a very short time one variety enormously exceeds another.[335]

It cannot therefore be denied that selection is at least able to eliminate unpractical mutations. But, so say the opponents, it cannot produce anything positive.

Here we come to the second objection. A small change, it is said, for example in the wing colour of a butterfly, cannot be of the slightest use to it, even when it is a move in the direction of mimi-cry. For a bird which does not eat the immune species will not take the slightest notice of that slight change but eat its possessor just as freely as an unprotected companion. In order that such a protection may be effective, a very high degree of similarity must be present from the start, and this could not possibly be acquired as the result of a single mutation. But even if we suppose such a mutation to take place it would be of no use, since it would be completely washed out in a short time by general crossing. It must be admitted that the first argument in the case in question is very attractive. But that it cannot be generally true is shown by a case such, for example, as the formation of a web for swim-ming. Here even a small variation would already give a small advantage, and selection could do the rest, if the second argument can also be refuted.

There are three arguments which can be brought against panmixy. We may first of all assume that mutants which re-semble one another will be slightly more fertile amongst themselves than with somewhat different forms, and this has been proved in several cases by statistics.[336] The infertility, which usually occurs in the case of hybrids between different Linnaean species may be taken as the extreme case of an increasing difficulty in the formation of hybrids. In the second place, it is not only imagin-able, but probable, that favourable mutants frequently become isolated in space from the parent form. In the case of animals we may assume a direct association, but apart from that, wind, water, birds, etc., may carry off mutated seed and thus cause the production of a local stock of the new variety. M. Wagner has even developed a complete migration theory of descent on this basis, a great deal of which is doubtless correct. But thirdly, we may on the basis of Mendelism, make almost the contrary assertion under certain conditions, namely that a single favourable mutant has a great prospect of holding its own, since the quality in question will appear again and again in its descendant.[337]

In conclusion a word concerning mimicry. Scarcely any other problem of the theory of descent has been the subject of so much recent controversy, and some find salvation in the theory of selection, while others reject it altogether. It is first of all certain that a great many of the cases originally regarded as mimicry are greatly in need of confirmation as to whether the likenesses in question really produce a protective effect. In this respect Heikertinger's sharp criticism[338] has brought forward much new relevant material, while on the other side, Wasmann continues to defend firmly the cases of ant mimicry which he himself has carefully examined.

Heikertinger's main argument is that we observe quite generally in nature a mutual adaptation of animals and their booty to one another, so that enemies specially adapted find out animals and eat them in spite of evil-smelling, supposed protective, substances. Though offensive to us or other animals, such protective arrangements have no value (according to Heikertinger) against the enemies which are specially adapted to a given booty, and they do not come into question as regards other animals which might be imitated by them, since these do not attack the animal in question. Furthermore, Heikertinger and other critics[338] have also pointed out that the upholders of the mimicry hypothesis have in many cases failed to bring any proof that the enemies of the booty animal in question really make use of the sense upon which the mimicry is supposed to work. Thus for example, the famous imitation of wasps by flies must be absolutely useless as a protection against spiders, since the latter, as experiments prove, do not depend upon the sense of sight.

It is extraordinarily difficult for an outsider to form any kind of judgment as long as the specialists differ in this way. Cases such as the famous Indian leaf butterfly *Callima paralecta*, the twig grasshopper, and many others appear so directly convincing, that one is obliged to say that it cannot be the result of chance.

Hhs

There must certainly be some inner connection between the form of the leaf and the butterfly. And what else can this be than that the butterfly is to be protected by the form which it has assumed ? Heikertinger himself admits this, but nevertheless requires us rather to abstain from any explanation than put forward hypotheses which cannot be supported by experiments. This is certainly true, but on the other side, it is also true that science is not a mere question of recording facts but of providing explanations.[339] How are we to arrive at such explanations without the formation of hypotheses ?

We may certainly explain numerous cases of simple protective colouring quite easily, without calling in selection, by the hypothesis of 'seeking' or of 'trying' and 'remembering.' The hare may perhaps lie down in the furrow, because he has learned in the course of development that his enemies then notice him least, the latter being numerous and on the watch for him with all possible senses (in this case Heikertinger's limitation of the relationship between enemy and booty is certainly not to the point). There is much in favour of the idea that the hare is conscious of this protection, so far as an animal can be conscious of anything. But this explanation does not hold in the case of the white polar bear, which certainly did not migrate to the north because they were white. Here we must rather assume that direct action (aberration due to cold with subsequent natural selection) is the most plausible explanation.[340]

But in the case of the famous examples mentioned above, this explanation is quite useless, and we are then really only left a choice between resignation, or natural selection, or vitalistic 'explanations,' according to which 'nature,' or the 'élan vital' or something else of the kind once devoted itself, in a sort of playful way, to painting on the back of a butterfly's wings the exact copy of withered leaf with its stalk and ribs, this having then led the fortunate possessor of this 'sport of nature' to hasten to make use of it by sitting on the plant so as to resemble a leaf. Or the drawing was actually given to it by one of those 'entelechial' powers, in order that it should be thoroughly well protected, though here again it is difficult to see why the same power should not give its enemies the same power of finding it. Or are we to suppose that it takes less care of the enemies ?

As in so many other cases, we must leave these debatable questions open for the present. The problem is not who is right but how much is each right. But we must state expressly that, after all that has been said, the rôle of selection is much greater than the sharp opponents of Darwinism were inclined to allow. The objection so frequently made, that selection could not produce anything new but only get rid of unsuitable material, misses the point. For the selection theory has never asserted that selection produces variations; Darwin, on the contrary, assumed the presence of variations as an undoubted fact, as all subsequent theorists of selection have done. Selectionists merely maintain that, among many variations having no relation to the needs of the organism, there may be some that are adapted to the need, and that these are maintained by being preferred in selection, and that a new species may come about in this way. We can see no reason why this should not happen. But if it does happen, selection is then the decisive factor, and it is not justifiable to reproach it with assuming the most important factor, namely variation. The fact that ultra-Darwinists such as Haeckel have quite unreasonably raised it to the level of a cosmic principle has nothing to do with the point.

Biology has thus quite recently been returning to the idea of selection to a greater extent than one would have supposed possible ten or twenty years ago.[341] The decisive rejection of every kind of Lamarckism by the modern theory of heredity, has had most to do with this. The author must also admit that, as time goes on, he becomes more and more convinced of the inadequacy of Lamarckian principles and hence takes up an essentially different position in this edition from that in the last. The case of Kammerer has had something to do with this, though a slip of this kind should not be regarded as of any importance. An assertion may be correct even if it is supported with bad arguments. But Lamarckism has been so badly beaten along the whole line, that apart from the bad impression created by such occurrences, there is not much left of it. This leaves only selection remaining as a real scientific explanation of descent, since the preformist theories last discussed are no explanation, but only a restatement of the problem in another form.

We might conclude our discussion of the principle of selection at this point, but for the fact that certain philosophical consequences, which rightly or wrongly have been linked up with it, did not require our attention for a moment. The resistance originally met in all religious and idealistic circles by the whole doctrine of descent (whereby, as we have already said, descent and Darwinism were hastily thrown into one pot) has to-day ceased to exist, at least in Germany, although there exist a few ecclesiastics who would, if they could, apply similar means to those applied in America to prevent the spread of the doctrine of descent, at least in schools.[342] These circles are, however only small conventicles, and official theology of both confessions has spoken with fair unanimity in favour of the reconciliation of the theory of descent with religious belief, Wasmann having done most for this on the Catholic side and Dennert on the Protestant.

In these circles, however, they are far from adopting the same neutrality towards the factor problem. On the contrary, there is a strong distaste for the principle of selection, and a very strong preference for psycholamarckist and orthogenetic theories. Darwinism is accused of delivering the marvellous organic world over to blind chance, and hence to be difficult to reconcile with the divine government of the world.[343] How false this point of view is, can be seen from the simple consideration that the whole world is full of chance of this kind, apart from any selectionist explanation of the formation of species. If hundreds of thousands of germs must perish before the goal is reached, that is just as 'dysteleological' as the Darwinian chance which demands that hundreds of variants shall perish until one is the useful new adaptation. If the latter is a contradiction of religious belief, so is the former.

We are here confronted with the real fundamental problem of present-day religion, the question of how all that is apparently senseless and worthless in the world is to be reconciled with a divine government of the world, the problem of theodicy. It is not rendered any easier or any more difficult when Darwinian chance is added to already existing difficulties. So all that is correct in the whole resistance to Darwinism is simply the consideration that it is responsible for the loss of one of the supports of the primitive physico-theological proof of God's existence, which draws from single useful adaptations observed in nature the conclusion, that a God must be behind them, who has made them so particularly effective. It is very questionable whether this is a real loss to a high form of religion.

Such tendencies have of course nothing to do with science; it has to decide as far as possible how we may suppose the development of species to have taken place. If, to-day, a certain resignation has entered into scientific biology, since it has become evident that we are very far removed from this goal and have a great deal to clear up before we dare to tackle this problem successfully, such a momentary and temporary state of affairs should not mislead us as to the necessity for undertaking the task. It is true that the biologists of to-day hardly think of wasting their time with genealogical trees in the manner of Haeckel, only to see later that they have built on sand.

We are clearly conscious to-day that what the genius of Darwin left us was not a finished solution, but a great and broad problem which will give work to generations of research. If this was earlier forgotten in the first flush of excitement over the new field of knowledge, nothing more happened than is often observed in all parallel cases in the history of science. We have first vague and indefinite ideas, curiously compounded of clear knowledge and all kinds of mystical shadows. Then comes a genius, a Copernicus, Newton or Darwin, and a science suddenly begins to grow out of the chaos of opinions. Then, however, things inevitably go to an extreme.

The genius, the pioneer, is chosen as a leader, even where his intuitions are no longer applicable. For a long time all goes well, the shaky structure is supported with all kinds of subsidiary hypotheses; but finally even that is no further use, and a critical reaction begins, at the height of which a great deal of what is still of value is swept away. Finally, when this reaction has subsided, what was real in the first work of genius begins to show its fertility in further insight, and sober investigation, free from all extreme views, can march straight forward to its goal. In the theory of descent, we are still not quite at the end of the first period of criticism, but we have already probably passed its maximum. Everywhere we see the inception of investigations which have neither a dogmatic nor a critical tendency, but are solely inspired to find out the truth. Future progress will depend upon such work.

What was said here of the theory of descent in general is true, to quite a particular degree, of one special problem, which really forms the reason for the bitterness of the controversy, namely, the descent of man. A certain inward necessity has resulted in all philosophical ideas of the universe, which have played a part in the doctrine of descent, receiving their crassest expression in the conclusions which they have drawn regarding man. Dogmatism and criticism have been hardest at work in this field, and it is certainly not too much to say that one half of what has been spoken and written concerning descent, has been devoted to this special field. The problem of mankind is of course the point at which scientific knowledge passes over into philosophy of life, which, after all, must always be at the same time a view of our own personal position.

There can be no question that the application of the theory of descent to mankind means a revolution in thought equally as great as that produced 300 years ago by the abandonment of the Ptolemaic for the Copernican system in astronomy. Whether this results in our whole view of the universe being put on a completely new basis, is another question, which cannot be easily answered in a few words. It must appear daring for us at the present day, standing in the midst as we do of this revolutionary process, to attempt to come to any decision as to its more far reaching consequences. History shows us that an objective judgment concerning such fundamental questions is only possible for later generations. It must nevertheless be attempted if our view of the world is not to remain an impossible and quite unsatisfactory torso. We must therefore conclude our discussion of the philosophy of nature by examining man himself, although we shall at many points have to cross the boundary between philosophy of nature and philosophy of culture.

For man is 'a wanderer between two worlds,' on the one side in nature, on the other in civilisation, which has a similar relationship to purely biological nature as that between the latter and physics. It is a plainly marked higher step, which always remains recognisable, even if we are compelled to assume continuous transition. We are dealing here also, with new 'emergents,' and in this case with some of an entirely different character, namely psychical products such as science, law, morality, religion, etc., along with civilised activity in the narrower sense, technology, economic life, the formation of society, etc. All these raise mankind above the level of the animal world, into a realm beyond biology, the realm of values.

PART IV

NATURE AND MAN

THE PECULIAR position of man as part of living nature has been realised plainly by the leading spirits of all civilised nations. We only need to recall the well-known verses of Sophocles and Pindar,* or the Eighth Psalm. As opposed to this, primitive peoples frequently show a fairly close contact with the animal world as regards their life feelings. The distance between man and animals, although present, does not here become so clearly conscious, the animal being humanised in much the same way as by our own children. We have in Totemism the religious form of this view of the animal. It also appears to have affected the higher religions for a very long time, for its traces can be found almost everywhere in them.

Our world of Christian Western civilisation, on the contrary, has always been under the influence of a decisive refusal to connect man in any way closely with the lower parts of nature, and particularly with animals. This refusal has its roots in Judaism, where an ethical monotheism, unique in the history of religion, developed into a definite opposition to the continual recurring tendencies in the people towards actions characteristic of savage cults (Baal, Astarte, etc.). Since the fundamental doctrine of Christianity itself assumes a still higher valuation of mankind than that assumed in the religion of the prophets, it is not surprising that in the history of the Christian church, the bond between man and nature very soon breaks entirely, all the more readily because the development of the late Greek world into Neoplatonism was a development in the same direction.

*φύσις κέρατα ταύροις,
πόδας δέδωκεν ἵπποις,
ποδωκίην λαγῴοις
. . . .
. . . .
τοῖς δ'ἄνδρασι φρόνημα.

487

This view held that it was man's true task to sever as far as possible all connection between the body, that is the whole natural foundation of his life, and the soul, which was supposed to come from a 'higher world,' the task being to bring the soul back to its true destiny, from which it had been drawn aside by entering the world of matter. Although the church frequently condemned gnostic doctrine having this tendency, and nominally attained to the dogma that, on the whole, creation is 'good' (according to Genesis i., 31), history shows that this had practically little influence on the course of development.

It was only modern science that brought a serious change. Nevertheless, the period of 'enlightenment' which followed in some respects again increased the contrast between man and nature. Man, as the only reasoning being, appeared to that time so far above all the rest of creation, that German Idealism exhibited scarcely less opposition than the church, when Darwin's works suggested, for the first time, the serious possibility of a true connexion between man and animal. Whereas for us the waves of this stormy quarrel have already calmed down to a certain extent, the example of America shows us how strongly the opposing forces are at work under the apparent calm of the surface.

It is now our task to trace out this problem in its various ramifications and arrive if possible at a calm and objective position. The problem contains many particular questions; the first being that concerning the genetic connexion between mankind and subhuman nature. The second concerns the natural biological conditions controlling his higher activities both as regards the higher individual and the social organisation; and the third inquires as to the effect that man, on his part, exerts upon nature. Our first question is thus that of man's beginning.

I

THE ORIGIN OF MAN

Although this is not, as Haeckel imagined, 'the question of questions,' it is still of enormous importance for every view of life, and without a careful consideration of it, no satisfactory conclusions can be reached.

Our knowledge concerning the origin of our own race, as we must unfortunately admit from the start, is very scanty. The results obtained by history and pre-history, by geology and palae-ontology, by ethnography, ethnology and philology, form to-day an enormous mass of knowledge as compared with the position of anthropology a hundred years ago. But it is really very little, merely a small beginning, compared with what we ought to know, to be able to say that we are fully clear concerning the origin of our race. The fact that we are clear about this from the start does not afford any reason for resignation, tinged either with want of courage or even secret satisfaction. What Copernicus taught us was but the beginning of a real investigation of cosmic structure, a tiny fraction of the problems which still occupy astronomy. And yet it was of more importance than everything the astro-nomers have since discovered or ever will discover. For it was the first step, and a most difficult step, towards a natural point of view, free from the old prejudices of our sense impressions. This first step, this emancipation from the old naïve, geocentric idea, counted for more than all the following steps put together, for it made all the others possible.

Something similar to this took place in the last century concern-ing organic nature. The knowledge that the origin and growth of this world, and the events in it, have taken place according to laws which can be investigated, and still are so taking place, and that the human race also belongs in this great sequence of organic development, will be inseparable in all future time from the name of Darwin, even if his teaching gives place in many respects to

489

further knowledge, and even though he may, like Copernicus, have had many forerunners. They did not carry general conviction, but this one man, the 'Newton of the blade of grass,' called for by Kant, carried the day and gave the intellectual life of his time a new direction. Whatever his personal merits may have been, history rightly connects the whole transformation with his name.

What makes true knowledge so difficult in astronomy, is the smallness of our extent in space, the fact that we are tied to this earth, which thus appears to us as the only stationary body, around which everything else moves. What impedes progress in biology is our narrow limitation as regards time. Humanity only knows its own race and the world of plants and animals, from their present state and from historical traditions covering a few thousand years. What is that as compared with geological time? How should we come upon the idea of gradual transformation, since everything has been the same as it now is 'within human memory.' In both cases a supposed religious objection, the strongest of all possible tendencies, was added.

The history of the controversy concerning the Copernican system is well known. In the biological and anthropological discussion, the opposition of the church, even to-day, does not appear so incomprehensible as in the former case. We will come back to the reason for this later. In any case, it is clear that there could be no question of a scientific pre-history of mankind, as long as people were completely satisfied with the possession of a documentary biblical history of creation, according to which man had been created about six thousand years ago as a single pair of individuals. Everything worth knowing was already told us there. To have forced mankind, that is to say, the educated European world, into at least a profound revision of this simple minded traditional view, is perhaps the most important aspect of the whole of modern anthropological investigation. For if this whole traditional structure gave way only at a single point, the result was an absolute necessity for revising it completely and fundamentally, and that left the road open to simple practical investigations. We shall turn to this later, but we will now first of all see what real result has come out of it.[344]

The most important question is that of the geological age of mankind. That mankind is the last being which has appeared on earth, probably the very last, can hardly be doubted according to all discovery. The chief question is whether we have to assume the existence of mankind in the last epoch but one of the earth's history, the Tertiary, or whether he first appeared in the last epoch of all, the Quaternary or Diluvial. Investigators have generally inclined to the first supposition from the start, but actual investigation has so far obstinately refused to produce any proof of man's existence in the Tertiary. Nevertheless, a certain kind of document has been known for a long time, which has been regarded on many sides as a proof of tertiary man. This is the 'eolith' (meaning stone of the dawn, the dawn of human existence). These are pieces of flint, the edges of which are cut in a peculiar rough manner and sharpened, so that the whole gives an impression of intentional preparation.

Stones of this kind have been found in large quantities in the tertiary deposits of France, Belgium and England. But Rutot, who first investigated them closely and who regarded them as artificial products and hence as proofs of the existence of tertiary man, was not able to carry conviction. Later on, the aspect of matters was altered by the work of Klaatsch, who travelled about Australia, originally with the object of making comparative racial investigations. He found there primitive flint tools among the savages, not only exactly like eoliths, but even in some cases still more primitive. Klaatsch was also able to study on the spot the simple technique used by the natives to produce these tools. All that he saw led him and many other investigators to the conviction that eoliths must after all be regarded as artificial products, and hence as a proof of the existence of tertiary man.

But the opposition to this hypothesis is by no means dead. It is, after all, very strange that not a single skeleton should have been found among so many eoliths to remove all question of doubt. It cannot also be denied that the peculiarities of the eoliths can be explained by natural forces. Under sudden strong pressure the edges of the flint split off in the way seen in the case of the eoliths, and the technique of the Australians is nothing but an application of this fact. It has been observed that pieces of glass which, by falling on a roadway, become trodden on, give the finest examples of miniature eoliths (Sarasin and others).

We must therefore avoid deciding this question here. The fact that eoliths can be formed by natural means certainly does not exclude the possibility that they could be produced by artificial means, as the Australians prove. But the converse is also the case. Let us therefore leave this question for the moment. It is in any case not admissible to say that science has proved the non-existence of tertiary man. It has not proved with certainty that he existed (see page 330), but every day may bring the discovery of a skeleton and therefore of proof. The reason why we are continually brought back to the supposition that our race goes back to this period is the state of affairs revealed to us in our discoveries of diluvial man. Mankind appears to be, already in these oldest remains, at such a high point of bodily and mental development, that if we adopt the view of development and not the supposition of a 'new act of creation' (see below), a long previous history must be postulated for this development to be reached and that can only lie in the Tertiary. But we must not omit to remark that there are also anthropologists who would place the origin of man not much farther back than the last phases of the Tertiary. We have therefore for the present to rely upon the human remains found in the Diluvium, and will there of course seek for the oldest evidences of our racial past. Apart from some doubtful finds in South America and Central Asia, we have to consider six discoveries :—the Javanese *Pithecanthropus*,[345] the Heidelberg, Piltdown, Rhodesian, and Pekin remains, and the so-called Taungs Skull. The order is that of discovery.

The most famous is the find of the Dutch military Doctor Dubois, made in 1891 in the neighbourhood of the village of Trinil in Java. First the roof of a skull and a molar tooth were found : later a thighbone and a second molar. Haeckel, and with him the whole scientific world of his time, enthusiastic as it was for evolution, saw in these remains the long sought-for 'missing link.' But they were immediately subjected to sharp criticism, both as to whether the parts found in different places really belonged together, and also as regards their interpretation. Only a part of the anthropologists adhered to their interpretation as a link between ape and man. Most regarded the creature in question as a blind branch of development, which had arisen before the evolution of actual man, and some even took it to be merely a particularly large ape (gibbon). Both these views gained in weight when it was later shown that true men had lived along with the Pitecanthropus at the same geological period. How dangerous it is to draw final conclusions in branches of science still in the full flux of development, has been very well illustrated in this case. For the remains of primitive man found in the neighbourhood of Pekin have shown that, there also, creatures of the same or a similar kind lived. The so-called *Sinanthropus Pekinensis* is, apart from a few unessential points, identical with the Javanese *Pithecanthropus*, and since the first, of which considerably more fragments, including the teeth, were found, was undoubtedly a member of the *Hominidæ*, the latter also must be the same. The famous Javanese 'missing link' must therefore, after all, be regarded as an ancestor of man.

It is not unnecessary to remark that the gibbon hypothesis was particularly stressed by such critics as deemed it necessary, in the interests of Christianity, to protest against any direct connection between man and the animals. From this side also, there was much ridicule poured upon those who, with Haeckel, saw in Pithecanthropus a true link between man and ape. Now the laugh is on the other side, and only the fact that the whole battle around evolution has died away prevents this fact being common knowledge. Naturally, the truth is that religion is in no way affected by the particular manner in which God may have caused man to come into existence. We shall return to this question later.

Regarding the Pekin discoveries,[347] we should add that the fairly numerous fragments found (teeth, parts of lower jaw, occiput) belong to a number of separate individuals, and in many respects remind us of the chimpanzee, but also allow some relations with the Piltdown fragments to be recognised. The age of the remains is in this case exceptionally easily fixed from numerous fossils which accompany them; this gives the date as the beginning of the Diluvium. The search is being continued, and it is to be hoped that a completion of the discovery will advance matters considerably.

The discovery at Piltdown in Sussex was at first very much contested.[346] While the discoverers (Woodward and Dawson, 1913) supposed it to be the remains of an unknown, late Tertiary or early Diluvial, primitive man (*Eoanthropus*), most well-known investigators at first considered that the very incomplete skull and the lower jaw did not belong together, but that the first belonged to a special race of primitive man (*Homo Dawsoni*) and the latter, including the upper molar tooth also found, belonged to a fossil chimpanzee. During the war, a second discovery at the same spot (1917), the same parts being found together, has considerably increased the probability of the original interpretations. If it is really correct, this is certainly the oldest specimen of mankind known, and it is noteworthy that we have the characters of man and chimpanzee, in a quite peculiar combination.

While the three finds so far mentioned are without doubt early Diluvial or late Tertiary (there is naturally no sharp line of division between these two periods), the geological age of a further important find, the Rhodesian or Broken Hill skull,[348] is fairly uncertain. The place of discovery, the silver mine Broken Hill, is in South Africa. The skull was discovered by miners in 1921, and first examined with care by the English anthropologist Woodward. Its most striking characteristic is the very unusually large eye sockets, in which respect, as in several others, it is fairly similar to a gorilla skull. It would be taken as such, but for the fact that it carries a completely human set of teeth. Since the skull was accompanied by a number of animal bones of undoubtedly recent species, it is necessary for the present to exercise the greatest care in interpreting it. In certain points it shows similarities with the Neanderthal (see below) so that many anthropologists regard it as the African representative of that race. Altogether, not very much can be done with this find.

A further important South African discovery was made in 1924 at Taungs in Bechuanaland, in the course of blasting in a chalk quarry during the building of a railway. The parts found certainly belong to a young creature, a full set of milk teeth and the first permanent molar having been found. In the case of a human being, this would correspond to an age of six years. Of the skull, the whole facial bones are there; the brain cavity is missing, but a complete cast in chalk of it was found, so that a reconstruction of the whole skull does not present the slightest difficulty. The eye sockets are very large as in the gorilla, but the whole form of the skull reminds us strongly of the chimpanzee. It would certainly be ascribed to this animal, but for the fact that certain characteristics are distinctly human. Hence the name given to it by Dart of Johannesburg, who was the first to examine it – *Australopithecus Africanus* – has been contested, since we are perhaps not dealing with a real *Pithecus*, that is, ape. In certain respects the *Australopithecus* or 'Taungskina' resembles the

Pithecanthropus and the Pekin man, but in others it is closely related to the Piltdown skull. A decision is hard to make. We are, as Kleinschmidt rightly remarks in this connection, in the position of a butterfly collector, who possesses for the moment only the caterpillar of a new tropical species.

To these five remarkable finds we must add a sixth, perhaps the most remarkable of all – the famous Heidelberg lower jaw, found in 1907 in a sandpit in Mauer near Heidelberg.[349] As figure 77 (after page 496) shows, this jaw is entirely apelike in its massiveness, in the breadth of the upward branch, in the almost right-angled junction of the latter, and in the very flat *Incisura semilunaris* (the half-moon shaped depression at its upper end) – the likeness again being closest to the chimpanzee – but the jaw is furnished with a completely human set of teeth, and the curve in which they are set, seen from above, exhibits that almost complete regularity which is characteristic of the human jaw. In the case of apes, this curve shows a distinct break at the eye-teeth. In other respects also, the skull is completely human, and hence the name *Homo Heidelbergensis* given to it by its discoverer Schoetensack, seems justified. Unfortunately, this jaw is the sole relic we have of this creature, which is quite unique. The age of the remains is about the first or second intermediate period, and thus in the first third of the Diluvium. It may thus be estimated at about 200,000 years, or perhaps considerably more. In any case, it is considerably older than the relics of primitive man which we shall now consider, and which date from the middle Diluvium, and later.

We now come to a period of which we are able to form a much clearer picture, than we can of the earlier period represented by the few relics we have just spoken of. We have evidence derived from a large number of excavations carried out in the most various parts of Europe (France, Southern England, Sweden, Belgium, Croatia etc.). We know for certain that these men lived together with the reindeer, rhinoceros and mammoth during the period when Europe was partially covered with ice, and that they lived by hunting these and other animals. In the earlier geological layers, we first meet with a race which, in accordance with the first place in which it was discovered, is now generally known as *Neanderthal man* (figure 78). This race was responsible for the oldest flint instruments (wedges), which are recognisable with certainty as of human origin.

The first stages of culture are designated by the name Prechellean, upon which follow the Chellean and the Acheulean (the names are derived from the places where certain finds have been made); and further the Mousterian, etc. In the second part of the Old Stone Age there then appears, alongside the Neanderthal man and later superseding him, a new race, so-called Aurignac man, and then other races, for example, the so-called Cro-Magnon, which approached present day man more closely (compare the two figures 79 and 80, on Plates overleaf). The picture of the Neanderthal man – *Homo mousteriensis* – is not quite right according to later investigations, since the parts of the skull were not put together correctly at first. In later reconstructions[350] the prognathic (underhung) jaw is less prominent). The origin of this race

FIG. 78
Skull of a Neanderthal child from Le
Moustier (from Hauser-Klaatsch)

is unknown. There is apparently no connection between the Heidelberg and Neanderthal man, nor between the latter and the Aurignac type. The Cro-Magnon was originally considered by many scientists to be a cross between the latter. But this hypothesis is mostly given up.

Hauser, who discovered the two skulls figured here in the valley of the Vezeré, believes that Aurignac man came into Europe at the end of the glacial epoch from the East, and then either conquered or drove out Neanderthal man, and possibly ate him. But the remains earlier supposed to be those of a cannibal feast in Krapina, in Croatia, have recently been regarded as doubtful, and it is not even certain that Neanderthal and other bones are mixed up together. Other scientists[351] suppose Aurignac man to have come over from Africa since he is mostly found in Western Europe, but the remains discovered are much too scanty for anything certain to be said on this point with certainty.

The Cro-Magnon race is regarded by quite a number of present-day investigators[352] as the direct forerunner either of the Nordic,

PLATE XI

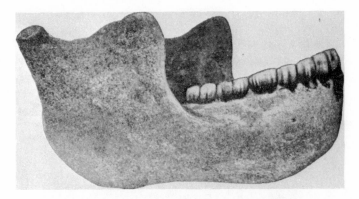

FIG. 77. Lower Jaw from Mauer, near Heidelberg

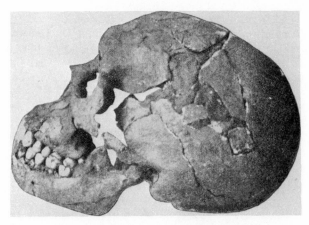

FIG. 79. Skull of a young Neanderthal man from Le Moustier
(from Hauser-Klaatsch)

PLATE XII

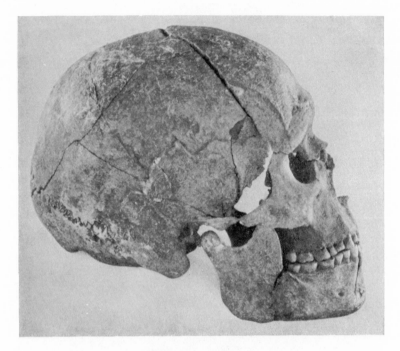

FIG. 80. *Homo Aurignacensis Hauseri,* taken from Dr. Hauser's book *Der Mensch vor* 100,000 *Jahren.* Thüringische Verlagsanstalt, Jena

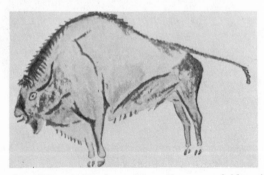

FIG. 81. Bison bellowing. From the cave of Altamiva

or of some race closely allied to it, but to be distinguished from it ('Dalic' or 'Falic'), but this is also by no means certain. We are thus obliged to confess that the discoveries from the Diluvium certainly give us some idea of mankind's condition in those days, but are not in any way sufficient to tell us whence they came or what happened to them. What became of the descendants of Neanderthal men ? And what preceded Aurignac man, who obviously represented a very high type ? And whence did the individual races arise, distinguished by modern anthropology under the names nordic, alpine, mediterranean, near-asiatic, etc. ? We cannot answer any of these questions to-day and it is not even certain that we shall ever be able to answer them, since very few remains have been preserved, and of these only a very small fraction has been found.

We may further inquire what the culture of those diluvial men or their predecessors was like, even if the question of eoliths cannot be answered for the present. We know nothing at all in this respect of Heidelberg and Piltdown man. Perhaps we may once more be fortunate enough to find not only a complete skeleton of the former but also some tools. This would bring us a great deal further.

Concerning the culture of Neanderthal man, on the other hand we already have a fairly clear picture. The youth of Le Moustier, dug up by Hauser, had undoubtedly been buried ceremoniously. It is difficult to resist the further conclusion that human beings practising such customs must already have had a kind of belief in an existence after death, such as is general among primitive savages in our day. The fact that they knew the use of tools, although in a very primitive form, is proved by the discoveries of flints. Even fire seems already to have been known to them, as appears from part of the bones of animals eaten being burned.

The cultural achievements of the Aurignac men were already much higher. They were in all probability responsible for the well known cave paintings, the magnificent artistic qualities of which never cease to surprise us. These paintings are found as a rule in inaccessible or almost inaccessible parts of the caves in question ; they were in all probability not entirely playful, but had probably some religious significance (such as weapon magic). The truth to nature of the drawing, and also of the sculptures on mammoth ivory, is all the more surprising (see figure 81, on opp. Plate).

I1s

It is vastly superior to the products of later times (Neolithic), which are not superior to those of the existing Bushmen,[353] but it is again completely uncertain how these cultures developed. We find the peoples in question in possession of them, but where and how they arose we know as little as we know concerning the race itself.

If we now conclude this discussion of the history of our race by asking what science is able to tell us concerning the relationship between mankind and animals, we must admit that it can tell us almost nothing of a positive character and make only a few negative statements with certainty. One of these is that there can be no question of the descent of man from the apes most closely related to him. The chief reason for this statement lies in the fact that the teeth are in the case of mankind without doubt less specialised in type than those of the gorilla or the chimpanzee. The teeth of the latter might be descended from those of man, but not conversely.

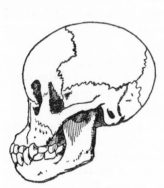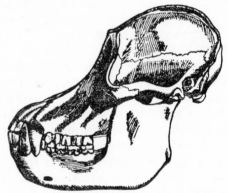

FIG. 82.   FIG. 83.

Skull of a young Orang   Skull of an old Orang

The biogenetic law mentioned on page 445 also points in the same direction. The skull of an orang (figures 82, 83), when young has the closest resemblance to a child's skull, whereas that of the fully grown orang possesses the enormous set of teeth which is the chief distinction between ape and man. If the development of the individual gives us a picture of that of the species, the progenitors of the orang must have been more like man than is the orang of to-day. It is hence generally assumed that the original 'Primates' were more like present day man in the form of the head than the great apes of to-day.   While one fraction retained forms more primitive and hence more capable of development, and was thus enabled to develop into man, another part entered a blind alley by developing a large lower jaw with powerful canines, and produced the present day anthropoids.

While this line of thought would lead us to assume a polyphyletic origin for man, this view would certainly be opposed to that of numerous anthropologists, who are firmly attached to the idea of a monophyletic origin, mainly on account of the unlimited fertility of the different races among themselves, but also on account of the very close bodily resemblance.[354] As regards this latter, it is true, certain very curious experimental results have become known recently which appear to show that human blood may be divided in respect of its composition into four distinct types known as A, B, AB, and O.[355]   This state of affairs was first suspected as the result of blood transfusion in medicine, where along with some astonishing successes, catastrophes were also known to occur.   It was then found that it is not permissible to inject into a human being the blood of any other human being, but only that of the same group or of the group O.   Whether these differences in blood, which space does not allow us to discuss here, have an origin in some way connected with the history of the race is quite uncertain.

The causes which effected the transition to man are just as uncertain as is the question of polyphyly.   Since man is the only living animal which shows in its bodily construction no special adaptation for the struggle for existence – his body possesses neither weapons for attack or defence, nor special organs for flight[356] – it has been concluded that the transition from animal to man can only have taken place in an environment relatively free from danger, and also in a mild climate, which facts, according to Klaatsch, exhibit (?) 'an unmistakable parallel to the well-known story of Paradise in Genesis, and the myth of the Golden Age.'

Most anthropologists are of the opinion that this transition took place in a steppe country analogous to our present day prairie, and very many actually find in the transition from a tree life to a steppe life the cause of man giving up a semi-erect walk for one completely erect, whereby both head and hand were set free for upward development. However that may be, anyone who does not assume that man fell from heaven in a finished state must assume some kind of a transition, and it is further certain that even in the case of a monophyletic origin, a separation of mankind into different races took place very early. It is frequently assumed that the present-day dwarf races are to be regarded as the undeveloped residue of early man, but there are many objections to be made to this view. A beginning with smaller forms would otherwise correspond to a very general rule of the development of species.

Very many different answers have been given to the question as to where, assuming a monophyletic origin, we are to look for the cradle of humanity. Almost all parts of the world have been put forward as such. Many people are of the opinion that humanity was first developed in the famous lost continent of Atlantis, or at least went through its main development there, whereby an explanation is obtained for the absence of all palaeont-ological traces previous to the Heidelberg and Piltdown period, that is in the Tertiary.[357] But these hypotheses are just as uncertain as most others in this field. Taking it all in all, we know very little at the present time, indeed practically nothing, concerning the development of our own species. Neither the problem of genealogy nor that of factors have found anything resembling a sufficient solution in our own case.

The sole matter of agreement between all scientists (with perhaps a single exception, Professor Fleischmann of Erlangen) is that man has descended from the animal kingdom; and we must expressly point out that this is also true of Dacqué, whose writings have recently attracted so much attention.[358] Although Dacqué, as we have already stated above, postulates a special development of man through a series including the whole animal world, and speaks in this sense of a 'Noachite,' 'Adamite' etc., man, he nevertheless expressly states that these 'men' were externally similar to amphibians, higher mammals etc. Their humanity was only potential and they can hardly have made use of it in practice, for they have not left behind any traces of human powers. It seems to give Dacqué pleasure to describe such beings as men, because their great-grand-children finally became so. An ordinary scientist would call such a being an animal, even if it carried the man 'in its knapsack' ten times over.

Still more uncertain than the transition from animal to man is the further question as to the more distant ancestors of man. We have absolutely no information on this point. It was once believed that something similar to a human hand had been found in the traces of an animal, the *chirotherium* (hand animal), which lived in the palaeozoic and mesozoic periods, and that this animal might therefore be regarded as an ancestor of man. But this hypothesis (first stated by Snell and later taken up again by Klaatsch) rests upon too weak a basis ever to be taken seriously by science. It has already been mentioned that Steinmann, on the other hand, wished to regard the metareptile living in the Permian, as the ancestor of the later primates (see page 478). It will be readily understood that empirically-minded scientists, and many others who believe themselves to be interested in this matter remaining a 'mystery,' make game of suppositions of this kind, though this is unjust as long as the authors of the hypotheses in question are themselves conscious of the fact that they are nothing but ingenious and unproved notions.

We may now inquire whether this picture of the result of modern anthropology, which is obviously wanting in very many directions, allows of any conclusion of a general philosophical character to be drawn. In the first place, it is clear that any kind of dogmatism in the midst of this tangle of hypotheses is objectionable, and that every unprejudiced judge can only take the view that we must wait and go further into the matter. Sufficient caution has certainly not been shown at times in this respect (compare pages 455, 484). 'The authors of the theory of Descent,' says Kohlbrügge,[359] quite rightly, 'have hoped to stride forward with seven-league boots over a path, that will undoubtedly need centuries of intensive investigation.' It is entirely to be condemned, when the public is not merely not enlightened as regards the incompleteness of all investigations up to date, but even industriously given the illusion that everything of importance is already settled.

But it is equally unjustifiable when this incompleteness, with its many points open to criticism, is used to discredit the whole investigation and represent it as a collection of arbitrary hypotheses, situated entirely in the air, and merely the product of anti-religious and atheistical tendencies in their authors. For it is no longer the case to-day, that any arbitrary construction purporting to be a history of man, and supposed to serve religious interests, can be brought into agreement with scientific truth. The fact that mankind is older than the 6,000 years of the Old Testament is as certain as any other scientific fact. At the very least 40,000 to 50,000 years have passed since Neanderthal man lived, and at least twice as many since the time of the Heidelberg man, and probably a great many more.

It follows of necessity that the history of the Patriarchs in the Old Testament must have contained purely mythical elements to a large extent at least (which fact has naturally been known to theology for a long time, for other reasons) and that we therefore have complete freedom for scientific investigations, even from this point of view. Our fathers were mistaken in believing that the Bible formed an infallible document concerning scientific and historical questions. The Bible contains matter of religious value – how far this is true of its various parts is a matter for theology – but there was hardly such a thing as science at the time when it was written, at least as far as its authors were concerned, and from an historical point of view it cannot be treated in any other way than other documents. It may naturally contain a great deal of historical matter, which does not concern us here. But it is certain that the story of the six days of Creation and of Paradise, is not history, but myth; that is, religious truth in the form of an historical narration.

The attempt to find in certain results of anthropological investigation, for example the supposed development of man in a mild climate and in an environment free from danger, confirmation of Biblical stories, or even (according to Dacqué) of actual reminiscences of humanity is, in the author's opinion, useless and impracticable, and from the religious point of view injurious; for they assist in thoughtless retention of a blind dogma of inspiration, which is nevertheless untenable, instead of gradually releasing mankind from it. Dacqué's method of the supposed interpretation of palaeontological facts by means of human myths may be used to prove anything. Fancies of this kind help neither science nor religion but merely injure both.

If theology, at least that of all varieties of Protestantism, has long ago freed itself from the idea of the literal inspiration of the Bible (this is unfortunately not true of the orthodox laity), the controversy concerning the problem of humanity is, nevertheless, by no means entirely abolished. For in reality, it was much less the formal principle of Christian doctrine, the authority of the Bible, about which the whole controversy arose but rather the material principles of Christian belief, which are to a certain extent not reconcilable with the experimental results and justifiable hypotheses of science. We shall have to go into this side of the matter later. The question is the quite general one of the evaluation of the human mind generally in relation to the natural (biological) foundations of its being, and this problem interests not Christian dogma alone, but the whole of cultural philosophy, and finally, all sciences of mind, and especially psychology. We must regard the problem of the origin of man, from this side as well, since it is also the kernel of the problem of civilisation.

II

NATURE AND CULTURE

As distinguished from animals, mankind possesses language, morality, law, religion, custom, art, and science, and also the use of tools and technology, social economy, and social life. Of all these things, we find among the higher animals at the most a few suggestions, and these are so generally known that I do not need to state them; we need only refer to the Australian leaf birds with their artistic instincts, and the 'states' of the ants and termites. These suggestions are however nothing more than suggestions, for what is the chief characteristic of man, namely that he carries on in freedom a continual development in all these matters, that his legal and political forms and his art etc., are perpetually being transformed and, at a variable rate, growing; all this is entirely absent among animals. They stand still at the point once reached. Every young spider spins its net in exactly the same stereotyped manner as the old spider, which it has never known, every song bird builds its nest in exactly the same manner as thousands of its ancestors before it. Even when such an instinct (see below) is really changed on account of changed environment, the new state of affairs is just as invariable as the old.

With mankind, on the other hand, change is typical. Upon this depends its endless power of adaptation, which has made it lord of the whole earth. Although it knows tradition and customs, inherited from its ancestors, these are something entirely different from animal instinct, it is not unconditionally bound by them, but can change them at any time, once it makes up its mind to it. Whereas the animal is obliged to go the way which its instinct determines with the same necessity that water runs down hill.

This confinement of the animal to stereotyped behaviour corresponds to its limitations to concrete single experiences. We have no real proof that an animal has any power of forming an abstract concept, a general law, or a generally valid principle, as regards any field. The dogs and horses which extract cube roots are therefore more than suspicious from the start. Why in all the world, if all these animals possess such powers, have they never made use of them? A dog certainly associates in a short time a whipping with the theft of a sausage, but it never thinks of forming a general notion of theft, no matter how many such experiences it may have, let alone a general notion of good and evil. This want of the power of abstraction is connected with the absence of speech, since the word is the materialised concept. Anthropoid apes, as has so often been observed, are not devoid of the organs of speech, that is, of the tongue and larynx; they do not speak simply because they have nothing to say. However differentiated animal sounds may be – a cat alone can express a large number of different moods and desires by its mew – the animal is devoid both of articulate speech and conceptual thought.

It is also devoid of the power of considered action in the full sense of the word, that is, the power of acting efficiently in face of completely new situations, of inventing new tools to meet new needs, and of building up that great power of action upon nature which we admire to-day as technology. At this point the contrast is not quite so sharp as in other directions. There can be no doubt that higher animals, for example, dogs, elephants, cats etc., are capable of actions which do not depend on mere inherited habit, not upon trial and error (personal experience therefore), but are the result of forming initially a useful combination of certain circumstances.

W. Köhler has proved this with certainty with some of his chimpanzees.[360] These arrived, although with some difficulty, at the right idea on their own account, that, for example, a box might be placed beneath an apple and used to assist in reaching it; they therefore use it in the truest sense of the word as a tool. They also arrived on their own account at the notion of using sticks for similar purposes.[360] When a dog's master in going out shuts the door against it, it often occurs to it to run round as quickly as possible to another exit in order to see if it cannot follow its master in that way. Such actions may justifiably be ascribed to thought, but even the arts of the most primitive people are beyond the highest efforts of animals. An animal reaches in this respect to about the stage of a twelve to eighteen months old child, at the very best.

With this, we have recounted the most important points which distinguish man from animals, but naturally a great deal might be added. Altogether, the conclusion is, that the beginnings of all human faculties are to be found in animals, but that in order to turn such a faculty into the power of a civilised human being, a certain something must be added. This something is what is usually termed mind. It is obviously related to the low psychical life of animals in the same way that the organic is related to the inorganic. One does not exclude the other, but includes it and brings it into a higher and more comprehensive region. In other words, we have here as we have already mentioned at the beginning of this chapter a new emergent of the development of life upon earth. All this is so clear that no further account of it is necessary. But it is much more difficult to answer the question in what this new thing ultimately consists and how we are to imagine its appearance from the animal kingdom.

A comprehensive and perfect definition of mind can hardly be given. It is certain that self-consciousness (the ego, and the feeling of possessing freedom of will) must be regarded as two fundamental components of this higher human existence. A child becomes a human being in the full sense of the word at the moment when it first uses the word 'I will' fully consciously. Child psychology[361] also teaches us that this rise to a higher stage does not take place suddenly and finally, but that it is only at first in certain moments and on special occasions that the consciousness, hitherto sunken in purely instinctive life, becomes aware of itself, only to sink back again quickly, and often not to reach the same point again for weeks. Only quite gradually do such single experiences build up a continuous, conscious, and plainly recollected connection and one day the little creature becomes a new 'personality.'

If we may apply to this matter the ontogenetic-phylogenetic parallel, we must conclude that the development of mankind took place in a similar way. At first single highly gifted individuals may have shown signs of higher mental power. Special occasions, for example great common danger to the herd, may have been the direct occasion for such mind power breaking through. But this again disappears, and only in the course of many thousands of years will such a 'super-consciousness' (see also below) have gradually appeared more frequently and for a longer time. This hypothesis must at least be made by any one who is not willing to assume that a radical mutation, which in this case would be nothing but a genuine act of creation at a certain time (with all the objections which can be raised to such an idea), suddenly caused the production of a human being in a finished state, from man's immediate animal progenitor.

We have also good grounds for supposing that the most important part in this process was played by speech. Conceptual thought did not develop before language, or after it, but simultaneously with it. Thinking is actually inward speech, and can be shown to be always accompanied in this way, indeed most people always think so loudly that actual movements take place, as we shall see below in another connection.[362] The believers in a temporal act of creation are accustomed to object at this point that thinking must have existed before language could have come into existence as its tool. The answer is, that as long as language did not exist thought was impossible, for it is only possible to think in a language. The converse is also true, as we have seen above, in considering animal language. It follows of necessity that both must have developed together. There are logically some grounds for giving priority to thought, but this purely logical priority must not be turned into a priority in time. Furthermore, we must not forget that it is not language alone that has produced the effect, but that the use of tools, that is conscious action, has had as much to do with the development of man as conscious thought.

When we take the last factor into consideration, and then inquire what the process of the development of language, and hence of abstract thought may have been, we must admit that we are completely ignorant and shall probably remain so. For it is clear in the nature of things, that we shall never be able to learn anything with certainty regarding these beginnings, since writing was certainly a much later discovery than language, while the latter naturally leaves no relics behind. We have not the slightest idea whether, or how, the Heidelberg or Piltdown man spoke. The conclusions which have been drawn from the development of the roots of the muscles on the jaw bones are and remain uncertain hypotheses.

The most we can hope for is some light from an investigation of present-day primitive peoples. We learn from these that the language of gesture is much more used by them than by us, and it is clear that it is a matter of indifference whether an optical or acoustical sign is used for a certain idea such as water, enemy, or buffalo. We may therefore suppose that at first only certain concrete objects of everyday importance, such as those mentioned, received such signs, which were transmitted by sound or gesture to others, the beginning being made by pictorial representation as regards sight and imitative sound as regards hearing (for example the mew of the cat).[1] This process must have taken a great deal of time, and new signs were only added quite gradually. The whole process of cultural development shows everywhere, without doubt, the typical character of an exponential function; at first it goes quite slowly, then more quickly as more is accomplished, and finally it moves at express speed. The reason for this is naturally that every step forward in any region is fertile in results for the other regions.

If we prolong this function backwards towards prehistoric and historical times of which we know something, we can form some conception of how slowly things must have moved at first. This will be true for language in particular and hence for conscious thought. That the beginnings of technology lie in the development of the first simple tools, such as clubs and flints, is so obvious that it is hardly necessary to give any reasons for it. The most important invention, fire, certainly came somewhat later, but then formed the starting point for the whole further development. One of the most important stages on the further road was the invention of the first simple vessels, which led from the use of natural hollow bodies, such as coconut shells and animal skulls, to home-made clay vessels and baskets.

Where and when these inventions were made is beyond our knowledge. They appeared in the later Stone Age, and at once became the instruments of further cultural, and particularly artistic, progress. The products of technology and art allow us

[1] Reference should be made to Sir R. Paget's theory of the development of speech from internal oral gesture. Deliberate movements of the hands and arms tend to be accompanied by movements of the tongue, etc. A child's first attempts to write exhibits this fact amusingly. The exhaled and voiced air gives rise to different sounds accordingly. See Paget, *Human Speech*. (London, 1930.) [*Trans.*]

to follow the process called by Wundt 'heterogony of purposes'; the gradual liberation of the technical and aesthetic ideals from direct utility. Even in the oldest cultures which show clay vessels in any form, we already find on them a simple ornament, which without doubt merely satisfied an aesthetic need, and the same is true of all tools of prehistoric man.

The carrying out of the technical ideal, on the other hand, the striving after fitness for a purpose for its own sake, is of recent date, and has only been clearly envisaged by modern Western civilisation. The technician of to-day builds his machines, buildings etc., as fitly designed as possible, even when a direct necessity is not present. It goes against the grain with him to waste energy, material, or human power. He builds the foundations of a railway embankment, a water-way, a bridge, as strong as is necessary to carry the load, but no stronger. Ancient architects, on the other hand, built as strongly as they were able, according to the principle that double sewing holds better. Here, therefore, something useful was being constructed, but utility as such was not yet the highest ideal; other ends were in view.

Matters are quite analogous in the development of the sciences. These also, at first, serve a practical need. Geometry arose from surveying and architecture, physics from technical problems, astronomy from the necessity for calculating the calendar etc. The development of an ideal of pure science, of the discovery of truth for truth's sake, is, as far as we are able to learn from history, a very late development. Although we cannot subscribe to the traditional idea that the Greeks were the inventors of pure science in this sense, since it is at least doubtful whether this ideal did not become conscious in Egypt, India and even in Sumerian Mesopotamia long before the great period of Greek science, it is nevertheless certain that this scientific ideal is of very late date as compared with the total age of humanity.

The most difficult question to answer is that concerning the beginnings of social life, and of the ideas of law, morality and religion. Certain social institutions are already found in the case of many animals, to mention only insect states, herds and their leaders, animal 'marriages,' the care of offspring.[363] There can hardly be any doubt that mankind possessed a first and simplest social connection, that of the family, from the start, for the great apes nearest related to him show well developed (monogamic) family life, and the love of the mothers for their young is proverbial. Only a very foolish view, derogatory to nature and only apparently religious, can find anything undignified in the fact that love between parents and children and married people is based upon our prehuman nature. Religion cannot do better than consider them sacred for the very reason that they are so deeply rooted in nature.

The same is also true of the social instincts, which man must also have had like the great apes, before his development to self-consciousness. He has been a political animal (*zoon politikon*) in the true sense of the word from the very beginning. But in these matters also, heterogony has been subsequently displayed; rules of behaviour originally followed purely instinctively became 'principles of action' by reason of their coming into consciousness. Many such principles finally produced a general moral principle, the command to do good and avoid evil. When, how, and where this first happened is again undiscoverable. We have no idea, for example, how far the Neanderthal or even the Heidelberg man was acquainted with moral or legal principles. Social rules certainly exist in the case of herd animals, and individuals who offend against them are relentlessly killed or driven out of the community (solitary buffaloes or elephants).

The still more difficult question of the origin of religion cannot be discussed here, for the result would be altogether too open to attack unless of sufficient length.[364] We need only mention that numerous recent investigators no longer adhere to the series which was once very generally assumed, namely, the order of development: animism, fetichism, totemism, polytheism, henotheism (monolatry), monotheism or pantheism. They regard, as more probable in the beginning, an indefinite belief in a mysterious power dwelling in all sorts of things, the 'mana' which is later succeeded by animistic and totemistic ideas, fetishism being a degenerate form, which branched off from the line of upward development. But we know in this matter also nothing with certainty. Religion is the only higher faculty of mankind, of which we cannot find any beginning in the animal kingdom. It is obvious only that it is closely connected with the 'occult' faculties of spiritual life (see below), and hence we may find in certain animal feasts, for example in the well authenticated 'charming' of birds by snakes, a certain point of contact with the human faculty of grasping the *mysterium tremendum fascinosum* (Otto).

There is a great deal more that could be said on all these matters besides these few short remarks. But we cannot go into the subject any further here, and can only make two statements by way of conclusion. It is certainly quite wrong to represent prehistoric man, as is often done, as three-quarters animal,[365] without articulate language, instigated only by lust of the chase and knowing only the right of the stronger; falling in herds upon bears, mammoths, bison, and then fighting over the booty. There must have been friendly relationships, common games, family life and social feeling even in those days; and just as primitive savages to-day are mostly big children, the same will have been true of mankind at that time. But on the other hand, the opposition to all exaggeration of the idea of development must not be driven so far as to refuse to admit any kind of development, and pretend that there was no serious difference between prehistoric man and ourselves.

The specifically human character must have developed slowly. Or are we to suppose that mankind, after having been produced at a certain time by mutation, in other words, by an act of creation, in a finished state, also suddenly received language, artistic powers, morality etc. ? What language did he speak ? Possibly Hebrew ? This view is obviously out of the question. But if this be admitted, and it be granted that everything must have developed slowly, it may nevertheless be said that development could not have taken place unless the potentiality were there, and so cannot be explained from the animal point of view. We may reply that this argument consists simply in projecting backwards what exists to-day as potentialities into the past. This does not bring us a single step forward in our understanding of things, just as little as does the life-force of the older vitalism. Anyone who admits a gradual development in all these respects has no reason for imagining more in the earlier stages in the form of potentialities than actually appears in the characteristics of these stages.

The same is true, as was pointed out, concerning Dacqué's Noachitic and Adamitic men, which were not men at all, but amphibians or reptiles or something else. This is merely playing with words in order to cover up the conflict that certainly exists between the idea of development and certain ideas hitherto held on the religious side, which however are found on closer examination not to be necessary to true religion. The dignity of mankind, which naturally cannot give up religion without giving up itself, does not depend on the manner in which mankind came about, but upon what it now is.

The mistake made by the pure evolutionists was to imagine it necessary to stamp mankind as merely a higher animal, when it was found that he was derived from the animals, and the mistake was made by their opponents on the religious side of believing this, and hence of opposing the idea of development itself, instead of rejecting the false conclusions drawn from it. Man has grown from the animal kingdom, but he is now much more than an animal. He stands with his feet on the earth, but his head reaches to heaven. He is really 'the wanderer between two worlds.' If we adhere quite clearly to this result we are now equipped to discuss the second group of questions, the question of the relationship of the natural foundations of human existence at the present time to man's personal high intellectual powers and to the realm of cultural values, which he has built up with their aid. We will first consider the biological foundations of mental life.

III

BRAIN, SOUL, AND CONSCIOUSNESS

We now proceed from what has been said in general above concerning the problem of body and soul, and will enlarge this with particular reference to human, spiritual and mental life.

Man has the heaviest brain of all animals (absolutely; and also, when we exclude very small animals, relatively), a fact known to every school child to-day. It is evident that this is closely connected with his mental powers, since we are otherwise able to prove throughout the whole of the animal kingdom increase in mental capacity with the development of the brain. The cerebrum, which undoubtedly is responsible for the higher intellectual activities, is a late acquisition in the history of the species; it was preceded by the cerebellum, and this again by the prolongation of the spinal cord called the *medulla oblongata*. At this point a particularly important investigation deserves mention, due to the anthropologist, Kleinschmidt[366], who started life as a theologian.

By careful comparison of the brain cavities of various recent and fossil human races, in particular the Neanderthal, Kleinschmidt showed that it is an error to draw conclusions concerning the mental qualities from the external form of the skull. In the case of the Neanderthal man for example, the strong projection of the bone over the eyes does not arise out of a special form of the brain cavity, but is simply the result of a particularly strong development of the sinuses (the hollows in the bone under the eye-brows, which are connected with the nose, and often give rise to unpleasant headaches). The two annexed figures show better what is meant than any further description would do. According to Kleinschmidt's results, the differences between single races of man in the form of the brain are extremely small. But it is nevertheless true that the average brain capacity of the Negro, Blackfellow, etc., as compared with the European, or of the Neanderthal as compared with the Aurignac, is smaller.[367]

Concerning the functions of the various parts of the brain, and the question of the localisation of the individual human faculties, for example, speech, hearing etc., we have thousands of special

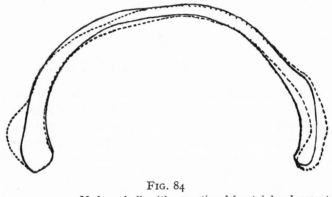

FIG. 84

—————— Modern skull, with exceptional frontal development
·················· Neanderthal skull, with strong frontal development

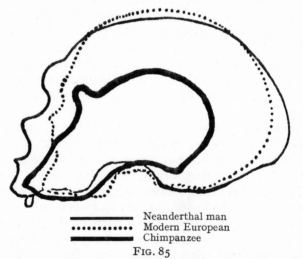

—————— Neanderthal man
············ Modern European
—————— Chimpanzee

FIG. 85

Actual relative sizes of skulls.　From Kleinschmidt's *Formenkreislehre* (1926).

investigations, the results of which, though extremely interesting, cannot be described here.[368] The most valuable material is given by pathological anatomy, then by experiments with animals, which are unfortunately unavoidable in this connection, and then by experimental psychology.

We know to-day from these investigations, that there is no strict localisation of single mental operations in the brain – as was assumed by Gall and others, but that certain areas (which nevertheless run into one another) are normally the bearers of certain faculties, for example, the association of concepts with speech and with writing. Nevertheless, such localisation cannot be proved for the highest mental operations, while it is on the other hand certain that they are in some way bound up with the correct functioning of the brain. In the war, injury to, for example, the moral feelings or logical powers as the result of brain damage was often observed, without any appearance of the failure of important lower senses. The patients in question began to tell lies, to steal, and the like, or were unable to draw sensible logical conclusions from given data.

It is therefore certain that the highest spiritual and mental activities depend to a high degree upon the brain, but it is nevertheless unquestionable that when the brain is damaged, other parts of it are able to take upon themselves the functions of the damaged part after a time, as already mentioned above. In face of these facts, two views are still fairly directly opposed to one another; one regards the whole of our mental and spiritual operations as pure 'functions' of the brain, which come into being with it, disappear with it, and are completely dependent upon it. According to the other view, the brain stands in the same relation to the mind, as the instrument to the player. The latter is naturally unable to play when the instrument is broken, but he himself is not the instrument or any function of it, but its guide and master.

It need hardly be said that most scientists and medical men lean to the first view, which has a strong resemblance to the materialism of former times, while the other is more in favour with vitalists, idealistic philosophers, and theologians. The author is of the opinion that it is useless to discuss such matters at the present time, since we are absolutely unable to say anything about the details of the psycho-physical problem (see above). It is probable that neither is the mind a mere function of the brain nor the brain a mere tool of the mind, but that both are functions of an unknown 'third,' and are therefore connected with one another in some way – how, we do not know.

The matter is still more mysterious to-day – although the

beginning of the way to a solution may lie here – because we have no longer to reckon only with our conscious mental life on the one hand and with our bodies on the other, but with a third factor, namely the 'unconscious mind,' which at one time was bitterly opposed by the extreme representatives of both the materialistic and the spiritualistic positions. On account of the fundamental and important rôle which these matters occupy in psychology and in the whole of the philosophy of nature, we must discuss them at greater length.[369]

Psychology believed for many decades that it would be able to discover the secrets of spiritual life by means of 'experimental methods.' Much interesting matter resulted, but taking it all together, we are unfortunately obliged to admit that the enormous amount of trouble taken has hardly brought a due reward. All these investigations concerning associations, obstructions, reflexes, etc., have not brought us any nearer to the actual problem, and psychology finally lost sight of its true problem, the soul or mind itself, entirely. Now modern psychology has made the important discovery of the unconscious life of the mind, which was first declared by many learned men to be a contradiction in terms (since 'mental' was supposed to be identical in meaning with 'conscious') but it is nevertheless generally accepted to-day, since the assumption is supported by cogent reasons.

The most important facts are those concerning psycho-analysis and posthypnotic suggestion. Psycho-analysis – if we omit here its pansexualist application by Freud – proves that what are called repressed complexes exist, which may influence the actions of a person for years and even for his whole life, without his being conscious of these deep-seated forces. Only by means of a careful 'deep analysis,' in particular with the aid of the analysis of dreams, can these complexes, which are hidden in the unconscious, be dragged to light and got rid of, whereupon the disturbances to normal life disappear. The evasion, that we are here dealing as in all cases of the unconscious merely with physical dispositions, is not admissible, for physical dispositions are obviously incapable of setting going every time, in innumerable instances, precisely those reactions which result from the repressed complex and the momentary situation. Combinations of this kind can only be understood as mental processes; a process of deduction, consideration, choice, etc., must have taken place.

In short, nothing but purely mental processes can have preceded the corresponding action.

These processes, however, take place in the unconscious, and it is only their result that comes to light. The matter of dream symbolism seems to be quite analogous. When the repressed feelings, wishes, etc., so often take the most extraordinary forms of expression in dreams – forms which can only be interpreted by a trained expert – mental activity is necessary for this transformation, which simply cannot be ascribed to brain molecules or electrical fields of force. Dreams in this respect often act in an artistic manner, and artistic achievements are usually regarded as of a mental or spiritual nature.

More convincing still than the psycho-analytical phenomena are those of post-hypnotic suggestion. Say the hypnotist gives the hypnotised subject a quite general order, to tell the first acquaintance that he meets next day at three o'clock something agreeable relating to his working life. The subject actually carries out this order, but is nevertheless highly surprised afterwards on being told that his kind remark was not a free effort of will, but forced upon him posthypnotically. The only possible conclusion is, that all mental processes, from the general order to the particular form in the given case, take place unconsciously. On the other hand, it is quite unimaginable how a purely material 'historical' reaction basis, could possibly achieve such effects which might have an infinite variety of forms, according to which acquaintance the subject happens to meet. It is quite impossible to avoid the assumption of unconscious mental processes in this case. But if it is necessary to postulate them in this case it is then simplest to make use of them for explanatory purposes in other generally known cases, where for example we are unable to find the solution of a problem in spite of earnest thought, but afterwards find that the solution comes into our mind when otherwise occupied. Here also everything becomes clear as soon as we assume that the unconscious has gone on working while the waking mind was busy with other things. On the other hand, to cling to the alternatives, conscious or bodily, must lead, here also, to the most absurd accessory hypotheses.

Further reasons for the assumption of unconscious mental life are given by the facts of multiple personality (in bad pathological cases, so-called demoniacal possession), automatisms (table-turning, automatic writing, etc.), in short, the whole field of those phenomena usually designated by the words occultism or spiritualism, which are not really occult in the narrower sense, but are easily subject to psychological explanation on the assumption of unconscious mental phenomena. They will be found in any good text book of so-called parapsychology,[370] and they are recognised by the whole of science to-day as real effects, and must not be confused in this respect with occult matters in the narrower sense, such as telepathy, clairvoyance, etc., the existence of which is denied or at any rate much doubted, on many sides. We will speak of these matters at length in a short time. For the present, what we have said is sufficient to give some grounds for the conclusion drawn above, that the supposition of unconscious mental life can no longer be avoided. As a matter of fact, it is made by far and away the majority of investigators in this field, and we must first consider of what use it is to us in our discussion of the fundamental problem of the relation between body and soul.

We must first of all realise that, if we are no longer able to avoid this assumption, the whole mental life of animals is brought into a new light. In this connection we must first recollect the extraordinary achievements of animals often designated by the much misused word, instinct. An instinctive action is defined as one which, on the one hand, is obviously adapted to a quite definite purpose which may often be only attainable by highly complicated means; but, on the other hand, cannot possibly be traced back to experience, since it is performed without the animals ever having tried to perform the action previously, or ever having seen it performed. Thus, for example, the yucca moth already mentioned above, and flies, perform quite complicated actions in the flower of the palm lily and of the fig respectively without any previous knowledge whatever of the position of affairs, or of the final result of their actions (which they do not survive to see). Nor has a spider which spins its first web immediately upon emerging from the egg, ever seen anything of the kind.

In the insect kingdom, these achievements are quite particularly remarkable, and one may even say that this branch of the animal kingdom has developed farthest in this respect. But they are never entirely absent, and even man possesses some few instincts, such as that of sucking in the new-born child, although man represents the peak of the contrary type of development, that in the direction of free mental activity. An essential feature of instinct is that it always proceeds in the same manner. An animal acts under its influence as if under irresistible compulsion as has been shown in many interesting experiments with hymenoptera (wasps and bees). But, on the other hand, this automatism does again not prevent the drive, for the satisfaction of which the instinct exists, seeking another mode of expression when the external circumstances are changed. There are caterpillars, for example, which roll up certain leaves from quite definite points, usually the tip, and spin them over to form a cocoon.[371] These actions always take place in the same way. But if the tip is cut off all leaves available for the caterpillar it can be compelled to perform the business in a different manner, with the production of a cocoon. In this case, the course of the action is changed, but in such a way as to accord with the original purpose. In this respect also, instinct resembles a hypnotic command, which likewise, when deprived of the normal means of its execution, finds some other method.

We can hardly be mistaken in further supposing that the 'tropisms' described above, regarded from the psychological standpoint, are fundamentally of the same character. From these, however, there would appear to be a fairly continuous bridge to the simple reactions of a physico-chemical system, determined by law. Might we perhaps venture upon the hypothesis, that physical causality on the one hand, and conscious free action of man on the other, are only the extreme ends of a series, in the middle of which we find the phenomena just discussed, and that therefore, we are possibly dealing with a similar case as that of waves and corpuscles in modern physics? At first they appear to us absolute opposites; but perhaps they are not contradictory, but contrary opposites, between which there is a continuous series of intermediates. At present we do not know whether this is a possible way to solve the whole problem, nevertheless much else tells in its favour, as we shall see shortly. But we must first say a few words concerning the occult phenomena just mentioned.

Occultism as a science is quite a recent development. As a superstition and pseudo-religion, as a magic art and 'medicine,' it is probably as old as humanity itself.[372] If we first of all refrain from all criticism, we can divide what is sometimes called to-day 'parapsychology" instead of occultism, into the following five main groups of phenomena (real or supposed).

The first group is formed by the phenomena of the unconscious shortly referred to above, in particular therefore suggestion and hypnosis, automatism, cryptomnesia and cryptaesthesia (that is, abnormally increased power of memory and perception), schizophrenia and multiple personality, and some others which we will omit.

The second contains the 'parapsychical' phenomena in the narrower sense; that is, telepathy, clairvoyance and its branch, so-called 'psychometry' (clairvoyance directed by objects), prophecy (clairvoyance as regards the future).

The third group is formed by what are called paraphysical phenomena; telekinesis (motion at a distance), materialisation, ghosts, and the like; the fourth by the so-called spiritualistic phenomena in the narrower sense (communication with supposed super-human intelligences) and the fifth by astrology and other varieties of cosmological secret sciences, in which must be reckoned theosophy and anthroposophy. With the latter the circle is closed, for from this point we are led immediately back to the phenomena of the first group, in particular, schizophreny; the mystics of all shades belong with their ecstasies to the first group, but also on the other hand in the main to the fifth group.

Unfortunately, space will not allow us to discuss all these matters in a way sufficient to do justice to the subject, a knowledge of which is absolutely essential if one is to get as clear a view as possible concerning the final questions of the philosophy of nature. I must therefore expressly ask the reader at this point to consult a good modern description of the matter,[370] and must limit myself here to a few short indications of the most important points.

As regards the first group, the super-normal powers of the unconscious, we are chiefly interested in the cryptomnesic and cryptaesthetic phenomena. It is certain that the unconscious retains not only the memory of things long ago forgotten, but also of things which were perceived with some sense organ or another, but not 'apperceived,' that is, not made clear to consciousness; amongst others, those images which were only received upon the edge of the retina, and never consciously. The case of the servant girl is famous, who suddenly spoke Hebrew when in a fever. It was proved that she had been in service twenty years before with a clergyman who was accustomed to walk up and down his room for half an hour and read aloud from a Hebrew bible while the girl cleaned the room. In other cases[373] unexceptionable proof was given that persons had noted matters which they could only have derived from a newspaper lying in a café which they had glanced at as they passed, but about which they had not the least knowledge.

The well-known fact, often confirmed, that in moments of the greatest danger, the events of the whole life with their smallest details, pass before the inward eye like a film, depends upon this entirely unexpected memory power of the unconscious. The best proofs of hyperaesthesia are given by certain experiments on sensitive persons, which may of course be easily falsified by unconscious suggestion and perhaps by telepathy. The most famous of these are the investigations of Chowrin,[374] whose subject was able to decipher writing by feeling with the hand through seven sheets of writing paper laid upon the top of it, and curiously enough was also able to recognise the colours of papers contained in closely sealed cardboard boxes, in the same manner. In this case, the fact that the power diminished with increase in the number of the sheets of paper is in favour of telepathy and true clairvoyance having been excluded.

The phenomena of hypnosis and suggestion are generally well known to-day. We need only remark that in these conditions, as in the case of the very closely allied autohypnosis or trance, we are dealing with a more or less complete elimination of the guiding consciousness, in such a way that either the unconscious is able to act freely, or a guidance is taken over by a foreign consciousness. The hyperaesthesia of 'mediums' (that is to say, persons with a strong tendency towards elimination of the consciousness) has the effect of causing them to react to the very slightest hint of the experimenter. And thus in telepathic and other series of experiments numerous sources of error arise in this way. It is certain, for example, that such persons are able to read off the reversed image of playing cards held in the hand and reflected in the eye of the experimenter, and are thus liable to simulate clairvoyance or telepathy. Likewise, they are able to perceive and interpret the quite involuntary whispers or movements of whispering which most people make when thinking intensely about a certain matter, which may again simulate telepathic powers.[375]

It occasionally happens that narrower and more durable complexes are formed within the unconscious, which represents the total sum of all ideas and wishes which have been purposely repressed as unethical in the course of the individual life. This may further lead to the complete formation of a second personality, which generally alternates with the normal personality, one not knowing what the other has said and done. In less common cases, however, both personalities exist together, and cause extremely painful states of mental strain. The case of Miss Beauchamp is well known, whose two personalities were finally welded together by Dr. Prince, who also described the case.

High degrees of schizophrenia of this kind must have played a decisive part in the witch hunts of the Middle Ages, but they are also present in the case of our present-day mediums, magicians, and ecstatics. Only, on account of the changed conditions, the split-off personalities give themselves out to-day as the spirits of the dead, while in those days they figured as devils and evil spirits.

This subterranean world exists more or less in everyone, and only dreams, and possibly hypnotic experiments, throw any light into it. If these depths are analysed in this way, or by means of the 'automatisms' (table-turning, automatic writing, etc.) a striking fact is the peculiar power of the unconscious to make use of all kinds of symbolic expressions. The complexes of wishes or ideas which are repressed in waking life, in particular those of an erotic description, form themselves in dreams (also under anæsthetics) into many very strange symbols, and the very remarkable phenomenon called transposition takes place. A group of ideas which, for example, were originally of a visual nature (things seen, etc.) is revealed in an acoustic form, for example as conversation, but nevertheless betrays its real nature by the identity of the contents. This peculiarity of the unconscious, which has often been observed is the reason for many apparent occult phenomena (and also, by the way, gives a new and striking proof against the purely physiological theory of the unconscious, since such a transposition undoubtedly represents the result of mental work).

We now come to the occult phenomena themselves, and will take first the second group, the parapsychical phenomena telepathy and clairvoyance, with their various sub-divisions. Twenty years ago, the line was drawn very decisively by orthodox science between the phenomena just discussed, which were regarded as genuine, and the latter phenomena which were regarded as false. But quite recently a change of opinion has taken place, and the second group also is now regarded as genuine in principle (although not in every individual case). In other words, it is agreed that there is such a thing as true knowledge of the content of another person's mental life, which is not transmitted in the usual way by the known senses. Even such critical investigators as Lehmann, Dessoir and Baerwald, admit to-day the existence of genuine telepathy. Only a few recalcitrants such as Moll, Bruhn and others, still strive to hold the old position, but they may be regarded as already defeated, since the number of proofs of the other view is too great. Further information will be found in all later books on parapsychology, even those written from a highly critical point of view.[370]

Instead of many instances, we will only give the most famous, the story of the watch. After the American medium, Mrs. Piper, had been in England for a few days, she was investigated by a committee which she had never seen before, and one of these, the physicist, Oliver Lodge, handed her a gold watch. Mrs. Piper declared that the watch had belonged to a deceased uncle of Lodge's (Uncle Jeremiah) who had had an accident once while swimming with his brother, who was long dead and who had not been known to Lodge; that the boy had owned a gun and a stuffed snake, and had once, together with his brother and other boys, beaten a cat to death on Smith's field. These statements could all be verified with the expenditure of great trouble. It is impossible to suppose that Mrs. Piper, already knowing in America that Lodge would be present among the investigators, employed an army of private detectives to find out all about Lodge's uncle. This hypothesis is of no value when we consider how much trouble private investigators, who knew all the family relationships, had to confirm these stories by questioning old people at the home town of the uncle, etc. Mrs. Piper would have had to be a Croesus in order to be able to perform this feat, not only in this but in very many other cases.

We are obliged to assume that a genuine occult phenomenon happened, that is to say, something that cannot be explained by our present methods. Mrs. Piper, like all mediums, believed that the spirit of the dead uncle had been talking through her. For science, however, one of two suppositions is sufficient, either that the stories were buried in Lodge's unconscious mind from the days of his childhood, and that Mrs. Piper took them telepathically from this source, or that there is really such a thing as genuine clairvoyance, that is, a view of the past orientated by a certain object. (In occult literature this kind of clairvoyance is called psychometry.)

In the present case it is difficult to decide between the two hypotheses. The fact that Lodge knew nothing of the stories does not prove anything, on account of the hypermnesia of the unconscious. Similar remarkable results have been recently produced by Doctor Pagenstecher, a medical man of German origin from Mexico, with his medium, Maria Reyes de Z. Here again the question as to whether telepathy or true clairvoyance is the explanation is difficult to decide for the present.[370]

Critical investigators, such as Baerwald, are inclined to-day, in view of the impossibility of denying such really occult phenomena, to admit telepathy, but to continue to reject everything else as supersensual. Telepathy would then be explicable by known physical processes, since wireless telegraphy proves the possibility of the conveyance of intelligible pictures by means of electric waves, even through empty space. It is supposed that something similar may lie behind telepathy. It might be assumed that the transmitter and receiver, which in any case are usually people who are spiritually 'in tune,' represent a certain analogy to the transmitter and receiver in wireless.

The author believes that there is serious objection to be made to the hypothesis. If telepathy, the existence of which can hardly be denied any longer, depends upon physical transmission of energy, it would certainly have the characteristic possessed by all transmissions of energy through space that it would decrease in intensity with the square of the distance. Nothing of this kind appears in the cases which have been reported, assuming their genuineness. The communication between two minds or brains takes place, so it appears, with equal clarity at any distance, spontaneous cases of this kind being reported as happening at a distance of many thousands of kilometres across the ocean; and there is certainly no sign of higher intensity when the distance is very small.

To this objection we must add a second. The supporters of the view in question are compelled to interpret all cases which on the face of them suggest true clairvoyance (telepathy being apparently impossible) in this peculiar manner, which often leads to very forced hypotheses. One of the most striking cases of this kind is that of the Loretto chapel, which will be found in Baerwald. The English archaeologist, Bligh Bond, who had the task of excavating at an abbey destroyed during the Reformation and partly buried, conceived the idea of asking a well-known medium to describe, by means of automatic writing, the position and character of the remains which he might expect to find. The 'spirits' called up by this medium, those of the one time abbots, etc., not merely gave a description (of the Edgar Chapel: *Tr.*) afterwards confirmed in detail, but also an indication that close to the abbey another building would be found buried, the Loretto chapel. Bligh Bond sought for this twice without result, but finally found it after the war in 1919, and again it proved to be exactly as described by the medium.

It is clear that this case is inexplicable on a natural basis, that is to say, by means which we are accustomed to imagine. Baerwald wishes to refer it to telepathy; he thinks that Bligh Bond may perhaps once have had in his hand an old document, or something of the kind, which he had now long forgotten, but in which these things were described – the medium having drawn this knowledge from his unconscious mind. The possibility is certainly imaginable, but it does not appear very probable to the author. For a detailed description as given by the medium requires that Bligh Bond did not merely read the documents or plan in question once and hastily, without taking them into his waking consciousness. Even the unconscious cannot take in more than is offered to it. But the knowledge in question would have necessitated the reading of several pages, or even a whole volume. That could only occur consciously, but then the supposition that a professional archaeologist would forget such a thing is extremely improbable.

It must furthermore not be forgotten that the first excavation was already correctly foreshadowed. Bligh Bond must therefore have found out about both monuments in some way or another from such documents. It appears much simpler and more obvious in this case to suppose that the medium was actually able, by reason of the questions put by Bligh Bond, to feel his way as it were, in some way unknown to us, to the things themselves; in other words the supposition of true clairvoyance, in this case spatial.

Before turning to the question of the consequences which follow on an eventual recognition of this whole group of phenomena, we will say first a few words concerning the other three groups of phenomena. During the last decades, it is less the parapsychical, than the paraphysical phenomena which have been the subject of controversy, particularly since their chief supporter, Schrenck-Notzing, succeeded in gaining over a large number of well-known German intellectuals, who became more or less confirmed supporters of his ideas. Though this controversy is not yet decided, the author feels justified in saying that a very extended study of the literature[376] has not so far been able to convince him that these phenomena are, like those of the first group, to be regarded as established facts.

On the contrary, a study of this kind produces the uncomfortable feeling of walking on a swamp, in which healthy commonsense is liable to sink. The whole arrangement of the experiments in question, above all their production, almost without exception, in dark rooms, the frequent cases in which mediums have been proved guilty of various kinds of fraudulent practices, and the notoriously uncritical and prejudiced attitude of many of the most active investigators, all produce the impression that we have no real guarantee against error and conscious or unconscious deceit by the medium. Quite different and much less exceptionable means of objective determination must be used, such as photography or cinematography with invisible light. It would lead us too far to give reasons for all this at this point, and hence I must refer the reader to the special literature.

The only part of this group, which seems to me to have something in it, is that of the spontaneous appearance of ghosts and phantoms. But here, as was shown by the detailed investigation and criticism of Baerwald, it is so extremely difficult to decide between really new phenomena and telepathy, that I do not feel qualified to draw any final conclusion. I think that any investigator who approaches the subject without prejudice will come to the same conclusion. A strong argument in favour of the telepathic explanation of the phenomena in question, is that persons who experience them together often perceive quite different things; some see something, others hear something, and so on. This entirely corresponds to what was called above 'transposition,' in connexion with complexes of sensations perceived or produced unconsciously. The fact also, that other persons present (namely those without any 'gifts' of 'mediumship') see nothing at all, tells in favour of this interpretation. But it must not be denied that a great deal can be said against it.

The whole hypothesis would be refuted at a single blow if it should really be proved, as appears in a few cases, that the phantoms in question are able to effect mechanical motions. But I should not like to assert that the reports at present available are sufficient to prove this, though many who have worked on the subject, for example, the philosophers T. K. Oesterreich and A. Messer, already do so. The total results in this field do not allow of a verdict at present. It is to be hoped that new and more methodical methods will soon enable us to see more clearly. The matter available at present is, in my opinion, ninety per cent. of it completely worthless, and the remainder uncertain. I am obliged, as I have said, to state this conclusion here without being able to give further reasons for it, and hence would ask the reader once more to look through the relevant literature for himself.

We now come to the fourth and fifth group, the phenomena which claim to prove communication with another world. As far as I can see, the great majority, at least of German scientific occultists, reject these phenomena, or at any rate place a large question mark against them. Oesterreich has quite correctly pointed out that it is fundamentally impossible to demonstrate a truly spiritualistic interpretation of the phenomena in question. Whatever the superhuman intelligences in question may tell us must, if it be taken seriously, first of all be confirmed in some way. But this can only be done by asking questions of witnesses, or by finding out something about non-living objects. But both of these may always be interpreted as telepathy or true clairvoyance.

It is difficult to meet this argument. Neither the so-called 'minutes,' or the 'cross correspondences' [377] equally often brought forward, can avoid the difficulty. Only one way, as I can see would lead to a proof of the existence of independent intelligences beyond those of man: if, namely, the mediums were to produce revelations of an artistic, ethical or scientific description which were not only far beyond the powers of the mediums themselves, but also of all living contemporaries, who might otherwise be responsible for them by means of telepathy.

There can be no possible question of this.[378] On the contrary, the communications 'from the other side' are, for the most part, ridiculously devoid of any interest, and quite trivial. The so-called spirit has left his cuff-links somewhere, or locked an important document concerning land in a distant drawer and now tells the heirs through a medium. It is true that the efforts of mediums occasionally exceed normal powers in artistic production, and also in the matter of language, but the investigation of these phenomena has shown that we are undoubtedly dealing with quite remarkable feats of the unconscious. But these are exceptions; the greater part is simply ridiculous rubbish, which only quite unusually simple-minded people could regard as a revelation from another world.

With this we have also supplied our criticism of the fifth group. There is nothing contained in the 'revelations' of the theosophists and anthroposophists which has not been said as well or better elsewhere, although we readily agree that it is often said very effectively. We may simply remark that what is new is not true, and what is true is not new. There is certainly nothing to be said against Steiner's in part excellent ethical suggestions and directions for self-development. They obviously depend upon a profound knowledge of human nature, but in order to arrive at this, there is no need to be a clairvoyant, but simply an expert in the human soul, such as have been produced in almost all higher religions.[379] But all that Steiner has produced in the field of the specifically 'occult,' his marvellous *Chronicle of Akasha* and his *Knowledge of Higher Worlds* is a collection of purely arbitrary statements. Their 'confirmation' by his blindly devoted disciples proves absolutely nothing, since it represents no more than a variant of the well-known phenomenon, that a hypnotised person will take a cabbage for a pineapple, and eat it.

Far worse than all this is of course astrology, and allied hocus pocus,[380] such as the 'Sidereal pendulum,' cabbalistic mysticism, etc. It is not worth while discussing these subjects in a book devoted to scientific investigations. I am quite unable to understand how Dacqué, in other respects a serious investigator, can defend such things.   In all the numerous astrological publications sent to me for review, I have never been able to find anything but mere nonsense.

It seems to me as though my ears are filled
With chatter of a hundred thousand fools. (*Faust*)

We now come, after having already spent too long over the statement of fact, to the sole question which is really the matter under discussion, namely, what bearing all these modern discoveries in 'deep psychology' and 'scientific occultism' have upon the fundamental problem of the psycho-physical relation, and in particular, the problem of human mental life.   It is easy to overestimate the importance of the results in question, but also equally easy to under-estimate it.

We must first turn to the question, already alluded to, of a possible physical explanation of telepathy, which we have already rejected above.   Against it, in addition to what has already been said there are two further objections: first, the extremely low possible energy of such a radiation, if it exists at all, from the telepathic sender.   Our broadcasting stations find it difficult to transmit intelligibly over a distance of a few hundred miles by the use of some kilowatts.   On the other hand, we have a sender whose energy, if it exists at all, is certainly only a fraction of a millionth of a watt, and is yet intelligible over the whole of the earth, from Hong Kong to Mexico, or from India to England (these cases are proved).   This is not merely improbable in itself, but becomes totally impossible when we remember that the complete continuity of the electro-magnetic radiation of such minute amounts of energy, though commonly assumed, is very doubtful according to the quantum theory.

There is a second reason.   Even if we were to admit the existence of such waves[381] and also assumed that two brains may be tuned to one another with respect to them, what would be the use of such a physical tuning?   According to actual experience, it is not a matter of this at all, but of a purely ideal correspondence of interests, wishes, etc., and these are not physical wave lengths. Further, how are we to imagine that electrical waves of this kind transmit the thoughts in question?   The wireless wave effects it by having oscillations of frequencies corresponding to musical notes, superimposed upon it in the way that we know.   It

Lls

therefore transmits sounds in the first instance, and they only acquire a sense because they produce words which have a meaning. For this purpose, a complicated technique of sending is necessary. How are the 'brain waves' to produce the same effect ?

We need to supply a whole number of new hypotheses, in order to give any kind of plausible explanation. In any case we know absolutely nothing about the whole matter from this point of view, but are merely inventing *ad hoc*, simply in order to hold on to a physical theory at all costs. To the author this does not appear to be the right way at all. Psychological phenomena should be understood first of all on a psychological basis, that is to say, they should be as far as possible arranged to give wide psychological relationships. Baerwald himself refuses to attempt to refer the productions of the unconscious directly to physical laws, and he is quite right in doing so, even if he may also think that such a thing may be possible later.

Let us therefore ask ourselves whether the matters we are discussing, both clairvoyance and telepathy, may not perhaps be naturally brought into a more general psychological connexion. This appears to the author to be the case. We need only remember what was said above concerning organic 'wholes,' and concerning the hypothesis of a spiritual connexion transcending the individual, to see that in this way telepathy and clairvoyance may be understood in a general way even if they cannot be actually explained. Since there is no doubt that primitive races have a much stronger inborn 'mediumship' than the more civilised Europeans, we can only conclude that self-consciousness is the last and latest spiritual acquisition, and it would follow from this that man through his unconscious may extend into those 'super' or perhaps more correctly 'sub' individual regions, of which we spoke in connection with animal psychology (see page 423).

But if, as we there saw, the limits of so-called individuals in these domains become just as unsharp as they have long appeared to the eye of the physicist; or in other words, if the unquestionable universal inter-connexion of nature has for its counterpart a similar connexion in the spiritual world, we may perhaps be able to make the matter clearer by means of the following picture. The upper or self-consciousness is like the fairly narrow, brightly illuminated, surface of the sea upon which the light from a lighthouse falls, while the unconscious is the dark and unfathomable depths which lie below it.   Or the consciousness, the ego, is the single wave upon this sea of spiritual happening, the unconscious, on the other hand, the sea itself; down below there are no single waves any longer, but only water itself.

Telepathy would then be regarded as a faculty possessed by certain people who are able, more easily than the vast majority, to descend into the depths of the unconscious (or better, perhaps, to fetch up something from it).   When in these depths, they are able as it were to feel their way to regions quite inaccessible as a rule to the consciousness (the ego), namely complexes belonging to another 'individual' (telepathy), or even to a dead collection of events, not generally regarded as having any spiritual nature (clairvoyance).   Indeed, clairvoyance in time, which is prophecy,[382] might also be included by this means, since it would be a matter of indifference from the point of view of the Minkowski universe, whether this feeling-of-the-way took place in the 'time-like' or 'space-like' direction of the world.

Hypotheses such as these, when measured according to the science of thirty years ago, are incredible, but when examined more closely they lose their strangeness.   We must realise that the universal physical connexion of the Cosmos was likewise once merely a vague idea, which has only acquired life and content by its special development in the latest physics.   In the same way, the idea of panpsychism might also gradually develop from a mere philosophical programme into a well-founded scientific doctrine, the question simply being whether the facts give it sufficient support.   This appears to the author, after considering the whole present position of biological, psychological (including animal) and occult research, to be the case in so high a degree that he felt no hesitation in giving the hypothesis so large a space.

It should be remarked that Eduard von Hauptmann, who was fifty years before his time in this matter, expressed the idea in full detail and with the most serious criticism,[383] at a time when the whole of science was entangled in materialistic dogma on the one side, and in uselessly gyrating around Kant's work on the other.

Kant, by the way, also interested himself at times in the problem of occultism, as his book, *Dreams of a Seer of Spirits*, proves. This happened under the impression produced by the works of the well-known Swedish Bishop Swedenborg, who, according to all reports, must have been an unusually gifted medium.[384] Kant, it is true, did not reach any positive position, the material at his disposal being too defective to render this possible.

Occultist investigation first gained firmer ground by the valuable work of the English Society for Psychical Research. It is unfortunate that in Germany this position of quiet objectivity has not yet been reached. For the present, occultism in Germany is still a matter of parties, so that there are pro-occultists and anti-occultists; this is about the same as if there were pro-physicists and anti-physicists. Let us hope that we shall soon reach a point at which there are only investigators.

Let us now see what further consequences can be drawn from our hypothesis, always remaining conscious that we are upon uncertain ground and that we can only make a few loose suppositions. If such a thing as true prevision of the future exists – and there are a number of well attested cases, in which every other interpretation leads to violence to the fact; I may mention the Holt and Genthes cases[385] – we cannot avoid the recognition of a certain determinism. In some way or another then, it must be possible to know, or at least to make a probable guess concerning the future state of the world as a consequence of the present, and that not only by all the means of science known to us, but also by direct intuition on the part of the mediums.[386] This once more brings us to the problem of time, to which we have referred several times already, and also to the problem of the will, which is particularly interesting at this point. How are we to regard these problems from the standpoint of a panpsychical view of the world, when we include what modern physics, with its doctrines of relativity and the quantum, have taught us about the fundamentals of material existence?

As we have already seen, it is not improbable that physics will be able to eliminate the concept of substance from its system completely before very long, all its laws being then reduced to formal, arithmetical relations of an indifferent Something. We have already mentioned Eddington's view that this something might, according to its nature, be the psychical and we stated that in our opinion this was meant, not in an epistemological, conscientialist sense but in a metaphysical, spiritualist sense. The whole material world would, therefore, be only a formal constitution of this psychical, in which it appears for one of its own parts, namely the individual soul.

If we take this as the basis of our further considerations, we are first confronted with the question as to the application of the time form to this metaphysical background of the world, the universal-psychical. Erich Becher, who was himself a believer in this idea, held on, as we have already mentioned above, to the transcendental validity of time. But the epistemological reasons which he gave for this are not convincing, as the author believes himself to have demonstrated in another place,[387] and as one of Becher's scholars, A. Wenzl, who has recently worked upon the problem[387] admits.

The real ground for Becher's position was, as the author already suspected, his attitude in the problem of body and soul. Becher was a believer in the interaction theory, and since he saw in causality a direct succession in time, he was obliged to leave the time factor in the psychical, which he regarded as the last reality of the world, and which appears to us in part directly as such, and in part also in the (mathematical) form of matter. But since we have already decided above (see page 244), on account of the theory of relativity, to admit no difference in this respect as regards space and time, but to regard both as mere forms of appearance of a transcendental order, the further question arises how this is to be made to agree with the panpsychical hypothesis just discussed.

In order to answer this question, we must first ask whether there can be such a thing as the psychical if time be eliminated. Direct impression causes us to answer no. For our own spiritual being is unquestionably in time, and not in space. A sensation, an action lasts just so many seconds, but does not occupy a space of so many cubic inches. This is the reason why it is so much more difficult to abstract from time than from space.

But there are psychical existences which are independent of time. Among these are the theoretical truths, such as the laws of pure arithmetic, but above all, all values. All these are 'valid,' they have no duration, they have not come into existence, all that has come into existence is their temporal expression in man, but the validity of an arithmetical law is obviously independent of the time and manner of its discovery. Here we have the spiritual – for such laws cannot be called material – the being of which stands outside time (it is the great achievement of the *Gegenstandstheorie* that it has made this fact clear as opposed to all psychologism)[388] and can therefore serve us as a characteristic example for the timelessly psychical.

But Becher would perhaps say that although this is true for logic and for values, the spiritual, in the connexion we are discussing, also exhibits itself as acting, and that the concept of action, as well as that of aim (in other words the teleology of organic nature) contain the time concept. This objection is perhaps met by a comparison. Let us consider for a moment the work of an artist, which is always used as a figure of the divine creative power, and let it be a composer whose work we consider. In his work we have, it is true, time succession contained as an essential of its constitution; melody, rhythm, and dynamic are nothing without the time in which they take place; only harmony is beyond time, it does not happen, but simply exists.

But it is a generally known fact that another art produces quite similar effects to music, namely architecture. Our Gothic cathedrals have been called 'frozen music,' and it is true that the lines of such buildings have an effect similar to melody, although they do not take place in sequence of time, but simply exist. The comparison may help us in imagining a *telos*, an action that is without time. What in melody is significant succession, is, in the curves of the cathedral, significant form, which the consciousness, in 'travelling over it' (see above), experiences in a manner quite analogous to the manner in which it experiences melody. In this way we may perhaps form some notion of the way in which the universal psychical is effective, without bringing it down into time.

But it is in this connexion that the problem of freedom is to be considered, which as we know, is usually formulated by asking whether I can act at a given moment in one way or another. As soon as we adopt Minkowski's point of view, we recognise that we are dealing with none other than the problem already known to us, that of the contingency of the world. We saw above that this may possibly be abolished in view of the general 'thusness' of the world, its subjection to laws, which would mean that it could be resolved into pure arithmetic, namely, an arithmetic of a statistical kind (see above page 216). But we also saw that this more than ever results in the 'historical prime structure' of the world, that is to say its peculiar state here and now, becoming completely contingent. And we further saw (see page 217 f.) that, according to the new physics, it is no longer permissible to regard this historical prime structure as completely determined by two (generally infinitely close) world sections $t=t_1$, and $t=t_2$, but that we must rather spread the freedom thus left for the initial state over the whole 'world,' the division of which into time and space does not appear necessary in any case.

If we continue this line of thought with special reference to our present problem, the result is that the statistical laws of physics (and only this kind of physics remains) are related to the assumed action of the psychical which really finally constitutes the world, as, for example, the suicide or marriage statistics are related to these actions regarded singly. These actions may be regarded as completely free, or in any case statistics has nothing to do with their causality taken singly, any more than physics (according to the quantum theory) is interested in the possible sub-atomic causality. Such action would be, regarded timelessly, nothing but a simple 'placing,' and freedom would mean nothing more than that no such logical connexion exists between these placings as would imply that when two (see above) were given, all the others would be given.

On the contrary, within certain limits given by the statistical rule but not clearly defined, a certain freedom would exist for each single placing. If we then at the same time bear in mind the hierarchy of the teleological view of nature discussed on pages 398 and 432, according to which the lower always forms a member or organ of the higher, we may perhaps form an approximate idea of the true position of affairs.

I am inclined to suppose that Kant also had something similar in mind, when he formulated his doctrine of the 'empirically unfree,' but 'intelligibly free' characters, and that it was only his rejection of all metaphysical speculation which prevented him from filling his simple 'postulate' with detailed content. He could hardly act otherwise in view of the classical mechanistic world-picture of physics which was present to his mind.

In this sense, the freedom would be directly comparable to the freedom of the artist who gives his work form within the rules imposed by his art. Only that in our case, the form-giving world-will is divided up into innumerable individual wills, which in their arrangement in steps, form the Cosmos. We shall return to the problem of evil, which is directly connected with this, later; at this point we are dealing with the problem of freedom as such.

We now see to what answer our question, propounded at the beginning of this discussion leads us, the problem namely as to what a possible proof of true prophetic clairvoyance would mean. The answer is that clairvoyance would prove nothing more than our present day rational or scientific prediction of the future also proves, namely that what happens in the world always follows rules. This, of course does not mean that the clairvoyant arrives at his knowledge by the same means that we use in arriving at rational conclusions. He simply knows it, just as we know our own spiritual world; we have it, we do not discover it. The fact that he has this more comprehensive knowledge proves, concerning the regularity of world events, no more and no less than our own direct knowledge of ourselves. The psychology of association proves that rules also hold for our minds. Another person would be able, on the basis of this knowledge, to predict of us at least certain psychical processes. Nevertheless, this does not compel us to accept absolute determinism.

When the author therefore rejects the Becher-Wenzl thesis that time relations must be ascribed to the noumena, he nevertheless agrees with Wenzl that the laws of the Minkowski world, in other words, world-mathematics, only imply a framework law within which free placing exists, or in other words, that a contingency of the whole world process in all its parts is possible.

We must at this point once again remark, that it is certainly not a matter of chance when the only being in the world known to us to be consciously in possession of the law of causality, is also precisely the being which feels itself to be free. In truth, the feelings of freedom and the need for causality are obviously only two sides of one and the same set of facts. Man seeks for reasons either in himself or elsewhere, the first implying the consciousness of freedom, the second the demand for causality. For this reason alone, the usual proofs of determinism are tricks, they resemble the snake which eats itself up by starting with its own tail. We must, however, once more emphasise the fact, stated at the conclusion of our study of physics, that all considerations of this kind do not alter the dependence of the physical course of events upon statistical regularity, to an extent which, in practice, amounts to absolute calculability.

It is perhaps allowable to form the further hypothesis, that the strictness of this regularity diminishes progressively in the series of successively superimposed voluntary actions; that for example human actions can only be calculated with a considerably less degree of probability than those of animals, and these again with a less probability than those say of the unicellular organisms. In this way we should arrive at a series of stages of freedom, at the highest point of which the World Being itself would stand with perfect freedom, since its action, the world namely, only happens once, thus of course excluding entirely the rules of statistics.

These considerations have, it is true, brought us right into the middle of metaphysics, and many of our readers have perhaps already thought that we have been untrue to our original plan. The author nevertheless regarded it as necessary to discuss these ideas, the general lines of which were already in his mind in earlier editions, somewhat more in detail in the present edition, since times have changed as regards our taste for metaphysics, and hence what might once have appeared a virtue, might now be felt as a deficiency.

He may also be reproached with the fact that occultism is now given a much larger space than was originally the case. It must be said in reply that this is not the fault of the author, but lies in the nature of the matter, since these questions call for detailed consideration at present. Our whole world picture is enlarged to such an extent by these results – even if it is only telepathy that is really true – that we are obliged to take notice of them whether we like or not. The rôle of the spiritual has been enlarged to a degree previously unimagined, and this fact cannot be any longer denied even by the most confirmed mechanist. It is quite certainly no longer permissible to regard it merely as a peculiar accompaniment of certain central nerve organs, and something of no interest to a serious scientist.

We have already said above (page 410), clearly and plainly, and we repeat here, that spiritual phenomena have just as much right to be regarded as a part of science as bodily, since both are undoubtedly equally real. Mere behaviourism simply results in our closing the road to the most important problems. As long as we are willing to reckon biology as a science, psychology also must, at least in part, be added. Biology without psychology is a torso, for living beings experience sensation, and act.

The objection that this leads to the whole of human civilised activity being a part of science, since it also depends upon spiritual activity, has no cogency. For in it we have a new emergent, and it has been the custom to designate the sciences dealing with these matters as the sciences of mind or culture; we can see no reason for altering this procedure. It is of course, true that psychology, along with its natural science side, has also a moral science side, just as biology has, along with its psychological and vitalistic side, a physico-chemical side. The natural-scientific side of psychology obviously affords the foundation for a quite general science of psychology, and not only one especially relating to human beings (the usual psychology) and to animals, but one comprising the whole realm of life, that is to say, a description of the spiritual side of the whole of natural life.

This problem has hardly yet been seriously attacked, and perhaps can only be further developed when the single psychologies, particularly those of the lower animals, shall have produced more definite results. We must learn that, in this field as in physics, the successful construction of a general theoretical basis cannot succeed at the first attempt, but that we shall have to strive for it for a long time to come. But the programme can already be outlined without any doubt at all, and all the positivist dogmas in the world will not prevent it. It is characteristic that certain positivist circles are already again opposing all attempts of this kind, just as once they opposed every attempt at a consistent theory of physics, by branding in both cases all attempts as metaphysical, and striving to ban them from science.

We are willing to recognise careful criticism thus put forward as necessary and commendable, but we reject the fundamental position upon which this negation is based. This is what Goethe satirised in his well-known lines concerning 'Learned Gentlemen.' 'The world is deep, and deeper than our daily thought' (Nietzsche). He who forgets this finally loses all contact with coming scientific development. The only possibility of progress lies in the combination of the soberest possible criticism with the most open-minded readiness to consider seriously everything really new. In this sense, modern scientific occultism also deserves consideration.

We cannot be too careful, but we can equally abstain from dogmatism.[389]

With this we must make an end of the subject and consider another side of the connexion of human civilisation with the natural biological conditions of man's existence. This side has been touched upon several times above, but it has been left to this point for more detailed consideration.

IV

HEREDITY AND CULTURE

We remark at the beginning that this theme is naturally divided
into two parts. The first deals with the question as to what influence the differences between human races, which are of course
each the bearers of certain heredity characteristics, have on their
cultural achievements; and the second is the more general question as to how the cultural efficiency within single peoples or
races, is divided between the various racial strains. Race hygiene,
or eugenics, seeks to draw practical consequences from the second
group of facts. We must, however, when using the first of these
two terms, bear in mind that 'race' is used in a different sense
than in anthropology, in which this word implies such differences
as Negro and Indian, Nordic and Alpine; whereas in so called race
hygiene no such considerations enter, at least directly. First
therefore we will take:

(a) *Race and Culture*

Here we are dealing with one of the most controversial subjects
of the present day and hence must be doubly careful; but the
author felt that it was not feasible in present circumstances to
evade the subject entirely.[390]

The definition of the concept of a race of mankind is very
difficult to formulate in a way not open to objection. We understand by the term in general a group of human beings, which is
plainly distinguished from other similar groups by bodily and
perhaps mental characteristics, these characteristics being transmitted by heredity in a constant manner. For the latter reason,
the Mulattos are not a constant race, since their offspring Mendelise; whereas say, the Mongols, the Malays, the Papuans, the
Hottentots, the Bushmen, etc., are without doubt recognisable
races, whose characteristic complexes of racial peculiarities always
reappear in their pure-bred offspring.

Before we pass on to the more difficult cases, and especially the Europeans, we must indulge in a short epistemological discussion. In mathematics and for the most part in theoretical physics, which of course serve as examples to the other sciences of the formation of exact concepts and judgments, we deal with sharply defined concepts, that is to say, with such the extent and content of which can be stated exactly, and which hence permit us to say with certainty in every single case, whether the case falls within the range of the concept.

In the other sciences, on the other hand, and especially in all 'existential' sciences (see above), which have to deal with the multiplicity of real life, the stipulation of absolutely sharp concepts does not bring us any further; not only, as is generally thought, for the present, but quite fundamentally, as we shall see immediately. It is wrong in principle to suppose that these other sciences are incapable of yielding any genuine knowledge, since 'in every science there is only as much real science as there is mathematics to be found in it.' Kant was never more wrong in all his works than when he wrote this sentence.[391]

In truth, in the whole of the science of reality, knowledge is only reached by working with approximate concepts, that is with concepts, the extent of which can only be determined within certain limits. A very simple and striking example is found in astronomy. When the astronomer sees at one point on a photograph of the heavens a very large number of stars very close together, perhaps so dense in the middle that they cannot be distinguished singly from one another, he calls this 'a star cluster' (see page 273) and is convinced that it is by no means merely a question of the images of a number of stars coming together by chance for reasons of perspective, but of a real configuration which must be treated as a unit made up of many stars.

We have seen above how important are the conclusions which can be built up upon such observations. Nevertheless, it is by no means possible to define in this case quite sharply, what single stars belong to the clusters in question. Single stars, especially those at the edge, may also appear as part of the cluster merely for reasons of perspective. But this does not prevent any sensible person from regarding the cluster as a reality. Exactly the same is true in other sciences and also everywhere in biology. The concepts of plant and animal, for example, are equally impossible to define with absolute sharpness, for as we know, the limits become uncertain in the lower groups of both kingdoms (particularly the unicellular organisms).

Nevertheless, there is no more erroneous phrase in the theory of cognition than the one which states such concepts to be merely aids to thought economy, used by us to orientate ourselves approximately. It is absolutely certain that nature herself forces upon us this sub-division of organic beings into two kingdoms with exactly the same evidence as the concept of the star cluster just mentioned. Hence the concepts of the animal kingdom and the vegetable kingdom are descriptions of real and not merely imagined units. We have not read them into nature, but have been compelled by her to accept them from her (see also above, page 232).

All this leads to the necessity, in the empirical sciences, at least in their special branches (page 249), not merely to define concepts in the manner customary in mathematics, but frequently to obtain these by a method which I should like to describe as 'typological.' We must first plunge, so to speak, right into the middle of the group of objects in question, and by examining its most notable and most 'typical' representatives, acquire a clear idea of the characteristics of this type as compared with another, just as the astronomer when he sees two star clusters close together on the plate would first fix the mean positions of both upon the stellar sphere, but would leave it doubtful whether a single star lying on the boundary of the two clusters belonged to one or the other, or neither.

He who wishes to know what the difference between a plant or an animal is, must not first study the limiting cases, such as the stationary cœlenterates, or the amoeboid stages of the mycetozoa, but a lion or a bee on the one hand, a beech tree or shave-grass on the other. He will then find it easy to see that a stationary position, the power of assimilating inorganic material, the possession of cell walls made of cellulose, are typical characteristics of a plant, whereas mobility, the development of special sense organs, the assimilation of already assimilated material, etc., are characteristics of the animal, which by no means prevents single groups of plants and animals from being devoid of one or more of these 'typical' characteristics. No one will doubt that the coral, in spite of its stationary existence, is an animal, nor that the mycetozoa in spite of their amoeboid stage, and the *Mimosa pudica* in spite of its short reaction time, are plants, since in all other respects they belong to these classes.

Only in cases where a real mixture of almost all typical characteristics takes place (as with many unicellular organisms) will any doubt exist, and in such cases we shall have no alternative but to give them special treatment. Experience tells us that the number of these is practically always very small, this only being another expression for the above-mentioned law formulated by Zilsel for the empirical sciences: 'we may neglect certain cases and still know something.'

Let us now turn this epistemological discovery to account in our problem of the human races. It is quite evident that, from the typological point of view, such races as the Indians and the Mongols, the Papuans or the Hottentots really exist. But then we must regard as equally real such races as the Nordic, Alpine, and others in Europe, the near-Asiatic, far-Asiatic and so forth in Asia Minor, only that in the latter case it is considerably more difficult than in the first to classify the numerous limiting cases and mixed forms. In many cases we are even obliged to give up the attempt.

The understanding of this problem, in itself so difficult, is rendered more difficult by the fact that political passions come into play. On the one hand, everything which makes it difficult to define the concept of race, is intentionally sought for, in order finally to arrive at the conclusion that the whole concept really has no meaning and would best be abolished entirely.[392] On the other side, the concept of *race* is mixed up with that of *people* in the most confused manner; or worse even, it is confounded with community of language. The result is a hopeless confusion of all judgments, since in the end no one has the least idea what is the meaning of the terms which he is actually using. Unfortunately, our higher schools with their hopelessly unbiological, almost purely linguistic and historical, training, are chiefly to be blamed.

A people (*Volk*) is a group of human beings bound together by history and a common culture, generally upon a common racial foundation, and generally also having a common language. It is open to doubt whether nations such as the Swiss, which have several national languages, may be regarded as peoples at all.

Conquered nations may even adopt the language of their conquerers, and yet remain a people just the same. Two different peoples may have the same racial composition, whereas conversely a single people, such as the Germans, for example, may consist of several different races.

The rule is, however, that the differences between the peoples rest upon different racial mixture. While, for example, there are components of like race in both the German and the French peoples, the mixture in the two cases is still very different, and there can be no doubt that the national differences depend for the most part upon this difference in race (along with this we have naturally the difference between the historical development of the two countries, which was different, not always merely for the reason that different mixtures of races were concerned; geographical, climatic, and other conditions played a part). Anyone who does not keep these facts clearly in mind, and, for example, continues to describe the Jews as a race, will never understand the matter properly. The Jews are a people, having, naturally, a different racial composition from that of the German or English.

Is it allowable to speak of races at all within so fluctuating a population as that of Europe? The answer to this is given by Mendelism, which teaches us that heredity characters are not changed as such by cross breeding, but only appear in new combinations. We will take a very simple example from botany. If we cross two of Johannsen's pure lines of beans (see page 458) we get on investigating the resulting stock a curve of variation (see page 457), which is composed of the two single curves by addition of the ordinates. If the peaks of the two curves are sufficiently far apart the resulting curve will have two peaks (figure 86), and this, conversely, is a proof for anyone investigating this stock of beans, that there are two pure lines in it.

This example shows us the ways and means open to the investigator of races to discover the presence of single races even among a very mixed population, and even to isolate them; naturally a very difficult matter in the case of man, in which experiment is not possible. It is necessary in this case to wait for a happy chance which produces one or other pure racial lines. Investigations of this kind have shown without doubt that in Europe, and particularly in Germany – we will leave the remainder of the world out of the question – we are dealing with at least four, if not five or six, different races. These are the Nordic, the Alpine, the Mediterranean, the Dinaric, and possibly the East Baltic and another branch of the Nordic, the 'Falic' or 'Dalic' race already mentioned on page 497.

We have further to add that, in the case of the Jews, we have the near-Asiatic (related to the Dinaric) and the Oriental race,

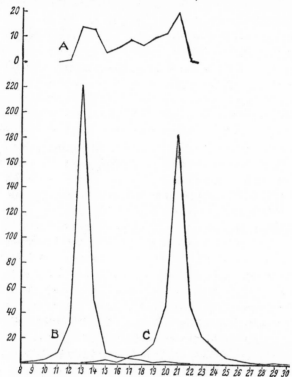

FIG. 86.
A, variation curve with two maxima, resulting from two single variation curves B and C (below).

together with some Negroid element; while in the East we have pure Mongolian additions (the East Baltic race is rejected by many investigators, being regarded as a mixture of Mongols with Nordics). These results are now placed fairly safely beyond doubt, although as regards details there is a great deal that is still uncertain. We do not need to give a description of the racial characteristics of these races since books on this subject are everywhere available to-day.

Mms

The only question that interests us is how far cultures depend upon race. This obviously depends upon another question, namely, as to whether psychical race differences, corresponding to the undoubted bodily race differences, really exist, and if so, whether they are strong enough to produce a serious difference in the nature, and also in the degree, of cultural achievement.

Regarding the first question, it can hardly be doubted that psychical race differences do exist. This already follows *a priori* from the fact that psychical qualities altogether are inheritable, as we have already seen above (page 389), and it is also self-evident that the strong gift for music of the Slavs, and the much smaller one of North European population, and also the 'hot blood' of the Southerner as compared with the calm of the Northerner, are mainly racial; that is to say, inherited and not determined by environment. Even anyone who rejects Günther's description of the psychical characteristics of European nations as one-sided and prejudiced, must recognise that such differences certainly exist, though they may be much more difficult to take hold of than the bodily differences.

But if this is the case, we cannot possibly deny that the nature and degrees of cultural achievement must also be determined, at least in part, by these hereditary psychical characteristics, in other words, by racial influences. How far, along with this, we must take into account the influence of environment (climate, geographical position, neighbours, flora and fauna, etc.), and how far besides the historical succession of events has its own autonomy (since cultural values are to a certain extent independently developing wholes, see below) naturally remain matters for further investigation. In any case it would be wrong, as for instance Boas and von Luschan[392] do, to push the racial factor completely aside with the excuse that it cannot be grasped with scientific exactness (because the limits are fluid, as we saw above).

This would be not to see the wood for trees. History itself shows us in the most concrete manner that great cultural achievements are the work of a few highly gifted races and peoples, i.e. mixtures of races. It is also quite evident that among these, those whose leading element was composed of the Nordic race, occupy the first place; but we may expressly point out that a mixture with a Northern upper layer appears to be a particularly favourable ground for culture, more favourable than a practically pure Nordic people such as we still find to-day in certain parts of Scandinavia.

To what extent this race was concerned in the ancient cultures of Asia Minor is now difficult to determine, since we know nothing for certain concerning the origin of the earliest of these nations, the Sumerian. There can be no doubt that enormous racial migrations took place in prehistoric times, and we must further recollect that cultural elevation itself is a very relative conception, and that trifles may often prevent an otherwise highly gifted people from not getting beyond a certain stage, which it might easily have much surpassed, given its inborn qualities, under more favourable conditions; whereas on the other hand, favourable historical conjunctions may cause a people to be placed in the first rank of civilisation, without having done anything on its own account to justify this. This is true, for example, of a considerable fraction of the present South European population, which is racially completely degenerate, but is able to live culturally upon its neighbours and its tradition.

The fact that the high endowments which our own forefathers must without doubt have possessed, only came very late to fruition, depended, conversely, on a few obstacles only: in the first place the tendency of the Germans to live in solitude, and secondly, on the rejection of writing for religious reasons, apart from the fact that suitable materials were not available in ancient Germany.

The idea that the importation of the 'treasures' of culture in Roman form during post-Carolingian times made in a few generations, out of a nation of rude barbarians, a nation of such great culture as the German already was in the Middle Ages, is absurd. Even if the greatest possible concessions are made to Lamarckism, we should be dealing with a complete biological impossibility. These gifts must have long been present in the genes before the first emissary of the Frankish kings, or even the first Roman entered Germany.[393]

We shall not deal any further with these matters, since we are engaged on the philosophy of science, and not that of culture. Our chief reason for mentioning them is because it is here that we see most clearly how great is the importance of scientific biological knowledge for cultural questions. The complete neglect of science in the curricula of the higher schools in our country has contributed in no small a degree to the fact that very large sections of educated people have totally inadequate conceptions of the driving factors in history. For not a few, 'culture' still means much the same thing as literature and art, at most with the additions of ethics and religion; as though science, technology, economics, etc., did not also belong to it, and as though the nation is only cultured when it produces works of literature.

Furthermore, a class educated in this way is without any insight into the biological conditioning of cultural achievement. It is ruled by a blind 'cultural' Lamarckism, as I may call it, which twists the knowledge of the autonomy of cultural development into the completely false notion that the autonomously developing cultural products of themselves produce the human beings belonging to them. In view of the enormous importance of this question for the whole future life of our people, I must go into the matter in a little more detail.

It cannot be denied, that, as a result of the 'heterogony of purpose' above referred to, the cultural achievements arising from concrete vital necessity have made themselves independent in the course of the development of humanity, and now follow their own laws of development. This is seen most clearly in science itself. Physics, for example, is driven forward to-day only by the problems arising out of itself, as compared with which all racial or national differences are of a completely secondary nature. There is no German or French or Japanese physics, but only a science of physics. At the most the selection, made by individuals from the great departments of physics as a whole, and the manner in which they present the selected material, is variable and dependent upon nationality. We only need to take up an English, a German, or a French text book of the same subject, such as electricity or modern atomic physics, to notice immediately national differences, but this has no effect on the content, nor upon the further development of physics.

For the same reason, it is pure nonsense to attempt to connect the relative arrest of physics after Maxwell with economic conditions in European countries. This stationary condition depended in the main upon purely internal reasons; a pause for digestion, so to speak, was necessary before new matter could be attempted. We therefore completely reject Marx's idea that all ideal cultural developments are simply products, or rather by-products, of economic developments. Their real driving forces are in themselves, at least as soon as they have already reached a certain height. It is this that is the autonomy of cultural development spoken of above, the new emergent as compared with the purely biological type, primitive human nature.

While recognising this clearly, and retaining it firmly, we must also emphasise the fundamental error of believing in consequence that these cultural goods create on their own account their human bearers and producers, and that when scientific, religious, artistic, political, and technical development, having reached a certain point, demand the creation of a further stage, the human beings will then always be available, indeed, will have been produced by the previous development, ready for the accomplishment of the new move forward. Demand certainly produces supply, but only when the endowment is there to render it possible. Where there is nothing of the kind, the most intense need cannot produce anything new. If mathematical gifts are not present in any serious degree in a people, it will not produce any mathematics, even if placed ten times over in a position in which mathematics would be very useful.

Of what use is their great historical tradition to the Greeks and Italians of to-day? Is it seriously believed that the intensive cultivation of this tradition at the present day can compensate for the racial degeneration which has taken place there, without ceasing, since the time of the great migrations from the North? Phenotypically, a great statesman such as Mussolini may be able to raise his nation for a time. But if he wishes to produce any permament result, his first care must be to improve its genotype. It would appear that Mussolini is the first of all European statesmen who has actually grasped this problem clearly.

Our philological and literary education in Germany, has, on the contrary, encouraged the belief that we should elevate the German nation by feeding it with German idealism, with Christianity, with social measures or hygiene, or suchlike, if only all these cultural advantages could act long and intensively enough. This is what I have described as cultural Lamarckism. What is the actual truth of the matter? This brings us to the second question raised above, the theme of:

(b) *Race Hygiene.*[394]

Here the word race, as we must once more expressly point out, is no longer to be understood in the sense of the anthropological races already mentioned, but as a sum of inherited characters which distinguish any group of mankind. This thereby abolishes the political flavour of the word, which many find objectionable. We are now dealing simply with the question of how the endowments for the various branches of culture are distributed in a people, and how this distribution changes in a good or bad sense, or can be changed; it being for the present a matter of complete indifference whether the people in question are blonde or brunette, long-headed or round-headed, etc. These questions will interest those for whom the word race in the anthropological sense is unsympathetic.

The first fact to be mentioned is this, that *there are inherited differences in cultural endowment, that is, in the capacity to create cultural values, among the members of a people, and these inherited endowments are on the average proportional to the social level of the various layers.* Although it frequently happens that individuals of high social position may be stupid or devoid of energy, or morally degraded, and that individuals of lower social position may be the opposite of all this – taking the average, the rule is correct, as is proved by the extensive statistics of school children collected by Hartnacke and Kramer, Duff and Thompson, and others, and above all, by the gigantic American army statistics. The percentage of highly gifted children (class 1) may be taken as 50% in the academic classes, and that of the poorly endowed (class 5) as 5% in the same classes, whereas these figures are almost exactly reversed for unskilled labourers. There exist also clearly recognisable differences between these latter and the class of skilled labourers, as has also been proved by statistics of school-children.[395] From this we see that the truth of the rule is not open to doubt. Besides all this, the inheritance of valuable as well as undesirable social endowments has been rendered certain by numerous other investigations, among which are Wood's statistics of Royal houses, and the notorious genealogical trees of Kallikak and Juke.

A second equally undeniable fact is that in all present-day civilised nations, the relative increase in population is inversely proportional to the social level.[1] For example, the numbers of children, according to German statistics by Dresel and Fries, (1921 to 1922) are shown by the following table.[396] The first column gives the class, the second the number of children per hundred families, the third and fourth the number which died before the age of sixteen and the last column the number remaining.

Social Class	Children per 100 families	Died before 16 years		Remaining
Academic . . .	270.7	5.4% =	14.6	256.1
Officials and Teachers .	304.0	6.1% =	18.5	285.5
Free Professions .	320.8	8.1% =	26.0	294.8
Commercial classes	341.0	9.0% =	30.7	310.3
Hand Workers . .	383.1	15.6% =	59.7	323.4
Employees . . .	431.9	16.3% =	70.1	361.8
Labourers . . .	596.4	20.7% =	123	473.4

These two facts prove with mathematical certainty that:

The civilised European nations are allowing their leading cultural elements to die out.

How quickly such a process takes place is shown by the following table.[396] If two groups A and B, are present in equal proportions in such a population, and group A has an annual excess of births of 1% while group B has an annual deficit of 1%, the relative proportions are:

After years	Group A %	Group B %
0	50	50
10	56.5	43.5
20	62.3	37.7
50	74.2	25.8
100	88.7	11.3
140	94.8	5.2

If B, for example, denotes a group with better inherited qualities, this means that in 140 years, that is to say, in about four generations, the nation has lost its endowment, at present amounting to 50%, to the point of leaving only 5%; indeed, a late age of marriage, even if the number of children remains the same, itself

---

[1] This statement is now no longer true for Germany, for the relative increase has become almost equal in all social classes with the sole exception of that of asocial individuals, which alone to-day has a fertility sufficient to maintain the stock, namely more than 5 children per marriage, the average number in all other classes having been reduced to 1.2 to 1.8. This result is, of course, still more disastrous than the former. (Note by the Author.)

produces a considerable displacément.   If group A has an average generation of twenty-five years while group B has one of thirty years, after 100 years B has already fallen to 33% and after 300 years to 11%.   As both causes actually combine, the fall in the well-endowed group is in reality still more catastrophic.

In this connection we now realise how important such purely theoretical investigations as those concerning the doctrine of selection and Lamarckism in Part III, may be; for the life and prosperity not only of our own nation, but of all civilised European nations, depends upon the measures which they are willing to take against these processes which are destroying their racial health. Such measures must of necessity depend upon insight into the causes of the processes.   The good will to help is blind, if it is not guided by true insight.   A physician can only be of use when he has first carefully investigated the cause of disease and when he is quite clear what his remedies can effect.

We are standing to-day beside the sick bed of the European civilised nations; the disease is racial degeneration, the questions to be answered are: firstly, what is the cause of it, and, secondly, what remedies exist ?   If we do not succeed in time in getting rid of the many fatal errors prevalent regarding both questions in wide circles, and particularly in those of the governing powers, the patient will die under our hands in spite of all precautions.

It is first of all an error to suppose that the chief cause of degeneration is direct injury to the germ cells; for example, by alcohol and the like.   It may be assumed, although it is not certain, that this is a contributing cause.   But it is certainly false to ascribe to it the chief part.   On the contrary, *racial degeneration would proceed almost undiminished on account of the present-day 'negative selection,' even if from to-morrow not a single drop more alcohol were drunk, or any further transmission of sexual disease took place.*

It is, secondly, erroneous to maintain, as socialists generally do, that only bad environment produces anti-social types, and that these would disappear if the environment were improved. Anti-social elements are undesirable Mendelian combinations, or mutations, they are not produced by the environment but at the most bred by it, and statistics prove that present-day environmental conditions are breeding them in the most dangerous manner.   For although such elements in earlier times usually produced a large number of children,[397] these were once greatly reduced by high infant mortality, whereas to-day we take every kind of so-called 'social and hygienic' precaution to keep alive as many as possible of such children with an hereditary taint, to bring them up as well as possible, and 'fit them to lead a useful life.'

This means to say that they are put into the position of being able to transmit their bad inheritance to as many descendants as possible.[398] The Lamarckian error prevents governments from seeing that it is absolutely necessary, while we cannot again drop the social measures for ethical reasons, to compensate their undesirable results by corresponding measures which limit the increase in these elements. In actual fact, *such measures operate by no means socially, but rather anti-socially;* they are born of pure individualism, which regards society as nothing but a utilitarian association of individuals with the object of helping one another as much as possible, thus completely losing sight of the fate of the whole, of the *societas* itself.

The whole of our present-day social, hygienic, and educational measures are dictated by this false fundamental position: *they forward the well-being of the individuals at present living at the cost of the whole of the nation, and of human civilisation.* We spend millions every year on our hospitals, asylums, reformatories; on institutions for infant welfare, support of unmarried mothers, the unemployed, etc., and have no means remaining to assist really gifted people without means, real invalids, and altogether the really valuable elements. Men and women of insight have long ago pointed out the folly of this before the War and Revolution. But it has since become very much worse instead of better.

It is further a dangerous error when both branches of the Church repeatedly declare, that only the increasing disregard by the people of religion is the cause of degeneration, and hence that the only remedy is to reinforce religion. It is certainly true that one of the conditions for national regeneration is moral reconstruction, for this must create the will to betterment. But it is an error when it is believed in religious circles that this good alone, without any knowledge of the true means, will produce the necessary effect. How often does good will towards a patient make use of the wrong means ! With all the good will in the world we cannot change the order of nature or, to use religious language, the order of divine creation. But this tells us implacably that, if those of low natural endowment continue to increase in numbers faster than those of high endowment, the end must come sooner or later.

The moral will must therefore be directed towards insuring that the right precautions are at last taken, and that everyone plays his part in seeing that they are carried out. In this sense the motto for the churches should be: *hic Rhodus, hic salta*. It is very much to be welcomed that a man such as the Jesuit Father Muckermann gives these demands his unqualified support. Unfortunately, neither his own co-religionists nor the majority in the Evangelical church supports him with anything like sufficient energy ; on the contrary, he is often enough sharply opposed.[399]

Even in the otherwise excellent book by Titius, *Nature and God*, all understanding of the immeasurable importance of this matter is wanting. The reason is that, in these circles, psycho-Lamarckism has been quite generally adopted (this is the case with Titius), and this is expected to save the situation on the basis of the idea that it is the mind that forms the body.

These people do not ever ask themselves whether, even if psycho-Lamarckism is to be regarded as a tenable theory of the origin of species (see above), there is any practical sense in expecting an improvement by such means, since it is established that what might perhaps be attained by these means in a hundred generations, would be more than lost in a single generation of negative selection. As compared with the rate at which the latter has operated in recent times, the Lamarckian influence of the environment is negligible, apart from the fact that it is quite uncertain whether it exists at all.

A further widespread error is the opinion that the whole process will regulate itself, inasmuch as when the degenerate 'upper ten thousand' do not produce enough children, new leaders in sufficient numbers will arise from the inexhaustible reservoir of the healthy common people. The truth is that this process, which is at the present time in full swing and is yielding many more 'leaders' than are necessary, does not resemble interest upon capital, but rather the ever more rapid consumption of the capital itself.

The system of a free road for talent has, under present conditions, the inevitable result that those who rise from the ranks arrive in a very short time at the same point as those who were already on top, namely, the one- and two-children system, which again leads to the elimination of their stock (for the maintenance of a stock demands at least 3.6 children per family). The governing classes and the people gaze entranced at the apparently brilliant success of this system – naturally it is brilliant from the phenotypical point of view – but, on account of their immeasurable ignorance of biology, have no notion that they are like the future bankrupt who builds himself a wonderful house by using up his working capital.

The unlimited advancement of the talented would only be a matter for rejoicing if the number of children in the higher social layers were the same, or approximately the same, as in the layers from which the new talent was recruited; it would even be very desirable if, in the latter, the birth rate were smaller than in the former. But as matters are, the result is merely to cause the race to degenerate still more rapidly.

We may regard this as enough of criticism, though a great deal more could be said, and ask what is to be done ? The answer can only be that everything must be done that promises to put an end to negative selection, or if possible to transform it into a positive selection. Since the social and ethical measures which act in a negative sense cannot to-day be abolished[400] – from the standpoint of individual ethics they are necessary – only two possibilities remain: first, to arrange that, in spite of these institutions, the increase in numbers of low-grade stock is limited; and, secondly and conversely, to take fiscal, administrative, educational, and other means to render possible again a larger number of children in the case of stocks with better heredity, while at the same time educating these classes in a sense of their duty to society.

Whether such a transformation is still possible may certainly be very much doubted. All that the theorist can do is to declare that he has done his duty, when he for his part has stated clearly the true position of affairs, and to demand that this knowledge shall be carried to as wide a public as possible, in which respect a small beginning has at last been made by introducing the science of heredity into the teaching in the secondary schools. If nothing can change matters, the 'Decline of the West' (Spengler) is as inevitable as that of the peoples of antiquity.

We cannot leave this subject without devoting a word or two

to the question of the biology of peoples, which was popularised
by the writings of Spengler, and has recently been further and
brilliantly developed in the writings of H. Piper.[401] According
to both authors, cultures and civilisation in general are subject
to the same laws of awakening to life, development, ripening,
ageing, and dying, as individuals. Both attempt to deduce from
this fact quite general laws of the biology of peoples, and suggest
that the immediate future of our own civilisation can be prophesied with the same certainty that we are able to feel in the
case of organic individuals.

The scientific philosopher cannot avoid taking up a position
towards these assertions at the risk of being reprimanded for intruding in a province not his own. For these 'culture-biologists'
make use of biological, that is to say scientific, concepts and laws;
and the scientists may therefore feel justified in examining the
question whether this method leads to judgments that hold water.
The answer must be in the negative. This extension of biological
'laws' – which are not even laws, but simple rules with exceptions
already known to us – to the development of civilisation is
inadmissible for one reason alone, since it depends upon a
purely outward analogy between two objects in reality totally
different.

Furthermore, it cannot be carried out, since every rule, even if
we do not demand very great strictness from it, nevertheless assumes a not too small number of single cases, in which the rule
is exemplified; whereas civilisations do not exist in this number,
or at least not such of which we know the history with sufficient
certainty. The extension is also unpermissible for a third reason,
namely that the case which most interests us, our own civilisation, is a special case with unique conditions, which have certainly
never been realised before, so that even if a considerable number
of other civilisations have followed the rule, the latter still remains
doubtful in our case.

I will now give shortly the most necessary arguments on these
points. Firstly, in dealing with the life, ageing, and death of the
biological individual, it is a matter, as we have seen above, of an
extremely complicated interaction of internal and external factors,
which only takes place with practical regularity in the usual
manner, because in the process the influences hostile to life generally increase gradually. We have seen that in certain suitable

cases, experimental interference with the normal process may result in death being at any rate postponed for a very long time, and that there is therefore such a thing as potential immortality in the case of the individual, although it never is realised in actual nature. But the development of civilisations depends upon the peoples which bear them. The peoples for their part again consist of individuals and the actual ageing of the culture is the degeneration of the people.

It is impossible to see any reason why a civilisation should stand still or even die, so long as enough culturally capable, that is energetic, active, and creative individuals are present. If we compare these individuals with the single cell of a multicellular organism, as in fact is done by the biologists of civilisation, we may compare the process, taking place in every people, of the continual renewal of the stock with each generation, to the continual replacement in the living body of dying cells by new ones produced by cell division.

This alone gives us a fundamental difference. In the organism, these cell divisions gradually become less and less, and the new generation of cells (with the exception of the germ cells), obviously becomes less resistant as time goes on (presumably owing to a continual increased disturbance of its chemical processes by the products of degeneration of the other cells). At the same time, these cells remain genotypically what they are, being only injured phenotypically.

The degeneration of a people, on the other hand, consists in the cells, in this case the individuals, being replaced by others that are genotypically less valuable. The two processes therefore have an entirely different fundamental character as regards this point, however similar they may be in their external character.

For this reason alone it is quite impossible to extend a rule found in the one case to the other case. This is also true for another reason, namely because the number of civilisations known to us from history is much too small to allow of the deduction of a rule, let alone a law. Both Spengler and Piper obviously had in mind, in the first place, the fate of the Graeco-Roman culture, and perhaps the Egyptian or ancient Persian, concerning the history of which we are more or less well informed. But the short-lived flowering of Arabian civilisation at the time of the Crusades already fits very badly into the scheme, as does also the contrary case of the extremely ancient Chinese civilisation with its sleep of almost a thousand years duration.[402]

But even admitted that a number of cases similar to these could be found, what would be the use of it when we are compelled to admit that there are still a much larger number of which we know absolutely nothing as regards their history. In the interior of Africa, in North and South America, it is certain that a whole series of, in part, very high civilisations have existed, of which we know nothing more than what a few poor material relics can perhaps tell us (Mayan, Pre-Incan, etc.).

If we must apply a biological rule, we can at most say that *far more civilisations have perished from external injury, by accident so to speak, than by the weakness of age, exactly as in the case of individuals;* but if we had only seen half a dozen individuals die of actual old age, we should hardly be justified in deducing a general rule.

Piper, in a discussion of this question,[403] has accused me of applying the measure of the formation of scientific concepts unjustifiably to historical matters. My reply to this objection, also raised by others, is that we are not dealing with a rule specific to natural science, but with a quite general logical rule. Whoever seeks to set up general rules, whether historical or scientific, is obliged to base them on a sufficient number of cases. From two, three or four single cases, one should never deduce a law.

But even if we admit such a rule – and it certainly is no more – the third question would still remain, whether this rule can be applied to the present case, namely, our own civilisation. Here we must say that this is completely impossible, since our civilisation depends upon conditions which have never been even approximately fulfilled in the whole history of humanity. It is unnecessary to support this in detail by reference to international intercourse, international relations, etc.

I need here only refer to a point often overlooked, but in reality the most important of all: *no civilisation, nor humanity anywhere, has ever possessed the knowledge that we possess, which alone enables a complete insight into the deeper causes of cultural processes, and in particular into the decline of peoples.* The science of heredity is, as a science, only twenty-five years old. It is true that ancient civilised peoples and their leaders, such as Lycurgus, had some practical knowledge of it,[404] and animal and plant breeding on a purely empirical basis have of course existed from the earliest times. But it is one thing to apply such knowledge purely practically, and another to possess it in the full light of clear theoretical insight.

The phenomenal development of our technology as compared with the primitive technology even of the most advanced ancient peoples, shows clearly what a difference it makes for us to know the laws of physics, and chemistry, of which they merely guessed a small fraction. Although there is yet no such thing as a real biotechnology, apart from a few beginnings such as plant breeding on the basis of Mendelian laws, it is as certain that it will one day come into being, as that the technology of to-day has followed upon physics.

How can one maintain in the face of this situation, that our civilisation must follow exactly the same laws as all the others? In the history of the world of organisms, a rule was followed for millions of years that new forms with new adaptations of a bodily kind continually arose while old forms died out. Nevertheless, one day man appeared, and with him something that had never before existed, a being capable of creating culture. Who will guarantee that a similar thing has not happened in the appearance of our civilisation, namely, a civilisation capable of turning potential into actual immortality? How are we to know?

Let it be noted that I am not upholding the blind faith of the apostle of progress in a perpetual forward development. Such a thing does not happen by itself, just as little as a hydroid polyp remains perpetually alive without interference by an experimenter. It is only on the basis of a clear insight into the true causes of cultural decay that a resolution can be formed, and lead to a successful elimination of these causes and hence of decay itself. *The people that is the first to make this resolution, will, unless every sign is misleading, rule the world.*

An idea of this kind may appear too audacious to pious minds. But it must be said that it really means nothing but what takes place in every individual life, namely the fact that the time comes when a person's fate is in his own hands, when, in religious language, God makes him responsible for what is to happen with his life. Piety does not consist in declining this responsibility and looking on in fanatical submission, but in taking it courageously and self-reliantly, in the belief that God helps those who help themselves. This at least is the attitude of European Christian piety, though not that of the Indian who longs for Nirvana.

But even those who reject such a line of thought completely, must recognise that our case is quite unique in history, and hence in no way justifies its being subjected to a doubtful rule deduced from other history. In this connection it should be plainly stated that the originators of such historical constructions in which biological concepts are made use of, would do better to use their biology, instead of in the construction of vague analogies, in investigating a case where it is really necessary, namely the actual cause of the decay of the historical civilisations. But their biology fails them just at the most essential point.

It must once more be emphasised that the ideas here expressed in no way contradict the view that the process of civilisation possesses its own internal law, or if it be preferred, may also be regarded as a free and unique placing (*Setzung*) within the possible limits. I only wish to show that between it and the biological foundations of our human existence very close relations nevertheless exist, and that it is in no way justifiable to deduce, from the autonomy of culture, its complete independence of its biological relations, or even a magic power of manufacturing its own biological foundations. The most fanatical vitalist would not assert that the entelechy is able, not merely to guide the physico-chemical processes in the organism, but even to create the material for them.[405] Everyone admits that a plant can only grow when the necessary food, light, etc., are present in forms which suit the species. In the same way we must recognise that cultures a stage higher do not create their own bearers, but can only direct, and therefore that the degeneration of these bearers leads implacably and irrevocably to the destruction of the culture itself.

Our comparison also shows us that, just as not all plants can live on any soil, so not all aspects of culture can find a home in every human race, but that there is such a thing, in a metaphorical sense, as desert soil and fruitful soil for cultures. This renders extremely doubtful the value of attempts, inspired by a blind belief in the omnipotence of civilisation, to treat Negroes and white men, Mongols and Indians as equally capable of civilisation, and to endow them all without distinction with the 'blessings' of our European civilisation.

A university on the German model is as appropriate, from this point of view, in a Negro state such as Haiti, as a *Victoria regia* on dry sand, which has to be watered all day long, or an edelweiss on heavy marshy soil. The childish belief that if this blessing were continued for a sufficient length of time, the people forming the group in question would gradually become suited to it, is on the same level as a belief that Europeans who live long enough in the Tropics will finally become black, or the belief of our feminists that to educate women like men for a sufficiently long time will finally make them exactly alike. Anyone with the smallest idea of the fundamental laws of biology can only pass over such cultural Lamarckism with the proverb that though you expel nature with a fork, she will always return.

Now that we have shortly reviewed the natural biological conditions of civilisation, we may consider the opposite problem, that of the effect of civilised man upon nature. The problem falls naturally into two parts. On the one hand we deal with inorganic, and on the other with organic nature. The first leads us to the philosophy of technology, the second being the question of the protection of nature and all that hangs together with it. We will deal with this later.

V

PHILOSOPHY OF TECHNOLOGY

At the head of our discussion concerning technology[406] we place the following statement:

Technology is not, as is still generally imagined, merely 'applied science' (physics and chemistry), but is a cultural field for itself standing on an equality with art, science, and religious ethics, as the fourth realm of values.

It would be more correct to say instead of the fourth, the first, since technology, as we have already seen, came into the world with the first tool made by primitive man, and thus became his first guide on the upward journey. For this very reason it possesses the deepest roots in its natural soil, and hence it is not remarkable that the categorical imperative of technology came, in an abstract form into man's consciousness later than the other three – the true, the beautiful and the good. We have only been conscious of it for about 150 years, and even to-day it is not fully recognised. The highest technical ideal is, as we already remarked on page 509, *fitness for a purpose,** and not only fitness within the limit of certain other aims, but fitness for its own sake, just as we seek in science truth for its own sake (and not with a view to its practical application), and in art, the beautiful for its own sake. For the true technologist, an ill-designed machine, an ill-adapted and unsuitable building are just as detestable as bad and insincere art for an artist; or, for a serious investigator, impertinent and empty chatter about serious matters; or, for the person of moral sensibility, the low and evil in any form. The German technician has coined the word *tinnef* for the smaller offences against his categorical imperative, but no name has yet been found for technical crimes on a larger scale, as the idea is still too new.

* *Zweckmässig,* for which no single word exists in English. Literally, "fitted to its purpose."

The 'categorical imperative' of technology runs as follows: everything made for any purpose is to be made in such a way that it is fitted to this purpose, within given limits, in the best possible manner; that is to say, the best possible result is to be obtained in the construction with the least possible means.   It is necessary to define the ideas of the 'best possible result,' and the 'least possible means' somewhat more closely, since they are capable of being misunderstood in innumerable ways, some of which are dangerous.

The worst of these misunderstandings is the view that these descriptions refer to the 'economic' side of the matter, or to put it more coarsely and plainly, the financial side.   When the modern technique of bridge-building sets out to ensure for a bridge, a prescribed degree of stability with the expenditure of the least possible means, it is not intended that it should cost as little as possible, but that iron is not to be wasted, rivets are not to be excessive in number, and unnecessary labour is not to be expended in building it, since all these offend against the technical ideal in itself.

The appearance of mere reference to the economic advantage, results from the fact that among the purposes which almost always guide the technician, the greatest possible economic return is represented.   The fact that this is a secondary matter and not part of the technical ideal as such, is seen from the fact that cases exist or might exist in which pecuniary advantage is entirely out of the question, whereas the technical ideal remains in full force.   I have elsewhere given in this connection the example of the building of an organ to be paid for by a rich patron, to whom money is of no importance.   He would nevertheless expect the organ builder entrusted with the work to construct an instrument best suited for a given space and for the musical end in view, and the builder would offend against this technical ideal if he built too large a structure for the space, or if he inserted stops which would never be used, and so on; in short, if he expended his means uselessly, the word means not being used in the sense of money, but rather of musical means (organ pipes, registers, etc.).   His offence would not consist of spending money unnecessarily, but, on the popular phrase, in using cannons to shoot at sparrows.

This example, which is no doubt not very common in technical practice, shows us clearly exactly what the essence of the matter is. If a technician, as is usually the case, is given the task of constructing a machine in such a way that its owner is able to earn the maximum of money by its means, it is not the fault of technology, but that of the purpose defined from outside, that the work of the technician is done in the service of Mammon. The actual purpose of technology is always given to it from outside, and its 'imperative' only demands that this purpose should be fulfilled in the most suitable fashion.

When our technology is still at work on the problem of transforming heat into work in a manner better than that possible with our present-day steam and other heat engines (steam turbines, explosion engines, etc.) this is not directly done in order to cheapen the production of energy, but first of all because it is an end in itself to increase the thermal efficiency of a heat engine as much as possible. If the problem set is to transform work into heat, then this must be done in such a way that the greatest possible fraction of the heat is so transformed (this fraction being limited, as explained above by the principle of entropy). The ideal of the designer of such machines is therefore the efficiency of the so-called Carnot cycle, the ideal process which delivers the greatest theoretical efficiency. His construction is technically the more perfect, the nearer he attains to this ideal, but the most perfect construction technically is not necessarily the most advantageous from a financial point of view. In this case the technical ideal competes with the financial ideal and a compromise must be arrived at.

This explains sufficiently what in technology is meant by the *principle of economy*. It is a misfortune in our day, that the only meaning of this principle which is worthy of humanity is transformed into a naked matter of pecuniary advantage which excludes any higher meaning. Truly understood, this principle places man directly alongside creative nature, which works in the widest sense according to the principle of the least expenditure of force, as, for example, in the rigidity of the plant, or of the animal bony framework, which are far in advance of our technical powers, in the honeycomb, and in many other cases. We can hardly assert that in these cases it is a matter of money.

On the other hand, it must be admitted that the technical accomplishment of living nature is by no means everywhere so exemplary, as enthusiastic admirers of it, for example Francé,[407] have often asserted. There are many things in living beings which the present day technician can do much better; Helmholtz long ago pointed this out as regards the mammalian eye, and it is obviously also the case as regards dyes. Nature has not approached the results of modern chemistry in this field. Altogether, there are many 'dysteleologies' in nature, and we shall have to discuss this point at greater length presently.

Another problem of the philosophy of technology, apart from that of the true nature of its fundamental values, inquires whether for each technical problem (understanding here as every- where within given means) there exists one or several solutions of equal value. When we walk through a museum, such as the Science Museum in South Kensington, and follow the history of a technical problem, such as that of the transformation of heat into work, we are at once convinced at the first glance that un- limited progress is not only possible, but actual, and that hence there can be no sense in inquiring as to the final best solution of a technical problem. But when we consider the matter further, this blind belief in progress gives way to doubts. For what is straightway interpreted as technical progress is often enough actually nothing of the kind, but, in the main, progress in the selection of available means. For instance, the progress repre- sented by the Diesel motor as compared with the steam engine does not in reality depend upon a technical improvement, but upon the fact that the introduction of paraffin oil for lighting brought into industry the subtances contained in crude rock oil.

It may of course be said that to seize upon a new means of this kind is in itself a technical achievement, since technology quite generally consists in the choice of the correct means to attain a given end. This is true, and in so far we may, as we have already done above, regard the explosion engine as a more perfect technical solution than the steam engine. If the idea of perfection is made so inclusive it is certainly more difficult to realise that there may be such a thing as an upper limit, an optimum, of the technician's powers, since in most cases no one can foresee what new means may become available as time goes on, principally derived from other branches of technology. We have already compared, in this sense, the progress of technology to an exponential function. But if for a moment we regard a collection of means, namely those at present known, as given, we may ask whether there is such a thing as the best possible construction within these limits. Or is here also progress without limit possible?

The history of technology shows us plainly, at least if we take certain examples, quite another picture; a plain asymptotic approach to a final condition representing the best that can be obtained with available means. A typical example for this is the development of the bicycle, the fundamental type of which is forty years old. It has only been developed in detail, and in this also has been stable for a long time already. Thus the pneumatic tyre is likewise about 35 years old, tangential spokes not much newer, though the back-pedalling brake is considerably more recent.

We are aware here of a process which is almost in every respect the exact analogy to the organic differentiation of the germ, which, as we saw above, also proceeds in such a way that a narrower specialisation takes place, and the cell material, at first 'totipotent,' becomes more and more determined (see page 347). We may even say, that if it were permissible to regard the production of a living individual as due to the activity of an intelligent constructor (as certain vitalistic theories of a semi-metaphysical character have done) this constructor proceeds in a manner quite analogous to human technology; both are aiming at an optimal construction, which in the organism, as it would seem, is mostly realised almost completely, while only appearing in human technical work in a few simple cases, since in this case the enlargement of the available means masks the purely constructive progress.

But if we assume this as generally valid, it would clearly mean that the solution of every technical problem is given, or pre-existent, in nature, just as is the solution of a mathematical problem; or its solutions, when the circle of means is enlarged.[408] Looked at from this point of view, Dessauer is not far wrong in saying that *technology shows us, in a humanly intelligible form, what the world process is from the standpoint of Platonism: the realisation of the idea, that is the passage of such an idea from the transcendental timeless realm of pure 'validity' into the world of time, space, and matter.*

In this sense, technology according to Dessauer is continued creation, and in fact it cannot be denied that what it creates is really something entirely new, and thus 'emergent' in the sense frequently alluded to above. It is absolutely certain that in the whole of nature, a watch or an electricity station would never have come into being without man and his mind. These are something entirely new, to which nothing analogous exists anywhere, even in living nature. They give us a picture, and without doubt the best picture, of the creative process, which is behind all creation in nature.

This process, which takes place in nature unconsciously, is continued consciously in man, to which statement we must add, in order to avoid misunderstanding, that in nature, apart from man, it is unconscious only from the standpoint of the single created individual. Whether or not it is consciously known to a highest consciousness (which would of course also be conscious of all human creative processes, since it would comprehend the whole of creation) is an entirely different question.

When we regard technology in this way, we shall find it easiest to form a just and pertinent judgment concerning one of the most difficult questions of the present day, the question of the value or otherwise of technology as a part of culture. I do not need to describe the position, since it is generally known. The enormous over-estimation of technical progress in the second half of the last century has been followed by a reaction, which sees in technology pretty well the root of all evil at the present time. It is supposed to be responsible for our life having become so over-burdened and over-driven, so noisy and restless, so desperately strenuous, so strange to nature and God. It, or the industry allied to it, is supposed to have cooped up mankind in the great towns, where everything that is unhealthy flourishes, and everything that is healthy withers, to have loosened the ties of men with their native soil by the modern ease of transport, and thus uprooted them, to have accustomed mankind, by the extension of the means of communication (newspaper, telegraph, wireless) to live only externally, to give ear to every sensation, and thereby to lose their own souls, and so on and so forth.

What are we to say to these accusations, which are by no means directed against technology alone, but quite generally against the whole process of the rationalisation of our life by human culture as a whole, as expressed by Dacqué in his book *Nature and Soul*, which has already been mentioned several times? If our

culture is to continue, a solution to this problem must be found,
which we are naturally not daring enough to attempt in this
space. We will simply try to find some common ground on
which the two parties now opposing one another with complete
absence of mutual understanding, the admirers of technical process and rationalisation on the one hand, the eulogists of the
'good old times' and of 'closeness to life and nature' on the other,
can at least enter into negotiations. We will first discuss
technology in the narrower sense, and then consider the more
general problem later on.

We must first of all admit that the whole of our technical and
industrial development was actually and historically unavoidable,
as regards the purely scientific and technical aspects. It is here
that what has been said above concerning the autonomy of cultural development comes into force; if a certain field of culture
has once flourished up to a certain point, further development
takes place mainly according to laws inherent in the field itself.
When once science and technology had reached the point arrived
at in 1800, further development was inevitable. It could only
have been stopped by external violence, which, according to our
very conception of higher mental culture, is inadmissible, and
from which European nations have suffered sufficiently in the
course of centuries to make them resolve that it should never
again be made use of.

In the second place, there can be no objection from the ethical
or the religious point of view, since 'every creature of God is
good,' also therefore the new creation which is taking place in
technology through the human mind (see above). A watch or a
wireless set is a work of creation just as admirable as the leaf of a
plant or a human body; the fact that one was created consciously
and the other unconsciously, is of secondary importance.

It must further be expressly stated that the technical products
in question are not, as is often maintained, produced from motives
of gain, but very often have been regarded by the inventors
themselves as gifts to humanity, by which they hoped to make
life easier for their fellow creatures, and bring content and happiness into their lives. It is true that matters have often turned
out very differently from the expectations of the inventors. Often
enough the invention of a single machine has deprived whole
masses of people of their accustomed means of support and thus

brought need, exile, and despair to thousands. But just as often, an invention which has been furiously opposed by certain sections of the people because they feared serious economic injury from it, has on the contrary been a blessing to them. The same thing happens with technical inventions as with great political or military decisions:

> *Rejoice not overmuch!*
> *The Rulers of our fate are jealous gods,*
> *Our over-hasty joy invades their rights,*
> *The seed it is that we lay in their hands,*
> *Time tells if joy or sorrow be the harvest.*
>
> WALLENSTEIN

However this may be, the total result is certainly a great economic advance for mankind, for it is quite certain that, but for technical progress (in the widest sense of the word, including rational agriculture) the European countries could hardly support a third of their present population. It is further an unquestionable fact, that the average standard of living in these civilised countries is, in spite of deplorable details, very much higher than that of 70 to a 100 years ago, to say nothing of several centuries.

In a house, for example, such as that of Albrecht Dürer in Nürnberg, the simplest labourer's family of to-day would feel exceedingly uncomfortable, and when this argument is countered by pointing to the wretched housing of the working class in our great towns, we may reply that such a comparison is only applicable when it is made between similar social classes in different centuries. There was scarcely such a thing as a proletariat, in our present day sense, in past times. But as far as anything similar existed the conditions of housing of it were certainly not better but much worse, than those of present-day industrial workers.

We need only consider what an enormous improvement modern methods of lighting and heating have brought about, to feel distinctly doubtful whether the numerous upholders of days gone-by would really be seriously willing to change places with those times. And those are merely external matters, although it is true that a great deal of our comfort depends upon them. Many other things are much more important, although their possession is quite natural to us to-day; we should miss them very greatly if we had to do without them suddenly. Our means of transport permit us to move about, and to visit our relations and friends in other districts frequently. Our means of communication enable us to keep in continual touch with our friends, and to give them when required a piece of news very rapidly. The suppression of terrible plagues, such as cholera and other infectious diseases, and the painless conduct of operations are other examples. Most people feel towards these things as they do towards unselfish and helpful friends; as long as we have them we do not think of them; only when they are gone are they missed.

Taking it all in all, technology is thus able to present us with a very large credit as opposed to the evils with which it has been debited. And to this must be added the greatest counter argument of all; that *the regrettable results of the industrialisation of our life are for the most part not the direct fault of technology, but of economic developments*, which have been connected with certain technical achievements. It is at least theoretically imaginable that far-seeing statesmen might have averted these consequences by guiding the economic and political development in good time in the right direction, without necessarily leaving the inventions in question in the main unused. It is true that at this point we need to make two limitations which are to the disadvantage of technology. It cannot be said of all unpleasant sides of the modern technical process, that they are caused by wrong economic and social organisation. Many are actually inseparable from the process itself.

Thus the foundation of our whole technology, the winning of coal, is an operation which, even when it is carried out with the latest machinery, is far from being a pleasant occupation, and one which the great majority of those who cannot praise progress sufficiently, would very probably refuse for themselves. But there are also many other positions in our modern industry where danger and even injury to the workman are unavoidable, as in sulphuric acid factories, coke ovens, etc. This is one of the limitations which we have to make, and the other is the answer to the counter argument usually made from the technical side, namely, that the matters in question only prove that the replacement of human labour by the machine has not gone far enough, for all such work must be finally taken over by machines.

The answer to this is that, to make the argument really

convincing, we need a further investigation of a question which I propose to call the 'convergence question' of technology. This is *the question whether, as we replace human power more and more by machines, the ratio of undesirable work really becomes continually smaller.* For we must not forget that every machine intended to take over a certain process of work from hand labour, must be made by a number of other processes of work, which, it is true, are for the most part also machine processes, but nevertheless also need human attention, which, even if it is not always directly injurious, is at least mentally injurious and mechanical. It is not so simple to decide whether this procedure, taken as a whole, represents a convergent or a divergent series; that is to say, whether the ratio in question always approaches an indefinitely small value, or whether it remains stable, or whether it becomes greater in the course of development.

This brings us to the most difficult part of the whole problem, the *effect of technology upon the spiritual and mental life.* The objections which have been raised in this respect may be divided into two groups. On the one hand, technology, together with the other considerations given above, is made quite generally responsible for the mechanisation of our daily work (not only that of the worker in industry, in which case it is quite obvious).

On the other hand, it is accused of showering the external goods of civilisation upon mankind, which goods, since they are really merely owned, but not taken true inward possession of, merely represent a varnish of civilisation, but not real culture, and thus take man away from a true and higher cultural good.

The first objection undoubtedly deals with the most question-able point of the whole of modern development. There can be no doubt that the modern 'soullessness' of work, which is a con-sequence of the division of labour carried to an extremity, is a terrible source of unhappiness for all those who possess a need for deeper culture. This fact does not prevent a considerable number of more primitive minds finding work of the most me-chanical possible description very agreeable, since they have no desire for more intelligent labour.

At this point, as has been generally agreed, we have the most serious problem of the future. We must leave it to those who are fitted to seek ways and means to lead us out of this difficulty. But as regards the other argument, that higher culture is being smothered by purely external culture, and by the all too easy appropriation of technical achievements, the very persons most to be accused of this are usually those who raise the accusation. Educated people of to-day (even when they are really educated) usually behave towards the technical things which they use every day in exactly the same way as the vulgar rich man, living in a villa built by the best architect in town, behaves towards the literary or artistic treasures which are to be found in it. He owns them all in a purely external fashion because he is able to pay for them, but it is not he who is master of the things, but the things that are master of him.

In the case of the vulgarian, this attitude is generally recognised and blamed, whereas as regards our technical achievements, no one finds anything wrong, simply because almost every one is equally to be blamed. Whoever wishes to effect improvement in this respect should first see to it that the natural sciences and technology do not remain an unknown territory for the spiritual leaders of our people, whereupon the mastery of things over men will cease of itself.

The objection that the mere superfluity of technical products is smothering man to-day is foolish. Otherwise one might raise the same objection to art or anything else; it is naturally true in certain cases, but this is likewise true of all cultural values. If this accusation thus rebounds upon those who make it, there is nevertheless something of truth in the statement that the way of life which results from technology may do much inward injury. I am not thinking here so much of the mad pace of modern life, since this is not only to a small extent a direct result of technology, but for the most part a result of the economic development. I refer rather to *the overloading of our minds with impressions of all kinds,* which are more or less a direct result of the means of travel and communication rendered possible by technology. But a great part of this is not to be laid to the door of technology, but to that of the greed of gain, which plays upon the desire for sensation without scruple, in order to make money. A wiser law might have already prevented a great deal of injury.

The fact remains that this by itself is not sufficient to solve the problem of how humanity is to deal with the enormous mass of new impressions which it is continually receiving. This is really a very serious biological problem, which needs taking the more seriously the more clearly we realise that cultural Lamarckism is impossible. Most apostles of progress endeavour to represent the matter as if the accommodation of mankind to this altered world were quite a simple matter, which might be completed in one or two generations. They start from the simple belief in the fixation by heredity of the influences of environment. We have seen above what is the truth of this matter, and do not need to repeat what was there said. As regards the present problem, it follows that in any case it is well worth the consideration of everyone suited to the task, and that we cannot possibly be satisfied with the superficial statement that humanity will get used to it as it has got used to so many other things.

It is true that man has the widest range of possible existence of all living beings, and is able to exist, thanks to his gift for invention, under conditions of the greatest possible variety. But *est modus in rebus, sunt certi denique fines.* We have already pointed out that Alpine plants cannot be made to grow in a marsh, and aquatic animals in a desert. It is true also for man that his power of adaptation is subject to certain biological limits, which can only be overcome by a long process of selection, which would last many hundred years. A sensible cultural policy cannot seriously consider this, but must regard present limits as fixed, and the question we are considering here is whether these limits are not already exceeded in our modern life as regards what human nerves are able to master.

A particularly important side of this subject is the question of noise, which quite recently has attained importance through the increase in motor traffic. We can only draw attention to these questions, and point out that they can only be solved on the basis of a true biological view. Mere talk of progress is just as useless as helpless praise of the good old times. We have before us the highly serious question as to *how developments, which are undoubtedly necessary from an historical point of view, and cannot be brought to a standstill, are to be made to conform to the elementary spiritual needs of mankind, which themselves cannot be abolished by simple decree.* If this problem cannot be satisfactorily solved, the result will be in the end self-destruction of our culture, since the foundation will be removed upon which alone it can flourish: namely, the existence of healthy human beings who enjoy their labour.

The questions discussed in the last section have now brought us to the more general problem alluded to above, the problem, namely, of how the reconstruction of natural conditions of life and of nature herself, brought about by rational thought, is to be evaluated. Technology is only a part of this process, which comprehends and penetrates the whole of human history. Mankind has not only made use of external inorganic nature, and changed it to a large extent for its own purposes. It has also domesticated animals and plants, exterminated others, and has finally emancipated its own conditions and laws of life to a large extent from the rule of mere drives and instincts, and thus, so to speak, rationalised itself as well. In our time we are confronted with the serious question whether this whole development is really a healthy or an unhealthy one; whether we are to welcome it, or to condemn and oppose it as far as possible. What right has man really to transform nature in this way? Or is this perhaps an idle question; do sense or right come into it at all, but rather natural necessity?

VI

THE PROTECTION OF NATURE

Mankind has actually exercised control over nature for many thousands of years before it (or rather individual thinkers) ever became conscious of the peculiar rôle of man in creation. As far as we can see to-day, this idea only appears at the highest point of cultural development, but then leads directly to the question of the right to this mastery. This question is then answered everywhere in connection with religion, and we are concerned here with two of these answers, the Jewish-Christian and the Hindu, these two representing the opposite poles of possible attitudes.

The Hindu doctrine is so generally known that we do not need to expound it here at any length. It denies every kind of prior right of man in relation to other living beings, and only allows him a single priority, that of seeing through the great delusion of the will to live, which is the source of all evil, and of finding his own salvation and that of the world in Nirvana. In strict contrast with this, we have the teaching of the 8th Psalm: 'For thou hast made him a little lower than the angels, and hast crowned him with glory and honour. Thou madest him to have dominion over the works of thy hands; thou hast put all things under his feet: all sheep and oxen, yea, and the beasts of the field; the fowl of the air, and the fish of the sea, and whatsoever passeth through the paths of the seas.'

There can thus be no doubt that the Jews taught the sovereign rule of Man over Nature, and regarded it as a privilege given to man by God. Whoever has studied the history of this people and its religion more closely (but in the real construction, and not in the traditional one taken over incautiously by Christianity from the Old Testament) knows that this view of the rôle of mankind is closely connected with the whole religious development of Judaism. For it shows us a perpetual battle by the priestly and prophetic leaders against tendencies in the people to nature cults (the service of Baal and Astarte), which persisted with great tenacity and continually broke out afresh.

In fully conscious contrast to all such tendencies of the 'peoples' (*gojim*, heathen), the priest stood for a strict ethical monotheism, which rose in personalities such as Jeremiah, the two Jesaiahs, Amos, Micah and Hosea, to a quite extraordinary ethical and religious height, but later on the other hand, solidified to a pure religion of law in post-exilian times; but in both cases it is directed against every kind of bond between man and all the rest of nature, since religion, and hence the whole sense of human existence, was limited to the moral relationship between man and God as the Giver of the moral Law.   In this one-sided attitude we have the unique world-historical meaning of the Jewish prophetic religion, but also its limit, which is most sharply expressed in the prohibition of the graven image (the First Commandment).

For the world of Western civilisation, this development has become of primary importance, since Christianity adopted from Judaism this complete exclusion of extra-human nature from the religious attitude, and this came about the more easily since the Jewish doctrine here met on common ground with Hellenism, which was otherwise entirely alien to it.   In the latter (in Neo-Platonism), the doctrine of the unique value of the spiritual and mental as against the merely natural was maintained for quite other reasons.

The two factors together resulted in the well-known words of Christ in Mark viii., 36, being interpreted in the very narrow sense, that the whole remaining world (that is the extra-human creation) was of no account as compared with the infinite value of a single human soul.   The whole of creation, seen from this point of view, was merely the setting for human history, and in particular the history of human salvation, and was finally to disappear completely in the end and give place to a 'Kingdom of God,' consisting only of saved souls.

The fact that such ideas have maintained their pre-eminence in both branches of the Christian church to the present day is obvious; although, since Christianity reached the Teutons, individual illuminated spirits, chief of whom is St. Francis, but also the German Mystics of the Middle Ages, Luther, etc., have protested against treating the whole of extra-human creation as valueless and finally negligible.

The revolution which has taken place in our time in all European countries, is to be referred mainly to the rise of natural science, which of course would not have developed to such an extent, if the aptitude for it had not existed in the leading European races. The Nordic aptitude for penetrating into infinite distance, and the wide mental range natural to it, on the one hand, and the strong tendency of the Alpine race on the other hand to kind and friendly self-immersion in the whole life of its immediate surroundings, resulting in a very friendly attitude towards animals,[409] have been chiefly responsible for this development.

There must be few educated people in civilised Europe to-day who would seriously support the relentless mastery of man, although we find on many sides strong doubts and objections regarding such movements as nature protection societies, anti-vivisectionism, vegetarianism, and the like, since exaggeration and fanaticism are feared, not without justification.[410] A procedure such as that adopted by American hunters towards the vast herds of bison, which were exterminated in a few decades merely for the sake of their tongues, would certainly be condemned unanimously to-day by the whole world, though it is true that it would be less the moral than the economic point of view which would settle the question.

This last remark brings us to the decisive question of the final justification for the demand, so general to-day, for a protection of nature. This demand, and the circles which stand for it, illustrate what very often happens in ethical and religious movements. They rise spontaneously from motives which at first are more matters of feeling than conscious reason, and are usually quite unclear regarding their final basis. Our scientific circles, who are fortunately so active on behalf of nature protection of every kind to-day, are not aware that, since we are dealing with a judgment of value, they are of necessity leaving their own element and entering upon the field of ethics and religion. They usually persuade themselves that their demands are a natural result of scientific knowledge as such. But it is easy to see that this is wrong.

What does scientific knowledge, pure and simple, prove? Only this: that there is a limitless abundance of forms of life (and also of inorganic matter) upon the earth, and that man is the latest in this series, and a member particularly highly and

Oos

peculiarly developed, (namely as brain animal), and perhaps also that it was wrong to believe, as in earlier times, that the whole of creation was there only for the sake of man (see also above, page 279). This is not only true as regards astronomical distances, but also as regards the earth, and in particular the development before the coming of man.

But as we have already said, it is a mistake to suppose that this gives sufficient grounds for demanding the protection of nature. The most that can be deduced from this line of thought is, that within the teleological structure of the whole of nature, discussed above, which may equally well be regarded theoretically as the causal structure, man does not occupy the position, from the point of view of theoretical logic, which was wrongly ascribed to him in former times. It is obvious that only a small part of creation, in the general biological sense is 'made for him,' as the fly is for the spider, the deer or the sheep for the wolf, or the grass again for the deer or the sheep. But all this in no way allows us to conclude that man has not the right to use, entirely as he pleases, whatever in creation is accessible to him. On the contrary, we might equally well conclude from the whole of the facts of science, that in all nature the right of the stronger holds without any modification, and hence that there is no reason for making an exception in the case of man, and putting moral hindrances in the way of his exploiting the remainder of nature. And the representatives of this somewhat brutal view would probably add that what the others (the protectors of nature) put forward as a moral demand, is probably nothing more than human egoism in another form. They would represent it as the opinion that, for the sake of human well-being, certain natural conditions should not be too much changed (see above). This, they would say, is a matter open to discussion, but it is pure imagination to pretend that the whole depends upon moral reasons; these moral reasons and moral demands are on all fours with others of the kind, being ultimately only disguised human egoism.

This latter statement shows us that the defenders of the 'right of the stronger' realise, that the whole problem is nothing but a part of the general problem of the foundation of ethics. In this respect, the protection of nature has no priority over ordinary ethics as between human beings themselves. In every such ethical discussion we are confronted with the same dilemma, that both the egoistic and also the altruistic or social motives can be regarded as products of natural development, and hence that no reason can be given for preferring one to the other. There are then only two fundamental possible positions, which, however, may be differentiated into very many different formulations. Either we refuse fundamentally to see in altruistic instincts,[411] which are unquestionably present already in the animal world and fully conscious in man, something of high and absolute value, which man or other creature does not make, but merely seizes hold of (just as with theoretical truth, see above). Or this is recognised, just as we recognise in all research not the production of truth, but merely the grasping of a truth already present, although we may of course hold very different opinions concerning the way in which it is done.

In the first case, if we are to be logical (as for example the German Monists are in this matter) we can only see in moral precepts the common agreement of certain social organisations, which have made such precepts as they consider suited to their purposes. This results in contradictions, such as an organisation enforcing a pacificist ideal with absolute tyranny, as is happening in Russia, and as happened in the French Revolution, when the guillotine was used to enforce liberty, equality, and fraternity.

In the second case, it is necessary to take a step which cannot be logically analysed, that of believing in an inner necessity of the world order, which stands above man and existed before him, according to which man had to attain to a moral ideal, because it in some way corresponded to the final object of the existence of the world, an object which we cannot, it is true, entirely comprehend, but are able to feel and experience within ourselves as a source of happiness and freedom. This position may be otherwise formulated either abstractly and philosophically, or in the form of a simple childish belief in God's commandments from Sinai. The essential point is the same: the humble reverence of man in the face of a higher power which was there before him and will be after him.

There is no need to enlarge upon the fact that the whole of German idealism lends its aid in this point to the religions of all peoples and times, and that practically all the great thinkers of humanity have taken this view. Only the materialism of modern times, and the logical positive Monism of to-day, have developed the position which denies all belief, with complete logic. After this has happened, only two alternatives exist. Which of these the author of this book stands for, does not need further explanation. We shall return at greater length to the whole problem of value, the deepest problem of all philosophy, later on.

If we now take our stand upon belief (not in the dogmatic sense of the Churches, but as explained above), it will be clear upon what the final justification of nature protection is based. If we regard the whole world as the act of a Final Reason and Will, and regard ourselves as part of this vast process, there arises in us directly the consciousness of our responsibility towards this Final Cause and its aims, and we feel bound to use all our scientific and historical knowledge in this sense, for in all knowledge God (to retain the ordinary religious terminology) reveals a small part of His plan and thus lays upon the recipient of the knowledge a certain duty. When now a highest value, namely that God's will is to be done, is fixed, all the theoretical views of a biological and teleological kind discussed above take on binding power. If we are responsible to God with every breath of our existence, we are also responsible with reference to all our fellow creatures, and doubtless even with reference to inorganic nature, as far as this is subject to our will.

There would then be nothing which in principle would be excepted from this responsibility. How we are to deduce from such a general view rules or directions for behaviour in special cases will, it is true, need special consideration; in all such cases the words 'seek and ye shall find' hold good. We have learned from history, and are no longer of the naïve opinion that we possess God's will in a finished and certified code in human institutions, and only need to look it up mechanically, or inquire, when we want to know what we are to do. On the contrary, every period and every nation is repeatedly confronted with new problems, and one of ours is obviously *the burning question as to how the little left by human civilisation from original creation, can be preserved, without civilisation being exposed to unbearable hindrances.*

In a more general way, this question may be stated thus: how can and must man reconcile generally his higher cultural values with other and lower values, which also have their justification? Can these values be arranged in order, and, if so, how much of this order can man discover? And to what extent has he the

THE PROTECTION OF NATURE

right or the duty to sacrifice the lower, or even the higher values?

This brings us to the kernel of the whole question. It has recently given rise to much public discussion in connection with the activity of a single distinguished man, the theologian, musician, and missionary doctor, Albert Schweitzer. Since the author has discussed this matter with him fully in another connexion,[412] and since a really full statement of the question could not be given in this place, it must suffice to refer the reader to the discussion here and to formulate shortly its result as the author sees it.

He rejects Schweitzer's fundamental position that there is no natural order of life values, or that man, at the least, cannot discover it, and that we can only therefore take upon ourselves at the most according to a subjective judgment, and 'with fear and trembling,' to destroy other life, even with the object of preserving life. Schweitzer went so far that he wished to regard his destruction of trypanosomes in order to save the life of his Negroes in Lamberene, not as an ethical duty, but as a free decision.in full consciousness of evil being done. According to him, there is only a single fundamental command in all ethics, the respect for the will of other life, which is supposed to come directly to the consciousness of man with its own will to live.

The author denies, on the one hand, that the latter is actually the case, and on the other, that it is right to formulate the highest commands of ethics in this way. It is not the life will of other single creatures as such, but the whole will, that is God's will, which must be regarded as the highest ethical law, and this may, as the whole of creation shows, demand the sacrifice of lower life in favour of higher. Man has not the right to claim to know and act better than the Creator himself in this respect. He has merely to ensure that life is not destroyed unnecessarily ('life' being here taken in a wider sense than the biological; the inorganic forms must be included, as Schweitzer himself also thinks).

For this reason, man has not merely the right, but actually the duty, not only of killing the trypanosomes in question in order to save the Negroes (it is obvious that Schweitzer himself, in spite of his theories feels this to be his duty otherwise he would not act in this way); he has also the right and duty to re-form both inorganic and organic nature to any extent for the sake of the higher goods of culture, which are also obviously a part of the world plan. In actual fact, almost everything done in past centuries in this respect produces by no means the impression of being injurious and senseless. We see the harmony in an old watermill in a valley, in the path on a mountain, in the domesticated animal in the household, in the steppe transformed into a meadow, and so on.

It is only the most recent overwhelming development of technology and the relentless energy of modern life in all its branches that have resulted in the catastrophic changes in both our inorganic and our organic environment, which have made the idea of nature protection so important to-day. When we visit towns such as those in the Black Country, and find them in every respect unlovely and inharmonious, ugly, smoky, and misshapen, having grown out of the earth in a night, we may well feel that man has here exceeded his rights, and disturbed creation in a brutally egoistic fashion which will bring its revenge in one way or another. The same is true of the extinction of many plants and animals resulting often from sheer greed of gain; we need only think of the wonderful Kaori pine of the South Seas, the dodo, and the great auk of the Arctic Seas, which will probably be followed by the whales in a very short time, and then of our own beautiful beasts of prey such as the lynx and the wild cat. It is a good thing that the governments have a word to say in the name of the whole of humanity, and against destructive greed, and thereby come to the rescue of extra-human creation; unfortunately much too little is being done in this respect. If the progress of civilisation is bought with the complete destruction of forms of life which can never be replaced, this is not progress, but impoverishment and retrocession. What is still to be saved must be saved now at the last moment.

The transformation of nature by civilisation concerns not only extra-human creations, but also humanity itself. Our own natural conditions of life have also been completely transformed as compared with the state of primitive man, and the process still continues. Here also the question arises how far the process is good and desirable, and in what respect it is to be regarded as false and injurious. It has led in part to very curious and sometimes fanatical plans and doctrines for the 'reform of living.' Some reject all animal food, others even the cooking of vegetable food, a third party wishes to get rid of clothing entirely, a fourth will only drink water, a fifth only live in huts in a wood, etc., etc.

Since Rousseau's *Émile*, innumerable novels have appeared which have described the 'return to nature.' *The Blue Lagoon* and *Tarzan* are examples of recent popular successes. In all of these, all modern modes of life are totally rejected, since they soften mankind, entangle them in all kinds of unnatural habits, and have caused great moral conflicts, which are destroying civilised man of to-day. A particularly acute sign of this question is the custom, now becoming general in civilised nations, of birth control, or more generally speaking, the rationalisation of sexual life. This in particular has aroused the most furious opposition of large circles of people which can by no means be dismissed with the term reactionary. What can be said fundamentally in regard to all these matters?

First of all, what has already been stated above as fundamental principle in such cases. It is just as wrong to hold blindly to tradition, or to any external authority whatever, as it is to doubt the existence of any objective ethical guidance in such cases. Here also there is such a thing as good and evil, but it is not as simple as the traditionalist supposes to decide what is good and what is evil in particular cases.

Let us first consider the more superficial changes in our conditions of life, which have already been discussed in another connection (see above, page 571). It is certain that, for example the use of larger quantities of soap, the introduction of more delicate cooking, the consumption of good wine, sleeping in modern spring beds, the possession of central heating, etc., etc., easily become habits so firmly ingrained, that people are unhappy when they must suddenly do without one or all these conveniences. This problem was one which affected soldiers in the war almost more than any other. When modern education in sport lays great stress on training in these respects it is obviously moving in the right direction, for when young we can accustom ourself to any, even the most primitive, way of living, but this does not hold for older people. Hence no greater kindness can be done to youth than to educate it with the greatest simplicity, though under the healthiest conditions that circumstances permit.

But such a principle is more difficult to carry out than to state. Taking it all in all, it cannot be avoided that, in the future, each new generation will grow up under somewhat more comfortable conditions than the previous generation, this being the nature of technical progress. But there is a limit to this rule. It does not mean to say that this process will grow according to a linear, or even an exponential function (arithmetical or geometrical series). We may rather imagine, and we ought at least to strive, that it will approach an upper limit, for when a person has a healthy dwelling, sufficient but not luxurious food, good lighting and heating, and suitable clothing, he has everything that is necessary for external comfort, and it is by no means necessary that he should increase further his demands in these respects.

The blind belief in perpetual progress may lead here also to erroneous conclusions. It is very much to be doubted whether in 2,000 years, although technical progress may make great advances in that time, the average dwelling will appear much more luxurious than a modern, medium-sized, well-furnished house. We shall no doubt find a number of technical achievements unknown to us, everyone will no doubt have a flat roof for his private airplane, food may be obtained by a postal tube connection from a central warehouse by pressing a button, telephones and television apparatus will be in every room.* But no one can use more than one bed to sleep in, one bath to wash in, one stove to heat with, one cupboard to hold possessions, and more than this will not be found in a room.

If by that time people have become sensible (which it is to be hoped will be the case), they will say to themselves that in all these respects, also as regards clothing, the simplest and most useful is also the best and most beautiful. The chief responsibility for the unhealthy excess of luxury which continually reappears in human development is borne by the vagaries of fashion and the endeavour, which is associated with them, to outdo one's neighbour. In this respect the feminine population have a great deal to learn. The old fairy story of the fisherman and his wife contains much wisdom. If this cause on the one hand, and the greed of gain which speculates upon it on the other, could be kept in check, we can imagine a state of saturation in civilisation, perhaps not very far ahead of us, which would then change as little from a fundamental point of view, as the solution when once found, of a technical problem. But the blind clamour for progress

*H. G. Wells has made numerous suggestions of this kind, some since realised. [*Trans.*]

on one side continually raises on the other a groundless fear of a development of luxury without end.

The problem of the change in our conditions of life does not appear so simple when regarded from the point of view of the bodily and mental states and processes of human individuals. There can be no question that rationalisation has made extraordinary advances in this respect also. How enormous is the difference between the eating and the drinking, the sleeping and the walking, the love and sexual life of the modern European, as compared with that of primitive people ! And this is not merely a result of custom and law, which themselves of course have grown unconsciously for the most part like organic structures, and have not been created artificially and rationally, but is also as the outcome of rational consideration, mainly of hygienic, but also of social aesthetic and other kinds.

We can easily understand the doubt arising, particularly with reference to this side of modern civilisation, whether the limit of what is biologically endurable has not long been past. From the complaints of dentists concerning the degeneration of our teeth, through those of oculists concerning the increase of short sightedness as the result of school work, to those of religious and ethical circles concerning the increasing rationalisation of sexual life, we have an immense number of gloomy prophesies concerning modern civilisation. What is there true or false about them ?

The first thing that is true above all is that an unavoidable consequence of easier conditions of life and less struggle for existence is the lessening and weakening of positive selection. For example, the increase in short-sightedness is in all probability by no means due, as hitherto believed by Lamarckism, to the bad effect on the eyes of school work (which would be an inheritance of acquired characteristics in the most perfect form) but to the fact that myopia (short-sightedness, or lengthening of the eye axis) is an inherited characteristic, which is less of a hindrance than formerly to those afflicted with it, since there are enough occupations which they can take up. Hence they can reproduce themselves to an extent not sensibly less than the average, and their characteristic defect spreads in a population in accordance with Mendel's laws, since they are no longer subjected, as in earlier times, to continual elimination. Whether the tendency is dominantly or recessively inherited is not certain but there are probably several different biotypes (genotypical combinations) causing an externally similar diseased state, and this explains the contradictions, hitherto not cleared up as to the results of various investigators.[413]

The position is similar with regard to the tendency to the spread of tuberculosis and many other diseases. The easier matters are made with regard to the disabilities for the individual, phenotypically, the more certain it is that the genotypical spreading is facilitated, since negative selection with regard to the disability takes place. There can therefore be no doubt that we may speak of a very serious danger in this respect to civilised man through civilisation itself (see also above), a danger which can only be met by very intensive race hygiene. Unfortunately, the usual complaints concerning the injurious effects of the softness of modern life obscure almost hopelessly all clear views. The clubs for sport and hard training, which are ever growing in number, and teachers, educators, the clergy, etc., are almost always fixed adherents of simple Lamarckism, and hence believe that the weak offspring of the present generation can be turned by sport and training into a more energetic and vigorous race; whereas they entirely overlook the more important factor of selection.

But much worse than this is another undesirable result of the present mania for sport as regards young women, for just those who are in body and character the most vigorous and valuable, do not marry early enough and hence are excluded, together with their healthy and valuable inheritance, from the reproduction of the race, either wholly or in part, since they have too few children in order not to be obliged to give up their pursuits. It is absolutely necessary to realise the *much praised improvement in the health of women through sport can only be actually of real use to the nation, if it causes these women to become more capable and willing to produce a larger number of healthy children of equally good inherited characters.* The fact that a mother climbs mountains or goes ski-ing does not make the slightest difference to the genotype of her children.

It is much more important that a woman who carries inheritance characters giving fitness for these bodily feats, should convey these powers to as large a number of children as possible, than that the same woman should actually develop these powers phenotypically.

Unfortunately, there is no sign whatever that this point of view of race hygiene is being taken into account in modern education. We are still, in this respect also, floating in the shallow waters of cultural Lamarckism.

It is still more difficult to come to a true judgment in the matter of the rationalisation of erotic and sexual life. Since it is impossible to deal here with this whole burning question[414] I will only say the following. We cannot regard it as a frivolous interference by man in the God-given order of nature, when he brings this side of his bodily and spiritual activity more and more, as time goes on, under the control of rational considerations. The justification for this, as far as it is a matter of rational ethical consideration or principles, has never been disputed by any sensible person, but always rather demanded by ethics itself.

But it is also obvious that considerations of a more practical kind also claim a certain right, and the whole history of love and marriage confirms this. Why should it therefore be regarded as immoral and 'unnatural' for two people, before entering on a marriage, to be clear and rational concerning the probable economic, hygienic, and personal consequences of the step, and hence refrain from it in certain circumstances, when they recognise that it will probably lead to more harm than good for themselves, for their children, their relatives, or society? The latter point of view in particular, is fortunately being put forward and finding increasing support by numerous social and ethical thinkers, among them the well-known race-hygienist, Muckermann, in Germany, and others in England.

The greatest objection is raised against the rationalisation of birth in marriage itself. But it is again not easy to see why this process should be judged fundamentally in any other way than the innumerable other processes of organic and also inorganic nature, which man has learned to regulate by means of his reason. If we are not to stamp the whole of human civilisation as unnatural – and that is impossible, since nature includes man, the brain-owning animal – it is impossible to draw any arbitrary dividing line. All that we can demand is that, as in every other case of the conquest of nature by man, this also must enter harmoniously into the whole sense of the cultural process, and not, as is also the case with so many other achievements of civilisation, fly directly in the face of the true sense of the process.

It follows from this that birth control as such is neither good nor bad, but, as in every such case, what is good or bad is what man makes out of this new possibility of controlling nature. He is able to make of it, as of everything else, a blessing or a curse. How the first is to be attained in this case, and the latter avoided, is a matter, the discussion of which would take us too far. It must suffice to give clear expression to the principle which alone, as I think, can form the basis of an objective and correct judgment, and do justice both to the fair demands of the evermoving development of civilisation, and to the eternal aims of all this development. [414]

This once more brings us to the problem so often touched upon, which must now be shortly discussed although it seems to take us away from our field of scientific philosophy, namely the problem of values.

VII

THE PROBLEM OF VALUES

This problem belongs, as we shall see, so closely to insight derived from science, that we have not merely the right but the duty, to enter upon it at the finish, since we would otherwise neglect one of the most essential tasks of scientific philosophy.

As regards a large number of points which bear on the question, a simple reference to what has been said before will suffice. The reader may be reminded that we saw in the technical fundamental value, that of fitness to purpose, a fourth fundamental value of equal status to the three other long recognised values, the True, the Good, and the Beautiful. To this should be added, that to these four categories a fifth is to be added, not in the opinion of all thinkers, but of a number (including the author) namely, the religious, with its fundamental value, the Holy.[415] This value would for religious people be placed above the others as necessarily including them all. But we do not need here to go further into the problem of the relationship of religion to the other four categories.

The individual judgments of value were in the course of human history, as we have already explained above (page 509), first made with reference to single concrete cases, and then only as single judgments. The aesthetic value must have been the first of all to have become independent in the sense of Wundt's heterogony mentioned above, for the oldest cultures known to us show surprisingly high forms of art. Nevertheless, even in this field a fully conscious art, in which the beautiful is pursued for its own sake and recognised as a problem in itself, only appears fairly late. When, where, how, and for what reasons, the actual process of spiritualisation took place, that is to say the detachment of the category in question from concrete cases and its complete emancipation, is for the most part outside our knowledge.

As far as it can be investigated, it is a subject for the history of civilisation. But this discipline must always remain conscious of the fact that it cannot give any reason for the setting up of the value, but merely show, after the event, how the value established and unfolded itself. This excludes all mere historicism, which erroneously confounds the question of value with the question of its historical origin.[416] The validity of mathematical and physical truths is completely independent of the way in which they were discovered. When they are once there, they hold, independently of all historical conditions. We have already pointed out that this leads to the rejection of Spengler's relativism. The cultural values are not only values for certain times, epochs, and peoples, but in each of them there is alongside the historical, racial, and other conditioned factors an unconditioned element, one simply valid, which is outside all relativity.

It is true that the historical approach teaches us with unmistakable plainness – and this is the kernel of truth in the relativist assertions of many writers to-day – that *it is fundamentally wrong for us to believe that we now hold in our hands, so to speak, the absolute and final truth, the last value, in any of these categories of value.* As soon as science presumes to say a final word, it becomes, as all experience shows, hasty dogmatism, and digs its own grave. For every scientific theory, as Einstein says, can be, and needs to be, merged in a higher and more comprehensive theory, and finally receives its limits from the latter.

It is obvious that the same is also true in the moral field. The history of ethics shows clearly how, in this case as well, lower truths are repeatedly merged into higher ones, and are combined with others to form over-riding judgments of value. It is thus and not otherwise that we are to understand the words of the Sermon on the Mount, 'Ye have heard that it hath been said . . . but I say unto you . . .' Every new insight of this kind does not mean the abolition of an earlier truth. The Newtonian theory of gravitation was not proved false by the Einsteinian, but now appears as a mere first approximation, whereas the more complete and comprehensive theory of Einstein gives us a second and closer. The relationship of the higher ethical judgments to the lower is to be regarded quite similarly, while in both cases direct errors are not impossible, which then cannot be comprehended, but merely refuted.

The incapacity of most people to distinguish between these two processes, which in the history of science are both obviously discernible, is the chief source of mere relativism, according to which nothing permanently valid exists in the midst of continual change and growth. It is like denying the existence of the sun because its rays are repeatedly darkened by clouds and sometimes appear to vanish almost completely. If it is asserted that there is in reality no sun at all but only occasional rays of light, which man as a matter of thought economy has accustomed himself to designate by the word 'sun' on account of certain relationships in space and time which occur repeatedly, the assertion cannot be refuted. We have seen above that this is actually the doctrine of modern positivism.

In the meantime, we others will not allow our pleasure in the real sun to be spoiled, nor our belief that, in spite of the relativistic darkening of its life, it will again shine through the clouds. With this naturally I mean no longer the physical sun, but the suns of the True, the Good, etc. It was the cardinal mistake of the Middle Ages to believe that all these values were laid down and settled in the written words of man, in human institutions, and so forth. This naïve realism (using the word in a transferred sense, although the narrow and epistemological sense also falls into the same category) had of necessity to be followed by an age of revolution, of release from external authority, and of the severest criticism of all tradition. For absolute truth, absolute value cannot be captured by human forms and formulae in any field whatever; all attempts of this kind fail repeatedly. Hence we were finally left with relativism pure and simple; that is, the dissolution of all belief in any kind of truth.

Kant's philosophy was nothing but an attempt, undertaken with magnificent power, to rescue the significance of values as such in the midst of this collapse, which Hume inaugurated and Kant himself actually brought to its final conclusion. It had to fail, and did fail, because Kant did not see, and also could not see as matters then stood, that the 'autonomous' personality, considered both theoretically and morally, or in other words, the knowing and evaluating mind, cannot guarantee this validity to be that of values beyond the individual, so long as it is merely regarded as pure personality. Whoever takes the subject as basis, will not arrive in the theory of knowledge at anything but 'absolute idealism' or illusionism, as the case may be, and in practical philosophy at anything but general relativism, it being practically indifferent whether the latter is supported upon single human individuals, upon collective individuals, such as Spengler's 'Cultural Soul,' or upon Kant's 'transcendental unity of apperception,' which is to be understood in the sense of a general human subject, or a general Ego.

In all three cases, values do not hold in themselves, but only for me or for the soul of my culture, or for man in general, just as in the theory of knowledge things are in no way there in themselves, but only as appearances to me. Kant's contemporaries, Fries, Schulze, and Jakobi, saw, and formulated clearly that only a super-personal principle could save the situation. The demand of the latter, stated by Geibel in his well-known words:

> *This is the end of philosophy,*
> *To know that we must believe.*

simply states the fact that there is no foundation for science or any other categories of values without the resolution to believe in the existence of the transcendental world of values. As regards the values of truth, this has already been stated at the close of our first section. All science depends upon the assumption that there is such a thing as truth. The only justification for relativism is that it is needed again and again to save us from relaxing into dogmatism and the Middle Ages.

At this point, we may draw from this statement, which is founded on the whole discussion in the first part of this book, the fundamentally important conclusion:

What is true in the field of the scientific knowledge of truth, may be transferred, by virtue of a justifiable analogy, to the other categories of values. That is to say, we have the right to believe in the existence of the Good, the Beautiful, etc., although here also, and even more strongly, we must oppose every kind of hasty dogmatism. In these other categories of value as well, the first assumption of all striving for truth is the belief that such striving is not in vain, not mere revolution about our own axis. Just as the whole of science without this assumption becomes meaningless, so also does the whole of ethics, of art, etc. This means the return from individualism to an objective point of view, from absolute idealism to realism, but to critical realism, that is to say, to a realism no longer subject to the delusion of simply taking the appearance for the thing in itself, the historical formulation of a theoretical, ethical, or other truth for this truth itself; a realism which on the contrary remains aware that it is the very characteristic of objective truth to transcend all purely human measure.[417]

Why is the old theory of physics now replaced by the new one? Simply because this is a step nearer the real truth, and not because one period regards the same matter in one way and another period in another. Properly understood, the history of striving after values with all its strange mistakes and vagaries, is a witness of the power of truth itself, for we should have no reason to recognise our relativities as such, if we were not driven to this recognition by the power of truth. We thus see before our mind's eye a new cultural epoch, in which the two opposing currents, the naïve realistic one of the Middle Ages, and the purely individualistic and relativistic one of the age of reason and history (18th and 19th century) are fused in an higher synthesis. It would be attractive to go further into the question of how this synthesis will work out in religion, but we must leave this aside, since it lies outside our actual theme. I have been unable to avoid this discussion since it had to form the foundation for what we have to say by way of conclusion, in the next chapter.

VIII

NATURE AND JUDGMENTS OF VALUE

By stating our theme thus, we imply our justification for including it in scientific philosophy. For a problem of the latter that cannot be avoided, is that of discovering how nature herself stands towards our human judgments of value; or in other words to the aesthetic, moral, and religious values, which are the highest possession of civilised man.

There are not a few, both on the scientific and also on the philosophical side, who regard this question as a delusion. For – so say both – nature as such is always free from values, all judgments of value are a specifically human and psychical matter, and only concern man. He can apply his judgments of value to nature as such just as little as, conversely, morality can be measured in cubic centimetres, or beauty in volts. Both are completely separate worlds, and the moral sciences must, in their own highest interest, oppose the natural scientist when he operates with ideas such as the 'art forms of nature' or the like. That is not his business, he may pronounce judgments of existence, but not of value. Conversely, the scientists themselves are accustomed to agree to this, since long historical experience has taught them, that the introduction of judgments of value into science has more than once dangerously obscured the view of research as regards objective fact.

Now I have of course no intention of disputing the correctness of the latter consideration. It is a commonplace to-day that the scientist as such has not to evaluate, but merely to get at the facts. But we have already seen above that these facts by no means include only causal relationships, but equally well teleological relationships, and while these again are also taken, for the moment, in a purely matter of fact sense (as for example in the statement already alluded to, that the fly serves as food for the spider), this whole teleological point of view leads naturally and quite logically to the idea of an universal teleological connection, in which all these single teleologies are contained only as components.

But from this thought it is only one step to the question behind it of the meaning of the whole itself, which question then brings us directly into the region of judgments of value. For it is obvious that the mental development of mankind, including its striving after values, stands in some kind of inward connexion with this final sense, since this development itself is a part of the whole, the meaning of which is under discussion. On the other hand, the scientist, as a human being, is not merely a scientist but also forms judgments of value like all other men, and this he does quite naturally also, in the first place, as regards those things with which he is mainly concerned.

The expression by great astronomers of the infinite elevation of spirit produced by the sight of the starry heavens and by immersion in its secrets is well known, as are also the innumerable exclamations of aesthetic admiration, and even ethical valuation of the wonders of the plant and animal world, from Haeckel's radiolariae to the complicated forms of animal pairing and care of offspring (see W. Bölsche). Are all these to be regarded as merely the private pleasures of the individual human beings, and as having no fundamental connexion with their work ? Or is there not after all a deeper and more necessary relation between the two; not only by way of the fact that to become an investigator requires a certain sense for such judgments of value, but also purely objectively ? That is to say, may it not lie in the nature of things, and not in our selves, that both are usually, or at least very frequently, found together ?

Against the assumption of an objective basis in nature herself for these judgments of value, the main objection that is made is that we can always find equally good grounds for negative as for positive judgments of value. In truth, we find at first sight quite as much that is meaningless as meaningful, quite as much that is beautiful as ugly, quite as much good as bad in nature; indeed it often appears as if the negative far outweighs the positive. This fact is the strongest support of relativism, which, as it appears, can appeal as regards the whole of nature to the idea expressed in the German proverb, 'What is an owl for one is a nightingale for another' ('One man's meat is another man's poison.'). We will now see how much truth there is in all this.

For this purpose it is best to begin with the aesthetic category. We have already pointed out that art is closely related to technology, and this very close relationship is seen most strikingly in architecture. And it is equally evident that what is simple and adapted to its end is generally also most satisfying from an aesthetic point of view, although it has repeatedly happened that this truth is forgotten, with the resulting production of impossibly overloaded styles. Now there cannot be the slightest doubt that nature herself produces technical values to an enormous extent. The technical wonders of a single blade of grass, of the trunk of a palm tree, of the blood circulation in an animal, of the eye, and so on and so forth, far exceed anything that human technology has been able to produce (see also above).

On the other hand, it is equally without a doubt, that sub-human nature already shows signs of the development of a certain striving after luxury. The innumerable decorative effects of colour, the song of birds, the numerous excess formations (see above), the play of young animals, and very much else, produces a strong impression of an ever-active power which, in a way closely similar to a child's play or the work of an artist, finds pleasure in pure creation for its own sake, and attempts it in every possible form. It is naturally wrong to attempt to introduce such considerations into the theory of descent or biology generally in place of scientific causal analysis.

But does this causal analysis, and the establishment of single teleological relationships, exclude the way of regarding nature which we are discussing? Just as those two exist of right side by side, so perhaps another approach to the problem may again be found. Who is to say that our power of understanding the world is exhausted by causal and teleological analysis? If our epistemological discussion led us to see that human capacity for knowledge on the one hand, and things-in-themselves on the other, must in all probability be in some way adapted to one another, since otherwise the fact of knowledge could not be satisfactorily understood (see above), it is surely not going too far to make the further supposition that our human faculty for aesthetic judgments concerning nature – and who has not at some time made use of this faculty? – corresponds in some way with certain objective sides of nature herself, and without her would be 'objectless' in the truest sense of the word. When Goethe says:

Had our eyes no kinship with the sun,
They had no power to see its light.
Did God's own power not dwell within ourselves,
How should Divinity enrapture us?

which is certainly true, this sentence also may be equally well reversed, if we allow ourselves to be guided by the analogy with knowledge: if the world were not flooded by the light of beauty, beings would never have existed with a sense for this beauty. How otherwise can an organ come into existence, which has no object? Just as the eye is there for light, the ear for sound waves, and the nose for chemical substances, so is our aesthetic sense there – for what?

Obviously for something which we no more generate by means of this sense itself than we do light, sound, and chemical substances by our other senses. Nor can we see the smallest reason why the investigator of nature should not have the right to make use also of this sense of his, as long as he understands how to keep what he perceives through it clearly separated from what his other senses (in the ordinary meaning of the term), teach him as regards knowledge. We likewise perceive in this a new and deeper foundation for what we have previously said concerning the protection of nature. The friends of nature among the investigators of nature – they are not all both of these at once – are quite right in demanding the preservation of the beauty of nature from a purely aesthetic point of view. It is objective, and not purely subjective values that are concerned.

But the relativist will say, that all this is none the less pure fancy. In reality there is nothing beautiful in itself, but only things and processes (or in the case of the positivists, elements in Mach's sense, that is phenomena), and only man brings aesthetic values into nature. Is this really the case? When I hear a symphony, is it only I that bring an aesthetic judgment of value into an otherwise meaningless volume of sound, or have not rather Beethoven, Mozart, or Brahms done the most, and the most essential to that end? And is the aesthetic impression of a grand Alpine landscape or of the seas fundamentally different from that of such a symphony or of a Gothic cathedral?

But it is said: naturally the aesthetic value of the symphony does not originate in Mr. Smith or Mr. Brown who listen to it, but in its creator, but that is explained by the fact that Mr. Smith and Beethoven are both specimens of the human species, and hence Mr. Smith, will read into or hear into such things the same or similar judgments of value, if in a less degree, as Beethoven or Goethe or Erwin von Steinbach put into them. Good, but the origin of the powerful, even overwhelming, impression of great natural scenery still awaits explanation. Who is here the Beethoven or Goethe? Or has every ordinary person now suddenly acquired the qualities of such men, and so is able to play the part of composer or poet towards these impressions on his own account?

Naturally I am aware of the answer given by the relativists to this statement. They point out that man has only gradually learnt to experience such beauties of nature aesthetically; that, for example, it was only in Goethe's time that the wonders of the Alps, which had hitherto been merely regarded with horror, were opened to general understanding, and that the same is true in our day of the magic of heath landscapes; and furthermore, that there are enough people who are left cold by all of it. This is quite true, but there are also enough people so stupid as to be incapable of learning from any book on physics or biology, and there are enough races who have progressed just as little towards any serious success in science as others have in the aesthetic comprehension of nature. Are we to conclude that science only exists for us who have it?

We saw that this supposition leads to absurdity and rejected it. The validity of scientific truths is quite independent of whether they are grasped, where they are grasped, and by whom they are grasped. Can it be otherwise with the aesthetic values? And is it not a much more satisfactory notion that humanity, which had to be slowly educated to a knowledge of truth, has also needed a long course of education in the other values, and will need to continue this course?

But at this point the artist himself may perhaps raise a weighty objection. Subjectivism and relativism are to-day a common opinion in his circles. We can speak to-day of an almost pathological fear, above all on the part of the wide artistic public, but also on the part of innumerable artists themselves, chiefly those of the second and third rank, a fear of talking of any kind of objective aesthetic forms or values. This was by no means always the case. On the contrary, the great artists of the past were almost all – with few exceptions – filled with the same feeling that also inspired the creators of the greatest scientific knowledge: the feeling that what was given to them to achieve was a gift from above, something that they laid hold of, and not really something which they themselves produced.

Language, which has been developed in connexion with purely concrete things, has unfortunately no adequate terms for this peculiar relationship, in which the very highest activity of the creative worker is finally revealed to himself as a pure vision. Goethe made the profound remark concerning Bach's music, that it was 'as if eternal harmony were conversing with itself' and in fact, this impression is surely that of everyone of any understanding for music, when listening to the works of Bach. How does it happen that no one then thinks of Bach, the badly paid cantor in St. Thomas's, the man with ill-behaved scholars, evilly disposed superiors, and innumerable children? What altogether is, in the case of such music, the man that 'created' it? Is he not perhaps really what he considered himself to be: a mere speaking tube, a means in the hands of a higher power creating through him?

These expressions are, as we have said, inadequate, for they deny all too strongly the intense activity of the artist concerned, which is always present at the same time. The truth of the matter is best expressed by saying quite directly that, in such men, a particle of the divine exhibits its activity, and that for this reason they are both at once: nothing and everything, an indifferent means and the highest creative power. I surely think that artists themselves have the greatest interest in preventing this highest view of their calling from being completely smothered in the waste of present-day mere relativism, and pursuit of originality at any price. What then are finally all the numerous would-be great, as compared with the few who really bring forth great things? The repeated and wearisome harping on the 'own,' the personal, is really, when examined closely, nothing but the poor cloak which a period fundamentally unproductive, because devoid of faith in the higher calling, wraps around its own nakedness and calls a royal mantle, whereas it is in truth really a beggar's ragged garment.

But if what we have said concerning art in the narrower sense is true, it cannot be too far-fetched a statement when we ascribe also to the aesthetic values of nature, what we have ascribed to the human artistic values; thus admitting that it is finally an objective side of nature that we grasp with our aesthetic sense, not one which we create. Only that the actual creative artist here is another than the conscious human mind, or perhaps more correctly: here artistic creation does not take place in the form of an individual consciousness, but in that of the still unconscious life of nature (see above pages 424, 531). But why should not this be just as capable of aesthetic, as of teleological achievement?

As Titius, whom we quoted above, quite rightly says, the whole inorganic world in its foundations is 'ruled by a symmetry, order and regularity,' which appeared to us almost incredible. T. K. Oesterreich states quite directly:[418] 'The world must be understood, not merely by the way, but quite seriously, as the work of art of a divine creative spirit . . . it is a part of its nature that even the most terrible may still bear aesthetic values . . . however far below the moral we may place the aesthetic, it is a value which permeates the world, one which takes a ruling position everywhere in creation, in the greatest as well as in the last structural refinement of the invisible' and he is also right when he adds: 'It is very difficult to believe that the whole was created by an absolutely good God (in view of evil, of which we shall speak immediately); but it is written, as it were, on the forehead of the world that it was created by a God who felt and acted artistically.'

This last sentence carries us away from the aesthetic judgment of the world to the ethical and religious, and there can be no doubt that these are the deepest problems of the whole of philosophy. They nevertheless belong, as I once more expressly emphasise, without question to scientific philosophy, which would indeed deprive itself of its truest and highest value, if it excluded them. For the final reason why persons of ripe and developed mentality pursue philosophy is that they desire to attain to a clear view concerning these final questions of our world view, as far as this is possible by the aid of our reason. Hence we will attack this last and most difficult task, the problem of evil with courage and determination, but as Frenssen said, 'With clenched teeth.' The poet sang of a world

Where every prospect pleases, and only man is vile

but in reality the whole living world, apart from man, is filled with pain and misery of all kinds.

Alongside much that is beautiful and harmonious, the terrible and cruel stands in every form and gradation. A moving description is given in a recent poem by W. Widmann,[419] and to go back to a somewhat earlier period, we may remind the reader of the well-known novel of Vischer, *Auch Einer*. It is naturally the simplest way to deal with the problem when, here again, we take our stand behind the positivist and relativist barriers, and say that everything, both beautiful and noble as well as terrible, only exists for humanity, which reads its own values into nature; the latter being neither one or the other in itself, but simply there, and necessarily as it is. But such mere rejection of every deeper inquiry will never satisfy permanently those who think more profoundly.

Precisely the same is true here as in the case discussed above on our aesthetic contemplation of nature: the plain and sober determination of the facts and their relations by science in no way excludes the further ethical and religious inquiry, but rather demands it as a necessary extension, if we are really to understand the world and not merely to analyse it theoretically (see above). Let us therefore attempt a short general view to begin with.[420] By the term 'evil' in the broadest sense we shall here understand everything which appears to us as if, according to our own standards of value, 'it ought not to be.' It is divided into two chief categories of physical and moral evil ('wickedness') but we are dealing here chiefly with the first, which itself may again be divided into the four divisions of dysteleology, bodily and mental suffering, and death.

We use the term dysteleology to describe a large number of arrangements, which contradict our ordinary standards of utility or good design (the technical ideal), or at least appear to contradict them, for example, the enormous waste of germs in living nature (a single living offspring for a million eggs), the apparently senseless destruction of life, particularly also valuable life, by geological catastrophes, the useless rudimentary organs and equally useless excessive developments (see above), and many such cases. This category of evil is comparatively the easiest to dispose of, for at least in very many cases we can show that a certain purpose, such as the maintenance of the species, is at any rate assured in this way, however clumsy it may seem. In many cases this explanation entirely fails, and taking it all in all, an unprejudiced view forces upon us the impression that Paulsen is right when he compares the utilitarian uses of nature, for example in the maintenance of species, to a hunter who fires off in a field a hundred thousand guns in every possible direction in order to shoot one hare.

As regards other physical evils, pain and death, a biological explanation can be given without difficulty. It is clear that pain causes living beings to avoid injurious influences. It serves as a warning; living beings like in other respects those we know, but devoid of the sense of pain, could never maintain themselves in the struggle for existence against those gifted with a sense of pain, they would die out in a very short time. It is equally evident that need has been the most important teacher in human civilised life, and still is so.

Death also, regarded biologically, is a necessary institution. For it strikes only the 'soma,' that is, according to our present-day view, the off-shoot existing at the moment of the germ plasm, which continues to reproduce itself quite independently (Weismann's germ path, see above). The soma may die, so long as the germ plasm lives on. Indeed the old must make way for the young, if any kind of progress – whether by selection or in any other way – is to come about. In accordance with this, the whole of living nature is based upon the mutual destruction of organisms, and in a definite territory there is actually kinetic biological equilibrium between the various kinds of plants and animals (compare page 401). A world in which 'the lion lies down with the lamb' is therefore a mere dream. The living world as known to us would be completely unsuited for its fulfilment. What in such a world would the lion do with his fangs, the snake with its poison, and the polyp with its prehensile tentacles?

We must however expressly state, that life according to the latest investigations is potentially immortal (see above page 360). Since there is no reason for regarding the organisms hitherto investigated as different from the others, we may conclude that death, though it actually strikes all organisms, is always and solely caused by the accumulation of injurious influences in the course of life (in which however we must also reckon internal changes which in normal cases produce death) but that above and beyond this rule, there is no secret and inward 'law of death.'

All this is naturally also true for man. But in his case and his alone we have to add to physical evil, moral evil – wickedness or sin. 'He distinguishes, chooses and judges,' (Goethe) but he does not himself act accordingly, and thus, in accordance with the measure of his own valuations, is guilty.   And this guilt increases already existing physical or natural evil in a monstrous fashion. *Homo homini lupus!* many more human beings are destroyed directly or indirectly by other human beings than by any of the powers of nature, quite apart from the innumerable tortures of every kind, both of body and soul, which human beings inflict upon one another.   We hardly need to mention that these facts are the chief source of all higher religions, and of numerous systems of philosophy of historical importance.

We must first be clear that we are dealing with a serious meta-physical problem, which cannot be abolished by a few cheap rationalisms.   Two ways of avoiding this problem are often attempted.   One is to point out the beautiful and pleasant side of natural life which undoubtedly exists alongside all this evil. Whoever experiences sympathetically the joy of all young life in nature, that of a playful kitten as well as of a human child, who-ever observes mother love, as it already exists in the touching self-sacrifice of animals, whoever hears the birds sing and rejoice on a lovely summer's morning, naturally arrives quite easily at the impression that only a dismal pessimist can talk of the evil in the world, and is inclined to rather cry with the poet:

God's in his Heaven – all's right with the world!

If he does not throw all philosophy overboard in such moments, he will find theoretical satisfaction in saying that unavoidable evil must simply be endured for the sake of all that is beautiful and pleasant.

But the same man may appear thirty years later as a worn out and tired creature, upon whom care and disease, or loss of his nearest and dearest, have set their stamp, he will then perhaps read us a lecture on the doctrine of Solomon, Schopenhauer, or Buddha, that all is vain.   'It were much better that nothing should come into existence . . . for all that exists is worthy to be destroyed' (Mephisto).   It is clear that the judgments in both cases depend upon the merely relative and subjective mood, and it is certain that no objective value is to be attached to them.

Whoever attempts seriously to think out, in a manner as far as possible free from the effects of his own temporary moods or per-sonal character, how far happiness or unhappiness, pleasure or pain, are distributed in the whole of creation, will always arrive at the result that the two, taking it all in all, about balance one another.   In any case he will have to admit that evil is real to an extent which does not allow us to pass it by as a subsidiary matter. Assuredly it must not lead us to under-estimate the positive and creative elements of life, but it is certainly not allowable to close our eyes before the dark side of nature.

Another and more acceptable way out is the heroic point of view as put in Felix Dahn's novel, *Julian the Apostate*, into the mouth of the German prince, Merovech Serapion, also in that of Teja in the *Fight for Rome*.   The hero says: Yes, the world is full of evils, they outweigh perhaps the good and beautiful, but for that very reason I stand firm and fight with them ever, even if I know that it is in vain.   *Si fractus illabatur orbis impavidum ferient ruinae.* There can be no question that from the practical point of view this attitude is in many cases the only one humanly worthy.   Teja's view of the world is theoretically identical with the Buddhistic. It is pessimism, but the practical consequence drawn by this representative of the Teutonic spirit is the opposite of that drawn by the Hindu with his quietism.   The latter finds salvation in the annihilation of the will to live, and the former in fighting to the death.

Such a view of the world has something heroic about it, but it easily leads to callousness and cruelty.   If nature is cruel, good, I will be the same !   Struggle for existence becomes the motto, and everyone is to pursue his own ends remorsely as best he can. Nietzsche's magnificent 'blonde beast' becomes all too easily the ideal man of the future.

This is not in itself necessary, for even a hero may be a human being of tender sensibility, meeting all beings with kindness and love – indeed, he may still remain a hero though sacrificing himself of his own free will for his brethren.   But then it is necessary that he should see evil also from another side (which, by the way, Dahn's Teja actually does, he having been more coloured by Christianity than he admits).   Such a one cannot be content to say that evil being there, it gives him a welcome opportunity of exhibiting his heroism.   He is obliged to experience it sympathetically in his heart; and then he cannot stop at that, but must take with him the deep conviction that it really 'ought not to be so' and that the problem is therefore not solved if individual heroes succeeded in conquering it.   What is to happen to the innumer-able creatures to whom such a power is not given ?   Animals that are timid by nature, or men who are not all heroes by nature, and cannot be so ?   Is there no solution to the problem for these ?

An especial variety of this 'heroic' solution is very widespread to-day in biological circles: the reference to the biological necessity of pain and death which we have pointed out above.   These are certainly facts, but what is the use of my explaining the biological necessity of pain to a person suffering terrible torture from cancer or kidney disease ?   Or to a mother at the grave of her dearest, the biological inevitability of death ?   The question behind it all remains completely untouched: why must the world be made in such a way that these bitter necessities exist in it ?

While these, so to speak natural attempts to evade the problem turn out to be untenable, matters are no better with certain other solutions which have grown up upon a specifically religious ground. These solutions are distinguished from the former by at least taking the problem seriously, and by really recognising the full weight of evil.   But the reasons they give cannot satisfy us.

One of these answers is the pedagogic.   God allows evil in order to discipline man by it.   We may reply to this assertion, which is much favoured in religious circles, that in far the great majority of cases such a pedagogic justification can at best be constructed after the event, and that in the second place, it not seldom leads to simple cruelty to others, as for example the assumption by a person that the disease and death of his wife or child are the result of God's discipline for him (who is naturally the person of chief importance).   Thirdly, it completely fails in the face of mass misfortunes, which obviously pass one by and strike another without any reason (war, earthquake, etc.).

The believer has certainly not merely the right but the duty to ask himself in the case of every such stroke of fate, what God now demands from him in this position; and in this sense he may also ask: why has God laid this upon me ?   But as soon as this question assumes the sense that the event in question has hap-pened simply for his education, piety turns into blasphemy, or at least naïve religious egoism, which has nothing to do with true religion.

But if the pedagogic answer is thus open to grave doubt, it must also be admitted on the other hand, that there is a considerable element of truth in it, inasmuch as need, as we have already said, is in fact the teacher of humanity, and has at all times been so. In this sense we may regard evil generally as a means of education in the hands of a cosmic power, but we must avoid giving this idea an individual turn. All that then remains of it is the quite general statement that, for a religious view of the world, it is self-evident that evil also serves God's purposes, and it then becomes, as Mephisto says, 'A part of that power which always wills evil and brings about good' and in the same sense the Lord himself says in the Prologue:

> *Hence this companion purposely I give,*
> *Who stirs, excites, and must, as devil, work.*

But it is clear that while this answer may produce a partial satisfaction, it does not remove the gigantic unreason of this whole pedagogic point of view. In innumerable cases need and evil bring about something entirely other than good, and often even wickedness. The German nation has by no means been improved as a result of its defeat, which was certainly undeserved in proportion to its guilt. On the contrary, it has become much worse, and would in all human probability have been much more likely to progress morally if it had been victorious in the beginning in 1914, when all good instincts were awake in our nation, and then gained a just peace.

In many individual human destinies also, the failure of the teaching of suffering is obvious. Innumerable people otherwise quite capable and valuable become, as a result of disease, embittered or hypochondriacal, and torture those nearest to them whom they once loved and cared for; and this fact is just as real as the other one that many become through such suffering, better, more patient, and more amiable. Is it logical to ascribe the benefit in the latter case to divine discipline, but the degeneration in the former to human weakness alone ? The truth is that both suffering and good fortune may make man better, but that both of them also make him worse.

We now come to the second religious solution, the doctrine that all physical evil is the consequence, and further the juridical consequence (punishment) of human sin. This doctrine is held even to-day by the majority of religious people among us as impregnable dogma. It is supported upon the teaching of the Apostle

Paul (in the Epistles to the Romans and Corinthians) who for his part went back to the Old Testament story of the Fall.   This solution brings us right into the middle of the metaphysical discussion of the problem, which without doubt is here grasped in all its keenness.

But at this point we are given an especially plain example of the extent to which the results of the natural sciences trench upon metaphysics.   It is modern natural science which must unconditionally veto this theory, at least in its common form.   While it cannot be denied that in many cases physical evil may come to human beings as the result of certain sins, it is quite impossible to generalise this experience over the whole of evil.   For in the first place, the whole of extra-human creation suffers in the same way, although perhaps to a somewhat less degree in single cases (on account of the lower level of consciousness), and this suffering came into the world millions of years before mankind.

In the second place, it is equally certain that man took over this suffering from his animal ancestors; he was mortal and sensitive to pain long before he could 'sin' in the moral sense (for he had not yet a free moral will).   Both reasons together make it impossible to maintain this doctrine in the usual manner.   (The question of how far we are to lay the responsibility for it upon Paul, may be left out of account here as a purely historical matter.   But it is certain in any case that he has almost always been understood in this sense).

If a 'Fall' retains any sense in face of modern knowledge of nature, it can only be this, that man in the moment when he acquired an ego, also became capable of the conflict between moral values already recognised by him, and his own lower instinct. This naturally happened already in dim primitive times, always and everywhere anew, and still happens anew in every growing child.   These lower and egoistic instincts are, however, nothing but a possession also common to the animals, and what is characteristically human is simply the wakening to consciousness of the conflict between them and the higher claims which have become autonomous.

If then this primitive form of the doctrine in question cannot be maintained, we must still pursue the question whether it may not perhaps contain a deeper core which is still valid to-day, and which it is only necessary to free from a too narrow, historically conditioned, complex of ideas in order to lay hold of a deeper truth. In actual fact, as soon as we state the question thus, this biblical doctrine touches the deepest wisdom of another people, which appears to have the highest aptitude for metaphysics of all mankind, the Hindu.

We have already seen above that the true cause (speaking metaphysically) of all evil lies in the division of the one world-will into innumerable part and individual wills. This is precisely the fundamental thought of all Hindu wisdom. The famous 'Veil of Maya' is nothing but the part-consciousness, which mistakenly believes itself to be the centre of its world, with its forms of thought and categories. If this veil is torn asunder, that is to say, if the illusion of ego is destroyed, the world is saved. It passes into Nirvana.

In this last doctrine of salvation, it is true, the Hindu philosophy takes precisely the opposite direction to Christianity, which finds the secret of the salvation of the world, not in the complete extinction of the individual will, but in the voluntary sacrifice inspired by love (since in love will is able to meet will without both being destroyed). But the assumption is common to both, that the source of both physical and moral evil or sin lies in the self-seeking of the individual will, and any other solution of the question but this is in fact impossible unless we renounce all metaphysics.

This assumption is also found everywhere in Christianity, where the influence of the Germanic spirit, racially allied to the Hindu, makes itself felt. It is chiefly seen in the case of the German mystics (Eckhart, Böhme). But it has always had difficulty in fighting the superior power of the Jewish juridical theory. We hardly need say that, on the other hand, the adoption of the conclusion drawn by the Hindu doctrine of salvation was impossible for the active European mind, with its fundamentally positive attitude towards life. The fact that such Buddhistic tendencies are spreading at present among ourselves is a sign of serious mental disease in our culture.

This brings the task of scientific philosophy to an end; indeed, the last sections have carried us far beyond it. Nevertheless, we could not omit them, since scientific philosophy, as we see, has very important contributions to make to this problem, which in itself belongs to the philosophy of religion. A further pursuit of it, would however, take us into the purely religious field. We should then have to investigate what conclusions regarding the idea of God are to be drawn from the position of the problem as stated. We must renounce any attempt to carry the matter further here, and can only refer the reader to the author's essay on 'Evil in the World from the point of view of Science and Religion.'

The fundamental question which arises at this point is why, if the world rests upon an Absolute (God), and a world could only come into being by the division of this One into many, this one Absolute did not prefer to be self-sufficient, but instead took upon Itself the existence of a world (these expressions being naturally mere images of eternal existences), the pain and joy of which It had Itself to share. This question is, as we see, the problem of contingency (page 214) as seen from the point of view of value. We have already seen that human understanding cannot solve the problem even purely theoretically. It is also unable to solve the problem of value connected with it, the so-called problem of theodicy, in a logically unexceptionable manner – this is self-evident, since the final values can just as little be founded logically as the final facts, since they themselves are the necessary presumptions of all judgments of being and value.

But although our mind, brooding over the problem, has to admit that, having investigated what is capable of investigation, it has no other choice than quietly to revere what cannot be investigated, mankind is not for this reason completely barred from entering this realm, concerning which Frennsen's fine-saying holds: 'We do not live for this life, but for the secret that lies behind it.'

Where the scientist and the philosopher have to be silent, the artist, the poet, and the prophet may still speak to us, indeed, this is their truest realm. What can no longer be stated in the ingenious phrases of human wisdom, may still be brought home to the sensitive spirit by the language of religious inspiration, art, and above all music; to such an extent that we may well be in doubt which way brings us nearest to the root of the matter. This revelation reaches its highest and purest degree when all three speak together, when the poet's gift comes to the help of the prophet, and both are brought by the language of music to a degree of expression infinitely beyond mere words.

Listen with your inner, and if possible with your external, ear, to the great chorus of Creation's rejoicing in Beethoven's Ninth

QQs

Symphony: hear Schubert's *Allmacht*, Haydn's Star symphony ! Or go to church when Brahms' immortal Requiem is being given, and let the eternal Song of Praise fill your ears: 'Lord, Thou art worthy of praise, and honour, and glory. For Thou hast created all things.' And go on Good Friday to hear Bach's St. Matthew Passion, and the Missa Solemnis and the B minor mass, the two greatest works produced by human, nay superhuman, inspiration. When you then hear the angels in Heaven singing 'Sanctus, sanctus Dominus, Deus Saboath' – with Beethoven in mystery far removed from all earthly things; with Bach in endless rejoicing, rising ever higher and higher – then you get a faint inkling of the reason why God did not remain God alone, but created a world with joy, life and love, but also with pain, death, and sin, and of where the solution of this contradiction is to be found. No philosopher in the world can tell you more about it than Bach, Beethoven, and Brahms.

THE PERIODIC SYSTEM OF THE ELEMENTS

I	II	III	IV	V	VI	VII	VIII	(O)
1 H 1,008								2 He 4,00
3 Li 6,94	4 Bo 9,02	5 B 10,82	6 C 12,00	7 N 14,01	8 O 16,00	9 F 19,0		10 Ne 20,2
11 Na 23,00	12 Mg 24,32	13 Al 26,97	14 Si 28,06	15 P 31,04	16 S 32,07	17 Cl 35,46		18 A 39,94
19 K 39,10	20 Ca 40,07	21 Sc 45,10	22 Ti 47,9	23 V 51,0	24 Cr 52,01	25 Mn 54,93	26 Fe 55,84 27 Co 58,97 28 Ni 58,69	
29 Cu 63,57	30 Zn 65,38	31 Ga 69,72	32 Ge 72,6	33 As 74,96	34 Se 79,2	35 Br 79,92		36 Kr 82,92
37 Rb 85,45	38 Sr 87,63	39 Y 89,93	40 Zr 91,25	41 Nb 93,5	42 Mo 96,0	43 Ma	44 Ru 101,7 45 Rh 102,9 46 Pd 106,7	
47 Ag 107,88	48 Cd 112,40	49 In 114,8	50 Sn 118,7	51 Sb 121,76	52 Te 127,5	53 I 126,92		54 X 130,2
55 Cs 132,81	56 Ba 137,37	Rare Earths 57—71	72 Hf 178,6	73 Ta 181,5	74 W 184,0	75 Re	76 Os 190,9 77 Ir 193,1 78 Pt 195,2	
79 Au 197,2	80 Hg 200,6	81 Tl 204,4	82 Pb 207,20	83 Bi 209,0	84 Po (210,0)	85 —		86 Em (222,0)
87 —	88 Ra 225,97	89 Ac (227)	90 Th 232,15	91 . Pa (230)	92 U 238,2			

Rare earths (57 – 71): La, Ce, Pr, Nd, (Il), Sm, Eu, Gd, Tb, Dy, Ho, Er, Tu, Yb, Lu. The number in front of the symbol of each element is its 'atomic number,' the number beneath is its atomic weight (isotopy not being taken into consideration). The Roman numerals at the top indicate the column number. Elements in the same column exhibit great similarity with one another.

NOTES

PART I

[1] The exact definition of an individual substance can only be given on the basis of thermo-dynamic laws which have been mainly investigated by Willard Gibbs. See any standard work on physical chemistry, e.g. *A System of Physical Chemistry*, by W. C. M. C. Lewis, Vol. ii (New York and London, 1925).

[2] The elements of chief practical importance are: – carbon, oxygen, hydrogen, nitrogen, sulphur, phosphorus, silicon, iron, magnesium, calcium, potassium, sodium, chlorine (these are all constituents of organic compounds), aluminium, copper, tin, zinc, lead, nickel, mercury; the noble metals, gold, silver, and platinum; and further, chromium, manganese, cobalt, arsenic, antimony, bismuth, iodine, bromine, fluorine, boron, barium, strontium, radium, uranium, thorium, cerium, tungsten, osmium, iridium, zirconium, helium, neon.

[3] See E. Cassirer, *Substance and Function* (London, 1923), also most of the general philosophical works given in the bibliographies at the end of this volume.

[4] On the history of the Atomic Theory, see: F. A. Lange, *The History of Materialism*, trans. by E. C. Thomas (London, 1925); J. W. Mellor, *A Comprehensive Treatise on Inorganic and Theoretical Chemistry*, Vol. i (London and New York, 1922); J. C. Gregory, *A Short History of Atomism* (London, 1931).

Recent developments in atomic physics have been described in a very large number of books, of which the following may be mentioned: E. N. Da C. Andrade, *The Structure of the Atom* (London, 1927); Lord Rutherford, J. C. Chadwick, and C. D. Ellis, *Radiations from Radioactive Substances* (Cambridge, 1930); A. E. Ruark and H. C. Urey, *Atoms, Molecules, and Quanta*, (New York and London, 1930); G. Gamov, *Constitution of Atomic Nuclei and Radioactivity* (Oxford, 1931).

[5] See text-books on the history of philosophy, e.g. Windelband's excellent description.

[6] The method of obtaining atomic weights may be found in all good text-books of inorganic chemistry.

See 'The Architecture of the Solid State,' by W. L. Bragg, *Nature*, Vol. 128, pp. 210, 248 (1931), with reference to the account of solids which follows in the text.

[7] *Naturwissenschaften*, 1926, No. 3, p. 43.

[8] For an account of the properties of thin films, see J. Alexander, *Colloid Chemistry*, Vol. i, Art. 'Surface Energy and Surface Tension,' by W. D. Haskins, pp. 226 ff.; H. Freundlich, *Colloid and Capillary Chemistry*, trans. by H. Stafford Hatfield (London, 1926), gives a very full account of the whole subject. There are numerous small treatises in English, including one by H. Freundlich; also *Soap Films*, by A. S. C. Laurence (London, 1929).

⁹ An account of Perrin's work on suspensions will be found in his book *Brownian Movement and Molecular Reality*, trans. by F. Soddy (London, 1910).

¹⁰ The calculation of the 'mean free path' and Loschmidt number may be found in J. Walker's *Introduction to Physical Chemistry*, Chs. x and xxiii (London, 1910), and in other books too numerous to mention.

¹¹ The figures are mainly derived from Dorn, 'Experimentelle Atomistik' in the volume *Physik* from *Kultur der Gegenwart* (Teubner, Leipzig, 1915).

¹² See Erich Becher, *Einführung in die Philosophie* (Munich, 1926). See N. Campbell, *Physics, the Elements*, p. 19 (Cambridge, 1920).

¹³ On the various stages of certainty in knowledge, see R. Richter, *Der Skeptizismus und seine Ueberwindung*.

¹⁴ Cf. W. Kodwiss, *Naturw. Wochenschrift*, 1919, Nos. 49 and 50.

¹⁵ I am glad to find that this statement meets with the assent of all eminent physicists; e.g. Bohr, *Naturwissenschaften*, 1930, No. 4, p. 73. [Ostwald withdrew his objection to atoms; see his Preface to his *Outlines of General Chemistry*, trans. by W. W. Taylor, London, 1912. The Preface is dated 1908.]

¹⁶ Adickes, *Kant contra Haeckel* (Berlin, 1906).

¹⁷ The most overwhelming criticism is that of Stallo (*The Concepts and Theories of Modern Physics*, London, 1872), who does not leave modern atomic notions a leg to stand upon.   Further illustrations of my remarks are to be found in Ostwald, *Grundriss der allgemeinen Chemie*, 2nd ed., 1890, pp. 3–6 [Eng. trans. *Outlines of General Chemistry*, London, 1890] ('What the real nature of matter may be remains unknown to us – and a matter of indifference').   Mach, in the *Mechanik*, 4th ed., p. 321 [Eng. trans. *The Science of Mechanics*, 4th ed., 1919] ('We can nowhere perceive atoms, they are, like all substances, creations of thought. . . . Though atoms may very well be suitable to represent a number of facts, the scientist who has taken to heart Newton's rules concerning philosophising will treat these theories as of provisional utility, and strive to substitute for them a more natural view').   Compare also Mach's *Wärmelehre*, pp. 429 and 430.

The attempt, mentioned in the text, to construct the Law of Multiple Proportions without the use of atoms, is given by Wald in Vol. ii of Ostwald's *Ann. der Naturphilosophie*.   For Boltzmann's useless protest see *Wied. Ann.* **60**, 231 (1897).

Further examples of a hypercritical attitude towards atomic theory are found in the case of Nernst (see text), Drude (*ibid.*), Riecke (Preface to his text-book of Physics), Grimsehl (Conclusion of the first edition of his text-book).   Dorn also cites, in the article referred to in Note 11, Hertz's view without contradicting it, although the whole article itself proves the contrary.

The most peculiar point of view is that of O. Lehmann, the discoverer of liquid crystals.   (In Vaihinger's *Annalen der Philosophie*, Vol. i, 'Das Als Ob in der Molekularphysik.')   After having stated in detail the number and nature of the reasons for assuming the real existence of molecules and atoms as an unavoidable conclusion (he says expressly that we are forced to make this statement, since the observations exclude any other interpretation) he suddenly concludes by talking about biology, and finds that in it the phenomena are not explicable on an atomic basis.   Instead of drawing the only possible conclusion that this lies in the nature of the relationship between physics and biology (if vitalism is right), he finds the way out through the 'As if' point of view in molecular physics, against which the whole of his initial statements form a complete protest.

As a further example of the extreme caution and reserve exhibited in recent times by eminent physicists before again accepting atoms as real, we may cite W. Voigt's words from the volume on Physics in *Kultur der Gegenwart* (p. 717).   'If the laws deduced on the basis of hypothetical ideas for visible processes prove to agree with experience, the *admissibility* of the fundamental idea is demonstrated.   If the laws for various classes of phenomena can be deduced from the same idea, we may even ascribe probability to it.   Such theories thus open up the possibility for us . . . of gaining an insight into a world not directly accessible to our senses.' Although it is later stated (p. 722) that 'the molecular structure both of matter and electricity must be taken as proved,' there is nowhere any expression of the opinion that this is all really much more important than anything that can be effected heuristically, and in the matter of thought economy. by the atomic theory.

Haas in his excellent *Introduction to Theoretical Physics* (trans. by T. Verschoyle, London, 1924), says that an experimental test of the theory, and possibly an experimental decision as to the correctness of its fundamental assumptions may be given.

Among the earliest to recognise, if somewhat conditionally, the real value to knowledge of hypotheses, were Volkmann (*Erkenntnistheoretische Grundzüge der Naturwissenschaften* and *Einführung in das Studium der Theoretischen Physik*), and Höfler and Poske (*ZS. f. ph. u. ch. Unt., Heft* 1, 1912). Among philosophers, Eduard von Hartmann first made a determined stand against the 'hypotheseophobia of modern physics' (*Die Weltanschauung der modernen Physik*, 2nd ed., Leipzig, 1902), and he was followed by Erich Becher, whose book, *Die philosophische Grundlagen der exakten Wissenschaften*, contains a devastating criticism of Mach's views.   Becher starts very cleverly from sound, in which we can most clearly recognise the law that a physical reality (vibratory motion) is at the bottom of the sensation (musical tone).   On the other side we have, amongst others, the founder of the philosophy of 'As If,' Vaihinger.   Though he distinguished sharply between 'Hypotheses' (= verifiable suppositions) and 'Fictions,' he classified the atom as belonging unquestionably to the latter.

The English reader should consult K. Pearson, *The Grammar of Science*, 3rd ed., London, 1911.   Pearson is the most eminent English disciple of Mach.

[18] See the previous note.

[19] 'A more exact understanding of this problem requires a knowledge of the Calculus, which Newton invented for this particular purpose.   See text-books of physics.

[20] See the authors referred to in Note 17.

[21] Such an attempt was made for example by Petzoldt in *Naturwissenschaften*, 1922, No. 10, p. 230.   Compare the reply to this by Thirring, in the same journal, p. 231.   Petzoldt maintained that Mach's opposition was not to the atomic theory as a fully justifiable physical hypothesis, but only to 'the metaphysical atom.'   But the actual wording of Mach's

statement shows that without a doubt, Mach fought the atom in physics for the reason that he regarded it as metaphysical. It is natural that this fact should be embarrassing for positivism and pragmatism. But unfortunately what was done cannot be undone.

[22] See text-books of physics. The gravitational constant is the force, measured in dynes, with which two masses of one gram each placed one centimetre apart, attract one another. According to the best recent determinations, its value is 0.0000000668. Recently the Italian physicist, Majorana, states that he has found an absorption of gravitation due to intervening matter. If this is true, all astronomical calculations would need considerable correction. If gravitation is a field action similar to electric and magnetic force, this assumption would be a probable one, but the absorption must in any case be very small, otherwise it would certainly have been already detected.

[23] A very sensible criticism of the whole formation of physical hypotheses and theories has been given quite recently by H. Feigel, *Theorie und Erfahrung in der Physik* (Karlsruhe, 1929). My only objection is that the author has not been able to free himself from Mach's 'Conscientialism' regarding the problem of reality, to the extent that he has freed himself from Mach's scepticism and pragmatism in the matter of the problem of truth. B. Russell, *Our Knowledge of the External World* (London, 1914); A. N. Whitehead, *An Inquiry Concerning the Principles of Natural Knowledge* (Camb., 1919); C. D. Broad, *Perception, Physics, and Reality* (London, 1914) and other books in the English bibliography.

[24] Concerning the true sense of Newton's words see Verweyen, *Phil. d. M.*, p. 9.

[25] I refer again to my description in Poske-Bavink, *Oberstufe der Physik*, pp. 16 *et seq.*

[26] See also article by Voss in the mathematical volume of *Kultur der Gegenwart*.
[See B. Russell, *Introduction to Mathematical Philosophy* (Cambridge, 1917); F. P. Ramsey, *The Foundations of Mathematics* (London, 1931).]

[27] See the histories of atomic theory given in Note 4, also Frost, *Bacon und die Naturphilosophie* (Munich, 1927).

[28] See Gehrcke, *Physik und Erkenntnistheorie* (Leipzig, 1921); Mach, *Erkenntnis und Irrtum*, p. 439; also *Space and Geometry*, trans. by T. J. McCormack (Chicago and London, 1906).

[29] Weyl, *Space, Time, and Matter*, trans. by H. Brose (London, 1922).

[30] Poincaré, *Science and Method*, trans. by Maitland (London, 1914); Dingler, *Die Grundlagen der Physik*; Schlick, *Raum und Zeit in der gegenwärtigen Physik*; and particularly Reichenbach, *Philosophie der Raum-Zeit-Lehre* (Berlin, 1929). See also references in the relativity bibliography below.

[31] See, for example, L. Lange, Note 32.

[32] The pre-relativist literature on the problem of motion is already almost unmanageable. A fundamental work is L. Lange's *Die geschichtliche Entwicklung des Bewegungbegriffs* (Leipzig, 1886). We may also mention Streintz, *Physikalische Grundlagen der Mechanik*; Mach's *Mechanik*.
Liebmann's *Analysis der Wirklichkeit*. Historically, Aster, *Raum und Zeit in der Geschichte der Philosophie* (Munich, 1925).

[33] Poske, *Vierteljahrschrift f. wiss. Phil.*, 1884; Mach, *Mechanik*, 4th ed., p. 279; Höfler, *Lehrbuch der Physik* (Braunschweig); Heymans, *Die Gesetze und Elemente des wiss. Denkens* (Berlin).

[34] Strictly speaking, this definition must be preceded by the experimental result, that the relationship between the accelerations produced in any two bodies by the influence of any kind of arrangements (charges of powder, stretched spring or the like) is independent of the nature of these arrangements, and hence only dependent upon a property of the two bodies themselves.

[35] See *Naturwissenschaften*, 1919, pp. 149, 327.

[36] I must here again refer the reader to the text-books of physics.

[37] Frost, see Note 27.

[38] E. Mach, *The Science of Mechanics*, 4th ed., trans. by T. J. McCormack (London, 1919).

[39] Volkmann, *Einführung*, p. 72.

[40] Bavink, 'Formalistisches und realistisches Definitionsverfahren in der Physik,' *ZS. f. ph. u. ch. Unterr.*, Vol. xxxi, No. 5, p. 161.

[41] Quite recently, the experimental investigation of the question whether gravitation is so strictly additive a property, that the effect upon one body upon others can be simply calculated by addition of the effects of its single parts, has again been taken up seriously. See Note 22.

[42] This view comes out quite naïvely also in Kant, for whom the *a priori* truth in pure science of the conservation of matter was true apodictically. (*Critique of Pure Reason*, N. K. Smith's translation, London, 1929, p. 212.) Particularly characteristic is the story, two pages further on, of the philosopher, where *weight* suddenly appears as quantity of substance. Cf. text, p. 215

[43] The literature is enormous. We may mention Sigwart and Wundt, *Logik*; Liebmann, *Analysis der Wirklickkeit*; Cohen, *Logik des reinen Erkennens*; Becher and Dingler, *Naturphilosophie*; König, *Die Entwicklung des Kausalproblems*; Wentscher, *Geschichte des Kausalproblems* (Leipzig, 1921); Verworn, *Ueber Kausale und Konditionale Weltanschauung*; Cornelius, *Einleitung in die Philosophie*; Kleinpeter, *Der Phänomenalismus*; K. Pearson, *Grammar of Science* (London, 1911); J. S. Mill, *System of Logic* (London, 1875); J. Ward, *Naturalism and Agnosticism* (London, 1907); A. Wolf, *Essentials of Scientific Method* (London, 1928).

[44] On this question see M. Schlick, 'Naturphilosophische Betrachtungen über das Kausalprinzip,' *Naturwissenschaften*, 1920, No. 24, p. 461.

[45] See Cornelius, *Einleitung in die Philosophie*.

[46] Curiously enough the French philosopher, Meyerson, has misunderstood my argument in the same way, and Poske also, with whom I once had a long discussion by letter.

[47] Thus quite recently again, Zilsel, *Naturwissenschaften*, 1927, p. 280.

[48] K. Meyer, *Die Entwicklung des Temperaturbegriffes* (Braunschweig, 1923).

[49] This is only apparently a purely nominalist definition. Compare Note 40.

[50] We cannot here go into the famous discussion between Leibniz and Descartes whether $\frac{1}{2}mv^2$ or mv (the momentum) is the measure of the 'force' which is conserved.

[51] The introduction of the name energy in the general sense is due chiefly to Kelvin and Rankine. Concerning the law of energy in general see Mach, *Geschichte und Wurzel des Satzes von der Erhaltung der Arbeit*; Planck, *Wissenschaft und Hypothese*, Vol. vi (Leipzig, 1913); Rey, *Die Theorie der Physik* (Leipzig, 1908); Poincaré, *Science and Hypothesis*, trans. by W. J. G. (London, 1905); Ostwald, *Naturphilosophie*.

[52] It has already been remarked that Ostwald's chief work on scientific philosophy is dedicated to Mach, and upholds a conventionalism similar to that of the latter in his 'analysis of sensation,' the classical work of this school.

[53] Even when such proofs of God's existence are not set up, it is at least attempted, e.g. by Sapper in his *Naturphilosophie*, to find in the physical domain a teleology allied to that in the biological.

[54] It is an open question whether the law of energy still holds in the interior of the atom. Compare pp. 196, 213 ff.

[55] The kinetic theory of gases gives us in a very simple manner, for example, the values for the specific heats of monatomic gases, and this leads directly to the law of Dulong and Petit for the atomic heat of solid elements. This, however, proves to be only true as the limiting value at high temperatures. The general law was first given by the quantum theory in a satisfactory manner. See the author's *Grundriss der Neueren Atomistik* (Leipzig, 1921); and on the latter point Valentiner, *Anwendungen der Quantenhypothese* (Braunschweig, 1920). A. Haas, *Quantum Chemistry*, trans. by L. W. Clodd (London, 1930).

[56] *Phys. Zeitschr.* 9, 465.

[57] The form of the second law is by no means definite. A large number of laws of more general and more special content, in some cases very different from one another, are included under it. A more detailed discussion would take us too far. See the essay of Schnippenkötter mentioned in Note 180.

[58] See text-books of physics, for example.

[59] The reader should not allow himself to be confused by the fact that water waves are transverse. These do not take place inside the water, but on its surface, which tends to assume a horizontal plane form on account of gravity. This gives the effect of a 'quasi-elastic' force, as if the surface consisted of a stretched rubber membrane. Even this view is not quite exact.

[60] It already appeared in the earlier optical theories of Fresnel and Neumann, that two magnitudes directed at right angles to one another (vectors) can be introduced with equal justification, one being interpreted at that time as an elastic displacement, the other as a twist. The relationship between the two has the same form as the Maxwell equations (p. 112). In the formalistic theory of light, the nature of these two magnitudes (the light vector) was left undetermined.

[61] Rupp, *Phys. Zeitschr.* 28, p. 290; Lewitsky, *ibid.*, p. 281.

[62] Regener, *Naturwissenschaften*, 1929, p. 183.

[63] It is characteristic of the obstinacy of positivist prejudice that this attempt is even repeated to-day, as for example by Frank of Prague in an essay which will be noticed at greater length later. (*Naturwissenschaften*, 1929, Nos. 50, 51.) Here the assertion is actually once more served up to us that the identity in nature of light and electro-magnetism really amounts

to no more than the formal agreement between the laws of propagation of electro-magnetic and optical waves. I can only refer the reader to the text in reply to this.

[64] The working-out of the atomic heat of the elements from the kinetic theory will be found in J. Walker's *Introduction to Physical Chemistry* (London, 1919), and numerous other text-books.

[65] The most important is given by the van't Hoff law. See, on the ionic theory of electrolytes, any text-book of Physical or Electro-Chemistry.

[66] $N\epsilon = 4.325$ Coulombs; taking N as 27.2×10^{18}, we get $\epsilon = 0.159 \times 10^{-18}$ Coul. See p. 130.

[67] Recent theories, of A. Werner, have led to some modification in our views concerning the ions, since we have now reasons for assuming that not simple charged atoms such as K^+ or H^+ are formed, but hydrated ions, that is, complexes of these atoms and water molecules. But this makes no difference in principle.

[68] In spite of the undoubted success of this corpuscular hypothesis of the cathode rays, there have always been physicists who have attempted to give an explanation of the observed phenomena on the basis of a continuity theory. The best known work of this kind is that of the Viennese physicist, Jaumann. Quite recently, as we shall see, the contrast between corpuscle and wave has in any case been obliterated to such an extent, that there would be no point in a discussion. We return to this matter in detail on p. 185 ff.

[69] See on this point J. J. Thomson, *Conduction of Electricity Through Gases* (Cambridge, 1928).

[70] On radio-activity and isotopes, see the book mentioned in Note 4 by Fajans, and also F. W. Aston, *Isotopes* (London, 1924).

[71] See Baer, 'Der Streit um das Elektron,' *Naturwissenschaften*, 1922, pp. 323, 344. Regener, *ib.*, 1926, No. 11; Millikan, *Physical Review* 26, 99.

[72] See Aston, *ib.* Note 70.

[73] Since K_{41} goes over into Ca_{41} with radiation of β-rays, and hence without change of weight, it should be possible to prove for calcium derived from minerals, containing very little of it together with much potassium, an atomic weight higher than the chemical atomic weight 40.01. A. and O. Frost actually believe that they have demonstrated this on the mineral microcline (having 11% K and 0.42% Ca), but this result is not as certain as Hönigschmidt's research on lead. See *Nature*, 125, 48, No. 3141.

[74] A general report on the position of the physics of the nucleus was given by C. D. Ellis in *Nature*, Vol. 129, p. 674 (1932), being a report of a Royal Society Symposium held on April 28th, 1932.

[75] According to the very plausible theory of Kossel (*Naturwissenschaften*, 1919, pp. 339, 360), the normal number of electrons in the outside shell of lithium is 8, corresponding perhaps to the number of corners of a cube, but perhaps again based on another reason (see p. 190). If now an alkali metal atom, say sodium for example, having only one electron on its outside shell, meets a halogen atom, say chlorine, which has seven there, the outside electron of sodium passes over to the outside shell of chlorine, making this into a noble gas shell (with 8 electrons, therefore saturated and incapable of further combination). This theory gives an excellent explanation of what are called 'heteropolar' compounds (those easily dissociated electrolytically) but fails in the case of 'homopolar,' such as water,

the carbon chains of organic chemistry, etc. The new wave mechanics (p. 185) has, here also, opened up many new possibilities.

[76] For works on Relativity, see Author's Bibliography, which contains many translated works; and the Supplementary Bibliography, containing original works in English.

[77] Further material regarding the experimental confirmation of the theory of relativity, in particular, regarding the 'Ether wind,' is to be found in the books cited in the last Note.

[78] If the motion of the two systems with respect to one another takes place in the X direction, and if the origins of co-ordinate coincide at the time $t = t' = 0$, the system of equations in question, the so-called Lorentz transformation, is given by $y' = y$, $z' = z$, $x' = \dfrac{x - vt}{\sqrt{1 - \beta^2}}$, $t' = \dfrac{t - \beta x/c}{\sqrt{1 - \beta^2}}$. In the last two equations we have the well-known paradoxes. But we may refer here particularly to a further point which is rarely mentioned in popular accounts of the relativity theory. When applied to the fundamental Maxwell-Lorentz equations (pp. 113 ff.), the Lorentz transformation gives the curious consequence, that the electro-dynamic (electromagnetic) and the electro-static phenomena likewise appear only as differences of standpoint. What appears as an electro-static field to an observer moving with the electric charges considered, and relatively at rest with them, appears to a stationary observer (or one moving relatively to the charges) as an electro-magnetic field. The coupling of the two field vectors (the electrical and magnetic force), which remained incomprehensible from the point of view of the absolute theory and the ether hypothesis, here appears, therefore, as conditioned by the purely kinetic transformation. Magnetism is not *produced* by moving electricity, it *is* electricity moving relatively to the observer, if we may formulate the result somewhat inexactly, but in a manner more easily comprehended. Conversely, the so-called 'induced electric voltage' is identical with the motion of the magnetic field, which, according to the old theory, caused it.

[79] The Minkowski world becomes euclidian, if we introduce instead of t, the imaginary co-ordinate cit. This makes all calculations much easier to grasp.

[80] In his famous thesis at Göttingen, 1854, republished recently by H. Weyl (Berlin, 1921).

[81] See Bavink, *Hauptfragen der Naturphil.* I, pp. 15, 53 (Berlin, 1928).

[82] Dingler, *Das Experiment* (Munich, 1928). See also Note 76.

[83] Speaking more generally, we can say that the state of the world at two different points of time must be given. In a special case, these might be infinitely close together, and this then amounts to our being given, not only the distribution of the quantities defining the state in space at this one moment, but also their differential co-efficients with respect to time (in Laplace's case, therefore, the components of the velocities). The question as to what takes the place of this requirement in Minkowski's four-dimensional world, has received little discussion. See Carnap's essay, mentioned on p. 151.

[84] Cf. F. Klein, *Vorlesungen über Differentialgeometrie* (Leipzig, 1907) : or any text-book of the Differential Calculus.

[85] Cf. M. Schlick, 'Naturphilosophische Betrachtungen über das Kausalprinzip' (*Naturwissenschaften*, 1920, p. 461); also Gerhards, see Note 153.

[86] The special theory of relativity does not make any essential alteration, but only makes the choice of the t axis arbitrary.

[87] Reichenbach, *Philosophie der Raum-Zeit-Lehre* (Berlin, 1929); also 'Die Kausalstruktur der Welt,' *Sitz.-Ber. der bayr. Akad.*, Munich, 1925.

[88] The calculation of the apparent increase in mass due to electric charge is to be found in O. W. Richardson's *The Electric Theory of Matter* (Cambridge, 1916), pp. 228 ff.

[89] In the calculation for more rapid motion, the difference between longitudinal and transverse mass must be taken more exactly into account. See Richardson, Note 88, p. 231.

[90] Lenard maintains, it is true, that on the basis of his theory of a primal ether, the observations also follow from an absolute theory. See P. Lenard, *Über Relativitätsprinzip, Aether, Gravitation* (Leipzig, 1920).

[91] The most important of these are the Doppler effect, the aberration of the fixed stars, and Fizeau's light-drag experiment. The apparent contradiction between the latter and the experiment of Michelson and Morley illustrates especially clearly the dilemma in which physics found itself before Einstein's time. See also Note 78.

[92] See Sommerfeld, *Atomic Structure and Spectral Lines*, trans. by H. L. Brose (London, 1923). The opponents of the relativity theory deny, it is true, that the determinations of Sommerfeld's fine-structure constant really gave the value required by the relativity theory. Gehrcke and Lau (*Ann. d. Ph.*, 1922, p. 388) claimed, on the contrary, to have obtained the very value demanded by the absolute theory; in the case of the hydrogen spectrum, the ratio between the two being $4 : 5$. But investigations by Lennan, Janicki, and others gave different values again, and Paschen finally showed that the hydrogen spectrum is for certain reasons unsuitable for an exact decision of this question. It must therefore be regarded for the present as unsettled.

[93] According to Seeliger, the phenomenon can be explained by a ring of cosmic dust which is also responsible for the Zodiacal light. Before Einstein, Gerber and Ritz had already attempted to deduce the motion of the perihelion in another way. But compare here, for example, von Laue (*Naturwissenschaften*, 1920, No. 37).

[94] The most recent results have again given deviations from the theoretical Einstein value.

[95] For further details see Gehrcke, *ib.* p. 75.

[96] *Zeitschr. f. Ph.*, Vol. i, p. 51. *Naturwissenschaften*, 1919, p. 629.

[97] The companion of Sirius has a much lower brightness than the main star, and must therefore, since it belongs otherwise to the same spectral type, have a much smaller surface. Since its mass can be calculated from the period of rotation, we apparently get for it an enormous density, which Eddington estimated at 50,000 times that of water. In order to make this comprehensible, Eddington took refuge in the hypothesis, that at this density the atomic nuclei are packed closely together without electron rings. The fairly large red displacement was thus not to be interpreted as a Doppler effect, but as an Einstein effect, very large on account of the enormous gravitational field. But Anding of Gotha has recently given a quite different and very plausible interpretation by supposing the companion to be itself a double star. The most recent observations of Ennis, of Johannesburg, and of van der Blos seemed to confirm this interpretation

with fair certainty, since they have actually seen the companion as a double star. There are naturally very serious objections to be brought against the assumption of a density of 50,000.

[98] A report on the present position was given by P. S. Epstein in *Naturwissenchaften*, 1929, p. 923. A conference recently took place between the investigators concerned, Miller, Lorentz, and Michelson, in Pasadena, at which all the pros and cons were debated. The result, regarding which Epstein reports as above, is not very encouraging to Miller's point of view.

[99] Courvoisier's papers appeared as follows: *Astr. Nachr.*, 1926, 226, p. 242; 1927, 230, p. 428; 1928, 234, p. 138. Shorter accounts in *Phys. ZS.*, 1927, p. 674, and *ZS. f. Geoph.*, 1928, 2, p. 49. This work is to be taken with great caution at present.

[100] For this reason success was obtained on the basis of the old mechanistic and formalistic theories by Ketteler and Helmholtz, and later, using the electron theory, by Drude.

[101] The elementary mathematics of the Zeemann effect will be found in A. Haas, *Introduction to Theoretical Physics*, and in most larger text-books of physics. See also G. Birtwistle, *The New Quantum Mechanics* (Cambridge, 1928).

The classical work on the subject is Voigt's *Magneto- und Elektrooptik* (Leipzig, 1908).

[102] The measurements of the wave lengths of X-rays have been brought to extraordinary accuracy, especially by the Swedish physicist Siegbahn, so that they can be given to-day almost as exactly as optical wave lengths (which are accurate to 7 figures). Quite recently, one of Siegbahn's pupils, Baecklin, has worked out a method of precision, which allows X-rays to be measured directly with ordinary gratings (*Naturwissenschaften*, 1929, No. 12, p. 201). The values obtained agree almost exactly with those calculated from crystal experiments. For the latter calculation we need to know the Loschmidt number, since we must use it to deduce the constant of the crystal grating. Conversely, the Baecklin experiment gives a new value for this number, and hence also for the elementary electric charge. These being $N = 26.94$ trillions, and $e = 4.793 \times 10^{-10}$ (instead of the Millikan value, see p. 131). This gives for the Planck quantum, $h = 6.591 \times 10^{-27}$.

[103] The law has the form $S\lambda = \dfrac{C_1 . \lambda^{-5}}{e^{c_2/\lambda T} - 1}$, where $C_1 = 8\pi ch$, $c_2 = ch/k$ and $3/2k$ is the amount, easily calculated from the Loschmidt number, by which the energy of a monatomic, ideal gas molecule is increased when the temperature rises by one degree. See Valentiner, *Die Grundlagen der Quantentheorie* (2nd ed.) and F. Reiche (Berlin, 1921), *The Quantum Theory*, trans. by H. S. Hatfield and H. L. Brose (London, 1930). A. Berthond, *The New Theories of Matter and the Atom*, trans. by E. and C. Paul (London, 1924), gives a fairly full, non-mathematical account of the whole subject. A complete account will be found excellently given in A. Haas, *Theoretical Physics*, trans. by T. Verschoyle (London, 1924).

[104] The chief point here is the thermal behaviour of substances in the neighbourhood of absolute zero. Specific heats become smaller and smaller as we approach it, the Law of Dulong and Petit, which is easily derived from the classical theory (see p. 84 and Note 55), thus ceases to hold, and we must apply Nernst's heat theorem, or the 'Third Law of

Thermodynamics': The specific heat of all bodies approaches the value zero at the absolute zero of temperature. From this follows the practical impossibility of reaching absolute zero (it can only be approached asymptotically). See *Vorträgen über die Kinetische Theorie der Materie*, by Planck, Debye, Nernst, etc. (Leipzig, 1914); A. Haas, *Quantum Chemistry*, trans. by L. W. Clodd (London, 1930). Any full-length modern text-book of Physical Chemistry deals with this matter.

[105] The elementary calculation of the relatively fine structure will be found in *The Atom and the Bohr Theory of its Structure*, by H. A. Kramers and H. Holst (London, 1923). The full theory will be found in A. Sommerfeld, *Atomic Structure and Spectral Lines*, trans. by H. L. Brose (London, 1923).

[106] Bohr's theory of the Periodic System will be found in the book by Kramers and Holst, Note 105.

Bohr's arrangement is given in the accompanying figure 87.

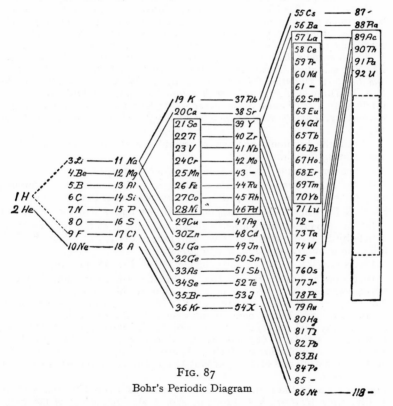

FIG. 87

Bohr's Periodic Diagram

Concerning the maximum possible number of chemical elements, see V. V. Narliker, *Nature*, Vol. 129, p. 402, 1932.

[107] An easily understood statement of the deduction of the law of

refraction for Einstein's emission theory will be found in an article by Flamm, 'Die neue Mechanik,' *Naturwissenschaften*, 1927, p. 569.

[108] New methods for determining h from Röntgen spectra are to be found in A. H. Compton, *X-Rays and Electrons* (New York, 1926).

[109] Dempster, *Phys. Rev.* 30, 644.

[110] For the Compton effect see Andrade's *Structure of the Atom*, pp. 688 ff.; A. H. Compton, *X-Rays and Electrons* (New York, 1926); (most elementary) C. G. Darwin, *The New Conceptions of Matter* (London, 1931).

[111] Bohr's condition in terms of wave mechanics is explained in A. Haas, *Wave Mechanics*, Ch. viii; A. Sommerfeld, *Wave Mechanics*, pp. 69 ff.

[112] Kikuchi's work on the diffraction of electrons is given in C. G. Darwin's book, see next Note.
See also G. P. Thomson's Royal Institution Lecture, 'Electron Optics,' reported in *Nature*, Vol. 129, p. 81 (1932).

[113] An account of these recent wave mechanics theories is given in C. G. Darwin's *The New Conceptions of Matter* (London, 1931). This is a book which assumes practically no previous knowledge of physics. The advanced student may be referred to the books on wave mechanics by Schrödinger, De Broglie, and G. P. Thomson, all published in 1931, and to J. J. Thomson's *Beyond the Electron* (Cambridge, 1931); G. Birtwistle, *The New Quantum Mechanics* (Cambridge, 1928); A. Haas, *Wave Mechanics and the New Quantum Theory*, trans. by L. W. Clodd (London, 1928); P. A. M. Dirac, *The Principles of Quantum Mechanics* (Oxford, 1930); F. A. Lindemann, *The Physical Significance of the Quantum Theory* (Oxford, 1932); Heisenberg, *The Physical Principles of the Quantum Theory* (Cambridge, 1930).

[114] For the harmonic analysis of sound see Harvey Fletcher, *Speech and Hearing* (London, 1929).

[115] See for example Jordan, *Naturwissenschaften*, 1927, p. 105.

[116] An explanation of Bohr's conditions by means of the 'spinning electron' is given in G. Birtwistle's *Quantum Mechanics* (see Note 113).
See C. G. Darwin, *The New Conceptions of Matter*, and *re* 'neutron,' Note 119.

[117] An account of Sommerfeld's electron theory of metals is given in an article by R. H. Fowler in *Nature*, Vol. 126, p. 611 (1930). See also J. C. M'Lennan, *Nature*, Vol. 129, p. 959 (1932).

[118] Gamov, *ZS. f. Ph.* 53, 601; Gurney, *Phys. Rev.* 33, 122, 127.

[119] Bothe and Kolhörster, 'Die Natur der Höhenstrahlung,' *Naturwissenschaften*, 1929, No. 17, p. 271. Against this view Das, *ib.*, 1929, No. 17, p. 543; Holmes, *Nature*, 123, 943 (who expresses the same doubts as the author). For a general discussion on cosmic radiation (Royal Society, May 14th, 1931) see *Nature*, Vol. 127, p. 859 (1931). See also *ib.*, H. Geiger's article, p. 785, and R. A. Millikan's, p. 167. See also J. H. Jeans, 'The Annihilation of Matter,' *Nature*, Vol. 128, p. 103 (1931), which deals with nature and origin of cosmic radiation, and also the expansion of the universe.

The existence of the neutron is now regarded as assured by the work of J. Chadwick (*Nature*, Vol. 129, pp. 129, 312); *Proceedings of the Royal Society*, A, Vol. 136, p. 693; N. Feather, 'Artificial Disintegration by Neutrons,' *Nature*, Vol. 130, p. 237, 1932; and *Proc. Royal Soc.*, June 1932. See also *Discovery*, Vol. xiii, p. 139 (1932).

[120] Estermann and Stern have recently obtained undoubted and clear interference pictures with positive rays (H and He) with LiF plates, in quantitative agreement with De Broglie's theory.

[121] *Naturwissenschaften*, 1928, No. 51, p. 1094 (Perles relationship).

[122] Eddington, *Proc. Roy. Soc. Lond.* 122, 358.

[123] Fürth, *Naturwissenschaften*, 1929, p. 688. See also Dirac's new attempt at explanation, *Proc. Roy. Soc. Lond.* 126, 360.

[124] Sommerfeld, *Planckheft der Naturwissenschaften*, 1929.

[125] Innumerable attempts in this direction have been made, e.g. Reichenbächer, *ZS. f. Ph.* 58, 402.

[126] Reichenbach has also arrived at quite a similar result (*Relativitätstheorie und Erkenntnis a priori*), although from another point of view; my standpoint is almost that of Bohr (*Naturwissenschaften*, 1930, No. 4); I recommend this profound address to the very special notice of the reader.

[127] *Critique of Pure Reason*: 'A philosopher was asked: How much does the smoke weigh? He answered: Deduct from the weight of the burnt wood, the weight of the ash and that gives you the weight of the smoke.' Here Kant naturally makes a chemical mistake; he did not know that the weight of the smoke also includes the weight of oxygen necessary for combustion. The correct form is: Weight of the smoke is equal to weight of the wood, plus weight of oxygen minus weight of ash. But this is of no importance for the present problem.

[128] Compare the references given in Notes 113 and 115, particularly the little book by Haas.

[129] This is also probably the essential sense of Weyl's words quoted in the former editions of this book: 'Physics is not at all concerned with the material content of reality; what forms the subject of its knowledge is simply a formal view or statement of it. . . . If it was the folly of the scholastic methods to wish to deduce real existence from purely formal premises, the philosophy which is called materialism is only a variety of scholasticism.' The objections which I raised elsewhere to this remark of Weyl's at the close of his book *Space, Time and Matter*, are not yet quite answered to my satisfaction, but I can the more readily lay them aside, since, as is shown further on in the text, the danger that mere formalism might suppress justifiable questions of existence has become less. Weyl's words are correct, as far as they refer to the last things of physical knowledge (see p. 203).

[130] Titius, *Natur und Gott*, Gottingen, 1926, p. 577. I take this opportunity of recommending this excellent book, which gives an extremely good description of the whole body of present-day science, and follows quite other roads than those previously taken by theological writers.

[131] This idea is also to be found in 'Über ein neues Grundgesetz, usw. (Leipzig, 1921, *Abh. d. sächs. Akad.* 38, 4.)

[132] The fact that simple validity is a form of being, as much as the existence of the world of bodies, which is usually regarded as the only form of being, has been very clearly shown by the more recent *Gegenstandstheorie*.

See, for example, Al. Müller's extremely clear statement in *Der Gegenstand der Mathematik mit besonderer Beziehung zur Relativitätstheorie* (Braunschweig, 1922).

RRS

133 Schottky, 'Das Kausalproblem der Quantentheorie,' *Naturwissenschaften*, 1921, Nos. 25, 26. Nernst, *ib.*, 1922, No. 21. Jordan, *ib.*, 1927, No. 5.

134 Planck's *A Survey of Physics*, translated by R. Jones and D. H. Williams (London, 1925), *Naturwissenschaften*, 1926, No. 13, pp. 249, 255. Quite recently, Planck has maintained humorously against Schrödinger, that his theory is deterministic (on the occasion of Schrödinger's admission to the Berlin Academy). See *Naturwissenschaften*, 1929, No. 37, p. 733. Schrödinger himself, on the other hand, has recently quite definitely expressed his adherence to the opinions of Heisenberg and Jordan. See his Inaugural Address, *Naturwissenschaften*, 1929, No. 1, 'Was ist ein Naturgesetz?'

135 Born, *Naturwissenschaften*, 1929, No. 7. This is a reply to Riezler's essay on 'Die Krise der Wirklichkeit' in 1928, Nos. 37, 38. The reader is also referred again to Bohr's essay mentioned in Note 126.

136 See Mahncke, *Leibniz und Goethe in der Harmonie ihrer Weltansichten* (Erfurt, 1925).

137 Sommerfeld, *Naturwissenschaften*, 1924, p. 1047. Franklin, *Science*, 60, 258.

138 Reichenbach, *ZS. f. Phys.* 53, Nos. 3 and 4 ('Stetige Wahrscheinlichkeitsfolgen'); also *ZS. f. Ang. Chemie*, 1929, No. 19, p. 457 ('Das Kausalproblem in der gegenwartigen Physik'). The first named paper contains a new, original, and very fertile formation of mathematical concepts, which is as well adapted to the new facts of 'continuous succession of probabilities,' as the ordinary calculus to càusal successions. Space forbids entering into it here, but its importance is emphasised.

139 A. Haas, *Wave Mechanics and the New Quantum Theory*, trans. by L. W. Clodd (London, 1928). See also C. G. Darwin, 'The Uncertainty Principle in Modern Physics,' *Nature*, Vol. 129, p. 746 (1932); A. Eddington, 'The Decline of Determinism,' *Nature*, Vol. 129, p. 333 (1932); Bohr, *Naturwissenschaften*, 1930, No. 4; De Broglie (see Note 111). The limit of accuracy is $h/2\pi$.

140 Haas, *ib.* (Note 139), p. 102.

141 See Note 113.

142 Cf. Planck, Note 134.

143 See Note 134.

144 See Note 135.

145 See Note 138.

146 This is the subject of a work by H. Bergmann, *Der Kampf um das Kausalgesetz in der jüngsten Physik* (Braunschweig, 1929). But I regard it as too closely adhering to Kantian apriorism.

147 This is actually done by Jordan in the article mentioned in Note 113, and by other positivists recently.

148 *Naturwissenschaften*, 1929, No. 3.

149 See especially Born, Note 135.

150 Dingler, *Grundlagen der Naturphilosophie* (Leipzig, 1913).

151 Chwolson's lecture has been published by Engelmann, and also, as regards its essential points, in the journal of the Keplerbund, *Unsere Welt*, Feb. 1911. Poincaré, *Dernières Pensées* (Paris, 1913).

[152] Haas, 'Die Axiomatik der Modernen Physik,' *Naturwissenschaften,* 1919, p. 744. Also *ib.,* 1920, p. 121.

[153] Cf. particularly M. Schlick, 'Naturphilosophische Beträchtungen über das Kausalprinzip,' *Naturwissenschaften,* 1920, p. 461. Also Gerhards, see Note 163.

[154] The Category of 'Order' appears to me – as opposed to Kant, but in agreement, for example, with Driesch – as the last and only remaining category of natural-scientific knowledge, but it has not only immanent (phenomenal), but also transcendental validity. See my *Hauptfragen der heutigen Naturphilosophie* (Berlin, 1928), Vol. i, p. 35; also this book, pp. 233 ff.

[155] Bohr, *Naturwissenschaften,* 1928, p. 245; *ib.,* 1929, p. 483.

[156] Likewise Reichenbach, *ib.,* Note 138.

[157] Planck, *Naturwissenschaften,* 1926, p. 249; Bohr, *ib.,* 1929, p. 245; *ib.,* 1930, p. 73. Riezler, *ib.,* Note 135.

[158] Cf. Bavink, *ib.,* Note 40.

[159] Carnap, *Physikalische Begriffsbildung* (Karlsruhe, 1926).

[160] This procedure is adopted, for example, by Pohl in his well-known and excellent *Einführung in die Elektrizitätslehre* (Berlin, 1929). I cannot follow him. If current strength = the reading of an instrument made at X's factory in Y, science ceases to be more than the mere tool of a handicraft.

[161] This is what I was told, as a student, in Felix Klein's lectures on differential geometry.

[162] Volkmann, *Erk. Grundzüge der Naturwissenschaft* (Leipzig, 1896). Galileo spoke of 'resolutive and compositive methods.' Cf. Frost, *Bacon und die Naturphilosophie,* see Note 27.

[163] E. Zilsel, *Das Anwendungsproblem* (Leipzig, 1916), pp. 103 and 169. An excellent statement of the same ·fundamental idea is found in W. Köhler's *Gestalt Psychology* (London, 1930), a book to which we shall refer later in another connection. See further Emil Müller, *Bedeutung und Wert mathematischer Erkenntnisse* (Vienna, 1917); H. Dingler, *Die Grundlagen der Physik* (Berlin, 1919); E. Gehrcke, *Physik in der Erkenntnistheorie* (Leipzig, 1921); Gerhards, 'Der mathematische Kern der Aussenweltshypothese,' *Naturwissenschaften,* 1922, Nos. 17, 18.

[164] Meinong, *Die Erfahrungsgrundlagen unseres Wissens* (Berlin).

[165] This view is maintained, for example, by Ph. Frank of Prague, in the essay – referred to in the text at great length further on (p. 217) – in *Naturwissenschaften,* 1929, p. 971, which is a reprint of a lecture given at a meeting of physicists in Prague in the previous year. This lecture was obviously in Sommerfeld's mind when attacking pragmatism.

[166] See *Wissenschaftliche Weltauffassung – Der Wiener Kreis – Veröffentlichungen des Vereins Ernst Mach* (Vienna, 1929).

[167] I refer once again to the profound statement of the 'forms of being' of the various 'kinds of objects' (*Gegenstandsarten*) in Alois Müller's *Einleitung in die Philosophie* (Bonn, 1925).

[168] Carnap, *Kantstudien,* 1923, No. 28.

[169] Typical examples of the formation of such 'divergent' theories are Stahl's phlogiston, Berzelius's electrochemistry, Black's material theory of heat, and others.

[170] Concerning critical realism and its foundations see Eduard von Hartmann, *Das Grundproblem der Erkenntnistheorie* and *Kritische Grundlegung des transzendentalen Realismus* (Leipzig, 1889 and 1875); Külpe, *Die Realisierung* (Leipzig, 1912 and 1920); *Einleitung in die Philosophie*, edited by A. Messer, 19th ed.; A. Messer, *Der kritische Realismus* (Karlsruhe, 1923); E. Becher, 'Naturphilosophie' in *Kultur der Gegenwart* (Leipzig, 1914); *Einführung in die Philosophie* (Munich, 1926); Bavink, *Hauptfragen der heutigen Naturphilosophie*, Vol. i (Berlin, 1928).

[171] See the two works referred to in Note 170.

[172] The same.

[173] Bavink, *Kantstudien*, 1927, Nos. 2 and 3.

[174] Driesch, *Ordnungslehre* (Jena).

[175] Winternitz, see Note 72.

[176] Bavink, see Note 154.

[177] See Note 166.

PART II

[178] E. Becher, *Geisteswissenschaften und Naturwissenschaften* (Munich, 1921).

[179] We may put forward, as a fourth law, that of the indestructibility of quantity of electricity, and as a fifth Nernst's heat theorem (unattainability of absolute zero, see p. 78). For the present, however, nothing can be done with these in cosmology. Besides, it is fairly certain that the first of these laws is included, with those of energy and mass, in a 'substance law.'

[180] A thorough discussion of the whole question is to be found in Schnippenkötter's *Der entropologische Gottesbeweis, die Geschichte einer apologetischen Irrung* (Bonn, 1920). This also gives the whole literature. 'The End of the World from the Standpoint of Mathematical Physics,' by A. Eddington, *Nature*, Vol. 127, p. 447, deals with entropy, and also with the Uncertainty principle. E. N. da C. Andrade, *The Mechanism of Nature* (London, 1930) deals with entropy.

[181] See L. v. Bertalanffy, 'Leben und Energetik,' *Unsere Welt*, 1929, 8.

[182] The work of Einstein and de Sitter is to be found in *Sitzungsberichte der preuss. Akad.*, 1917, 6, and Amsterdam, *Proc.*, 1917, Vol. 19, p. 527. The question of the shape of the universe and its limits is discussed in the following: *Nature*, Vol. 128, p. 699 (1931), art., 'The Evolution of the Universe,' discussion by J. Jeans, G. Lemaître, W. de Sitter, A. Eddington, R. A. Millikan, E. A. Milne, J. C. Smuts, E. W. Barnes, O. Lodge (all, of course, authorities in this field). See later, A. Eddington, 'The Expanding Universe,' *Nature*, Vol. 129, p. 421 (1932), a report of a Royal Institution lecture in which he gives his own theory of development, and of the relation between the cosmic constant and world curvature. Also E. A. Milne, 'The Expansion of the Universe,' *Nature*, Vol. 130, p. 9 (1932); J. H. Jeans, 'The Annihilation of Matter,' *Nature*, Vol. 128, p. 103 (1931). See further: M. Planck, *The Universe in the Light of Modern Physics* (London, 1931); A. Eddington, *The Nature of the Physical World* (Cambridge, 1928); and more popular, J. H. Jeans, *The Universe Around Us* (Cambridge, 1929), and *The Mysterious Universe* (Cambridge, 1930).

[183] Weyl, *Space, Time, and Matter*. See also Winternitz, Note 64.

[184] This point, in my opinion, gives the only possible starting point for a fresh discussion of the problem of the freedom of the will, both human and Divine. In most discussions, the usual concept of time is simply assumed. The question is asked whether Man or God could, at a given time, determine the course of world happening one way or another. This statement of the question is wrong from the start, as soon as we have realised that time possesses metaphysical reality neither from the subjective point of view (taken as epistemologically phenomenal) nor from the objective point of view; reality belongs only to that order of things which, projected in the 'time-like' direction, appears to us as time, in the 'space-like,' as space. Before the question can be discussed at all, it must first be translated, so to speak, into Minkowski's language. But then it obviously coincides with the problems of the contingency of the world, discussed on pp. 101, 214, 534, and also with the psychophysical problem (see pp. 404, 531). No one acquainted with the matter will maintain that the problem has hitherto found a satisfactory solution. Determinism fails to explain the feeling of freedom or voluntariness which accompanies all conscious actions of the will, while indeterminism is baffled by the unambiguity of physical causal connections, which are a matter of experience (see pp. 208, 535). The whole decidedly gives the impression of an intrinsically false statement of problem. But I do not feel capable of attempting a solution of the problem.

[184a] I only became acquainted after this book was in the press with the recent monumental work of Hermann Friedmann, *Die Welt der Formen* (Munich, 1930), which develops the 'system of morphological idealism' to a point of perfection which seems like a revelation in our time, until recently so hostile to metaphysics. I do not hesitate to rank this book along with the greatest achievements in the history of philosophy, such as Lotze's *Microcosmus* (London, 1886), Ed. v. Hartmann's *Philosophy of the Unconscious* (London, 1931), Schopenhauer's *World as Will and Idea* (London, 1883–6), etc. Friedmann's starting point is the contrast between the sense of touch and that of sight, and correspondingly, metrical and projective geometry (or geometry of position). He wishes, to put the matter shortly, to show that the reality concept of mechanistic materialism depends upon the one-sided notion that only that is 'real' which can be expressed in the language of the sense of touch ('haptically,' as Friedmann puts it). In truth, the domain of sight and hearing (the optical-acoustical) are much richer; it is in no way possible to translate everything existing in the latter domain into the language of touch, whereas every statement in the latter domain can be 'transformed' into one of the former. But 'optics' gives us only the qualities of form (*Gestalt*) as a primary experience, and the haptic-metrical must be derived from it by taking apart (analysis). From this point of view Friedmann then undertakes a truly magnificent synthesis of our whole knowledge, from mathematics and physics to art and religion, and one is again and again amazed at his fabulous grasp of the vast field of knowledge over which he ranges. I have only one objection to the book: the author does not do complete justice to the achievements of 'mechanism' in biology, and to some extent even in physics. Here – particularly in relation to Mendelism and allied matters – his feeling, no doubt originating in art, for the primacy of the form (*Gestalt*) over the elements (cf. Goethe's remark concerning *Teilen in der Hand*) carries him away to some extent. It is possible to hold on to this point of view even if one has a higher opinion of these investigations than Friedmann has. Generally,

one may describe the book as a revival, on the plane of modern knowledge, of Plato's work.

185 For recent views on the shape and structure of our universe of fixed stars, the reader is referred to Note 182, and also to J. H. Jeans, 'Beyond the Milky Way,' *Nature*, Vol. 128, p. 825 (1931); Russell, Dugan, and Stewart, *Astronomy* (London, 1926).

Shapley's theory, and his work on the problem of the δ Cepheids, will be found in the *Harvard Astronomical Bulletin*, 841, 861, 876, 881, etc. See also Eddington's article, 'Star,' in *Ency. Britt.*, 14th ed.

186 Hagen, *Naturwissenschaften*, 1921, p. 935.

187 Nernst, *Das Weltgebaude im Lichte der neueren Forschung* (Berlin, 1922); Eddington, *Stars and Atoms* (Oxford, 1927).

Recent articles on these matters, see Note 182; also H. Dingle, *Modern Astrophysics* (London, 1927); F. J. M. Stratton, *Astronomical Physics* (London, 1925).

188 A. Kühl, *Naturwissenschaften*, 1924, p. 1126. Recent discussion in English concerning the Martian canals will be found in *Popular Astronomy*. See also: Percival Lowell, *Mars and its Canals* (New York, 1911); Trumpler, 'Observations of Mars,' *Lick Obs. Bull.*, Vol. 13, p. 19.

189 A. R. Wallace, *Man's Place in the Universe*, 4th and enlarged ed. (London, 1904).

190 *Das Problem der Entwicklung unseres Planetensystems* (2nd ed., Berlin, 1921). See references in Notes 182, 186 for theories of the development of our planetary system; also J. H. Jeans's *Astronomy and Cosmogony* (Cambridge, 1928).

191 Riem, *Unsere Welt*, 1924, No. 1.

192 See references in Note 187.

193 For the Russell diagram, see Dingle, or Stratton, Note 187; Russell, Dugan, and Stewart, *Astronomy* (London, 1927); H. N. Russell's article, 'Stellar Evolution,' *Ency. Britt.*, 14th ed.

194 E. A. Milne, 'Stellar Structure and the Origin of Stellar Energy,' *Nature*, Vol. 127, p. 16; J. H. Jeans, 'Beyond the Milky Way,' *Nature*, Vol. 128, p. 825 (1931).

195 Nernst, see Note 183.

196 *Scientia*, Dec. 1928.

197 Holmes, *Phil. Mag.*, 1928, 1, 1055, considers that the most trustworthy method is the calculation from the lead content of uranium minerals. For pegmatite from the middle of the Precambrian he found about a milliard years. See H. Jeffreys, *The Earth* (Cambridge, 1929); A. Holmes, *The Age of the Earth* (Benn's Sixpenny series).

198 Concerning Wegener's continental theory see his *Origin of Continents and Oceans* (London, 1924). See also *Naturwissenschaften*, 1924, p. 34; 1925, pp. 77, 84, 669; 1929, p. 743.

English works on geology and palaeontology: J. Joly, *The Surface History of the Earth* (Oxford, 1925); A. Morley Davies, *Introduction to Palaeontology* (London, 1923); A. H. Swinnerton, *Outline of Palaeontology* (London, 1923); also H. G. Wells, *Outline of History* (London, 1930), and Wells and Huxley, *The Science of Life* (London, 1929).

199 See Note 14. See also the references given in the previous two Notes.

Wegener's theory makes certain assumptions concerning the nature of the earth's interior.

See H. Jeffreys, *The Earth, its Origin, History, and Physical Constitution* (London, 1924); A Holmes, *The Age of the Earth* (Benn's Sixpenny Library).

[200] M'Lennan and Shrum, in 1925, produced the green line by means of a discharge through a mixture of oxygen and helium or neon. See M'Lennan, *Proceedings of the Royal Institution*, 1926.

[201] Recent investigation concerning the state of the upper atmosphere will be found in articles by W. J. Humphreys, 'The Atmosphere: Origin and Composition,' *Scientific Monthly*, 1927; *Scientific American*, 1928; also in N. Shaw's *Manual of Meteorology* (Cambridge, 1926–30).

[202] A short account of the modern 'Polar Front' theory will be found in Napier Shaw's *Manual of Meteorology*, Vol. ii, pp. 422 ff. See also the same author's *The Air and its Ways* (Cambridge, 1923).

[203] An account of astrology from the historical and critical point of view in T. O. Wedel's *The Mediaeval Attitude towards Astrology* (Yale, 1920).

[204] The fact that so modern a scientist as Dacqué fails to apply this criticism, and even breaks a lance (*Natur und Seele*, Munich) for the reality of astrological superstition, is one of the worst of the many follies which characterise our times. Unfortunately, Verweyen has quite recently joined him.

PART III

[205] Investigations concerning the chemical complexity of the various constituents of cytoplasm have been made by, among others, the famous physiologist Abderhalden. If we assume with Emil Fischer that the proteins are built up from amino-acids, of which we know to-day more than 20, we already have an enormous number of possible different substances built up from these 20 constituents (when all 20 are employed in one molecule, $20^{20} : 2$, when only 19, 18, and so on are employed, the numbers are still enormous). See on this point Wells and Huxley, *The Science of Life*. Fox and Ramage (*Proc. Roy. Soc. B.*, Vol. 108, p. 157) have shown that iron and copper are both essential constituents of protoplasm, in addition to C, H, O, N, P, S, Na, K, Mg, Ca, and Cl, which are already known to be so.

Concerning vitamins, see *Vitamins: A Survey of Present Knowledge*, Medical Research Council Report, No. 167; Wells and Huxley, *The Science of Life*, pp. 636–9.

Space does not allow us to enter into the extremely interesting story of the discovery of vitamins, the famous explanation of the production of antirhachitic Vitamin D from ergosterol by Pohl and Windaus, the dangers of the present boom in vitamins, etc. It is not improbable that the 'hormones' mentioned later are directly connected chemically with the vitamins.

[206] Concerning the true mechanism of the process of assimilation, see later on in the text, p. 311, and also the literature mentioned in Note 214.

[207] Concerning the assimilation of nitrogen, see *Naturwissenschaften*, 1925, p. 21; 1927, p. 606; 1928, p. 457.

[208] Plenty of examples in any text-book of organic chemistry.

209 Concerning enzymes, see *Naturwissenschaften*, 1925, p. 937; 1927, p. 585.

210 See *Colloid and Capillary Chemistry*, by H. Freundlich, trans. by H. S. Hatfield (London, 1926); *Colloid Science applied to Biology*, a Faraday Society Symposium (London, 1931); J. J. Lloyd, 'Colloid Structure and its Biological Significance,' *Biological Reviews*, Vol. vii, p. 254 (1932).

211 For the synthesis of haemin, see *Naturwissenschaften*, 1929, p. 611. See also *Physiological Reviews*, Vol. x, p. 506, article by Anson and Mirsky.

212 Reinke, *Grundzüge der Biologie*, 1909.

213 The work of C. Neuberg and his school has become of fundamental importance for the problem of fermentation. See his report in *Natur-wissenschaften*, 1921, p. 334; 1923, pp. 657, 732; 1925, p. 980. See also M. Schoen, *The Problem of Fermentation* (London, 1928); A. Harden, *Alcoholic Fermentation* (London, 1923).

214 The problem of assimilation has been solved to a considerable extent, but not completely, by Willstätter and his pupils. See his account in *Naturwissenschaften*, 1925, p. 985; also Klein, *ib.*, 1925, p. 21.

215 Byk, 'Die Synthese der Molekularen Asymmetrie,' *Naturwissen-schaften*, 1925, p. 17.

216 Against this view see Spek, *Naturwissenschaften*, 1925, p. 893, who goes back to Bütschli's idea of a simple 'froth structure' for the cytoplasm. An excellent account of our present knowledge concerning the extremely complicated structure of the interior of the cell will be found in Francé, *Der Organismus* (Munich, 1928). The mere recital of the various structural elements, known to us to-day, of 'simple' cells, takes about a page. See also T. R. Parsons, *The Fundamentals of Biochemistry* (London, 1927); W. Seifriz, 'The Structure of Protoplasm,' *Biological Reviews*, Vol. iv, p. 76 (1929).

217 See Weidenreich, *Naturwissenschaften*, 1923, p. 485, and Francé's book mentioned in the last note.

218 The chief means for determining the individual elements of cell structure consists in dyeing them with certain organic colouring matters. These methods were brought to great perfection particularly by Ehrlich. In these investigations Ehrlich also showed that not only dyes, but also certain poisons, and medicaments (probably also hormones) are particularly strongly held (absorbed) by certain cells. It is on this fact that the 'chemo-therapy' of sleeping sickness (atoxyl, etc.) and of syphilis (salvarsan) depends.

219 See Miehe, *Biol. Zbl.*, 1923, 1. But later investigations of the d'Hérelle phenomenon, discussed further on, allow other possibilities to be envisaged.

See Wells and Huxley, *Science of Life* (London, 1929), p. 185; H. H. Dale, 'Biological Nature of the Viruses,' *Nature*, Vol. 128, p. 537 (1931); 'Problems of Filterable Viruses,' *ib.*, Vol. 129, p. 48 (1932); 'Vitamins and Viruses,' *Discovery*, xiii, p. 126 (1932).

220 Molisch, *Populäre biologische Vorträge* (Jena, 1920).

221 An account of the d'Hérelle phenomenon will be found in *A System of Bacteriology in Relation to Medicine*, published by the Medical Research Council, Vol. vii (London, 1930), where d'Hérelle's own ideas are given. Bechold's more cautious views will be found in the *Umschau* (edited by

him), 1930. I have here followed him in essentials, but the important practical meaning of the phenomenon does not quite receive full justice.

[222] Concerning micelle structure of cellulose fibres see Seifriz, Note 216. For an original experiment by Koltzoff see *Biol. Zbl.*, 1928, 6.

[223] The pseudocrystalline character of organic materials was recently proved by F. Rinne in experiments on spermatozoa. The Debye-Scherrer method gave clearly marked diffraction pictures (*Ber. der natf. Ges. Freiburg*, xxx, 1, 2).

[224] See W. Bayliss, *Principles of General Physiology* (London, 1921).

[225] Since this problem forms the subject of the whole of this part of the book, I content myself here with referring to a few recent authors who take a similar view: Dürken, *Die Hauptprobleme der Biologie* (Kempten, 1925); Ehrenberg, *Naturwissenschaften*, 1929, p. 777; and particularly L. v. Bertalannfy, *Kritische Theorie der Formbildung* (Jena, 1929), where further literature will be found.

[226] In the case of the d'Hérelle phenomenon they are of course dependent upon the presence of bacteria.

[227] Much material will be found in W. Bölsche's well-known book *Love-Life in Nature*, trans. by J. G. A. Skerl (London, 1924), which, however, is more imaginative than scientific both in its quite one-sidedly sectionistic attempts at explanation, and also in several other directions. Other works dealing with the question of sex in organic life are Wells and Huxley, *The Science of Life*; F. A. E. Crew, *Genetics of Sexuality in Animals* (Cambridge, 1927); T. H. Morgan, *Heredity and Sex* (2nd ed., Oxford, 1914).

[228] For this reason the ascaris was used in the title of Goldschmidt's excellent popular account of modern experimental biology, *Ascaris, eine Einführung in das Wissen vom Leben für jedermann* (Leipzig, 1922).

[229] Far and away the most material is to be found in W. Roux's journal, *Archiv für Entwicklungsmechanik*. We may also refer for a general account to Driesch's *Philosophie des Organischen* (Leipzig, 4th ed., 1928), and also to the books mentioned in Notes 228 and 232.

[230] Hence we now distinguish between regulation eggs and mosaic eggs; see for example Bertalannfy, Note 225, where a bibliography will also be found.

[231] Leeuwenhoeck (the inventor of the microscope) himself names a student at Leyden as discoverer of the spermatozoa. But he undoubtedly deserves the credit for having followed up the discovery and made it generally known.

[232] An excellent account of some particularly characteristic results in the field of regeneration will be found in Koelsch's *Verwandlungen des Lebens* (Zurich, 1919), also in Thesing's *Experimentelle Biologie II*, from Teubner's Collection *Natur und Geisteswelt*. A fuller account in Driesch, *Philosophie des Organischen*; Dürken, *Experimental-Zoologie* (Berlin, 1919); Dürken, *Experimental Analysis of Development*, trans. by H. G. Newth (London, 1932).

[233] H. Petersen, *Archiv für Entwicklungsmechanik*, Vol. 47, 1 and 2.

[234] Klebs' most important results will be found described in Dürken's book, Note 232; Klebs' original paper is in *Sitzber. d. Heidel. Akad.*, 1913, No. 5.

235 The literature on transplantation and implantation is of enormous extent. I refer the reader to the works mentioned in the previous notes, and also to the summaries by Ubisch and Braus, *Naturwissenschaften*, 1922, Nos. 12 and 20. Spemann, *Archiv für Entwicklungsmechanik*, 48, 1. See also the following Note. The reader is also referred to Dürken, Note 232.

236 The redeterminability of the transplanted part does not mean (as has recently been shown by the Spemann school: see Bautzmann, *Naturwissenschaften*, 1929, p. 818, for references) that the cells in question were completely indetermined. When transplanted into neutral territory (e.g. eye socket, abdominal tissue) they differentiate in a quite definite way according to their own inner laws. See Spemann, *Naturwissenschaften*, 1924, pp. 65, 1092; 1927, p. 946; 1929, p. 287. See also de Beer, *Biological Review*, ii (1927).

237 Concerning chimaeras, and the closely related 'xenias,' see text-books on genetics, Note 242. Also Taube, *Naturw. Wochenschr.*, 1922, No. 34; Goetzsch, *Naturwissenschaften*, 1922, Nos. 15, 36, 39, 41; 1923, No. 18. Issajew, *Biol. Zbl.*, 1923, 2. The latter seems apparently to have succeeded in producing a kind of cross by 'complantation.'

238 Concerning the culture of animal tissues apart from the organism see E. N. Willmer, 'Tissue Culture from the Standpoint of General Physiology,' *Biological Review*, iii, p. 271 (1928); T. S. P. Strangeways, *Technique of Tissue Culture in Vitro* (Cambridge, 1924); A. Fischer, *Tissue Culture* (London, 1925).

239 Concerning the cancer problem see *Reports of the Medical Research Council*; Cope, *Cancer: Civilisation Degeneration*; H. J. Paterson, *Lectures* (1925); *British Empire Cancer Campaign*; *Ninth Annual Report*, 1932; *Report of the International Conference on Cancer* (Bristol, 1928); *Cancer: International Contributions in Honour of James Ewing* (Philadelphia, 1932); *Ninth Scientific Report of the Imperial Cancer Research Fund* (London, 1930).

240 According to P. Weiss (*Naturwissenschaften*, 1928, p. 626; *Biol. Zbl.*, 1929, No. 6) the choice of nerve paths does not take place in the brain or spinal cord; a nerve impulse, representing a certain effort of will, is sent into all nerve paths, but only received by those nerve endings which are adapted to this specific form of excitation (analogy with the reception of wireless waves). This hypothesis would somewhat diminish the difficulty of the body-soul problem.

241 Concerning hormones, also of plants, see Haberlandt, *Biol. Zbl.*, 1922, 4; Koppanyi, *Naturw. Wochenschr.*, 1922, 42; Söding, *Unsere Welt*, 1928, p. 70. The activity of the heart is also regulated, according to Loewe (*Naturwissenschaften*, 1922, p. 52), by a special hormone produced in the heart itself. Concerning the chemistry of the hormones see Barger, *Naturwissenschaften*, 1928, p. 940.

See also C. Lovatt Evans, *Recent Advances in Physiology* (London, 1930).

Concerning the male sexual hormone, see Loewe and Voss, *Klin. Wochschr.*, 1927, 11; and the female sexual hormone, *Naturwissenschaften*, 1927, p. 417; 1928, 946, 952, 1088.

The latter, 'progynon,' has recently been prepared pure (by Butenandt, see *Forschungen und Fortschritte*, 1930, No. 1), also the thyroid hormone, thyroxin. See L. T. Hogben, *The Comparative Physiology of Internal*

Secretion (1927); S. Vincent, *Internal Secretion and the Ductless Glands* (London, 1924).

[242] The regulation of this whole system of interlocked processes is probably effected by the 'sympathetic' nerve system. (Cf. Hansen, *Naturwissenschaften*, 1928, p. 931.)

[243] Concerning the investigation of personality types see Baur-Fischer-Lenz, *Human Heredity* (London, 1931), trans. by E. and C. Paul.

[244] The fundamental paper by Gurvich appeared in *Arch. f. Entw. Mech.*, 1923, Vol. 100. See further *Biol. Zbl.*, 1929, 3, 49; *ib.*, 1929, 7 (general critical survey), and also the important paper by Stempell, *ib.*, 1929, 10. According to Reiter and Gabor (*Phys. Ber.*, 1929, 2), the active rays have a wave length of about 340 $\mu\mu$, but according to Frank (*Biol. Zbl.*, 1929, 3, 49) 200–240 $\mu\mu$. More exact determinations by Chariton, Frank, and Kannegiesser (Leningrad) gave the most powerful effects at 206 and 220 $\mu\mu$ (*Naturwissenschaften*, 1930, 19). It may be mentioned here as a curious fact, that the Soviet Government continually interferes with and spies upon a man like Gurvich, because he has given an account of (not advocated) Driesch's vitalism in his lectures. At the present time only atheistic materialism may be taught in Russia, hence vitalism is forbidden. Freedom of thought with a negative sign ! (From authentic reports which I have received.)

[245] The presence of photographically active rays was recently demonstrated by beautiful experiments with butterflies' wings.

[246] How far the independent capacity for life of single parts even of higher organisms may go in certain circumstances was recently demonstrated by the Russian scientist Bryuchenenko, by a rather horrible experiment performed on the freshly severed head of a dog.

[247] On the problem of death see Doflein, *Das Problem des Todes*; Korschelt, *Lebensdauer, Altern, und Tod* (Jena); Wells and Huxley, *The Science of Life*, p. 88; Raymond Pearl, *The Biology of Death* (Philadelphia and London, 1922).

[248] Steinach's paper is in *Archiv f. Entw. Mechanik*, Vol. 46, No. 4, 'Über Verjüngung durch Experimentelle Neubelebung der alternden Pubertätsdrüse.' Steinach has been very much attacked recently, particularly from the medical side. On the other hand, his results have been largely confirmed by Harms in Marburg. See also S. Voronoff, *Rejuvenation by Grafting* (London, 1925), and *The Study of Old Age and my Method of Rejuvenation* (London, 1928) by the same author.

[249] General works on heredity in English will be found in the Supplementary Bibliography at the end of this book.

[250] Compare Wells and Huxley, *The Science of Life*.

[251] Steinach regards the interstitial cells as the seat of the production of hormones (see p. 352); they are called by him the glands of puberty; but this is regarded with scepticism by other investigators, who look upon the genital cells themselves as the producers of hormones.

[252] It is only quite recently that we have reached a clear understanding of the complicated rhythm of this maturation of the ova in the human female, and its hormonic connection with changes in the mucous membrane of the uterus. See T. Pryde, *Recent Advances in Biochemistry* (London, 1930).

[253] Concerning the work of Correns see *Naturw. Wochschr.*, 1922, No. 1.

On sex determination see R. Goldschmidt, *The Mechanism and Physiology of S. D.* (London, 1923).

254 See the books referred to in Note 249, particularly Morgan.

255 Concerning the bearings of modern heredity upon medicine see R. R. Gates, *Heredity in Man* (London, 1929).

256 The criticism of Mendelism has never been quite silenced by its success. Haecker's text-book in particular adopts a very cautious and indecisive attitude. See also Herz, *Naturwissenschaften*, 1923, p. 833; Fick, *ib.*, 1925, p. 524, and Belar's reply, *ib.*, 1925, p. 717. Stieve also, *ib.*, 1927, p. 951, has expressed himself fairly sharply against the universality of Mendelism in genetics; we shall return to his criticism later. But all this criticism, including Wettstein's proof of cytoplasmic inheritance, has failed to shake the foundations of Mendelism, and has only strengthened them.

257 An article by Penners, *Naturwissenschaften*, 1922, Nos. 34 and 35, gives a good account of the development of the matter up till then, Wettstein's results are summarised in *Unterr. Blätt. f. Math. u. Naturw.*, 1928, p. 271; 1929, p. 11. Similar results were also obtained by Kühn (*Naturwissenschaften*, 1927, p. 735). Both proved that, when certain lower plants were crossed, the offspring were 'matroclinic' in certain respects; that is to say, they showed characteristics of the female parent, and departed from those of the male parent. As opposed to this, Nawaschin showed (*Naturwissenschaften*, 1927, p. 919) that this phenomenon does not occur with flowering plants. See Caroline Pellew, 'The Genetics of Unlike Reciprocal Hybrids,' *Biological Reviews*, Vol. 4, p. 209 (1929); Shearer, de Morgan, and Fuchs, *Phil. Trans.*, 1912.

258 The mode of inheritance in the case of infusoria seems to be quite different, the nucleus apparently taking no part. See *Der Naturforscher*, Vol. 11, Nos. 28 and 29.

259 See *Naturwissenschaften,* 1922, No. 33; *Umschau*, No. 42. Objections to Meisenheimer have been raised, amongst others, by Lenz, *Arch. f. Rassen- u. Gesellschaftsbiologie*, 1926, Vol. 18, No. 2.

See L. Hogben, *Genetic Principles in Medicine and Social Science*, London, 1931.

260 For these methods see Baur-Fischer-Lenz, *Human Heredity*, trans. by E. and C. Paul (London, 1931).

261 See Note 249.

262 See especially Haecker, *Naturwissenschaften*, 1927, p. 710; J. S. Huxley, *Problems of Relative Growth* (London, 1932).

263 Scientists with a leaning towards occultism, such as Oesterreich, Driesch, and others, have recently connected this phenomenon with the supposed phenomenon of 'materialisation.'

264 Bohr has rendered especial service by a clear statement also of this question (*Naturwissenschaften*, 1930, No. 7).

266 Goldschmidt's hypothesis (text, p. 383) is not very different from this.

267 See in this connection, besides Driesch, Fr. Kottje, 'Vitale Energie' (*Ann. d. Phil.*, 1927, Nos. 2 and 3).

See *The Philosophical Basis of Biology* (London, 1931), by J. S. Haldane;

O. Lodge, *Modern Scientific Ideas* (London, 1927); *The Idealisation of Mechanism*, 1930.

[268] See the authors named in Note 225, and Ungerer, Note 266; also Gradmann, *Naturwissenschaften*, 1930, Nos. 28 and 29.

[269] See Friederichs, *Naturwissenschaften*, 1927, pp. 153, 182; Thienemann, *ib.*, 1925, p. 589, and Gradmann (previous note). Concerning Heidenhain's 'Synthesiologie,' see *ib.*, 1926, p. 149. Also J. W. Bews, 'The Ecological View-point,' *Nature*, Vol. 128, p. 145 (1931); C. Elton, *Animal Ecology* (London, 1927).

[270] See Lloyd Morgan, *Emergent Evolution*.

[271] W. Köhler, *Gestalt Psychology* (London, 1930).

For Friedmann, see Note 184*a*.

[272] See Driesch, *The Science and Philosophy of the Organism*, 2nd ed. (London, 1929), and his pamphlet, *Das Ganze und die Summe* (Leipzig, 1922). See also the Supplementary Bibliography, under 'Vitalism.'

[273] Regarding Kant's attitude towards the problem of organic purpose, see Ungerer's *Die Teleologie Kants und ihre Bedeutung für die Logik der Biologie* (Berlin, 1921), which is otherwise well worth reading.

[274] Regarding animal psychology, see K. C. Schneider, *Tierpsychologisches Prcktikum* (Leipzig, 1912); and also B. Schmid, *Von der Aufgaben der Tierpsychologie* (Berlin, 1921); Ziegler, *Tierpsychologie* (Berlin, 1921); Kafka, 'Tierpsychologie,' from *Handb. d. vergl. Psych.*, i, 1 (Munich, 1922); Hempelmann, *Tierpsychologie vom Standpunkt des Biologen* (Leipzig, 1926); W. v. Buddenbrock, *Grundr. d. vergl. Psychol.* (Berlin, 1924); W. Köhler, *The Mentality of Apes* (London, 1927); S. Zuckerman, *The Social Lives of Apes and Monkeys* (London, 1931); Alverdes, *Social Life in the Animal World* (London, 1927); *Animal Psychology*, trans. by H. Stafford Hatfield (London, 1932).

[275] Concerning the problem of the colour sense of animals, and v. Frisch's investigations, see *Naturwissenschaften*, 1921, Nos. 29 and 37; 1923, p. 470; 1924, pp. 981, 988; 1927, pp. 321, 963; 1929, p. 221. Concerning the powers of communication of ants, see Eidmann, *ib.*, 1925, p. 126. See also H. M. Fox, *Blue Blood in Animals*.

[276] Concerning the curious nuptial dances of Bees, see *Umschau*, 1921, Nos. 36 and 37.

[277] See Note 270.

[278] Driesch, *Science and Philosophy of the Organism* (London, 1929). Bertalanffy expresses a similar adverse view (*ib.*, Note 225). Driesch, who of course himself introduced the hypothesis that the 'entelechies' are ultimately identical with his 'psychoids,' naturally only intends to set up a methodological first principle. But it may be asked whether the whole statement of the question is not spoiled from the start if we begin by 'acting as if' there were no such thing as the spiritual or psychical. A recent lecture by H. Gradmann of Erlangen (*Naturwissenschaften*, 1930, Nos. 28 and 29) is particularly characteristic of the attitude of most biologists. Gradmann points out quite correctly that between the living individuals, alone commonly described as organisms and their subdivisions (organs or cells) on the one hand, and the higher wholes (animal states), symbioses (organisations of animals and plants) on the other hand, every possible transition exists, and that hence there is no sense in making a particular factor (the entelechy) responsible for forming the whole out of the individuals, if we

could completely explain these more loosely connected aggregations by the fact that their parts act upon one another, or in opposition to one another, according to known laws. In order that a wood may be formed we only need a piece of land protected from cattle, and left to itself for several decades; as a result of the seed of plants, and various species of animals, reaching it by chance, a community, which we call a wood, will be formed automatically, without its being necessary to lay the responsibility upon a special 'Spirit of the Wood.' This is a perfectly sound argument against Driesch's vitalism, which replaces a causal explanation by the entelechy. But Gradmann never seems to realise that his conclusion can equally well be reversed; if we find at the end of a series of such wholes formed by chance the human individual – to which Gradmann will not refuse the possession of a spiritual unity – it is necessary for the same reason to go backwards from this point, in accordance with the very principle of continuity he so skilfully asserts, to the wholes which are below man, and extend the principle to the loosest unit. Surely this other path, downwards from mankind, belongs just as legitimately to scientific Biology as the first series from the cell or its parts upwards through *Pandorina, Volvox*, etc. Is not the only possible method for this other path the spiritual insight (*Einfühlung*) which Gradmann expressly denies? Gradmann's view, that is, the whole of biological mechanism with a materialistic tendency, would only be the complete truth, if at the end of its path we found, in place of man, only a material system of the highest complication. But instead, we have a living being which without a doubt is possessed of spiritual and mental qualities. This is a proof that the procedure quite certainly represents only half the truth. But so much it certainly is.

279 Bohr, see Note 264.

280 See Driesch, *ib.*, Note 272; and books on animal psychology, Note 274.

280a Concerning the mechanism of 'geotropism' see Loeb, *Forced Movements, Tropisms, and Animal Conduct* (New York, 1918).

281 See the report by André on Lux's apparatus, *Unsere Welt*, 1922, No. 8.

282 Mach (*Analysis of Sensation*) separates the elements in a somewhat different manner according to the various disciplines which he deals with. In my view the distinction here made is more accurate – but it really is of no importance.

283 *Analysis of Sensation.*

284 Külpe, *Die Realisierung*, Vols. 1 and 2; Hönigswald, *Zur Kritik der Machschen Philosophie* (Berlin, 1908). E. Becher (see bibliography) points out how much simpler the 'hypothesis of a real external world' is as compared with the mere element-relations of the positivists (*ib.*, p. 91).

285 Husserl, *Logische Untersuchungen*, i, p. 84.

286 Hönigswald, *ib.*

287 A fuller account will be found in E. Becher, *Gehirn und Seele* and *Einführung in die Philosophie*; Busse, *Leib und Seele*; Külpe, *Einleitung in die Philosophie*; Driesch, *Leib und Seele*. See Bibliography.

288 Driesch, in the books cited in Notes 265 and 287, gives a very thorough criticism of the theory of parallelism. But he also rejects the theory of interaction in its usual form. His main argument is the greater degree of complication of the psychical as compared with the physical. The law of arrangement of degrees of complexity prevents, in his opinion, our

being able ever to make a complete picture of the psychical from the corporeal. On the other hand, in his opinion the psychoid (the entelechy) fulfils all requirements in this respect. Naturally, for Driesch ascribes to this *deus ex machina* the very degree of complexity which he requires for this purpose. His psychoidal image of the spiritual is, in my opinion, a simple duplicate. The solution of the riddle is not thereby given, but simply passed on to the relationship between entelechy (psychoid), and body. As regards this, Driesch is satisfied with the bare statement that we must assume states of the brain, resulting from stimuli of the sense organs, to be unambiguously connected with those spiritual states which are called sensations or true perceptions. But since this sensational element never appears pure, but always as part of a whole and characterised by some ordered symbol, it is not necessary to assume that a definite brain-state is unambiguously assignable to this whole; we must suppose that the whole is given by brain-state and soul-state (obviously psychoid-state) taken together. Yes, but how? This is exactly the question which we have to solve.

289 Unfortunately, too little attention is paid in philosophical literature to the fact that the epistemological problem of reality is nothing but the problem of body and soul seen from the other side, the inside. The epistemologist will always tend, even when he is a critical realist like E. von Hartmann or Becher, to start from the immanent sensations, etc., and then proceed to found his hypothesis of the real external world. This, for instance, is Becher's method. The author does not regard this as correct, but would take the 'primordial experience' (*Urerlebnis*, Verveyen) as neutral, and the subject – object relationship as directly given and not further analysable. See however, regarding the question of the correct starting point, an article by Seiffert, *Unsere Welt*, 1926, Nos. 1 to 3.

290 Eddington, *Space, Time, and Gravitation*; see Bibliography, p. 654.

291 The 'hypothesis of a super-individual psyche' has found a clever advocate in E. Becher's book *Die fremddienliche Zweckmässigkeit der Pflanzengallen und die Hypothese* (Leipzig, 1917). Against his view see e.g. Frankenberg, *Biol. Zbl.*, 1929, 12, 48. I am no longer prepared to uphold my complete rejection of this hypothesis as in the last two editions, although I still share the doubts aroused by the possible introduction of 'lazy teleology.' The hypothesis may be correct, even if its introduction for the purpose of explaining certain mysterious phenomena connected with galls was a mistake. The foundation for it given in the text has obviously nothing to do with such 'lazy' teleology.

292 Hartmann's chief work on the theory of knowledge is the 'theory of categories.' An excellent account of the essentials of the matter is given in *Das Grundproblem der Erkenntnistheorie*, and *Die kritische Grundlegung des transcendentalen Realismus* (Leipzig). See also his summary, *System der Philosophie in Grundriss*, or that of his disciple Drews, *E. v. H.'s phil. System in Grundriss* (Heidelberg, 1902).

293 S. Arrhenius, *Worlds in the Making*, trans. by H. Borns (London and New York, 1908).

294 See Molisch (Note 220), p. 101. Recently, Miehe, *Biol. Zbl.* 1, 1923.

295 This fundamental idea of Fechner's is in my opinion inevitable, unless metaphysics is radically rejected. Becher also supported it. See in this connection his Introduction to Philosophy.

[295] A complete comprehension of this historical fact seems to me to be somewhat wanting in the otherwise excellent book of Friedmann (Note 184a), which quite definitely proclaims the return to Goethe's purely idealistic morphology, as opposed to the excessive mechanistic and temporal descent of forms. See Note 184a.

[296] On the history of the doctrine of descent see Radl, *History of Biological Theories*, trans. by E. J. Hatfield (Oxford, 1930); C. Singer, *A Short History of Biology* (Oxford, 1932).

[297] See Singer's book in the previous Note.

[298] How far we are to count Goethe as one of the originators of the doctrine of descent is somewhat doubtful, since we cannot be sure whether he regarded the development of 'Individual Species' from the primary type, as purely logical, or actually historical.

[299] The only strong opponent of the doctrine of descent among German zoologists is Fleischmann of Erlangen. As a super-empiricist, he rejects every kind of theorising which goes beyond the simple observation of facts and their connections. In this respect he carries out his own principles, for he has confined himself to morphological and anatomical investigations. He does not realise that his rejection of theorising would lead whole branches of science, for instance geology, to lose their right to exist, and that finally nothing would be left, to use Wasmann's neat expression, but 'well ordered magazines of facts,' for the sake of which no one would be willing to engage in scientific work. Fleischmann has published, in collaboration with the theologian Grützmacher, a series of lectures under the title, *Der Entwicklungsgedanke in der gegenwärtigen Natur- und Geisteswissenschaft* (Leipzig, 1923), in which he undertakes to refute the doctrine of descent. As a matter of fact, he continually confuses in the most hopeless fashion descent as such with its special forms, such as the theory of selection, and accuses present-day theorists of things which have been forgotten for the last thirty years by the whole science. A single glance at, for example, R. Hertwig's excellent account of descent in *Kultur der Gegenwart* disposes of Fleischmann's whole fight against an opponent that no longer exists. Of other biologists, only a few such as Petersen of Heidelberg, have occasionally expressed themselves very critically regarding descent (*Naturwissenschaften*, 1921, No. 49) because they have likewise adhered to extreme empiricism. But no one else but Fleischmann has rejected the whole doctrine.

[300] It has taken long enough for this view to be actually adopted by the specialists. The historical interconnection of the problems was all too effective. The chief credit for the new and precise statement of the problem belongs to Tschudlok's clear analysis in his *Deszendenzlehre* (Note 296).

[301] See Note 297. As regards what follows, see also Plate, *Die Abstammungslehre* (Jena, 1925).

[302] Rauthner, *Naturw. Wochschr.*, 1921, No. 10. (See also T. N. George, 'Palingenesis and Palaeontology,' *Biological Reviews*, Vol. viii (1933).)

[303] The fact that the vestiges in question may sometimes serve other purposes (reproduction) does not in any way tell against their rudimentary nature.

[304] See Wasmann, *Naturwissenschaften*, 1926, p. 504. Concerning blood groups, Breitner, *ib.*, 1928, p. 849.
See also H. Munro Fox, *Blue Blood in Animals*.

[305] An interesting and new attempt has been made by Tischler to reconstruct the genealogy of plants on the basis of the number of chromosomes (*Biol. Zentralbl.*, 1928, p. 6).

[306] See in this connection, for example, Hertwig, *Zur Abwehr des ethischen, sozialen, und politischen Darwinismus* (Jena, 1918), where further literature will be found. Concerning the dangers of a too hasty application of biological concepts and laws to human culture see Drexler, *Naturw. Wochschr.*, 1920, No. 42. An insight into these dangers does not abolish the necessity for fully appreciating the biological foundations of human culture. See below in the text, and Note 400. See also Wells and Huxley, *The Science of Life.*

[307] Regarding this question, see the books on genetics referred to in Note 249, and also Hennig, *Naturwissenschaften*, 1927, p. 260.

[308] This table is from Voss' *Moderne Pflanzenzuchtung und Darwinismus* (Detmold, 1928).

[309] Further examples will be found in the books on genetics, and in Touton, *Naturwissenschaften*, 1928, p. 685, and *ib.*, 1927, Nos. 15 and 25.

[310] The literature regarding the inheritance of acquired characteristics is limitless. I give as a tiny selection: R. Hertwig's article 'Abstammungslehre' in *Kultur der Gegenwart*; O. Hertwig's *Werden der Organismen*; Kammerer's reports on his experiments in *Umschau*, 1909, No. 50; 1911, p. 7; 1913, p. 45; Demoll, *Arch. f. Entw.-Mech.*, 1921, Vol. 47. And above all, Baur-Fischer-Lenz, *Human Heredity.*

There is, of course, an equally large literature in English. See the Supplementary Bibliography to this book.

One of the few English upholders of the hereditary transmission of acquired characters is E. W. MacBride. See his Royal Institution lecture reported in *Nature*, Vol. 127, p. 933 (1931); J. B. S. Haldane's R.I. lecture opposing his view, *ib.*, Vol. 129, pp. 817, 856; and a subsequent discussion between them in *Nature*, Vol. 129, p. 900, and Vol. 130, p. 20.

[311] Concerning Kammerer, see previous Note, and Note 314.

[312] This view has recently been chiefly represented by Gemünd (*Leben und Anpassung*, Bonn, 1926). According to him, Driesch's so-called 'prospective potencies' are in reality retrospective. The necessity for taking an 'historical' standpoint in biology has recently been strongly urged by Bertalanffy. See his *Kritische Theorie der Formenbildung* where other literature will be found; see also Hennig, Note 299.

[313] Pavlov's famous experiments on mice are to be interpreted in this sense. Pavlov trained mice to come for food at the sound of a bell, and states that he proved that their offspring learned this response considerably more quickly. According to Lenz, these experiments only proved, that Pavlov or his collaborators gradually learned better and better how to train mice.

[314] Regarding Kammerer's case, see *Nature*, Vol. 118, pp. 490, 636 (1926).

[315] See further Lenz, *Arch. f. Rassenhyg. u. Gesellschaftsbiol.*, Vol. 21, No. 3; also Bavink, *Unsere Welt*, 1929, No. 5.

[316] The same is true for the oft-quoted experiments of Standfuss and Fischer on the fruit-fly Drosophila, as Goldschmidt (*Biol. Zbl.*, 1929, p. 7) has shown. Concerning Tower's experiments see Baur-Fischer-Lenz, *Human Heredity*; V. Haecker, *Umwelt und Erbgut.*

Sss

317 According to the experiments of A. Blum, see Fetscher's work quoted in Note 394, and also the essay by Rudin, *Naturwissenschaften,* 1930, No. 13.

318 See for example Dürken, *Biol. Zbl.* 12, 48. The experiments of Lesage, *Naturwissenschaften,* 128, No. 11, are probably to be interpreted in the same way. Further examples in Lenz, *ib.*

319 The chief upholder of this argument in recent times is Demoll, see Note 310. See also Gessner, *Die Krisis in Darwinismus* (Leipzig, 1910).

320 Stieve has recently drawn frequent attention to this sensitiveness (*Naturwissenschaften,* 1927, 951), but unfortunately makes it the occasion of a quite unjustified attack on genetical theorists, and a defence of Lamarckist ideas. No modern geneticist maintains that the germ cells are absolutely unchangeable by, and insensitive to external influences. See text.

321 A summary of the little that is known with certainty is given by Kosswig, *Naturwissenschaften,* 1930, p. 561; see also *ib.*, 1930, p. 434.

322 *Ib.*, 1919, No. 45.

323 Bohr, *ib.*, 1930, No. 4.

324 See the books referred to in Note 116.

325 See the books referred to in Note 253 on the determination of sex.

326 Figures in Weismann, *ib.*

327 The literature on Darwinism is limitless. The great majority of present-day workers are very critical towards it. Its chief supporters at present are Plate (see Bibliography) and lately the mathematician Study (see *Naturwissenschaften,* 1920, 1921). The zoologist Hesse also regards the theory of selection favourably on the whole. It is rejected entirely by the de Vries school, and part of the Mendelians, and the Neo-Lamarckists (O. Hertwig, Driesch, Pauly, Reinke, Dennert, etc.). The best summary of the whole problem of descent is given, in my opinion, by R. Hertwig's article 'Abstammungslehre' in *Kultur der Gegenwart.* See Bibliographies.

[Mention should also be made of S. Butler's *Unconscious Memory*, originally published in 1880 (3rd ed., 1920), in which both Darwin's theory and his mode of presenting it are subject to penetrating criticism entirely ignored at the time, and still fully valid. *Tr.*]

328 See Weismann, *On Heredity* (Oxford, 1892).

329 Just, *Begriff und Bedeutung des Zufalls im organischen Geschehen,* Berlin, 1925.

330 A new experiment of this kind, together with a full statement of its basis, has recently been published by Kranichfeld in No. 31 of *Vorträge und Aufsätze zur Entwicklungsmechanik der Organismen* (edited by W. Roux, Berlin). The reader is referred to my review in *Unsere Welt,* 1922, No. 11, p. 261.

331 In his book *Urwelt, Sage, Menscheit* (Munich, 1928).

332 *Naturwissenschaften,* 1912, No. 8.

333 Kleinschmidt's new *Formenkreislehre* (Halle, 1926) can also not be acquitted of this mistake.

334 Heikertinger is prominent among the supporters of this idea, see Note 318.

335 Prell, *Naturwissenschaften,* 1924, p. 149; Beljajeff, *Biol. Zbl.*, 1927, 2.

336 See for example Goldschmidt, *Vererbungswissenschaft* (2nd ed., pp. 78 ff.).

[337] Regarding the question as to the final relationship between the parent form and the mutations after unrestricted interbreeding, a discussion took place some years ago between Peter, Kranichfeld, and Nachtsheim in the *Naturw. Wochschr.*, 1922, Nos. 39, 52, to which the reader is referred.

[338] Regarding mimicry, see Pieper's *Mimikry, Selektion, und Darwinismus* (Leyden, 1907); Study, *Naturwissenschaften*, 1919, Nos. 21 to 23; Dahl and Heikertinger, *Naturw. Wochschr.*, 1920, many issues, and also *Biol. Zbl.*, 1922, Nos. 10 and 11. Also Heikertinger, *Die Frage der Schutzanpassungen im Tierreich* (Karlsruhe, 1927), and numerous papers by the same author, e.g. *Biol. Zbl.*, 1925, p. 705; 1926, pp. 351, 503; 1927, p. 462; *ZS. f. Morph. u. Oekologie*, 1925, p. 598.
'Mimicry,' art. by G. D. H. Carpenter, *Sci. Pro.* xxvi, 609 (1932).

[339] This is the view urged by Wasmann against Fleischmann in the discussion on the doctrine of Descent fought out in the *Münchener Neuesten Nachrichten* some years ago, see Notes 299 and 342.

[340] W. Schultz has published important new experiments on direct colour change caused by cold (*Arch. f. Entw.-Mech.*, Vol. 47, Nos. 1 and 2).

[341] See in this connection H. Grunwald, *Unsere Welt*, 1926, No. 10; also Just, Note 329.

[342] How many 'unexploded mines' still lie about in this region may be seen most clearly from the discussion which took place on Descent in the *Münchener Neuesten Nachrichten* between Fleischmann, Wasmann, Hertwig, Dacqué, Kaup, Lenz, etc. See my summary, *Unsere Welt*, December 1925.
Regarding 'Fundamentalism,' the reader may be referred to *Fundamentalism versus Modernism* (1925), by J. W. Johnson; *American Inquisitors* (1928), by W. Lippmann; *New Learning and the Old Faith* (1928), by A. W. Robinson.

[343] Thus Dennert in his pamphlet *Weltbild und Weltanschauung* (Godesberg, 1909) which may be taken as an adequate expression of the general mood of the time in Christian Church circles. For a discussion of it see the author's statement in *Unsere Welt*, 1924, Nos. 2 and 3.

PART IV

[344] A general account of our present knowledge of primitive man will be found in H. Klaatsch's *The Evolution and Progress of Mankind* (London, 1923), and in the works given in the Supplementary Bibliography, notably the works by Keith and Read, and in more popular style, Wells' *Outline of History*.

[345] See H. Weinert, *Naturwissenschaften*, 1925, 188.

[346] H. Weinert, *Kosmos*, 1925, p. 386; W. Dietrich, *Unsere Welt*, 1917, 363.

[347] Weidenreich, *Naturwissenschaften*, 1930, 118; *Umschau*, 1930, No. 2; *Unsere Welt*, 1930, No. 4.

[348] See the report on this in *Naturwissenschaften*, 1922, No. 4. Opposite view, H. Reck, *Naturw. Wochschr.*, 1922, No. 5. See also *Naturwissenschaften*, 1929, 233.

349 While this book is in the press I learn that Freudenberg has found new prehistoric human remains near the site of the old discovery (teeth, parts of skull and face, shoulder blade) which he ascribes to a large ape-like creature called *Hemianthropus Osborni.* It is possible that they form, together with the lower jaw of Mauer, part of the same skeleton.

350 That of Weinert (*ZS. f. Ethnol.*, 1926; *Umschau*, 1926, 31) probably takes first place at the present time.

351 Bayer, *Umschau*, 1929, 17.

352 Kern, *Stammbaum und Artbild der Deutschen* (Munich, 1927).

353 Very rich material for the comparison of the technology and art of prehistoric man with that of primitive races and children will be found in the new large work by Dennert, *Das geistigen Erwachen des Urmenschen* (Weimar, 1929), with whose conclusion I cannot however agree in all respects. See also H. Eng, *Children's Drawings*, trans. by H. S. Hatfield, (London, 1931): full account and Bibliography.

354 [See F. G. Crookshank, *The Mongol in Our Midst,* a popular exposition (with full bibliography) of the case for a polyphyletic origin of man from three stems corresponding to the chimpanzee, orang, and gorilla in the apes, and the White (taken in a very wide sense), the Mongol, and the Negro in man. Most populations, particularly the European, are a mixture of all three stocks; the evidence is derived from a study of physiognomy, personality, and bodily attitude in normal people, and of the characteristics of congenital imbeciles, which are believed to be atavistic in nature ('Mongolism,' dementia praecox). The opponents of this view allege that both imbeciles and normal race differences are the results of differences in endocrine (hormone) balance. *Tr.*]

355 Breitner, *Naturwissenschaften*, 1928, 849.

356 P. Alsberg has laid particular stress on this point in a lecture given in 1922 at the Naturforscher-Versammlung and in his book *Das Menschheitsrätsel.*

357 Thus Dacqué regards as the scene of the main part of human development the lost continent of Godwana (*Urwelt, Sage, Menschheit,* see Note 331).

At night all cats are grey.

358 See, in addition to the work mentioned in the last note, the one given in Note 204, and also *Leben als Symbol*, all three published in Munich.

359 *Die morphologische Abstammung der Menschen, Kritische Studie über die neueren Hypothesen* (Stuttgart, 1908).

360 W. Köhler, *Abh. der preuss. Akad. der Wiss.*, 1917–18; *The Mentality of Apes* (London, 1925). See also Bierens de Haan, *Naturwissenschaften*, 1927, 481.

361 Ch. Bühler, *Die geistige Entwicklung des Kindes* (1919). See also Dennert and Eng, Note 353.

362 Watson's *Behaviorism* twists this fact into the monstrous assertion that thinking is nothing but this inward conversation with one's self.

363 A rich store of material will be found in Hesse-Doflein, *Tierbau und Tierleben* (Leipzig). See also Alverdes, *Social Life in the Animal World;* S. Zuckerman, *The Social Life of Apes and Monkeys.*

364 See F. B. Jevons, *An Introduction to the History of Religion,* 9th ed. (London, 1927), and other books in the Supplementary Bibliography.

[365] See Note 344, and C. K. Aldrich, *The Primitive Mind* (London, 1931); A. Machin, *The Ascent of Man* (London, 1925).

[366] O. Kleinschmidt, *Formenkreislehre*.

[367] The average skull capacity is as follows: –

Anthropoids	about	500–600 cb. cm.
Pithecanthropus	,,	800–900 ,, ,,
Neandertaler	,,	1200–1250 ,, ,,
Piltdown (Keith)	,,	1400– ? ,, ,,
Present-day European	,,	1500– ? ,, ,,

but the range of variation is very great, particularly in present-day man, so that positive conclusions can only be drawn with difficulty.

[368] See F. Tilney, *The Brain from Ape to Man* (London, 1920); C. J. Herrick, *An Introduction to Neurology* (Philadelphia and London, 1927).

[369] Regarding the reasons for assuming the existence of unconscious mental processes, see modern treatises on general psychology, e.g. C. K. Ogden, *The A B C of Psychology* (London, 1929).

[370] Lehmann, *Aberglaube und Zauberei* (Stuttgart, 1925); Baerwald, *Der Okkultismus in Urkunden* (Berlin, 1925) (almost entirely sceptical); Dessoir, *Vom Jenseits der Seele* (Stuttgart, 1917); Hennig, *Wunder und Wissenschaft* (Hamburg, 1904); T. K. Oesterreich, *Occultism and Modern Science* (London, 1923); Tischner, *Telepathy and Clairvoyance*, trans. by W. D. Hutchinson (London, 1925); G. Pagenstecher, *Recent Events Seership, a Study in Psychometry* (New York, 1923).

See also F. Podmore, *Modern Spiritualism, a History and a Criticism* (London, 1902), and *The Newer Spiritualism* (London, 1910); Richet, *Thirty Years of Psychical Research*, trans. by S. de Brath (London, 1923); Osty, *Supernormal Faculties in Man*, same trans. (London, 1923); Geley, *Clairvoyance and Materialisation*, same trans. (London, 1927); Morton Prince, *Dissociation of Personality* (New York, 1906); MacDougall, *Body and Mind*, 2nd ed. (London, 1913); W. Brown, *Mind and Personality* (London, 1926); W. F. Prince, *The Enchanted Boundary* (Boston, 1930); T. Besterman, *Some Modern Mediums* (London, 1930); C. E. B. Roberts, *The Truth about Spiritualism* (London, 1932).

[371] Prell, *Naturwissenschaften*, 1925, 652.

[372] See Lehmann, Note 370; Hauer, Note 364.

[373] See Lehmann and Dessoir, Note 370.

[374] Baerwald, *ib.*, Note 370.

[375] The extension given by Hennig to this hypothesis (Note 370) cannot in my opinion be justified.

[376] Concerning physical phenomena see Geley and Richet, Note 370.

[377] The term 'cross-correspondence' is used by psychical research workers to describe communications from 'spirits' received at various times, which only make sense when joined together.

[378] An exception to the rule of the triviality of supposed spirit communications is found in the remarkable poetical products of the medium Mrs. Curran who was examined by the very critical leader of American scientific occultism Dr. Prince (*The Case of Patience Worth*, Society of Psychical Research, Boston, 1917). The spirit calling itself Patience Worth produced a poem in pure English of the 17th and 18th centuries which could not possibly be credited to the medium. The case is unique of its kind.

[379] By way of criticism of Steiner's *Anthroposophy*, see Hauer, *Werden und Wesen der Anthroposophie* (Stuttgart, 1922). See also, by a disciple, G. Kaufmann, *Anthroposophy, an Introduction to the Works of Dr. Rudolph Steiner* (London, 1922).

[380] See text, p. 296, and Notes 197, 198.

[381] Certain experiments of an Italian, Cazzamelli, with wireless receivers have been regarded as a direct proof of the existence of such waves; they are reported in the *ZS. f. Parapsch.*, 1926, Nos. 1–3.

[382] Remarkable cases will be found in Mathieson's book (Note 370), particularly the case of Genthe, and also in Baerwald, *Okk. u. Spir.*, p. 226.

[383] In his *Philosophy of the Unconscious*, and also in his work *Animismus und Spiritismus*.

[384] See Lehmann, *Aberglaube und Zauberei*, and Note 356.

[385] See Note 382.

[386] At this point we are reminded of the oft-quoted presentiment of animals at the approach of accidents and catastrophies such as earthquakes. But the reports on this point are very contradictory. Volcanic eruptions and the like are reported without reference to their natural preliminary warnings.

[387] Bavink, *Kantstudien*, 1927, Nos. 2 and 3; Wenzl, *Das Naturwissenschaftliche Bild der Gegenwart* (Leipzig, 1929).

[388] See Al. Müller, *Einleitung in die Philosophie* (Bonn, 1925).

[389] In this view I am in agreement with the chief representatives of both contending parties.

[390] With reference to the following pages, see in particular the second part of Baur-Fischer-Lenz and also Günther, *Rassenkunde des deutschen Volkes*; Clauss, *Rasse und Seele* and *Von Seele und Antlitz der Rassen und Völker* (all Munich); the books of Boas and Luschan, quoted later. And above all, the excellent, profound, and very full work by L. Schemann, *Die Rasse in den Geisteswissenschaften*, so far two vols. (Munich). The second treats of the chief epochs and peoples of history as related to race, the first the part played by race as such in history.

[391] It might be objected here that the statistical method to be applied in the cases in question is also mathematics, although only 'the mathematics of approximation'; and that therefore Kant's law would still hold. But that would result in his original opinion being actually turned into its opposite, for mathematics was Kant's ideal because it gives us 'apodictically certain' results, and not merely the probability of correctness.

[392] We find bad examples of such tendentious work in Boas' *Kultur und Rasse*, and Luschan's *Völker, Sprachen, Rassen*, though the latter work contains a great deal of detailed valuable material. On the other hand, Günther's well-known book is also by no means free from exaggeration in the opposite direction, and in popular nationalistic literature these exaggerations often become intolerable, even when they are in the service of a good cause. Here we often have hopeless confusion between the notions of race, people, and language.

[393] The real level of the culture of our distant ancestors has been made clear in recent times, particularly by Kossina. The gold ornaments found prove that art at least was on a high level. Opinions are very much divided concerning astronomical achievements; Teudt's assertions –

Germanische Heiligtümer (Jena) – are mostly rejected by the specialists, but have recently been confirmed from another side (*Umschau*, 1930, No. 3). One thing at least is correct in them, namely that it was a grave mistake to proclaim Charlemagne as the bringer of culture to Germany. He destroyed much more culture than he built up. We owe it to him, and above all to his son Louis the Good, that the spiritual bond uniting us to our forefathers has been almost completely severed. He succeeded, where Rome has failed, in bringing Germany under Roman servitude. Concerning the question of the early Germanic culture, see Worth's monumental work *Der Anfang der Menschheit* (Jena).

[394] Concerning race hygiene, see Baur-Fischer-Lenz, *Human Heredity*, the standard work. See Supplementary Bibliography for further works in English on eugenics, and also Shepherd Dawson, 'Intelligence and Fertility,' *Nature*, 1932, Vol. 129, p. 191.

[395] See Prokein, Lenz and Fürst, *Arch. f. Rassen- und Ges.-Biol.*, Vol. 17, No. 4.

[396] These figures are taken from Fetscher, *Erbbiologie und Eugenik* (Berlin, 1927).

[397] Natural sterility only seems to appear in a few extremely degenerate families, but the elements thus eliminated only form a minute fraction of the bad stock which is continually increasing in relative proportion.

[398] See Supplementary Bibliography, 'Eugenics.'

[399] See Bavink, Note 394.

[400] The founders of eugenics in Germany, Ploetz, Schallmeyer, Tille, and others, at first aroused much opposition by their somewhat acrimonious remarks on this point. Thus the notorious East End of London was once described by this school as the source of England's health, because the unsocial types die out there in the course of a few generations. O. Hertwig protests against this view in his book quoted in Note 306, which, however, goes too far in the other direction, since it takes entirely the point of view of Lamarckism.

[401] O. Spengler, *The Decline of the West*, trans. by C. F. Atkinson; H. Piper, *Die Gesetze der Weltgeschichte* (Leipzig, 1929). Frobenius, in *Paideuma* (Munich, 1921), takes a similar view; his researches on the lost civilisations of Central Africa are well known. See also the Supplementary Bibliography.

[402] Piper has recently undertaken to show, in a further work, that the history of China and Japan agrees with his theory. I must leave criticism to specialists.

[403] *Unsere Welt*, 1928, 12; 1929, 10; 1930, 1.

[404] See Schemann, *ib.*, Vol. ii.

[405] Many recent vitalists, who incline to occultism, appear however to attribute even this power to the entelechy or 'psychoid,' inasmuch as they use it to explain 'materialisation' phenomena.

[406] *Philosophie der Technik*, by Dessauer (Bonn, 1927), and the book of the same title by Zschimmer (Jena, 1920). [No subject has received more discussion in England during the last two centuries, but I am unable to find any book which could be described as a philosophy of the subject. All our literature is strongly political, ethical, and propagandist, with the exception of some economists; it turns around

the question whether machines, and the 'Capitalist' system under which they have been developed, are good or bad for man, mentally, morally, and physically. While the question has played a large part in the controversy between individualism and socialism, the opponents of machine as opposed to hand labour are to be found in both camps. A few recent works representing different schools of thought will be found in the Supplementary Bibliography under 'Civilisation,' but the reader may be reminded of the names of Ruskin, W. Morris, S. Butler the Younger, Adam Smith, J. S. Mill, H. Spencer, Ed. Carpenter. *Tr.*]

407 Francé, *Die Pflanze als Erfinderin* (1920); *Das Gesetz des Lebens* (Leipzig, 1920). See S. Butler, *Life and Habit*, and J. M. Montmasson, *Invention and the Unconscious*, with a preface on the relation between human invention and biology by H. S. Hatfield (London, 1931).

408 Ph. Frank (see Note 165) would naturally object strongly to this view, for in his opinion the belief in such pre-existent solutions is the root of all evil in the 'philosophy of the schools.'

409 The excellent treatment of animals in the Swiss mountain villages strikes all travellers.

410 See a full discussion of the whole question in *Unsere Welt*, 1929, Nos. 3 and 9.

411 See, for example, Kropotkin, *Gegenseitige Hilfe in der Entwicklung* (Leipzig, 1903).

412 *Unsere Welt*, 1929, Nos. 1, 2, 6 and 7.

413 See further, Baur-Fischer-Lenz, *Human Heredity*, Note 394.

414 The vast literature in connection with this problem has recently been enriched by a notable lecture by W. Hellpach (Bielefelder Reichselterntagung, 1930, *Nord und Sud*, Vol. 53, No. 6).

415 See Otto's fundamental work *The Idea of the Holy*, trans. by J. W. Harvey (London, 1928).

416 The problem of historicism has been most clearly envisaged and developed in recent times by Troeltsch.

417 In this respect I agree entirely with A. Sternberg, *Idealismus und Kultur* (Berlin, 1924). See my essay 'Vom Relativen zum Absoluten,' *Unsere Welt*, 1925, Nos. 7 and 8.

418 Oesterreich, *Das Weltbild der Gegenwart* (Berlin, 1925), p. 313.

419 Widmann, *Der Heilige und die Tiere* (Halle, 1921).

420 Bavink, 'Das Übel in der Welt vom Standpunkte der Wissenschaft. und der Religion,' *Unsere Welt*, 1925, Nos. 1 and 3; reprint (Detmold, 1925).

BIBLIOGRAPHY

(See also Supplementary English Biography)

Becher, E., *Einführung in die Philosophie*. Munich, 1926.

Becher, E., *Geisteswissenschaften und Naturwissenschaften*. Munich, 1921.

Carnap, *Der logische Aufbau der Welt*. Berlin, 1928.

Cassirer, Art. 'Erkenntnistheorie und Logik' in *Jahrbücher der Philosophie*. Berlin, 1913. See bibliography there given.

Cohen, *Logik des reinen Erkennens*. Berlin, 1902.

Comte, A., *The Positive Philosophy*; trans. and condensed by H. Martineau. London, 1852.

Comte, A., *The Fundamental Principles of Positive Philosophy*, being the first two chapters of *Cours de Philosophie Positive*; trans. by P. Descours and H. Gordon Jones, with biographical preface by E. S. Beesly. London, 1905.

Cornelius, *Einleitung in die Philosophie*. Berlin and Leipzig, 1919.

Driesch, *Die Logik als Aufgabe*. Tübingen, 1913.

Driesch, *Ordnungslehre*. Jena, 1912.

Driesch, *Wissen und Denken*. Leipzig, 1923.

Driesch, *Man and the Universe*; trans. by W. H. Johnston. London, 1929.

Friedmann, H., *Die Welt der Formen*. Munich, 1930.

Geyser, J., *Grundlegung der Logik und der Erkenntnistheorie*. Münster, 1919.

Hartmann, Ed. von, *System der Philosophie in Grundriss*. Leipzig.

Hartmann, Ed. von, *Das Grundproblem der Erkenntnistheorie*. Leipzig, 1889.

Hartmann, Ed. von, *Kritische Grundlegung des transcendentalen Realismus*. Leipzig, 1875.

Hartmann, Ed. von, *Kategorienlehre*. Leipzig, 1896.

Hartmann, Ed. von, *Philosophy of the Unconscious*; trans. by W. C. Coupland. London, 1931.

Hartmann, Ed. von, *The Religion of the Future*; trans. by E. Dau. London, 1886.

Heymanns, *Einführung in die Metaphysik auf Grund der Erfahrung*. Leipzig, 1905.

Höfler, *Logik*. 1890.

Husserl, *Logische Untersuchungen*, 2 vols., 2nd ed. Halle, 1913.

Husserl, *Ideas: General Introduction to Pure Phenomenology*; trans. by W. R. Boyce Gibson. London, 1931.

Koppelmann, *Untersuchungen zur Logik der Gegenwart*. Berlin, 1913.

Külpe, *Die Realisierung*. 2 vols., Leipzig, 1912–1920.

Külpe, *Introduction to Philosophy*; trans. by W. S. Pilsbury and E. B. Titchener. London, 1897.

Liebmann, *Zur Analysis der Wirklichkeit*, 4th ed. Strasburg, 1911.

Lotze, R. H., *Grundzüge der Naturphilosophie*, 2nd ed. Leipzig, 1889.
(The above is not translated, but translations exist of the *Grundzuge* [Outlines] of Logic, Metaphysics, Practical Philosophy, Psychology, and Philosophy of Religion. See E. E. Thomas, *Lotze's Theory of Reality*, London, 1921; H. Jones, *A Critical Account of the Philosophy of Lotze*, London, 1895; E. Robins, *Some Problems of Lotze's Theory of Knowledge*, Cornell, 1900.)

Mach, E., *The Analysis of Sensation*. Chicago and London, 1914.

Mach, E., *Populär-wissenschaftliche Vorlesungen*, 3rd ed. Leipzig, 1903.

Mach, E., *Erkenntnis und Irrtum*, 2nd ed. Leipzig, 1906.

Meinong, *Untersuchungen zur Gegenstandstheorie und Psychologie*. Leipzig, 1904.

Meinong, *Ueber Annahmen*. Leipzig, 1902.

Meinong, *Die Erfahrungsgrundlagen unseres Wissens*. Berlin.

Meinong, 'Ueber Gegenstände höherer Ordnung' and numerous other articles in Ebbinghaus' *Zeitschrift für Psychologie*, 1889.

Messer, *Einführung in die Erkenntnistheorie*. Leipzig, 1909.

Müller, *Einleitung in die Philosophie*. Bonn, 1924.

Nelson, 'Ueber das sogenannte Erkenntnisproblem,' *Abhandlungen der Friesschen Schule*. N.F. ii. 4. Göttingen, 1908.

Paulsen, Fr., *Introduction to Philosophy*; trans. by F. Thilly. New York, 1911.

Rickert, *Kulturwissenschaft und Naturwissenschaft*. Tübingen, 1910.

Riehl, Al., *Philosophie der Gegenwart*. Leipzig, 1908.

Russell, Bertrand, *The Problems of Philosophy*. London, 1912.

Schlick, *Allgemeine Erkenntnislehre*. Berlin, 1918.

Sigwart, *Logik*, 2 vols., 3rd ed. Tübingen, 1904.

Vaihinger, H., *The Philosophy of 'As If'*; trans. by C. K. Ogden. London, 1924.

Verweyen, *Philosophie des Möglichen*. Leipzig, 1913.

Ziehen, *Erkenntnistheorie auf psychophysiologischer und physikalischer Grundlage*. Jena, 1913.
(See also: *Introduction to Physiological Psychology*; trans. by C. C. van Liew and O. W. Berger. London, 1895.)

Wundt, *Logik*. 2 vols., 3rd ed. Stuttgart, 1906 and 1907.

PHILOSOPHY OF SCIENCE, SCIENTIFIC EPISTEMOLOGY
(CHIEFLY RELATING TO PART I)

Apelt, *Theorie der Induktion*. Leipzig, 1854.

Aster, E. von, *Prinzipien der Erkenntnislehre*. Leipzig, 1913.

Atomic Theory. See Note 5.

Bauch, Br., *Die Idee*. Leipzig, 1926.

Bauch, Br., *Studien zur Philosophie der exacten Wissenschaften*. Heidelberg, 1911.

Bauch, Br., *Wahrheit, Wert, und Wirklichkeit*. Leipzig, 1923.

Bavink, *Hauptfragen der Naturphilosophie*. Berlin, 1928.

Becher, *Die Philosophischen Voraussetzungen der exacten Wissenschaft.* Leipzig, 1907.

Becher, Art. 'Naturphilosophie' from *Kultur der Gegenwart.* Leipzig, 1914.

Becher, *Weltgebäude, Weltgesetze, Weltentwicklung.* Berlin, 1915.

Becher, *Grundlagen und Grenzen des Naturerkennens.* Munich, 1928.

Becher, *Erkenntnistheorie und Metaphysik.* Berlin, 1925.

Bergmann, E., *Einführung in die Philosophie,* Vol. i. Breslau, 1926.

Bonola-Liebmann, *Die nichteuklidische Geometrie.* Leipzig, 1919.

Boutroux, E., *Das Wissenschaftsideal des Mathematikers.* Leipzig, 1927. (See also: *The Contingency of the Laws of Nature;* trans. by F. Rothwell, Chicago and London, 1916; *The Beyond that is Within,* trans. by S. Nield, London, 1912; *Natural Law in Science and Philosophy,* trans. by F. Rothwell, London, 1914; *Historical Studies in Philosophy,* London, 1912; Crawford, L. S., *The Philosophy of Emil Boutroux,* Cornell, 1924.)

Carnap, R., *Physikalische Begriffsbildung.* Karlsruhe, 1926.

Carnap, R., *Scheinprobleme in der Philosophie.* Berlin, 1928.

Classen, *Vorlesungen über moderne Naturphilosophen.* Hamburg, 1909.

Classen, *Die Prinzipien der Mechanik bei Herz und Boltzmann.* Hamburg, 1898.

Cohn, *Voraussetzungen und Ziele der Erkenntnis.* Leipzig, 1908.

Couturat, L., *Les Principes des Mathématiques.* Paris, 1905. (See also *The Algebra of Logic;* trans. by L. G. Robinson and preface by P. E. B. Jourdain. Chicago and London, 1914.)

Dingler, *Die Grundlagen der Naturphilosophie.* Leipzig, 1913.

Dingler, *Physik und Hypothese.* Berlin, 1921.

Dingler, *Die Grundlagen der Physik.* Leipzig and Berlin, 1923.

Drude, *Die Theorie in der Physik, Antrittsvorlesung.* Leipzig, 1895.

Du Bois-Reymond, P., *Metaphysik und Theorie der mathematischen Grundbegriffe.* Tübingen, 1882.

Duhem, 'L'Evolution de la Mécanique,' *Revue générale des Sciences,* 15th March, 1903.

Einstein, *Sidelights on Relativity:* 1. *Ether and Relativity;* 2. *Geometry and Experience;* trans. by G. B. Jeffery and W. Perrett. London, 1922.

Eisler, R., *Einführung in die Erkenntnistheorie.* Leipzig, 1925.

Enriques, F., *Problems of Science;* trans. by K. Royce. Chicago and London, 1914.

Erdmann, *Die Axiome der Geometrie.* Leipzig, 1877.

Erdmann, *Ueber Inhalt und Geltung des Kausalgesetzes.* Halle, 1905.

Exner, *Vorlesengen über die physikalischen Grundlagen der Naturwissenschaften.* Vienna, 1920.

Feigl, *Theorie und Erfahrung in der Physik.* Karlsruhe, 1929.

Frischeisen-Köhler, *Wissenschaft und Wirklichkeit.* Leipzig, 1912.

Frost, *Naturphilosophie.* Leipzig, 1910.

Gehrcke, *Physik und Erkenntnistheorie.* Leipzig, 1921.

Geissler, *Die Grundzüge und das Wesen des Unendlichen in Mathematik und Philosophie.* 1902.

Görland, *Die Hypothese* ('Wege zur Philosophie,' No. 4). Göttingen, 1911.

652 BIBLIOGRAPHY

Guthberlet, *Naturphilosophie*. Münster, 1913.

Haering, Th. H., *Philosophie der Naturwissenschaft*. Munich, 1923.

Hartmann, E. von, *Die Weltanschauung der modernen Physik*, 2nd ed. Leipzig, 1902.

Hartmann, E. von, *Grundriss der Naturphilosophie*. Sachsa, 1907.

Hartmann, Nic., *Grundzüge einer Metaphysik der Erkenntnis*. Berlin, 1925.

Helm, *Die Energetik nach ihrer geschichtlichen Entwicklung*. Leipzig, 1898.

Helmholtz, *Popular Lectures on Scientific Subjects*; trans. by E. Atkinson, with introduction by J. Tyndall, and an autobiography of the author. London, 1893.

Helmholtz, *Schriften zur Erkenntnistheorie* (edited by Hertz and Schlick). Berlin, 1921.

Hertz, P., *Ueber das Denken und seine Beziehung zur Anschauung*. Berlin, 1923.

Heymanns, *Die Gesetze und Elemente des wissenschaftlichen Denkens*. Leipzig, 1894.

Heymanns, 'Ueber Erklärungshypothesen und Erklären überhaupt,' *Annalen der Naturphilosophie*, i.

Hilbert, *Grundlagen der Geometrie*, 5th ed. Leipzig, 1921.

Höfler, *Studien zur gegenwartigen Philosophie der mathematischen Mechanik*. Leipzig, 1900.

Höfler, *Zur gegenwartigen Naturphilosophie*. Berlin, 1904.

Höfler. Many papers, particularly in the *Zeitschr. phys.-chem. Unters.* See also the 'Logische Anhang' of his *Lehrbuch der Physik*.

Hönigswald, *Zur Kritik der Machschen Philosophie*. Berlin, 1903.

Hönigswald, 'Naturphilosophie,' in *Jahrbuch der Philosophie*. Berlin, 1913. Further references are there given.

Jacoby, G., *Allgemeine Ontologie der Wirklichkeit*. Halle, 1925.

James, W., *Pragmatism*. Cambridge (Mass.), 1907.

Keyserling, H. Graf, *Prolegomena zur Naturphilosophie*. Munich, 1910. (See also *Creative Understanding*. New York and London, 1929.)

Klein, F., *Vorlesungen über Differentialgeometrie*. Leipzig, 1905.

Kleinpeter, *Die Erkenntnistheorie der Naturforschung der Gegenwart*. Leipzig, 1905.

Kleinpeter, 'Hertz' und Machs Auffassung der Physik,' *Archiv für systematische Philosophie*, v, 2; 1899.

Kleinpeter, 'Zur Formulierung des Trägheitsgesetzes,' *ib.* vi, 4; 1900.

Kleinpeter, *Der Phenomenalismus*. Leipzig, 1913.

König, *Die Entwicklung des Kausalproblems*, 2 vols. 1880 and 1890.

König, 'Kant und die Naturwissenschaft,' *Die Wissenschaft*, Vol. 22. Brunswick, 1907.

König, *Das Wesen der Materie*. Göttingen, 1911.

Külpe, *Erkenntnistheorie und Naturwissenschaft*. Paper read at the Naturforscherversammlung in Königsberg, 1910. Leipzig, 1911.

Lehmann, O., *Das 'Als Ob' in Molekularphysik*. See Note 5.

Lipps, *Naturwissenschaft und Weltanschauung*. Paper read at the Natur-forscherversammlung. Stuttgart, 1906.

Lipsius, *Naturphilosophie und Weltanschauung*. Leipzig, 1918.

Mach, E., *The Science of Mechanics*; trans. by T. J. McCormack, 4th ed. 1919.

Mach, E., *Prinzipien der Warmelehre*, 2nd ed. Leipzig, 1900.

Mach, E., *Die Geschichte und die Wurzel des Satzes von der Erhaltung der Arbeit*. Prague, 1872.
(See also *Space and Geometry in the Light of Physiological, Psychological and Physical Inquiry*; trans. by T. J. McCormack, 1906.)

Messer, A., *Der kritische Realismus*. Karlsruhe, 1923.

Messer, A., *Einführung in die Erkenntnistheorie*. Leipzig, 1921.

Müller, Al., *Wahrheit und Wirklichkeit*. Bonn, 1913.

Müller, Al., *Das Raumproblem*. Brunswick, 1911. See Note 32.

Natorp, *Die logische Grundlagen der exacten Wissenschaft*. Leipzig, 1910.

Ostwald, *Vorlesungen über Naturphilosophie*, 3rd ed. Leipzig, 1905.

Ostwald, *Vorträge und Abhandlungen*. Leipzig, 1904.

Ostwald. Numerous papers in various journals, chiefly in the *Annalen der Naturphilosophie*, of which he was editor.

Ostwald, *Grundriss der Naturphilosophie*. Leipzig, 1908.

Petzoldt, *Das Weltproblem vom Standpunkt des rel. Positivismus*, 2nd ed. Leipzig.

Petzoldt, 'Das Gesetz der Eindeutigkeit,' *Vierteljahrschrift für wissenschaftliche Philosophie*, xix, 146.

Pfordten, von der, *Vorfragen der Naturphilosophie*. Heidelberg, 1907.

Picard, *Das Wissen der Gegenwart in Mathematik und Naturwissenschaft*. Leipzig, 1913.

Planck, *Die Stellung der neueren Physik zur mechanischen Naturanschauung*. Leipzig, 1910.

Planck, 'Das Prinzip der Erhaltung der Energie,' *Wissenschaft und Hypothese*, 4th ed. Leipzig, 1921.

Planck, *Die Einheit des physikalischen Weltbildes*. Leipzig, 1909.

Planck, 'Zur Machschen Theorie der physikalischen Erkenntnis,' *Vierteljahrschrift für wissenschaftlichen Philosophie*, 1910.

Planck, *A Survey of Physics*; trans. by R. Jones and D. H. Williams. London, 1925.

Planck, *The Universe in the Light of Modern Physics*; trans. by W. H. Johnston. London, 1931.

Poincaré, H., *Science and Hypothesis*; trans. by W. J. G., with a preface by J. Larmor. London, 1905.

Poincaré, H., *Science and Method*; trans. by F. Maitland, with a preface by B. Russell. London, 1914.

Poincaré, H., *Dernières Pensées*. Paris, 1913.

Poincaré, H., *La Valeur de la Science*. Paris, 1904.

Poske, 'Die Hypothese in Wissenschaft und Unterricht,' in the *Zeitschr. f. phys. u. chem. Unterr.*, 1912, 1. This journal is edited by him, and contains numerous other of his essays.

Potonié, *Naturphilosophische Plaudereien*. Jena, 1913.

Reichenbach, *Philosophie der Raum-Zeit Lehre*. Berlin, 1928.

Rey, *La Théorie de la physique chez les physiciens contemporains*, 3rd ed. Paris, 1930.

Stallo, *The Concepts and Theories of Modern Physics*. London, 1872.

Study, *Die realistische Weltansicht und die Lehre vom Raum*. Brunswick, 1914.

Verweyen, *Naturphilosophie*. Leipzig, 1915.

Verworn, *Ueber kausale und konditionale Weltanschauung*. Bonn, 1912.

Volkmann, *Erkenntnistheoretische Grundzüge der Naturwissenschaft*. Leipzig, 1896.

Volkmann, *Einführung in das Studium der theoretischen Physik*. Leipzig, 1900.

Voss, 'Ueber die mathematische Erkenntnis' in *Kultur der Gegenwart*. Leipzig, 1914.

Weinstein, *Die philosophischen Grundlagen der Wissenschaft*. Leipzig, 1906.

Weyl, H., 'Philosophie der Mathematik und Naturwissenschaft' in Bäumler and Schröter's *Handbuch der Philosophie*, Part II A. Munich and Berlin, 1926.

Wien, *Vorträge*.

Wiener, *Die Erweiterung unserer Sinne*. Inaugural Dissertation. Leipzig, 1900.

Wiener, *Physik und Kulturentwicklung*. Leipzig, 1918.

Wundt, *Prinzipien der mechanischen Naturlehre*. Stuttgart, 1910.

RELATIVITY

(a) Scientific Accounts

The Principle of Relativity: a collection of original memoirs on the special and general theory, by H. A. Lorentz, A. Einstein, H. Minkowski, and H. Weyl, with notes by A. Sommerfeld; trans. by W. Perrett and E. B. Jeffrey. London, 1923.

Einstein, A., *Relativity: the Special and the General Theory*; trans. by R. W. Lawson. London, 1920.

Weyl, H., *Space-Time-Matter*; trans. by H. L. Brose. London, 1922.

Weyl, H., 'Gravitation und Elektrizität,' *Sitzgsber. der bad. Akad.*, 1918, p. 465; see also Haas (c).

Laue, M. von, *Das Relativitätsprinzip*. Brunswick, 1919. A mathematical formulation of the theory.

An easily understood introduction is given by the larger works of

Born, M., *Einstein's Theory of Relativity*; trans. by H. L. Brose. London, 1924.

Eddington, A. S., *Space, Time, and Gravitation*. Cambridge, 1920.

By way of introduction to the more difficult works of Weyl and Laue, the following may be recommended:

Kopff, A., *Grundzüge der Einsteinschen Relativitästheorie*. Leipzig, 1921.

Haas, A., *Theoretical Physics*, Vol. ii.

Haas, A., *The New Physics*; trans. by R. W. Lawson, 3rd ed. London, 1930.

(b) *Popular Accounts*

Einstein, A., *The Meaning of Relativity*; trans. by E. P. Adams. London, 1922. In my opinion, in many respects not a very happy statement, particularly on account of the emphasis laid upon the well-known paradoxes.

Bloch, W., *Einführung in die Relativitätstheorie*. Leipzig. An easily understood and clear account.

Freundlich, E., *The Foundations of Einstein's Theory of Gravitation*, with a preface by Einstein; trans. by H. L. Brose, with an introduction by H. H. Turner. London, 1924.

Schlick, M., *Space and Time in Contemporary Physics*; trans. by H. L. Brose. Oxford, 1920. In my opinion the best and philosophically most profound of the popular statements. Particularly recommended.

Kirchberger, P., *Was kann man ohne Mathematik von der Relativitätstheorie verstehen ?* Karlsruhe, 1920.

Laemel, R., *Introduction to the Theory of Relativity*. London, 1926.

Mie, G., *Die Einsteinsche Gravitationstheorie*. Leipzig, 1921. A peculiar, and in many respects new, point of view.

Thirring, E., *The Ideas of Einstein's Theory of Relativity*. 1921.

(c) *Philosophical Works relating to Relativity*

Winternitz, J., *Relativitätstheorie und Erkenntnislehre*. Leipzig, 1923. Describes almost all relevant questions in an excellently clear and penetrating manner, and takes up a very sensible attitude, which does justice both to rationalism and positivism.

Reichenbach, H., *Relativitätstheorie und Erkenntnis a priori*. Berlin, 1920. Draws attention to the opposition existing between the views of Kant and Einstein.

Reichenbach, H., *Philosophie der Raum-Zeit Lehre*. Berlin, 1929. See text, p. 156.

Sellien, E., 'Die erkenntnistheoretische Bedeutung der Relativitätstheorie,' supplement to *Kantstudien*, No. 48. Berlin, 1919.

Cassirer, E., *Zur Einsteinschen Relativitätstheorie*. Berlin, 1921. Sellien and Cassirer attempt to show that no contradiction exists between Kant and Einstein. The same view is taken by

Schneider, J., *Das Raum-Zeit Problem bei Kant und Einstein*. Berlin, 1921. Further literature in Sellien and Cassirer.

Petzoldt, J., *Die Stellung der Relativitätstheorie in der geistigen Entwicklung der Menschheit*. Tries to show that relativity confirms Mach's positivism.

Siebert, O. A., *Einstein's Relativitätstheorie und ihre kosmologische und philosophischen Konsequenzen*. Langensalza, 1921. Treats mainly of the metaphysical problems referred to on pp. 267 ff. of the text.

Haas, A., 'Die Axiomatik der modernen Physik' and 'Die Physik als geometrische Notwendigkeit,' in *Naturwissenschaften*, 1919, p. 744; 1920, p. 121, are excellent accounts of the fundamental unification of physics by the relativity theory.

Planck, M., 'Vom Relativen zum Absoluten,' *Naturwissenschaften*, 1925, Nos. 3 and 4. An excellent account of what is not 'relativistic' in the relativity theory, as opposed to the purely destructive view of the positivists.

Geiger, M., *Die philosophische Bedeutung der Relativitätstheorie.* Halle, 1921. The clearest and most accurate discussion known to me of the main problem.

Carnap, R., 'Der Raum,' supplement to the *Kantstudien,* No. 56. Berlin, 1922. See text, p. 152.

Dingler, H., *Kritische Bemerkungen zu den Grundlagen der Relativitätstheorie.* Leipzig, 1921.

Dingler, H., *Das Experiment.* Munich, 1928. See also Note 82.

Dingler, H., *Relativitätstheorie und Oekonomieprinzip.* Leipzig, 1922.

Dingler, H., *Physik und Hypothese.* Berlin, 1921.

Müller, Al., *Die philosophischen Probleme der Einsteinschen Relativitätstheorie.* Brunswick, 1922. 2nd ed. of the *Raumproblem.* See Note 27.

Müller, Al., *Der Gegenstand der Mathematik mit besonderer Brücksichtigung der Relativitätstheorie.* Brunswick, 1922.

(d) Opponents of Relativity

Wiechert, E., Art. 'Mechanik' in the volumes on physics of *Kultur der Gegenwart.* Leipzig, 1915. Only takes account of the special theory. Much of it now out of date.

Wiechert, E., *Der Aether im Weltbild der Physik.* Berlin, 1921.

Lenard, P., *Ueber Relativitätsprinzip, Aether, Gravitation.* Leipzig, 1920. Emphasises the realistic element in physics as against mathematical formalism.

Gehrcke, E., *Die Relativitätstheorie: eine wissenschaftliche Massensuggestion.* Berlin, 1920. A very fierce attack, particularly on the general theory.

Geissler, K., *Gemeinverständliche Widlegung des formalen Relativismus.* Directed more against philosophical relativism than against the physical relativity theory.

Ripke-Kühn, L., *Kant contra Einstein.* Erfurt, 1920. Attempt to confute Einstein on the basis of Kant's *Critique of Pure Reason,* assumed to be correct. Directed particularly against the mingling of kinematics and mechanics.

PART II

Arrhenius, Sv., *Lehrbuch der kosmischen Physik.* Leipzig, 1903.

Arrhenius, Sv., *Erde und Weltall.* Leipzig, 1926.

Arrhenius, Sv., *Der Lebenslauf der Planeten.* Leipzig, 1921.

Arrhenius, Sv., *Worlds in the Making;* trans. by H. Borns. London and New York, 1908.

'Astronomie' from *Kultur der Gegenwart.* Leipzig, 1915.

Auerbach, *Die Weltherrin und ihr Schatten.* Jena, 1913.

Classen, *Das Entropiegesetz.* Detmold, 1910.

Credner, *Elemente der Geologie,* 10th ed. Leipzig, 1916.

Dacqué, *Die Erdzeitalter.* Munich, 1930.

Gockel, *Schöpfungsgeschichtliche Theorien,* 2nd ed. Cologne, 1910.

Isenkrahe, *Energie, Entropie, Weltanfang, Weltende.* Trèves, 1910.

Meisel, *Wandlungen des Weltbildes und des Wissens von der Erde.* Stuttgart, 1913.

Newcomb-Engelmann, *Populäre Astronomie*. Leipzig, 1914.

Oppenheim, *Probleme der modernen Astronomie*. Leipzig, 1911.

Plassman, *Himmelskunde*. Freiburg, 1913.

Schnippenkötter, *Der entropologische Gottesbeweis*. Bonn, 1920. See Note 180.

Troels-Lund, T. F., *Livsbelysning*. Copenhagen, 1899.

Troels-Lund, T. F., German trans. *Himmelsbild und Weltanschauung in Wandel der Zeit*. Leipzig, 1929.

Wallace, A. R., *Man's Place in the Universe*, 4th ed., with an added chapter. London, 1904.

Wegener, A. L., *The Origin of Continents and Oceans*; trans. by J. G. A. Skerl. London, 1924.

Ziegler-Oppenheim, *Weltentstehung in Sage und Wissenschaft*. Leipzig, 1925.

PART III: BIOLOGY, DOCTRINE OF DESCENT

'Abstammungslehre,' 12 lectures by O. Abel and others. Jena, 1911.

'Abstammungslehre,' from *Kultur der Gegenwart*. Leipzig, 1914.

Baur, *Einführung in die experimentelle Vererbungslehre*, 2nd ed. Berlin, 1914.

Becher, E., *Gehirn und Seele*. Heidelberg, 1911.

Becher, S., 'Seele, Handlung in Zweckmässigkeit im Reich der Organismen,' *Annalen der Naturphilosophie* 10, 1911.

Bergson, H. L., *Creative Evolution*; trans. by A. Mitchell. London, 1911.

Bertalanffy, L. von, *Kritische Theorie der Formbildung*. Jena, 1929.

Boelsche, W., *Love-life in Nature*; trans. by C. Brown. London, 1931.

Driesch, *Grundprobleme der Physiologie*, 2nd ed. Leipzig, 1929.

Driesch, *Wirklichkeitslehre*, 3rd ed. Leipzig, 1930.

Driesch, *Das Ganze und die Summe*. Leipzig, 1921.

Driesch, *The Science and Philosophy of the Organism*, 2nd ed. London, 1929.

Driesch, *The History and Theory of Vitalism*; trans. by C. K. Ogden. London, 1914.

Driesch, *Mind and Body*; trans. by T. Besterman. London, 1927.

Dürken, *Experimental Analysis of Development*; trans. by H. G. Newth. London, 1932.

Dürken, *Allgemeine Abstammungslehre*. Berlin, 1923.

Dürken, *Hauptprobleme der Biologie*. Munich, 1925.

Eimer, *Die Entstehung der Arten*. Jena, 1888–1901.

Eisler, R., *Leib und Seele*. Leipzig, 1906.

Erdmann, B., *Wissenschaftliche Hypothesen über Leib und Seele*. Cologne, 1907.

Gemünd, *Leben und Anpassung*. Bonn, 1925.

Goldschmidt, *Einführung in die Vererbungswissenschaft*, 3rd ed. Leipzig, 1921.

Goldschmidt, *Ascaris* (Popular introduction to modern biology). Leipzig, 1922.

Tts

658 BIBLIOGRAPHY

Haberlandt, *Sinnesorgane im Pflanzenreich.* Leipzig, 1906.
Haecker, *Allgemeine Verebungslehre,* 2nd ed. Brunswick, 1912.
Hering, *Ueber das Gedächtnis als allgemeine Funktion der organischen Materie.* Vienna, 1870.
Hertwig, O., *Allgemeine Biologie,* 2nd ed. Jena, 1918.
Hertwig, O., *Das Werden der Organismen,* 2nd ed. 1918.
Hertwig, R., 'Abstammungslehre' in *Kultur der Gegenwart.* See above. The best inclusive description.
Hesse, *Abstammunglehre und Darwinismus,* 5th ed. Leipzig, 1918.
Hesse, Doflein, *Tierbau und Tierleben.* A magnificent work. 2 vols. Leipzig, 1910–14.
Jacobi, *Mimikry und verwanndte Erscheinungen.* Braunschweig, 1913.
Johannsen, 'Experimentelle Grundlagen der Descendenz-Theorie' in *Kultur der Gegenwart.* Leipzig, 1914.
Johannsen, *Vererbungslehre.* Leipzig, 1921.
Kern, *Des Problem des Lebens.* Berlin, 1909.
Klebs, *Wilkürliche Entwicklungsänderungen bei Pflanzen.* Jena, 1903.
Klebs, *Ueber das Verhältnis der Aussenwelt zur Entwicklung der Pflanzen.* Heidelberg, 1913 (Stz. ber. der Akad.).
Köhler, W., *Gestalt Psychology.* London, 1930.
König, *Das Leben, sein Ursprung und seine Entwicklung auf der Erde,* 2nd ed. Berlin, 1905.
Kraepelin, *Einführung in die Biologie,* 3rd ed. Leipzig, 1912.
Kropotkin, *Gegenseitige Hilfe im Tierreich.* Leipzig, 1904.
Leduc-Gradenwitz, *Das Leben in seinem Physikalisch-chemischen Zusammenhang.* Halle, 1912.
Lehmann, *Experimentelle Abstammung- und Vererbungslehre.* Leipzig, 1913.
Leppelmann, *Das Gesetz der Erhaltung der Energie und die verschiedenen Auffassungen der Wechselwirkung zwischen Leib und Seele.* Dissertation. Münster, 1915..
Löb, *Vorlesungen über die Dynamik der•Lebenserscheinungen.* Leipzig, 1906.
Lotsy, *Vorlesungen über Descendenztheorien.* Jena, 1908.
Molisch, *Populäre biologische Vortäge.* Jena, 1920.
Morgan, T. H., *The Physical Basis of Heredity.* Philadelphia and London, 1919.
'Palaeontology' by O. Abel, in 'Abstammungslehre' volume of *Kultur der Gegenwart.*
Pauly, *Darwinismus und Lamarckismus.* Munich, 1905.
Plate, *Selektionsprinzip und Probleme der Artenbildung,* 4th ed. Leipzig, 1913.
Plate, 'Descendenztheorie' in *Handwörterbuch der Naturwissenschaften* II, pp. 497 ff. Jena, 1912.
Plate, *Vererbungslehre, mit besonderer Berücksichtigung des Menschen.* Leipzig, 1913.
Plate, L., *Abstammungslehre.* Jena, 1925.
Reinke, *Einleitung in die theoretische Biologie,* 2nd ed. Berlin, 1911.

Reinke, *Grundzüge der Biologie.* Heilbronn, 1909.

Sapper, K., *Naturphilosophie (Philosophie des Organischen).* Breslau, 1928.

Schaxel, *Grundzüge der Theoriebildung in der Biologie.* Jena, 1922.

Schneider, *Einführung in die Descendenztheorie,* 2nd ed. Jena, 1911.

Schneider, *Tierpsychologisches Praktikum.* Leipzig, 1902.

Schultz, 'Philosophie des Organischen' in *Jahrbuch der Philosophie.* Berlin, 1913. Very detailed bibliography.

Semon, *Mneme.* London, 1921. (English translation.)

Semon, *Das Problem der Vererbung erworbener Eigenschaften.* Leipzig, 1912.

Stöhr, *Der Begriff des Lebens.* Heidelberg, 1909.

Thesing, *Experimentelle Biologie* II. Leipzig, 1911.

Tschulok, *Descendenzlehre.* Jena, 1922.

Verworn, *Die Mechanik des Geisteslebens.*

Verworn, *Die Anfänge der Kunst.* Jena, 1920.

Voss, *Moderne Pflanzenzüchtung und Darwinismus.* Detmold, 1913.

de Vries, *Mutation Theory.* London, 1910–11.

de Vries and Klebahn, *Species and Variety.*

Wasmann, E., *The Problem of Evolution,* 2nd ed. London, 1912.

Wasmann, E., *Modern Biology and Evolution,* 2nd ed. London, 1910. See against him Dahl, *Die Redeschlacht in Berlin.* Jena, 908 Also *Nat. Wochschr.,* 1908, pp. 49, 161.

Weismann, A., *Studies in the Theory of Descent.* London, 1882.

Wertheimer, M., *Ueber Gestalttheorie.* Reprint from Symposion I, Weltkreisverlag, Berlin-Schlachtensee, 1925.

Wertheimer, M., *Drei Abhandlunger zur Gestalttheorie.* *Ib.,* 1925.

Wolff, *Mechanismus und Vitalismus,* 2nd ed. Leipzig, 1905.

PART IV: MAN AND NATURE

Baur-Fischer-Lenz, *Human Heredity*; trans. by E. and C. Paul. London, 1931.

Dessauer, *Philosophie der Technik.* Bonn, 1927.

Friesen, J., *Mensch und Tier in Eiszeitalter,* 2nd ed. Leipzig, 1921.

Hauser, *Der Mensch vor 100,000 Jarhren.* Leipzig, 1917.

Heilborn, *Allgemeine Völkerkunde,* 2 vols. Leipzig, 1917.

Hoernes, *Urgeschichte der bildenden Kunst,* 2nd ed. Vienna, 1915.

Klaatsch-Heilborn, *Der Werdegang der Menschheit und die Entstehung der Kultur.* Berlin, 1921.

Köhler, *Mentality of Apes.* London, 1925.

Kohlbrugge, *Die morphologische Abstammung des Menschen.* Stuttgart, 1908.

Müller, Al., *Psychologie.* Bonn, 1927.

Piper, H., *Die Gesetz der Weltgeschichte.* Leipzig, 1929.

Ranke, *Der Mensch,* 2 vols., 3rd ed. Leipzig, 1912.

Reinhardt, *Der Mensch zur Eiszeit,* 2nd ed. Munich, 1908.

Weinert, H., *Menschen der Vorzeit*.　Stuttgart, 1930.
Werth, *Das Eiszeitalter*, 2nd ed.　Berlin, 1930.
Wundt, *Elements of Folk Psychology*.　London, 1916.
Zschimmer, *Philosophie der Technik*.　Jena, 1919.

GENERAL ON WORLD-VIEW

Apel, *Die Ueberwindung des Materialismus*.　Berlin, 1909.
Bavink, *Hauptfragen der Naturphilosophie*, Vol. II.　Berlin, 1928.
Beth, *Die Entwicklungsgedanke und das Christentum*.　Gr.-Lichterfelde, 1908.
Dennert, *Die Weltanschauung des modernen Naturforschers*.　Stuttgart, 1906.
Dennert, *Weltbild und Weltanschauung*.　Hamburg (Detmold), 1908.
Dessauer, Fr., *Leben, Natur, Religion*.　Bonn, 1924.
Drews, *Der Monismus*, 2 vols.　Jena, 1908.
Driesch, *Man and the Universe*.　London, 1929.
Driesch, *Ethical Principles in Theory and Practice*.　London, 1930.
Driesch, *The Possibility of Metaphysics*.　London, 1930.
Eucken, *The Truth of Religion*.　London, 1894.
Fechner, *Zendavesta*.　New Edition.　Leipzig, 1920.
Geyser, J., *Allgemeine Philosophie des Seins und der Natur*.　Münster, 1915.
Gomperz, *Weltanschauungslehre*.　Jena, 1905.
Gomperz, *Das Problem der Willensfreiheit*.　Jena, 1907.
Gutberlet, *Theodizee*, 4th ed.　Münster, 1909.
Haeckel, *The Riddle of the Universe*.　London, 1929.
Haeckel, *The History of Creation*, 4th ed.　London, 1892.
Jellinek, *Das Weltengeheimnis*.　Stuttgart, 1921.　(Beautifully produced, theosophical in tendency.)
Kern, *Weltanschauung*.　Berlin, 1900.
Koppelmann, *Kritik des Sittlichen Bewusstseins*.　Berlin, 1904.
Koppelmann, *Weltanschauungsfragen*.　*Ib.*
Lotze, *Microcosmus*.　London, 1885.
Mahncke, *Der Wille zur Ewigkeit*.　Halle, 1917.
Mauthner, *Der Atheismus und seine Geschichte im Abendlande*.　Stuttgart, 1920.
Messer, *Deutsche Wertphilosophie der Gegenwart*.　Leipzig, 1926.
Messer, *Glauben und Wissen, die Geschichte einer inneren Entwicklung*.　Munich, 1919.
Messer, *Natur und Geist. philosophische Aufsätze*.　Osterwieck, 1920.
Oesterreich, *Das Weltbild der Gegenwart*.　Berlin, 1920.
Otto, R., *Naturalism and Religion*.　London, 1907.
Otto, R., *The Idea of the Holy*.　Oxford, 1928.
Pfannküche, *Religion und Natur im Kampf und Frieden*.　Leipzig.
Reinke, *Die Welt als Tat.*, 6th ed.　Berlin, 1915.

Schmidt, W., *Der Kampf der Weltanschauungen.* Berlin, 1904.
Schwartz, *Grundfragen der Weltanschauung.* Leipzig, 1902.
Unold, *Organische und soziale Lebensgesetze.* Leipzig, 1906.
White, *Geschichte der Fehde zwischen Wissenschaft und Theologie.* Leipzig.

HISTORICAL

Auerbach, *Entwicklungsgeschichte der modernen Physik.* Berlin, 1923.
Boll-Bezold-Gundel, *Sternglaube und Sterndeutung.* Leipzig, 1928.
Dacqué, *Der Descendenzgedanke und seine Geschichte vom Altertum bis zur Neuzeit.* Munich, 1903.
Dannemann, *Die Naturwissenschaften in ihrer Entwicklung,* 4 vols. Leipzig, 1911–13.
Darmstaedter, *Naturforscher und Erfinder.* Bielefeld and Leipzig, 1926.
von Hartmann, *Geschichte der Metaphysik.* Sachsa, 1900.
Lange, F. A., *Geschichte des Materialismus,* 7th ed. Leipzig, 1902.
Lenard, P., *Great Men of Science.* London, 1933.
Nordenskiöld, *Geschichte der Biologie.* Jena, 1926.
Radl, *History of Biological Theories.* Oxford, 1930.
Siegel, *Geschichte der Naturphilosophie.* Leipzig, 1913.
Uhlmann, *Entwicklungs-Gedanke und Artbegriff.* Jena, 1923.
Wagner, *Geschichte des Lamarckismus.* Stuttgart, 1909.
Windelband, *History of Philosophy.* London, 1931.

SUPPLEMENTARY BIBLIOGRAPHY
OF ORIGINAL WORKS IN ENGLISH

This Bibliography does not attempt to be exhaustive, but only to give some works, in most cases recently published, which will afford further information, and also indicate further sources. No standard text-books, with one or two exceptions, have been included, as these are too numerous to make selection possible.

Mention should be made of the *Encyclopædia Britannica*, 14th edition, which contains articles by authorities in every field, with full bibliographies. Further, the Home University Library, Benn's Sixpenny Library, the Thinker's Library, the Rationalist Press Association Reprints, the 'Recent Advances' series and many other similar publishing undertakings supply books, often very cheaply, which deal with most of the subjects.

GENERAL PHILOSOPHY AND THEORY OF KNOWLEDGE

Broad, C. D., *Perception, Physics, and Reality*. Cambridge, 1914.

Broad, C. D. *Scientific Thought*. London, 1923.

Broad, C. D., *The Mind and its Place in Nature*. London, 1925.

Burtt, E. A., *Metaphysical Foundations of Modern Science* (with bibliography). London, 1925.

Davies, J. Landon, *Man and his Universe*. New York and London, 1930.

Dingle, H., *Science and Human Experience*. London, 1931.

Eddington, A. S., *The Nature of the Physical World*. Cambridge, 1928.

Eddington, A. S., *The Theory of Relativity and its Influence on Scientific Thought*. Oxford, 1922.

Haldane, J. S., *The Sciences and Philosophy*. London, 1929.

Huxley, Julian, *What Dare I Think ?* London, 1931.

Needham, J. (Edited by), *Science, Religion, and Reality*. A collection of essays by A. Balfour, Malinowski, Singer, Eddington, Inge, and others. London, 1925.

Ramsey, F. P., *The Foundations of Mathematics*. London, 1931.

Ritchie, A. D., *Scientific Method*. London, 1923.

Russell, Bertrand, *Analysis of Matter*. London, 1927.

Russell, Bertrand, *Analysis of Mind*. London, 1921.

Russell, Bertrand, *Mysticism and Logic*. London, 1932.

Russell, Bertrand, *Our Knowledge of the External World*. London, 1926.

Russell, Bertrand, *Sceptical Essays*. London, 1928.

Russell, Bertrand, *The Problems of Philosophy*. (Home University Library.) London, 1912.

Thurstone, L. L., *The Nature of Intelligence*. London, 1924.

Whitehead, A. N., *An Enquiry Concerning the Principles of Natural Knowledge*, 2nd ed. Cambridge, 1925.

Whitehead, A. N., *Process and Reality*. Cambridge, 1929.

Whitehead, A. N., *Science and the Modern World*. Cambridge, 1926.

RECENT ADVANCES IN PHYSICS AND CHEMISTRY

Andrade, E. N. Da C., *The Structure of the Atom*. London, 1927.

Andrade, E. N. Da C., *The Mechanism of Nature*. London, 1930.

Birtwistle, G., *The New Quantum Mechanics*. Cambridge, 1928.

Bolton, L., *Introduction to the Theory of Relativity*. London, 1921.

Compton, A. H., *X-Rays and Electrons*.

Darwin, C. G., *The New Conceptions of Matter*. London, 1931.

Dirac, P. A. M., *The Principles of Quantum Mechanics*. Oxford, 1930.

Eddington, A. S., *Space, Time, and Gravitation*. Cambridge, 1920.

Fraser, R. G. J., *Molecular Rays*. Cambridge, 1931.

Gunn, J. A., *The Problem of Time*. London, 1929.

Lindemann, F. A., *The Physical Significance of the Quantum Theory*. Oxford, 1932.

Newmann, F. H., *Recent Advances in Physics (non-atomic)*. London, 1932.

Russell, B., *The A B C of Relativity*. London, 1925.

Schoen, M., *The Problem of Fermentation*. London, 1928.

Sullivan, J. M. W., *The Bases of Modern Science*. London, 1932.

Thomson, G. P., *Wave Mechanics of Free Electrons*. New York, 1930.

Thomson, G. P., *The Atom*. London, 1930.

Thomson, J. J., *Beyond the Electron*. (Lecture.) Cambridge, 1928.

Whittaker, E. T., *A History of the Theories of Ether and Electricity*. London, 1910.

ASTRONOMY, ASTROPHYSICS, COSMOLOGY

Dingle, H., *Modern Astrophysics*. London, 1924.

Dyson, F., *Science of To-day and To-morrow*.

Eddington, A. S., *Stars and Atoms*. Oxford, 1927.

Gregory, R. A., *The Vault of Heaven*, 2nd ed. London, 1923.

Jeans, J. H., *Astronomy and Cosmology*. Cambridge, 1928. 2nd ed., 1929.

Jeans, J. H., *The Mysterious Universe*. Cambridge, 1930.

Jeans, J. H., *Theories of Cosmogony and Stellar Dynamics*. Cambridge, 1919.

Jeans, J. H., *The Universe Around Us*, 2nd ed. Cambridge, 1930.

Jeans, J. H., *The Stars in their Courses*. Cambridge, 1931.

Phillips, T. E. R., and Steavenson, W. H., *The Splendour of the Heavens*. London, 1923–6.

Russell, Duggan, and Stewart, *Astronomy*. 1926–7.

Stratton, F. J. M., *Astronomical Physics*. London, 1925.

Biology and Biochemistry

Barger, G., *Some Applications of Organic Chemistry to Biology and Medicine.* New York, 1930.

Bayliss, W., *Principles of General Physiology,* 4th ed.   London, 1924.

Browning, E., *The Vitamines.* London, 1931.

Burn, J. H., *The Chemistry of the Vitamines.* London, 1932.

Elton, C., *Animal Ecology and Evolution.* Oxford, 1930.

Faraday Society, *Discussion on Colloid Science applied to Biology.*   London, 1931.

Fine, J., *Filterable Virus Diseases in Man.* Edinburgh, 1932.

Fox, H. M., *Biology.* Cambridge, 1932.

Hogben, L. T., and Winton, F. R., *Introduction to Recent Advances in Comparative Physiology.* London, 1924.

Huxley, J. S., *The Individual in the Animal Kingdom.* Cambridge, 1912.

Huxley, J. S., and Haldane, J. B. S., *Animal Biology.* Oxford, 1927.

Keith, A., *Engines of the Human Body.* London, 1925.

Marshall, F. H. A., *The Physiology of Reproduction.* London, 1922.

Medical Research Council, *A System of Bacteriology in relation to Medicine.* Vol. vii. London, 1930.

Parsons, T. R., *The Fundamentals of Bio-Chemistry.* London, 1927.

Pearl, Raymond, *The Biology of Death.* Philadelphia, 1922.

Plimmer, R. H. A., *The Constitution of the Proteins.* London, 1912.

Russell, E. S., *The Study of Living Things.* London, 1924.

Schaefer, E. A., *The Endocrine Organs.* London, 1924.

Singer, C., *A Short History of Biology.* Oxford, 1931.

Starling, E. H., and others, *Action of Alcohol on Man.* London, 1923.

Stockard, C., *The Physical Basis of Personality.* London, 1931.

Wells, H. G., and Huxley, J. S., *The Science of Life.* London, 1931.

Genetics and Evolution

Bateson, W., *Principles of Heredity.* Cambridge, 1913.

Castle, W. E., *Genetics and Eugenics.* Cambridge (Mass.), 1922.

Fisher, R. A., *Genetic Theory of Natural Selection.*

Goldschmidt, R., *The Mechanism and Physiology of Sex Determination,* London, 1923.

Morgan, C. Lloyd, *Emergent Evolution.* London, 1922.

Morgan, T. H., *The Physical Basis of Heredity.* Philadelphia and London, 1919.

Morgan, T. H., *Evolution and Genetics.* Princeton, 1925.

Morgan, T. H., *Theory of the Gene.* Yale, 1928.

Newman, H. H., *Evolution, Genetics, and Eugenics.* Illinois, 1926.

Russell, E. S., *Interpretation of Development and Heredity.* Oxford, 1930.

Sinnott, E. W., and Dunn, L. C., *Principles of Genetics.* London, 1925.

Walter, H. E., *Genetics.* New York, 1923

EUGENICS

Carr-Saunders, A. M., *Eugenics* (Home Univ. Library). London, 1926.

Darwin, L., *The Need for Eugenic Reform.* London, 1926.

East, E. M., *Biology in Human Affairs.* New York, 1931.

Galton, F., *Hereditary Genius.* London, 1869 and 1914.

Hogben, L., *Genetic Principles in Medicine and Social Science.* London, 1931.

Holmes, S. J. A., *Biology of Eugenics.* Berkeley, 1924.

Holmes, S. J. A., *The Trend of the Race.* London, 1921.

Human Biology and Racial Welfare, by a number of authors. London, 1930.

Landman, J. H., *Human Sterilisation.* London, 1932.

McDowall, *Biology and Mankind.* Cambridge, 1931.

Newman, H. H., *Readings in Evolution, Genetics, and Eugenics.*

Pearson, K. (Editor), *The Treasury of Human Inheritance.* Memoirs of the Francis Galton Laboratory for National Eugenics. London, 1907.

VITALISM AND MECHANISM

Butler, S., *Life and Habit.* London, 1878 and 1910.

Haldane, J. S., *The Philosophical Basis of Biology.* London, 1931.

Hogben, L., *The Nature of Living Matter.* London, 1930.

Jennings, H. S., *The Biological Basis of Human Nature.*

Lodge, O., *Beyond Physics.* London, 1930.

Needham, J., *Man a Machine.* London, 1927.

Rignano, E., *Man not a Machine.* London, 1926.

Ritter, W. E., *The Unity of the Organism.* Boston, 1919.

Stout, G. F., *Mind and Matter.* Cambridge, 1931.

Smuts, J. C., *Holism and Evolution.* London, 1931.

PREHISTORIC MAN

Aldrich, C. K., *The Primitive Mind and Modern Civilisation.* London, 1931.

Early Man, his Origin, Development, and Culture. (Lectures delivered at the Royal Anthropological Institute by G. Elliot-Smith, A. Keith, and others.) London, 1931.

Elliot-Smith, G., *The Evolution of Man.* London, 1924.

Elliot-Smith, G., *Human History.* London, 1930.

Keith, A., *The Antiquity of Man,* 7th impr. London, 1929.

Keith, A., *New Discoveries relating to the Antiquity of Man.* London, 1931.

Machin, A., *The Ascent of Man.* 1925.

Peake, H., and Fleure, H. J., *The Corridors of Time.* Vols. i, ii, and iii. London, 1927.

Read, C., *The Origin of Man and his Superstitions.* 1920.

Read, C., *Ethnos, or the Problem of Race from a new point of view.* London, 1931.

Rivers, W. H. R., *Psychology and Ethnology.* London, 1926.

CIVILISATION

Drift of Civilisation, The, by a number of authors, including Wells, Russell, Einstein, Inge. London, 1930.

Elliot-Smith, G., *In the Beginning: the Origin of Civilisation.* London, 1932.

Elliot-Smith, G., *The Migrations of Early Culture.* Manchester, 1929.

Finer, H., *Theory and Practice of Modern Government.* London, 1932.

Florence, P. S., *The Statistical Method in Economics.* London, 1929.

Hammond, J. L. L. and B., *The Rise of Modern Industry,* 3rd ed. London, 1927.

Hatfield, H. S., *The Conquest of Thought by Invention.* London, 1930.

Montmasson, J. M., *Invention and the Unconscious.* (With a preface by H. S. Hatfield on biology and human invention.) London, 1931.

Petrie, W. M. F., *The Revolutions of Civilisation.* London and New York, 1911.

Rivers, W. H. R., *Psychology and Politics.* London, 1923.

Russell, B., *The Prospects of Industrial Civilisation.* London, 1923.

Vowles, H. P. and M. W., *The Quest of Power.* London, 1931.

Wells, H. G., *The Outline of History,* 5th revision. London, 1930.

Wells, H. G., *The Work, Wealth, and Happiness of Mankind.* London, 1931.

PHILOSOPHY AND RELIGION

Alexander, J., *Space, Time, and Deity.* London, 1920.

Bosanquet, B., *The Meeting of Extremes in Contemporary Philosophy.* London, 1921.

Bosanquet, B., *The Principle of Individuality and Value.* London, 1912.

Bosanquet, B., *The Value and Destiny of the Individual.* London, 1913.

Caird, E., *The Evolution of Religion.*

Crawley, E., *The Tree of Life.* London, 1905.

Crawley, E., *The Mystic Rose,* 4th ed. London, 1932.

Dampier-Whetham, W. C. D., *History of Science and its Relations with Philosophy and Religion,* 2nd ed. Cambridge, 1930.

Eddington, A. S., *Science and the Unseen World.* London, 1929.

Frazer, J. G., *The Golden Bough,* 3rd ed. London, 1911–15.

Frazer, J. G., *The Belief in Immortality.* London, 1913.

Frazer, J. G., *Adonis: a Study in the History of Oriental Religion.* London, 1932.

Jevons, F. B., *An Introduction to the History of Religion,* 2nd ed. London, 1902.

Needham, N. J. (Editor), *Science, Religion, and Reality.* A collection of essays by A. Balfour, Malinowski, Singer, Eddington, Inge, and many others. London, 1925.

Ogden, C. K., Richards, I. A., and Wood, J., *The Foundations of Aesthetics.* London, 1922.

Rivers, W. H. R., *Medicine, Magic, and Religion.* London, 1924.

Rivers, W. H. R., *Psychology and Ethnology*. London, 1926.

Russell, B., *The Scientific Outlook*. London, 1931.

Russell, B., *Sceptical Essays*. London, 1928.

Russell, B., *Why I am not a Christian*. London, 1927.

Sorley, W. R., *Moral Values and the Idea of God,* 3rd ed. Cambridge, 1924.

Tiele, *Elements of the Science of Religion.*

Tylor, *Primitive Culture*, 4th ed. London, 1903.

NAME INDEX

Uus

SUBJECT INDEX

HISTORY, PHILOSOPHY AND SOCIOLOGY OF SCIENCE

Classics, Staples and Precursors

An Arno Press Collection

Aliotta, [Antonio]. **The Idealistic Reaction Against Science.** 1914

Arago, [Dominique François Jean]. **Historical Eloge of James Watt.** 1839

Bavink, Bernhard. **The Natural Sciences.** 1932

Benjamin, Park. **A History of Electricity.** 1898

Bennett, Jesse Lee. **The Diffusion of Science.** 1942

[Bronfenbrenner], Ornstein, Martha. **The Role of Scientific Societies in the Seventeenth Century.** 1928

Bush, Vannevar. **Endless Horizons.** 1946

Campanella, Thomas. **The Defense of Galileo.** 1937

Carmichael, R. D. **The Logic of Discovery.** 1930

Caullery, Maurice. **French Science and its Principal Discoveries Since the Seventeenth Century.** [1934]

Caullery, Maurice. **Universities and Scientific Life in the United States.** 1922

Debates on the Decline of Science. 1975

de Beer, G. R. **Sir Hans Sloane and the British Museum.** 1953

Dissertations on the Progress of Knowledge. [1824].
2 vols. in one

Euler, [Leonard]. **Letters of Euler.** 1833. 2 vols. in one

Flint, Robert. **Philosophy as Scientia Scientiarum and a History of Classifications of the Sciences.** 1904

Forke, Alfred. **The World-Conception of the Chinese.** 1925

Frank, Philipp. **Modern Science and its Philosophy.** 1949

The Freedom of Science. 1975

George, William H. **The Scientist in Action.** 1936

Goodfield, G. J. **The Growth of Scientific Physiology.** 1960

Graves, Robert Perceval. **Life of Sir William Rowan Hamilton.** 3 vols. 1882

Haldane, J. B. S. **Science and Everyday Life.** 1940

Hall, Daniel, et al. **The Frustration of Science.** 1935

Halley, Edmond. **Correspondence and Papers of Edmond Halley.** 1932

Jones, Bence. **The Royal Institution.** 1871

Kaplan, Norman. **Science and Society.** 1965

Levy, H. **The Universe of Science.** 1933

Marchant, James. **Alfred Russel Wallace.** 1916

McKie, Douglas and Niels H. de V. Heathcote. **The Discovery of Specific and Latent Heats.** 1935

Montagu, M. F. Ashley. **Studies and Essays in the History of Science and Learning.** [1944]

Morgan, John. **A Discourse Upon the Institution of Medical Schools in America.** 1765

Mottelay, Paul Fleury. **Bibliographical History of Electricity and Magnetism Chronologically Arranged.** 1922

Muir, M. M. Pattison. **A History of Chemical Theories and Laws.** 1907

National Council of American-Soviet Friendship. **Science in Soviet Russia: Papers Presented at Congress of American-Soviet Friendship.** 1944

Needham, Joseph. **A History of Embryology.** 1959

Needham, Joseph and Walter Pagel. **Background to Modern Science.** 1940

Osborn, Henry Fairfield. **From the Greeks to Darwin.** 1929

Partington, J[ames] R[iddick]. **Origins and Development of Applied Chemistry.** 1935

Polanyi, M[ichael]. **The Contempt of Freedom.** 1940

Priestley, Joseph. **Disquisitions Relating to Matter and Spirit.** 1777

Ray, John. **The Correspondence of John Ray.** 1848

Richet, Charles. **The Natural History of a Savant.** 1927

Schuster, Arthur. **The Progress of Physics During 33 Years (1875-1908).** 1911

Science, Internationalism and War. 1975

Selye, Hans. **From Dream to Discovery: On Being a Scientist.** 1964

Singer, Charles. **Studies in the History and Method of Science.** 1917/1921. 2 vols. in one

Smith, Edward. **The Life of Sir Joseph Banks.** 1911

Snow, A. J. **Matter and Gravity in Newton's Physical Philosophy.** 1926

Somerville, Mary. **On the Connexion of the Physical Sciences.** 1846

Thomson, J. J. **Recollections and Reflections.** 1936

Thomson, Thomas. **The History of Chemistry.** 1830/31

Underwood, E. Ashworth. **Science, Medicine and History.** 2 vols. 1953

Visher, Stephen Sargent. **Scientists Starred 1903-1943 in American Men of Science.** 1947

Von Humboldt, Alexander. **Views of Nature: Or Contemplations on the Sublime Phenomena of Creation.** 1850

Von Meyer, Ernst. **A History of Chemistry from Earliest Times to the Present Day.** 1891

Walker, Helen M. **Studies in the History of Statistical Method.** 1929

Watson, David Lindsay. **Scientists Are Human.** 1938

Weld, Charles Richard. **A History of the Royal Society.** 1848. 2 vols. in one

Wilson, George. **The Life of the Honorable Henry Cavendish.** 1851